Proceedings of GeoShanghai 2018 International Conference: Rock Mechanics and Rock Engineering

Lianyang Zhang · Bruno Goncalves da Silva
Cheng Zhao
Editors

Proceedings of GeoShanghai 2018 International Conference: Rock Mechanics and Rock Engineering

Springer

Editors
Lianyang Zhang
Civil Engineering and Engineering Mechanics
University of Arizona
Tucson, AZ
USA

Cheng Zhao
Tongji University
Shanghai
China

Bruno Goncalves da Silva
New Jersey Institute of Technology
New Jersey, NJ
USA

ISBN 978-981-13-0112-4 ISBN 978-981-13-0113-1 (eBook)
https://doi.org/10.1007/978-981-13-0113-1

Library of Congress Control Number: 2018939621

Printed on acid-free paper

This Springer imprint is published by the registered company Springer Nature Singapore Pte Ltd.
part of Springer Nature
The registered company address is: 152 Beach Road, #21-01/04 Gateway East, Singapore 189721, Singapore

Preface

The 4th GeoShanghai International Conference was held on May 27–30, 2018, in Shanghai, China. GeoShanghai is a series of international conferences on geotechnical engineering held in Shanghai every four years. The conference was inaugurated in 2006 and was successfully held in 2010 and 2014, with more than 1200 participants in total. The conference offers a platform of sharing recent developments of the state-of-the-art and state-of-the-practice in geotechnical and geoenvironmental engineering. It has been organized by Tongji University in cooperation with the ASCE Geo-Institute, Transportation Research Board, and other cooperating organizations.

The proceedings of the 4th GeoShanghai International Conference include eight volumes of over 560 papers; all were peer-reviewed by at least two reviewers. The proceedings include Volumes 1: Fundamentals of Soil Behavior edited by Dr. Annan Zhou, Dr. Junliang Tao, Dr. Xiaoqiang Gu, and Dr. Liangbo Hu; Volume 2: Multi-physics Processes in Soil Mechanics and Advances in Geotechnical Testing edited by Dr. Liangbo Hu, Dr. Xiaoqiang Gu, Dr. Junliang Tao, and Dr. Annan Zhou; Volume 3: Rock Mechanics and Rock Engineering edited by Dr. Lianyang Zhang, Dr. Bruno Goncalves da Silva, and Dr. Cheng Zhao; Volume 4: Transportation Geotechnics and Pavement Engineering edited by Dr. Xianming Shi, Dr. Zhen Liu, and Dr. Jenny Liu; Volume 5: Tunneling and Underground Construction edited by Dr. Dongmei Zhang and Dr. Xin Huang; Volume 6: Advances in Soil Dynamics and Foundation Engineering edited by Dr. Tong Qiu, Dr. Binod Tiwari, and Dr. Zhen Zhang; Volume 7: Geoenvironment and Geohazards edited by Dr. Arvin Farid and Dr. Hongxin Chen; and Volume 8: Ground Improvement and Geosynthetics edited by Dr. Lin Li, Dr. Bora Cetin, and Dr. Xiaoming Yang. The proceedings also include six keynote papers presented at the conference, including "Tensile Strains in Geomembrane Landfill Liners" by Prof. Kerry Rowe, "Constitutive Modeling of the Cyclic Loading Response of Low Plasticity Fine-Grained Soils" by Prof. Ross Boulanger, "Induced Seismicity and Permeability Evolution in Gas Shales, CO_2 Storage and Deep Geothermal Energy" by Prof. Derek Elsworth, "Effects of Tunneling on Underground Infrastructures" by Prof. Maosong Huang, "Geotechnical Data Visualization and Modeling of Civil

Infrastructure Projects" by Prof. Anand Puppala, and "Probabilistic Assessment and Mapping of Liquefaction Hazard: from Site-specific Analysis to Regional Mapping" by Prof. Hsein Juang. The Technical Committee Chairs, Prof. Wenqi Ding and Prof. Xiong Zhang, the Conference General Secretary, Dr. Xiaoqiang Gu, the 20 editors of the 8 volumes and 422 reviewers, and all the authors contributed to the value and quality of the publications.

The Conference Organizing Committee thanks the members of the host organizations, Tongji University, Chinese Institution of Soil Mechanics and Geotechnical Engineering, and Shanghai Society of Civil Engineering, for their hard work and the members of International Advisory Committee, Conference Steering Committee, Technical Committee, Organizing Committee, and Local Organizing Committee for their strong support. We hope the proceedings will be valuable references to the geotechnical engineering community.

Shijin Feng
Conference Chair
Ming Xiao
Conference Co-chair

Organization

International Advisory Committee

Herve di Benedetto	University of Lyon, France
Antonio Bobet	Purdue University, USA
Jean-Louis Briaud	Texas A&M University, USA
Patrick Fox	Penn State University, USA
Edward Kavazanjian	Arizona State University, USA
Dov Leshchinsky	University of Illinois, USA
Wenhao Liang	China Railway Construction Corporation Limited, China
Robert L. Lytton	Texas A&M University, USA
Louay Mohammad	Louisiana State University, USA
Manfred Partle	KTH Royal Institute of Technology, Switzerland
Anand Puppala	University of Texas at Arlington, USA
Mark Randolph	University of Western Australia, Australia
Kenneth H. Stokoe	University of Texas at Austin, USA
Gioacchino (Cino) Viggiani	Université Joseph Fourier, France
Dennis T. Bergado	Asian Institute of Technology, Thailand
Malcolm Bolton	Cambridge University, UK
Yunmin Chen	Zhejiang University, China
Zuyu Chen	Tsinghua University, China
Jincai Gu	PLA, China
Yaoru Lu	Tongji University, China
Herbert Mang	Vienna University of Technology, Austria
Paul Mayne	Georgia Institute of Technology, USA
Stan Pietruszczak	McMaster University, Canada
Tom Papagiannakis	Washington State University, USA
Jun Sun	Tongji University, China

Scott Sloan	University of Newcastle, Australia
Hywel R. Thomas	Cardiff University, UK
Atsashi Yashima	Gifu University, Japan

Conference Steering Committee

Jie Han	University of Kansas, USA
Baoshan Huang	University of Tennessee, USA
Maosong Huang	Tongji University, China
Yongsheng Li	Tongji University, China
Linbin Wang	Virginia Tech, USA
Lianyang Zhang	University of Arizona, USA
Hehua Zhu	Tongji University, China

Technical Committee

Wenqi Ding (Chair)	Tongji University, China
Charles Aubeny	Texas A&M University, USA
Rifat Bulut	Oklahoma State University, USA
Geoff Chao	Asian Institute of Technology, Thailand
Jian Chu	Nanyang Technological University, Singapore
Eric Drumm	University of Tennessee, USA
Wen Deng	Missouri University of Science and Technology, USA
Arvin Farid	Boise State University, Idaho, USA
Xiaoming Huang	Southeast University, China
Woody Ju	University of California, Los Angeles, USA
Ben Leshchinsky	Oregon State University, Oregon, USA
Robert Liang	University of Dayton, Ohio, USA
Hoe I. Ling	Columbia University, USA
Guowei Ma	Hebei University of Technology, China
Roger W. Meier	University of Memphis, USA
Catherine O'Sullivan	Imperial College London, UK
Massimo Losa	University of Pisa, Italy
Angel Palomino	University of Tennessee, USA
Krishna Reddy	University of Illinois at Chicago, USA
Zhenyu Yin	Tongji University, China
ZhongqiYue	University of Hong Kong, China
Jianfu Shao	Université des Sciences et Technologies de Lille 1, France
Jonathan Stewart	University of California, Los Angeles, USA

Wei Wu	University of Natural Resources and Life Sciences, Austria
Jianhua Yin	The Hong Kong Polytechnic University, China
Guoping Zhang	University of Massachusetts, USA
Jianmin Zhang	Tsinghua University, China
Xiong Zhang (Co-chair)	Missouri University of Science and Technology, USA
Yun Bai	Tongji University, China
Jinchun Chai	Saga University, Japan
Cheng Chen	San Francisco State University, USA
Shengli Chen	Louisiana State University, USA
Yujun Cui	École Nationale des Ponts et Chaussees (ENPC), France
Mohammed Gabr	North Carolina State University, USA
Haiying Huang	Georgia Institute of Technology, USA
Laureano R. Hoyos	University of Texas at Arlington, USA
Liangbo Hu	University of Toledo, USA
Yang Hong	University of Oklahoma, USA
Minjing Jiang	Tongji University, China
Richard Kim	North Carolina State University, USA
Juanyu Liu	University of Alaska Fairbanks, USA
Matthew Mauldon	Virginia Tech., USA
Jianming Ling	Tongji University, China
Jorge Prozzi	University of Texas at Austin, USA
Daichao Sheng	University of Newcastle, Australia
Joseph Wartman	University of Washington, USA
Zhong Wu	Louisiana State University, USA
Dimitrios Zekkos	University of Michigan, USA
Feng Zhang	Nagoya Institute of Technology, Japan
Limin Zhang	Hong Kong University of Science and Technology, China
Zhongjie Zhang	Louisiana State University, USA
Annan Zhou	RMIT University, Australia
Fengshou Zhang	Tongji University, China

Organizing Committee

Shijin Feng (Chair)	Tongji University, China
Xiaojqiang Gu (Secretary General)	Tongji University, China
Wenqi Ding	Tongji University, China
Xiongyao Xie	Tongji University, China

Yujun Cui	École Nationale des Ponts et Chaussees (ENPC), France
Daichao Sheng	University of Newcastle, Australia
Kenichi Soga	University of California, Berkeley, USA
Weidong Wang	Shanghai Xian Dai Architectural Design (Group) Co., Ltd., China
Feng Zhang	Nagoya Institute of Technology, Japan
Yong Yuan	Tongji University, China
Weimin Ye	Tongji University, China
Ming Xiao (Co-chair)	Penn State University, USA
Yu Huang	Tongji University, China
Xiaojun Li	Tongji University, China
Xiong Zhang	Missouri University of Science and Technology, USA
Guenther Meschke	Ruhr-Universität Bochum, Germany
Erol Tutumluer	University of Illinois, Urbana—Champaign, USA
Jianming Zhang	Tsinghua University, China
Jianming Ling	Tongji University, China
Guowei Ma	Hebei University of Technology, Australia
Hongwei Huang	Tongji University, China

Local Organizing Committee

Shijin Feng (Chair)	Tongji University, China
Zixin Zhang	Tongji University, China
Jiangu Qian	Tongji University, China
Jianfeng Chen	Tongji University, China
Bao Chen	Tongji University, China
Yongchang Cai	Tongji University, China
Qianwei Xu	Tongji University, China
Qingzhao Zhang	Tongji University, China
Zhongyin Guo	Tongji University, China
Xin Huang	Tongji University, China
Fang Liu	Tongji University, China
Xiaoying Zhuang	Tongji University, China
Zhenming Shi	Tongji University, China
Zhiguo Yan	Tongji University, China
Dongming Zhang	Tongji University, China
Jie Zhang	Tongji University, China
Zhiyan Zhou	Tongji University, China
Xiaoqiang Gu (Secretary)	Tongji University, China
Lin Cong	Tongji University, China
Hongduo Zhao	Tongji University, China

Contents

About the Editors

Lianyang Zhang received his PhD in Geotechnical Engineering from MIT and currently holds the Delbert Lewis Distinguished Professorship at the University of Arizona.

Bruno Goncalves da Silva earned his PhD and MSc from MIT with a concentration in Geotechnical Engineering, and a BSc from Instituto Superior Tecnico, Portugal, with a focus in Structural Engineering. He is currently an Assistant Professor in the New Jersey Institute of Technology.

Cheng Zhao received his PhD from Nagasaki University with a concentration on Geotechnical Engineering. He has been a member in the Department of Geotechnical Engineering at the Tongji University since 2010 and is currently an Associate Professor.

Theoretical and Numerical Study of Rock Behavior

A Method of Surface Subsidence Prediction for Compressed Air Energy Storage in Salt Rock

Zirui Cheng[1] and Bin Ye[1,2(✉)]

[1] Tongji University, Siping Road 1239, Shanghai 200092, China
[2] State Key Laboratory of Geohazard Prevention and Geoenvironment Protection, Chengdu, China
yebin@tongji.edu.cn

Abstract. Renewable energy resources are usually intermittent and unstable. Compressed air energy storage (CAES) provides a good solution to address this problem. Underground air storage caverns are an important part of CAES. Salt rock is known for its excellent flexibility and its extremely low porosity and permeability. During the operation of salt rock underground gas storages, surface subsidence above underground gas caverns would be generated due to the volume shrinkage of salt rock caverns. In this paper, a method to predict the subsidence above salt rock storage cavern is proposed. Firstly, using the finite-difference program Flac3D, the volume loss of the cavern caused by creep shrinkage of salt rock was calculated. Secondly, Schober's model was applied for deformation prediction of storage caverns in salt rock. Finally, surface subsidence was calculated. The method was applied to the Jintan under-ground gas storage in salt rock, the surface subsidence for 20 years operation was calculated, and the impact of the facilities on the ground settlement was effectively accessed.

Keywords: Compressed Air Energy Storage · Salt rock
Subsidence prediction · Numerical simulation

1 Introduction

Power generation technologies with renewable energy sources, such as wind, solar, tidal and waves, have been developing rapidly in recent years. These renewable energies play an important role in mitigating global warming and preventing environmental pollution. However, the generating capacities of renewable energy systems are limited by naturally occurring conditions, such as that they only produce electricity intermittently with low power and energy densities [1]. Large-scale energy storage technologies convert electrical energy into a form that can be stored for later conversion into electrical energy as needed [2] and thus can overcome the problem of intermittency to make renewable energy sources more stable and reliable. Compressed air energy storage (CAES) is a commercial, utility-scale technology that is suitable for providing long-duration energy storage. Compared with other energy storage systems [3], CAES can provide a large-scale energy storage of hundreds-of-MW and has a relatively high efficiency (66%–82%) [4].

L. Zhang et al. (Eds.): GSIC 2018, *Proceedings of GeoShanghai 2018 International Conference: Rock Mechanics and Rock Engineering*, pp. 3–11, 2018.
https://doi.org/10.1007/978-981-13-0113-1_1

Underground storage caverns are the key component of CAES power plants that determine their operating efficiency and reliability. From the engineering perspective, salt formations are considered as suitable repository media since they are assumed to be impervious to fluids. Salt rock permeability can be lower than 10^{-20} m^2, and salt has excellent creep and damage self-healing behavior [5], which can ensure the tightness of the storage caverns. Therefore, CAES in salt caverns should be an ideal solution to ensure the safety, stability, and economic operation of the power network in these areas.

Even though rock salt has many advantages for compressed air storage, it still has some inevitable disadvantages, particularly the deformations even when subjected to very low deviatoric stress [6]. After leaching a cavern, the internal brine pressure or gas pressure is always lower than the in-situ stresses, thus larger deviatoric stresses develop in the vicinity of the cavern due to stresses redistribution and then continuous shrinkage of the volume will be induced. The induced deformation of the strata around and overlying the cavern gradually transfers to the ground surface and then results in vertical subsidence and horizontal movement.

Many studies have been launched to predict the subsidence above underground storage caverns. The main research methods consist of field measurements, theoretical methodology and numerical simulations. Bérest and Brouard [7] and Fokker and Visser [8] launched underground field subsidence monitoring for underground gas caverns. It indicates that the surface subsidence persists very long and is a very slow process and has proportional relation with the volume loss of the cavern. Sroka and Hejmanowski [9] improved the Knothe model and proposed a double parameter time dependent model for the subsidence prediction induced by solid mining. Hu et al. [10] established a new evaluation parameter model of Knothe function for surface subsidence prediction.

In this paper, using the finite-difference program Flac3D, the volume loss of the cavern caused by creep shrinkage of salt rock was calculated, then Schober's model was applied for deformation prediction of storage caverns in salt rock.

2 Method to Predict the Subsidence Above Salt Rock Storage Cavern

According to the studied launched by many researchers, the surface subsidence above underground gas caverns was mainly caused by the volume shrinkage of salt rock caverns. Therefore, the method to predict the subsidence above salt rock storage cavern can be divided two parts: method to calculate the volume loss of the cavern caused by creep shrinkage of salt rock and method to calculate the surface subsidence.

2.1 Method to Calculate the Volume Loss of Storage Cavern

The creeping shrinkage of a salt cavern is induced by the rheology of rock salt. The creep characteristic of salt rock is very complex, a large number of creep constitutive models have been developed. The creep of salt rock can be divided into two states:

initial creep state and steady creep state. The creep constitutive models can be expressed as:

$$\varepsilon = \varepsilon_i + \varepsilon_s \tag{1}$$

where ε is total creep, ε_i is initial creep and ε_s is steady creep. For the cavern under operating period, steady state creep phase is considered as the constitutive model of rock salt. Norton-Hoff (power) Law is used to describe the steady state creep of rock salt, that is

$$\dot{\varepsilon}_s = A\left(\frac{\sigma_1 - \sigma_3}{\sigma^*}\right)^n \tag{2}$$

where A and n are creep parameters, σ_1 and σ_3 are the maximum and the minimum stresses, respectively; σ^* is unit stress. In this paper, the volume loss of storage cavern was calculated by the creep deformation of salt rock, and the deformation was obtained by the finite-difference program Flac3D.

2.2 Method to Calculate the Surface Subsidence

Schober indicated that a mining space would be left after the underground seam was exploited, and surface subsidence was the result of the spread of mining space to the surface, this means the existence of mining space is the root cause of surface subsidence. The propagation of surface subsidence caused by exploitation of underground seam can be described by Fig. 1, where β is angle of influence range, r_u is influence radius, Z_0 is the height of the top of the salt cavern from the ground, Z_u is the height of the bottom of the salt cavern from the ground. The surface subsidence function caused by exploitation can be expressed as:

$$S(X, Y, Z) = \iiint_{\Omega} Q(\xi, \zeta, \eta) F(X, Y, Z) \mathrm{d}\Omega \tag{3}$$

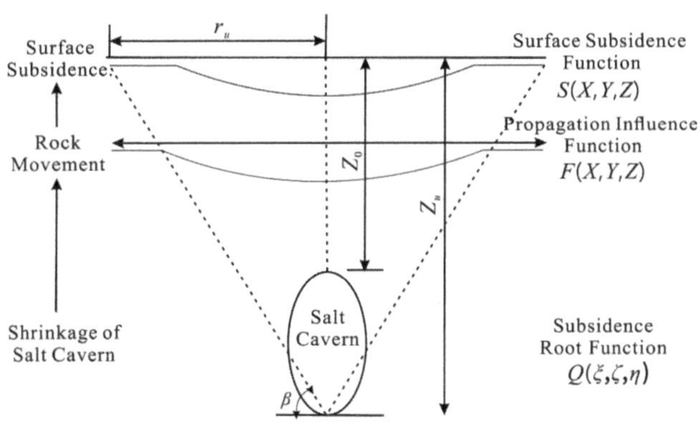

Fig. 1. Influence diagram of surface subsidence caused by volume shrinkage of salt rock underground gas storage

where $Q(\xi, \zeta, \eta)$ and $F(X, Y, Z)$ are the subsidence root function and propagation influence function, respectively, which will be expressed in the following text.

We assume that the initial volume of salt cavern is V_0 and the volume after the final surface subsidence is V_S^{End}. Therefore, the volume shrinkage of salt cavern λ can be expressed as:

$$\lambda = \frac{V_S^{End}}{V_0} = \frac{\frac{1}{3}\pi r_u^2 S_{\max}^{End}}{V_0} \tag{4}$$

where $r_u = Z_u \cdot \cot\beta$, $S_{\max}^{End}$ can be expressed as:

$$S_{\max}^{End} = \lambda \frac{3V_0}{\pi Z_u^2} \tan^2\beta \tag{5}$$

In this paper, Knothe function is used as propagation influence function:

$$F(X, Y, Z) = \frac{1}{r_z^2} \exp(-\pi \frac{r^2}{r_z^2}) \tag{6}$$

where r is horizontal distance of calculated point from the salt cavern, r_z is radius of scattering of exploitation influences. When considering the surface, $r_z = r_u = Z_u \cdot \cot\beta$.

According to the geometric shape of underground salt rock mining space, a suitable subsidence root function was selected. Underground salt cavern generally uses ellipsoidal structure, thus the salt cavern exploitation volume function used in this paper is:

$$\mathrm{d}V = \pi B^2 \left[1 - \frac{(Z - Z_u)^2}{A^2}\right] \mathrm{d}z \tag{7}$$

where A and B are the long axis and short axis of the ellipsoid, respectively.

Converts the coordinate system of Eq. (3), we can get:

$$S(d) = S_{\max}^{End} \int_{Z_0}^{Z_u} F(X, Y, Z) \mathrm{d}V \tag{8}$$

Submitting (5), (6) and (7) into (8), we can get the expression of surface subsidence induced by shrink of elliptic salt cavern:

$$S(d) = \lambda \frac{3V_0 \cdot B^2}{Z_u^2} \tan^2\beta \cdot \frac{1}{r_z^2} \exp(-\pi \frac{r^2}{r_z^2}) \int_{Z_0}^{Z_u} \left[1 - \frac{(Z - Z_u)^2}{A^2}\right] \mathrm{d}z \tag{9}$$

3 Application of the Schober Model for Surface Subsidence Prediction

3.1 Calculation Results of the Volume Loss of Storage Cavern

Finite-difference program Flac3D was used to calculate the volume loss of the cavern caused by creep shrinkage of salt rock. The thickness of the salt rock is assumed to be 250 m, the upper and lower cap rocks are mudstones, the depth of salt rock is 800 m. When calculating, the viscoplastic constitutive model Cpower was adopted in FLAC3D. This model combines the behavior of the viscoelastic two-component Norton power law and the Mohr–Coulomb elastoplastic models. The parameters of salt rock and mudstone were taken from the measured values of Jintan salt rock gas storage cavern, the specific parameters are shown in Table 1.

Table 1. Material parameters used for the numerical simulation

	E(GPa)	μ	φ	C(MPa)	σ_t(MPa)	ρ(kg/m^3)	A	n
Mudstones	28	0.2	50	2.2	1.5	2600	$1.55 \times 10^{-4.2}$	3.8
Salt rock	13.8	0.3	42	1.4	1	2300	$3.48 \times 10^{-5.9}$	4.35

The size of the salt cavern and the pressure of the gas storage are determined according to the operating parameters of a 50 MW proposed compressed air energy storage power station, the maximum storage pressure in the salt cavern is 10 MPa. The calculation mode is divided into four steps: firstly, initial stress balance; secondly, salt cavern excavation and stress redistribution; thirdly, an internal pressure of 10 MPa was applied; and finally, creep calculation.

Figure 2 presents the evolution of volume shrinkage ratio of salt cavern in 20 years when the storage pressure is 10 MPa. This figure indicates that the volume shrinkage ratio increases as operate time increases. In 20 years, the volume shrinkage ratio increases from 0 to 31.71%.

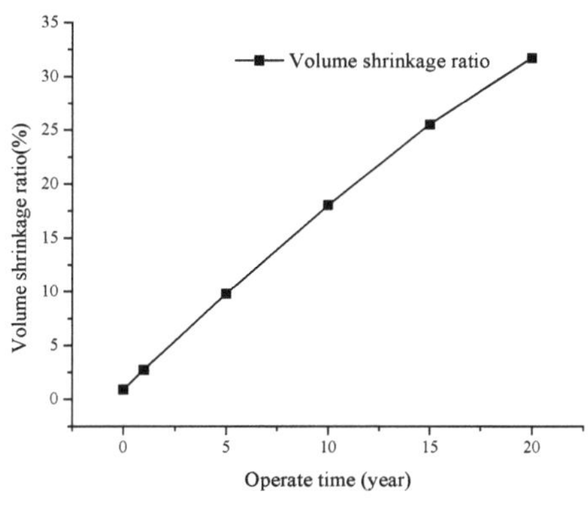

Fig. 2. Evolution of volume shrinkage ratio in 20 years

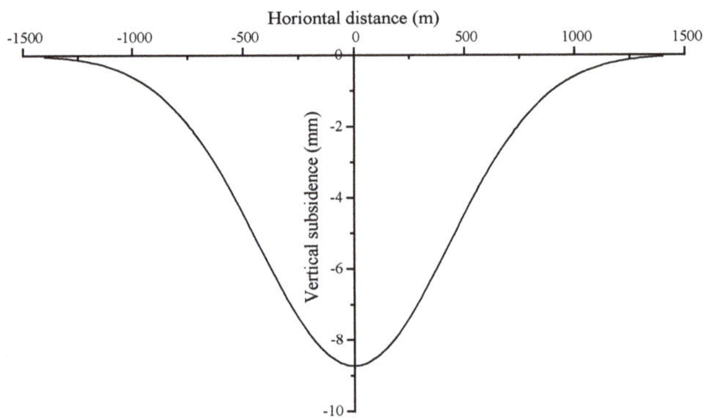

Fig. 3. Subsidence curve after 20 years' operation

3.2 Surface Subsidence Prediction

Submitting the volume shrinkage ratio obtained from numerical calculation into the expression of surface subsidence prediction (Eq. (9)), we can get surface subsidence induced by shrink of salt cavern.

Figure 3 presents the subsidence curve after 20 years' operation. The curve is funnel-shaped, it indicates that the range of surface subsidence induced by shrink of salt cavern is relatively large, the influence radius is about 1400 m, the maximum subsidence point is located above the storage cavern and the settlement is about 8.7 mm. Figure 4 shows the subsidence curves of different operating years. With the increase of operate time, the maximum subsidence increases. The influence radius of the range of surface subsidence is all about 1400 m, the farther from the center point, the smaller the difference between the settlement is.

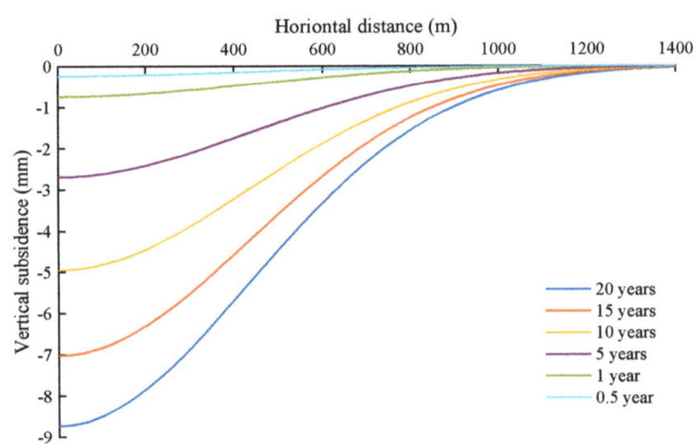

Fig. 4. Subsidence curves of different operating years

4 Influence of Different Storage Pressure on Surface Subsidence

Generally speaking, the storage pressure of a CAES underground storage cavern is variable, it is determined by the power of the CAES plant and stability of the underground storage cavern. Thus, the Influence of different storage pressure on surface subsidence is studied in this section. Three different storage pressures: 6 MPa, 10 MPa and 14 MPa are selected to calculate the volume shrinkage ratio and surface subsidence. The calculation results are shown as following.

Figure 5 presents the evolution of volume shrinkage ratio of salt cavern in 20a under three different storage pressures. It can be found that the volume shrinkage ratio decreases as the increasing of storage pressure. The volume shrinkage ratio after 20 years is 40.69% when storage pressure is 6 MPa, while the volume shrinkage ratio is 31.71% and 23.46 when storage pressure is 10 MPa and 14 MPa, respectively. This indicates that, deformation of the storage cavern will reduce due to the increased storage pressure.

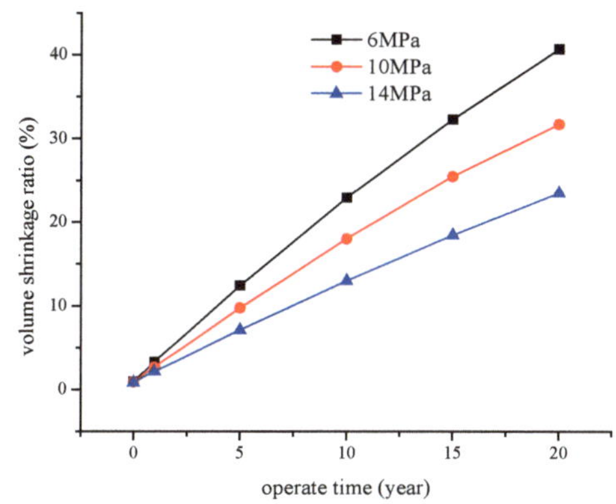

Fig. 5. Evolution of volume shrinkage ratio in 20 years under different storage pressure.

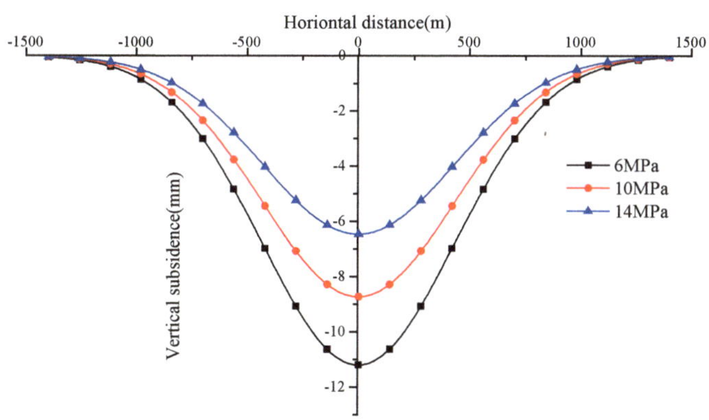

Fig. 6. Subsidence curve after 20 years' operation under different storage pressure

Figure 6 presents the subsidence curve after 20 years' operation under different storage pressure. The curves are all funnel-shaped, and the influence radius is about 1400 m, the maximum subsidence point is located above the storage cavern. The vertical subsidence decreases as the increasing of storage pressure. The maximum subsidence under a storage pressure of 6 MPa, 10 MPa and 14 MPa is 11.21 mm, 8.73 mm and 6.45 mm, respectively. This indicates that the surface subsidence can be reduced by increasing storage pressure.

5 Conclusion

In this paper, the surface subsidence is regarded as the result of the spread of the mining space to the surface and the subsidence is predicted by the integration of subsidence root function and propagation influence function, through these, the calculation method of surface subsidence caused by salt cavern shrinkage is established, and this method is applied to the Jintan underground gas storage in salt rock. The calculate results indicate that:

(1) This method can effectively predict the surface subsidence above salt caverns.
(2) The subsidence curve is funnel-shaped, the maximum subsidence point is located above the storage cavern, the influence radius of the range of surface subsidence is about 1400 m.
(3) With the increase of operate time, the maximum subsidence increases. After 20 years, the maximum subsidence reaches 8.7 mm.
(4) The surface subsidence can be reduced by increasing storage pressure.

References

1. Dopita, M., Williamson, R.: Clean energy Australia 2010. Technical report, Clean Energy Council, Australian Academy of Sciences (2010)
2. Chen, H., et al.: Progress in electrical energy storage system: a critical review. Prog. Nat. Sci. **19**(3), 291–312 (2009)
3. Hadjipaschalis, I., Poullikkas, A., Efthimiou, V.: Overview of current and future energy storage technologies for electric power applications. Renew. Sustain. Energy Rev. **13**(6), 1513–1522 (2009)
4. Succar, S., Williams, R.H.: Compressed air energy storage: theory, resources, and applications for wind power. Princeton Environmental Institute Report 8 (2008)
5. Huang, X.L., Yang, C.H., Li, Y.P.: Discussion on underground energy storage in salt cavern and its tightness evaluation method. In: Proceedings of 2009 International Conference on Management Science and Engineering, Beijing, China (2009)
6. Liu, W., Li, Y.P., Yang, C.H., Wang, B.W., Heng, S.: Analysis of physical and mechanical properties of impure salt rock. In: The 47th US Rock Mechanics/Geomechanics Symposium, vol. 2, pp. 1130–1139 (2013)
7. Bérest, P., Brouard, B.: Safety of salt caverns used for underground storage. Oil Gas Sci. Technol. **58**(3), 361–384 (2003)

8. Fokker, P.A., Visser, J.: Estimating the distribution of salt cavern squeeze using subsidence measurements. In: Proceedings of the 48th US Rock Mechanics/Geomechanics Symposium, Minneapolis, MN, USA, 1–4 June 2014
9. Sroka, A., Hejmanowski, R.: Subsidence prediction caused by the oil and gas development. In: Proceedings of the 12th FIG Symposium, Baden, Austria (2006)
10. Hu, Q.F., Cui, X.M., Kang, X.L., Lei, B., Ma, K.F., Li, L.C.: Impact of parameter on Knothe time function and its calculation model. Int. J. Rock Mech. Min. Sci. **78**, 19–26 (2015)

A New Method for Estimating Rock Fracture Toughness K_{Ic} from Strength Parameters

Jun Liu[1], Qi Zhang[1,2](✉), Guoliang Dai[1], and Xiao Wang[1]

[1] School of Civil Engineering, Southeast University, Nanjing, Jiangsu, China
zhangqi@seu.edu.cn

[2] State Key Laboratory for GeoMechanics and Deep Underground Engineering, China University of Mining and Technology, Xuzhou, Jiangsu, China

Abstract. Considering the difficulty in measuring the model I fracture toughness of rocks, this study aims to develop a quantitative relationship between model I fracture toughness and strength parameters such as the tensile strength, the uniaxial compressive strength, and the Hoek-Brown (H-B) strength parameter m_i. Based on the collected experimental data from the literature, a relationship between model I fracture toughness and tensile strength is derived statistically. The rock tensile strength can be determined using a closed-form relationship from the uniaxial compressive strength and the H-B strength parameter m_i, which can be easily measured. Combining the two relationships, a new method is proposed to determine the model I fracture toughness conveniently. Finally, collected experimental data is used to validate the proposed estimating method.

Keywords: Rock mode I fracture toughness · Uniaxial compressive strength Tensile strength · Hoek-Brown parameters m_i · Estimating method

1 Introduction

Brittle and semi-brittle fractures of rock are widespread geological structures of the crustal surfaces, and are important objects of researches in many fields, such as engineering geology, prospecting engineering, rock masses engineering, etc. The fracture toughness is a commonly mechanical index which is applied to characterize the resistance of instability expansion of inner cracks and the fracture damages of materials [1]. The evaluation of fracture toughness is a controlling factor for the rocks crack initiation, crack propagation and crack linkage [2]. Rocks are under the condition of compressive-shear stress commonly [3], and the compound compressive-shear fracture in crack propagation is categorized as the open mode, which is also known as mode I crack propagation [4, 5]. Other two crack propagation modes, which are slip mode (mode II) and tearing mode (mode III), are essentially the secondary phenomena during the expansion of mode I crack initiation [6]. Therefore, the measurement of mode I fracture toughness (K_{Ic}), which reflects the resistance to opening mode crack propagation of materials, is becoming concerned increasingly.

Several experimental methods for K_{Ic} such as three-point bend, short round bar, Brazilian disc, have been suggested by the International Society for Rock Mechanics

L. Zhang et al. (Eds.): GSIC 2018, *Proceedings of GeoShanghai 2018 International Conference: Rock Mechanics and Rock Engineering*, pp. 12–20, 2018.
https://doi.org/10.1007/978-981-13-0113-1_2

and Rock Engineering (ISRM) so far. The tests and analyses on static, dynamic fracture toughness have been carried out to determine K_{Ic} for different kinds of rocks [6]. Although the test procedures are standardized, the fracture toughness tests are still not widespread yet. That is because of the lengthy sample preparation time, premature failure of samples, and difficulties in obtaining the consistent notch dimensions of the tolerances specified [7]. The values of K_{Ic} obtained by different testing methods and sample types are often different. So as to seek an alternative method to determine K_{Ic}, some researchers have tried to obtain the correlations between K_{Ic} and rock tensile strength based on the mathematical statistical methods.

However, the rock tensile strength can be measured with tests, which are very difficult to operate and very discrete. Colak and Unlu established an empirical relationship between the H-B strength parameter m_i, the indirect tensile strength σ_{tB} and the uniaxial compressive strength σ_c [8]. m_i is one of the empirical parameters in H-B strength criterion which is the most widely used rock criterion. According to the engineering experience of laboratories and engineering sites, Hoek et al. proposed a relatively comprehensive method to determine m_i covering a variety of rock types [9]. Because it is easily to obtain m_i and σ_c, and the tensile strength σ_t can be derived conveniently.

Therefore, it is meaningful to seek a quantitative correlation between K_{Ic} and rock strength parameters to replace the experimental method. Based on experimental data from published literature, the relationship between K_{Ic} and σ_t is improved statistically. Then a new estimating relationship for σ_t is obtained by m_i and σ_c, which can be easily measured. Combined the two relationships, a new method for estimating K_{Ic} based on m_i and σ_c is proposed. Finally, collected experimental data is applied to validate the proposed estimating method.

2 A Relationship Between Rock Fracture Toughness and Tensile Strength

Based on the maximum circumferential tensile stress theory, the crack started to expand when the circumferential tensile stress reached the critical value K_{Ic} [10]. Whittaker et al. [11] concluded that the maximum principal stress theory can well explain the fracture characteristics of rocks. Deng et al. [6] applied the theory of maximum circumferential tensile stress to solve the mode I fracture problems. Theoretical studies proved a reliable relationship between K_{Ic} and σ_t is existed, which is expressed as follows:

$$K_{Ic} = \sigma_t \sqrt{2\pi r} \tag{1}$$

where K_{Ic} is the mode I fracture toughness (MPa•m$^{1/2}$). σ_t is the rock tensile strength (MPa). r is the crack propagation radius (m), which is a constant [10, 12].

Based on the experimental results of different rock types, a variety of empirical relations which can reflect the relationship between K_{Ic} and σ_t have been established. Table 1 lists some empirical relationships given by statistical methods. The experimental results show that there is a good linear relationship between K_{Ic} and σ_t.

Table 1. The empirical relationships of K_{Ic} and σ_t

Literatures	Equations
Deng et al. [6]	$\sigma_t = 8.23\ K_{Ic}$
Zhang [7]	$\sigma_t = 6.88\ K_{Ic}$
Whittaker et al. [11]	$\sigma_t = 9.35\ K_{Ic} - 2.53$

In order to improve the accuracy of the relationship, more experimental data [11, 13–15] are collected, as shown in Table 2.

Table 2. Collected testing data of K_{Ic} and σ_t

Rocks	K_{Ic} (MPa·m$^{1/2}$)	σ_t (MPa)	Rocks	K_{Ic} (MPa·m$^{1/2}$)	σ_t (MPa)
Marble [11]	2.680	17.30	Sandstone [15]	0.380	3.34
	2.260	15.40		0.370	3.30
	2.020	13.90		0.450	3.70
	1.700	12.10		0.460	3.45
	1.440	10.00		0.452	3.40
Sandstone [13]	0.490	3.62		0.425	3.16
	0.440	3.27		0.390	3.00
	0.418	3.17		0.385	2.96
	0.377	2.80		0.351	2.72
	0.345	2.57		0.333	2.69
	0.328	2.46		0.439	3.23
	0.298	2.16		0.416	3.08
Marble [14]	0.967	10.33		0.362	2.77
	1.482	4.59		0.329	2.61
Sandstone [15]	0.670	5.09		0.294	2.48
	0.280	2.65		0.276	2.43

According to the above collected data of K_{Ic} and σ_t, the linear function model is re-fitted and analyzed, and the lines of up and down envelopes will also be obtained. The newly fitted results are shown in Eq. (2) and Fig. 1. From the theoretical derivation of Eq. (1), when σ_t is 0, K_{Ic} should also be equal to 0. So the intercept of K_{Ic} and σ_t is set to 0.

$$\sigma_t = \alpha K_{Ic},\ \alpha = 6.872\ (4.58\text{–}10.65) \tag{2}$$

More than 90% of the data points in Fig. 1 scatter above the down envelope line and below the up envelope line. The line of slope 6.872 in Fig. 1 represents the average trend line with a good correlation coefficient maximally.

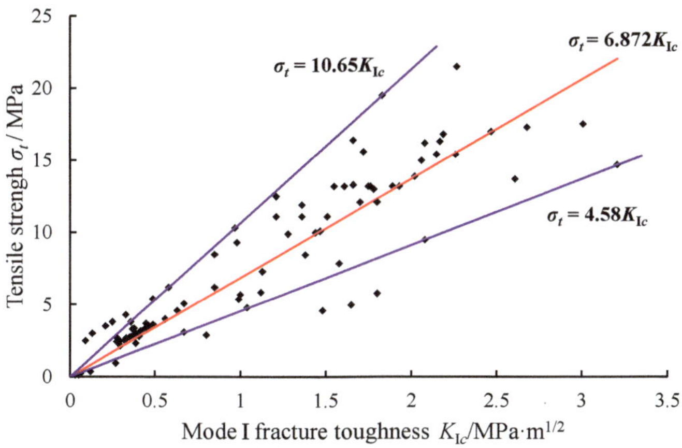

Fig. 1. Improved relationship between K_{Ic} and σ_t

3 A Relationship for Rock Tensile Strength Based on Uniaxial Compressive Strength and H-B Strength Parameter m_i

The empirical parameter m_i from the H-B strength criterion reflects the hardness degree of different types of rocks and can be determined by a guiding chart conveniently [9]. The accurate value of m_i requires a series of rock mechanics tests and the mathematical method is used to fit the rock triaxial compression experimental results. A series of studies of parameter m_i has been conducted, and the empirical relations between H-B strength parameter m_i and the indirect (or direct) tensile strength σ_{tB} (or σ_t), triaxial compression strength σ_c are concluded [8, 16–18].

Owing to that H-B strength parameter m_i is distributed between 2 and 35 approximately for the various rock types, the empirical equations proposed by Colak and Unlu [8] and Richards and Read [16] cannot predict effectively when m_i value is relatively larger or smaller. The empirical formula proposed by Cai [18] contains crack initiation strength σ_{ci}, which is a strength parameter obtained in the rock uniaxial compression test. Although the value of σ_{ci} can be measured using the acoustic emission equipment, σ_{ci} is not widely spread in engineering practice. Table 3 lists some of the empirical relations obtained by the statistical method.

From the empirical relationships in Table 3, it can be observed that different researchers have tried to derive the relationships for m_i based on σ_c/σ_t and σ_t/σ_c, but the specific equations are quite different. Similarly two coefficients a and b are introduced for σ_c/σ_t and σ_t/σ_c respectively, as shown in Eq. (3). A new estimating equation can reflect the relationship between m_i, the uniaxial compressive strength and tensile strength with better accuracy.

Table 3. The empirical relationships between m_i and rock strength parameters

Equations	Notes
$m_i = \sigma_t/\sigma_c - \sigma_c/\sigma_t$ [8]	σ_t is positive when the stress is compressive
$m_i \approx R = \sigma_c/\lvert\sigma_t\rvert$ [16]	m_i is consistent with the ratio of R
$m_i = 16\sigma_{tB}/\sigma_c - \sigma_c/\sigma_{tB}$ [17]	σ_{tB} is obtained by Brazilian disc
$m_i \approx 8$ or $12\sigma_c/\sigma_{ci}$ [18]	8 is valid for low confining pressure to tensile state, 12 is valid for high confining pressure; σ_{ci} is the crack initiation strength in rock uniaxial compression test

$$m_i = a\sigma_c/\sigma_t - b\sigma_t/\sigma_c \quad (3)$$

According to more 100 groups experimental data of the rock uniaxial compressive strength and tensile strength collected from [18–21], the Eq. (3) is fitted by the least square method. σ_t/σ_c and σ_c/σ_t are considered as the independent variables, and m_i is considered as the dependent variable. The fitting results are that a equals to 1.318 and b equals to 4.586. The new equation is expressed as follows.

$$m_i = 1.318\sigma_c/\sigma_t - 4.586\sigma_t/\sigma_c \quad (4)$$

Using the Eq. (4), σ_t can be determined indirectly by σ_c and m_i, while m_i and σ_c are easily measured. In order to determine a more accurate σ_t, several triaxial compression tests are required to obtain an exact value of m_i (Fig. 2).

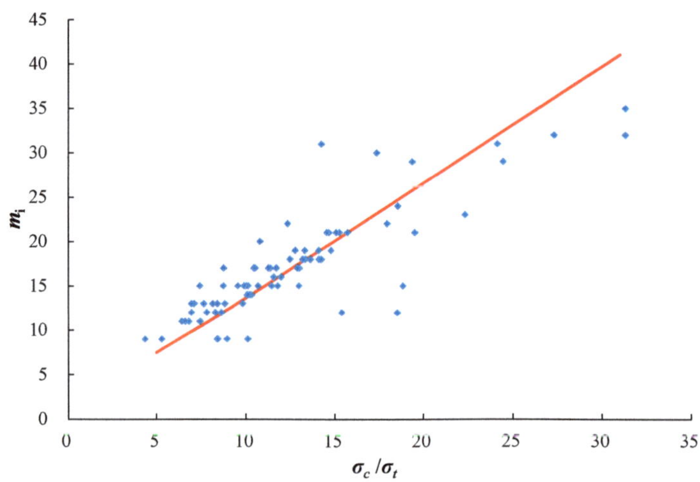

Fig. 2. The relationship between m_i, σ_c, and σ_t

4 Validation of a New Method for Estimating K_{Ic}

4.1 A New Method for Estimating K_{Ic}

Combined with the Eqs. (2) and (4) established in Sects. 2 and 3, a new method for estimating K_{Ic} simply from σ_c and m_i are proposed, as shown in Eq. (5).

$$K_{Ic} = -\sigma_c(m_i - \sqrt{m_i^2 + 25.601})/66.741 \tag{5}$$

With the new method, the estimated K_{Ic} can be determined conveniently according to Eq. (5). Only two input parameters σ_c and m_i are required, which are obtained by conducting the uniaxial compressive test or several rock triaxial compression tests.

4.2 Validation of the New Estimating Method

A number of testing data [6, 22–27] are applied to Eq. (5) to validate the accuracy. The empirical parameter m_i is determined according to the texture and mineral composition descriptions [9]. σ_t is calculated by substituting σ_c and m_i into Eq. (4), and then the value of K_{Ic} and the slope of the up and down envelope lines are calculated by substituting σ_t into Eq. (2). Meanwhile, calculation errors of K_{Ic} are listed and evaluated. Or σ_c and m_i are directly substituted into Eq. (5) to determine the value of K_{Ic}. The calculation results of the new estimating method are shown in Table 4. It can be concluded that the predicted values of the fracture toughness K_{Ic} estimated by Eq. (5) have relatively good precision. All the measured values are distributed in the intervals of the up and down envelope lines.

Table 4. Calculation results by the new estimating method

Rock type	σ_c (MPa)	Measured K_{Ic} (MPa·m$^{1/2}$)	m_i	Calculated K_{Ic} (MPa·m$^{1/2}$)	Error of K_{Ic} (%)
Sandstone [6]	56.31	0.37	21	0.507 (0.327–0.761)	37.03
	46.17	0.30	20	0.436 (0.281–0.654)	45.33
Ore [22]	53.10	1.56	7	1.303 (0.840–1.954)	−16.47
Marble [22]	62.70	1.59	8	1.377 (0.889–2.066)	−13.40
Granite [23]	150.00	0.734	35	0.818 (0.528–1.227)	11.44
Sandstone [23]	50.00	0.395	21	0.450 (0.291–0.676)	13.92
Migmatite [23]	60.00	0.320	32	0.357 (0.231–0.536)	11.56
Porphyritic granite [24]	177.00	0.88	34	0.993 (0.641–1.490)	12.84
	195.00	0.85	35	1.063 (0.686–0.595)	25.06
	220.00	0.80	35	1.199 (0.774–1.800)	49.88
	173.00	0.95	34	0.971 (0.626–1.456)	2.21

(*continued*)

Table 4. (*continued*)

Rock type	σ_c (MPa)	Measured K_{Ic} (MPa·m$^{1/2}$)	m_i	Calculated K_{Ic} (MPa·m$^{1/2}$)	Error of K_{Ic} (%)
Red sandstone [25]	67.24	0.80	19	0.667 (0.430–1.001)	−16.63
	84.98	0.95	20	0.802 (0.518–1.204)	−15.58
	98.88	1.11	21	0.890 (0.574–1.336)	−19.82
	99.49	1.15	21	0.896 (0.578–1.344)	−22.09
	103.79	1.09	21	0.935 (0.603–1.402)	−14.22
	90.29	1.08	19	0.896 (0.578–1.344)	−17.04
	66.36	0.60	18	0.694 (0.448–1.041)	15.67
Granite [26]	191.00	1.19	31	1.174 (0.757–1.761)	−1.34
Marble [26]	151.50	1.70	12	2.322 (1.499–3.485)	36.59
Sandstone [27]	70.00	0.68	21	0.630 (0.407–0.946)	−7.35

*Note: The up and down envelopes of the calculated K_{Ic} are listed in the brackets respectively.

5 Conclusions

In the paper, a new method for estimating K_{Ic} based on the empirical H-B strength parameter and uniaxial compressive strength is proposed. Some conclusions are listed as follows.

(1) With the relationship of $m_i = 1.318\sigma_c/\sigma_t - 4.856\sigma_t/\sigma_c$, rock tensile strength σ_t can be estimated by the H-B strength parameter m_i and uniaxial compressive strength, which can be easily measured.

(2) The empirical relationship between mode I fracture toughness K_{Ic} and rock tensile strength can be expressed as $\sigma_t = 6.872K_{Ic}$. The expressions of up and down envelope lines are also gave.

(3) Using the new estimating method of $K_{Ic} = -\sigma_c(m_i - \sqrt{m_i^2 + 25.601})/66.741$, K_{Ic} can be estimated directly by the H-B strength parameter m_i and the uniaxial compressive strength of the rocks with relatively good accuracy.

References

1. Wang Z.: Higher Fracture Mechanics. Science Press (2009)
2. Chen, J., Deng, J., Yuan, J.: Determination of fracture toughness of modes I and II of shale formation. Chin. J. Rock Mech. Eng. **34**(6), 1101–1105 (2015)
3. Fu, G., Li, H., Xue, X., Yin, G.: The mixed mode fracture criterion of brittle rock under compression - shear stress state. J. Chongqing Univ. Nat. Sci. Edn. **2**, 73–78 (1990)
4. Li, S.: Introduction to Rock Fracture Mechanics. Publishing House of University of Science and Technology China, Hefei (2010)
5. Pan, B., Tang, H.: The characteristics of the compression-shear fracture in rocks and the I-II composite fracture criterion. Earth Sci. J. China Univ. Geosci. **13**(4), 413–421 (1988)

6. Deng, H., Zhu, M., Li, J., Wang, Y., Luo, Q., Yuan, X.: Study of mode-I fracture toughness and its correlation with strength parameters of sandstone. Rock Soil Mech. **33**(12), 3585–3591 (2012)
7. Zhang, Z.: An empirical relation between mode I fracture toughness and the tensile strength of rock. Int. J. Rock Mech. Min. Sci. **39**(3), 401–406 (2002)
8. Colak, K., Unlu, T.: Effect of transverse anisotropy on the Hoek-Brown strength parameter 'm_i' for intact rocks. Int. J. Rock Mech. Min. Sci. **41**(6), 1045–1052 (2004)
9. Zhu, H., Zhang, Q., Zhang, L.: Review of research progresses and applications of Hoek-Brown strength criterion. J. Rock Soil Mech. Eng. **32**(10), 1945–1963 (2013)
10. Jiang, F.: Fracture Mechanics and Crack Growth. Beijing Institute of Aeronautics Press, Beijing (1984)
11. Whittaker B., Singh R., Sun G.: Rock Fracture Mechanics Principles, Design and Applications. Elsevier, Amsterdam (1992)
12. Li, H., Zhao, J., Li, T.: Triaxial compression tests on a granite at different strain rates and confining pressures. Int. J. Rock Mech. Min. Sci. **36**(8), 1057–1063 (1999)
13. Zhu, M., Deng, H., Zhou, S., Luo, Q., Cai, J.: Research on fracture toughness and tensile strength of sandstone under water-saturated and air-dried circulation. Yangtze River **43**(21), 63–69 (2012)
14. Wang, Q., Wu, L.: Determination of elastic modulus, tensile strength and fracture toughness of brittle rocks by using flattened Brazilian disk specimen-part II: experimental results. Chin. J. Rock Mechan. Eng. **23**(2), 199–204 (2004)
15. Deng, H., Li, J., Sun, X., Zhu, M., Yuan, X., Luo, Q.: Experimental research on fracture mechanical effect of sandstone under water corrosion. Chin. J. Rock Mech. Eng. **31**(7), 1342–1348 (2012)
16. Richards, L., Read, S.: A comparison of methods for determining m_i, the Hoek-Brown parameter for intact rock material. In: 45th U.S. Rock Mechanics/Geomechanics Symposium, San Francisco (2011)
17. Gercek, H.: Properties of failure envelopes and surfaces defined by the Hoek-Brown failure criterion. In: Proceedings of the 6th Regional Rock Mechanics Symposium, Konya, pp. 3–11 (2002)
18. Cai, M.: Practical estimates of tensile strength and Hoek-Brown strength parameter m_i of brittle rocks. Rock Mech. Rock Eng. **43**(2), 167–184 (2010)
19. Chen, Z., Wang, L., Jin, J., Huang, Q., Gong, J.: Comparative research of Brazilian splitting and uniaxial compression tests on three kinds of rock. Water Power **41**(11), 34–38 (2015)
20. Jiang W.: Comparative study on rocks in different tensile and compressive tests. Nanjing University, master thesis (2014)
21. Wang, Y., Xu, Q., Chai, H., Liu, L., Xia, Y., Wang, X.: Rock burst prediction in deep shaft based on RBF-AR model. J. Jilin Univ. **43**(6), 1943–1949 (2013)
22. Li, J., Gu, D., Cao, P., Wu, C.: Interrelated law between mode-I fracture toughness and compression strength of rock. J. Central South Univ. **40**(6), 1695–1699 (2009)
23. Zhu, C.: Discussion on compressive strength and fracture toughness of rock. Water Res. Hydropower Eng. **4**, 5–6 (1989)
24. Huang, Y., Xia, X.: The effects of physical properties on rock fracture toughness. Chin. J. Geotech. Eng. **9**(4), 91–96 (1987)
25. Wang Z.: Experimental study on fracture characteristics of rock high temperature shear (including mode II and III). Central South University, master thesis (2007)

26. Yin, X., Yan, Y., Liao, Y., Wu, J., Shi, Y., Li, H.: Experimental investigation on the measurement of facture toughness K_{Ic} of rocks using different methods. Chin. J. Rock Mech. Eng. **9**(4), 328–333 (1990)
27. Li, X., Tang, H.: Dynamic centrifugal modelling tests on toppling rock slopes. J. Geotech. Eng. **36**(4), 687–694 (2014)

A Numerical Study on Mechanical Behavior of Pre-fissured Sandstone Under Triaxial Compression by Three-Dimensional Distinct Element Method

Wenkai Zhu[1,2,3], Mingjing Jiang[1,2,3(✉)], Lei Li[1,2,3], and Tao Li[4]

[1] Department of Geotechnical Engineering, College of Civil Engineering, Tongji University, Shanghai 200092, China
mingjing.jiang@tongji.edu.cn
[2] Key Laboratory of Geotechnical and Underground Engineering of Ministry of Education, Tongji University, Shanghai 200092, China
[3] State Key Laboratory of Disaster Reduction in Civil Engineering, Tongji University, Shanghai 200092, China
[4] School of Civil Engineering, Tianjin University, Tianjin 300072, China

Abstract. Natural rock contains many complicated defects in different scales, resulting in reduction of strength and various special mechanical characteristics, e.g. discontinuous, anisotropic behavior etc. In order to investigate the strength and deformation properties of rock samples containing pre-existing single fissure, a series of three-dimensional (3D) numerical simulations of uniaxial compression and triaxial compression tests on the sandstone specimens were carried out with the Distinct Element Method (DEM), in which a new 3D bond contact model incorporating rolling and twisting resistances was implemented to reproduce the mechanical characteristics of sandstone during the tests. The influence of flaw inclination and confining pressure were investigated from macroscopic and microscopic perspectives. The numerical results demonstrate that the mechanical characteristics of the fissured sandstone samples are in good agreement with experimental results. The confining pressure has a remarkable effect on peak strength and deformation properties of pre-fissured sandstone samples. With confining pressure increasing, the peak strength and the deformation modulus increase. The pre-existing fissure reduces the strength compared to intact rock samples and the reduction extent of peak strength decreases with increasing fissure angle. The present research is helpful to improve the understanding of fracture mechanism of engineering rock and rock structures.

Keywords: Triaxial strength · Failure behavior · Sandstone
Fissure · DEM

1 Introduction

As a kind of complicated geological materials, natural rock contains many discontinuities such as joints, fissures, faults, etc., which significantly influence the strength, deformation, failure mode, crack propagation, etc. [1]. In order to deepen understanding

L. Zhang et al. (Eds.): GSIC 2018, *Proceedings of GeoShanghai 2018 International Conference: Rock Mechanics and Rock Engineering*, pp. 21–29, 2018.
https://doi.org/10.1007/978-981-13-0113-1_3

on mechanical behavior of fractured rock, efforts have been made in the past couple of decades. Among them, rock samples containing single fissure have been widely employed as a simplified and feasible way to experimentally study the effect of fissures on mechanical behavior of different kinds of rock samples, such as marble [2], sandstone [3] and limestone [4], or rock-like material as resin [5]. Fissure geometry is often treated as an important issue. Lee and Jeon [6] systematically studied the effect of fissure angle on crack initiation and coalescence of three types of samples (poly methyl meth acrylate, gypsum and granite) under uniaxial compression. Yang and Jing [7] investigated the strength failure and crack coalescence behavior of sandstone containing a single fissure with different lengths. But most of them are limited to uniaxial compression stress state.

It is widely acknowledged that rock material is always under three-dimensional stress naturally and under complex loading conditions in engineering practice. Therefore, it's important to investigate mechanical behavior of fissured samples under triaxial tests. Because of the limitation of experiment apparatus and complicated test process, only a few works have been reported [8, 9]. Although different observation methods, such as acoustic emission (AE) and X-ray computed tomography (CT), have been developed to great extent, essential cracking and damage process cannot be accurately acquired from physical tests. Distinct Element Method (DEM) is an available numerical method to provide macroscopic information such as mechanical behavior, and microscopic information at the grain scale level, which may help to improve the understanding of cracking and fracture mechanism of pre-fissured rock [10, 11].

This study aims to investigate the macro and micro-scale mechanical response of sandstone samples containing a single fissure with four inclination angles under uniaxial and triaxial compression. Four uniaxial tests and eight triaxial tests were carried out on the DEM samples using a novel 3D bond contact model incorporating rolling and twisting resistances. The stress-strain curves of samples under different confining pressures were analyzed in detail and the influence of the fissure angle on peak strength and deformation properties was presented. Finally, the growth of broken bonds was examined in both uniaxial and triaxial tests, which helps to explain the differences of macroscopic mechanical responses from microscopic points.

2 DEM Modeling

2.1 Contact Model

Rock is a typical cementitious geotechnical media which can be represented as an assembly of mineral grains and bond contacts between them. In the DEM, the mechanical behavior of rock is controlled by the properties of contacts between adjacent particles. The three-dimensional contact model used in this paper consists of two parts, particle contact and bond contact. A 3D particle contact model incorporating rolling and twisting resistances together with normal and tangential interaction has been developed by Jiang et al. [12]. The model was theoretically derived in a clear physical scenario which allows only two additional parameters, i.e. shape parameter β and local crushing parameter ζ_c to be introduced in order to incorporate complete rolling and twisting resistances. Figure 1 shows the mechanical responses of the particle contact

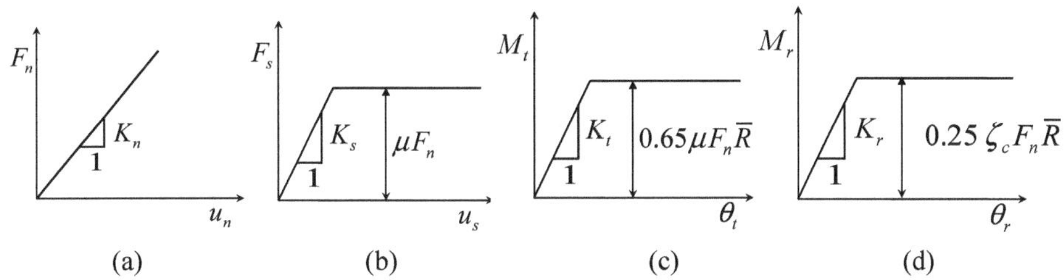

Fig. 1. Mechanical responses of the particle contact [12]: (a) normal direction, (b) tangential direction, (c) twisting direction, (d) rolling direction.

model in four directions where F_n, F_s, M_t and M_r are normal force, shear force, torque and rolling moment respectively. K_n, K_s, K_t and K_r denote normal, tangential, twisting and rolling stiffnesses respectively and the peak resistances in the shear, rolling and twisting directions are controlled by the normal force multiplied by corresponding coefficients.

Bond contact between adjacent particles is envisioned as an ideal elastic-brittle beam with its ends located at the particle centers. If the combined actions of shear force, rolling moment and torque exceed the failure envelops, the bond breaks with its behavior controlled by the model shown in Fig. 1. The bond failure criterion used in this paper was developed by Li et al. [13] by simplifying the 3D bond failure criterion proposed by Shen et al. [14]. The failure criterion can be expressed by a ellipsoid in the F_s-M_r-M_t space as

$$\left[\left(\frac{F_s}{R_s}\right)^2+\left(\frac{M_r}{R_r}\right)^2+\left(\frac{M_t}{R_t}\right)^2\right]=1 \tag{1}$$

where R_s denotes shear strength and $R_r(R_t)$ denotes rolling (torsion) strength of the bonded contact.

2.2 Generation of Pre-fissured Sandstone Samples

The DEM sandstone sample aims to match Berea Sandstone [15], so the particle size distribution used in the numerical assemblies was determined according to the micro-computed tomography images of Berea Sandstone from Bera et al. [16]. The sample with an initial void ratio of 0.8 consists of 40,000 particles and the length: width: height is 1:1:2.5. In order to make a homogeneous sample, the Multi-layer Under-compaction Method (UCM) originally proposed by Jiang et al. [17] was used. After being generated, the sample was "pre-consolidated" under a vertical pressure of 30 MPa by lowering the top rigid wall and lifting the bottom wall with side walls fixed until the equilibrium state was achieved. This process aims to simulate natural sedimentary process of sandstone under *K0* condition. Then, the microscopic bond parameters were assigned to the contact and adjacent particles were bonded together. Next, the sample was isotropically unloaded to the condition of sufficiently low

Table 1. Parameters utilized in the DEM simulation. [13]

Parameters	Values during sample generation	Values after sample generation
Particle effective modulus E (Pa)	3.0×10^{10}	3.0×10^{10}
Particle normal-to-shear stiffness ratio κ	1.5	1.5
Interparticle friction coefficient μ	0.3	0.3
Interparticle shape parameter β	1.5	1.5
Bond effective modulus $\bar{E}$ (Pa)	—	1.35×10^{10}
Bond normal-to-shear stiffness ratio $\bar{\kappa}$	—	1.5
Bond tensile strength σ_t (Pa)	—	3.5×10^7
Bond compressive strength σ_c (Pa)	—	1.4×10^9
Critical bond thickness g_c (m)	—	1.0×10^{-5}

confining pressure (1 kPa), prepared for subsequent tests. Microscopic parameters used during the sample generation and pre-consolidation are listed in Table 1.

When the equilibrium state was achieved after unloading, an open straight fissure with 0.15 mm aperture thickness was created by deleting particles that lay in the fissure position. The position of the fissure is illustrated in Fig. 2. The fissure inclination angle, α, is chosen to be 30°, 45°, 60° and 90° respectively in the study. The length of the fissure, c, is one third of the sample length.

Triaxial compressions of intact sandstone samples with the same parameters values used in this paper were simulated and compared with experimental results [15] in [13]. Figure 3 presents the numerical and experimental stress-strain curves of intact samples under conventional triaxial compression. It shows that the DEM and experimental results are in good agreement, i.e. main mechanical behavior of Berea Sandstone was reproduced.

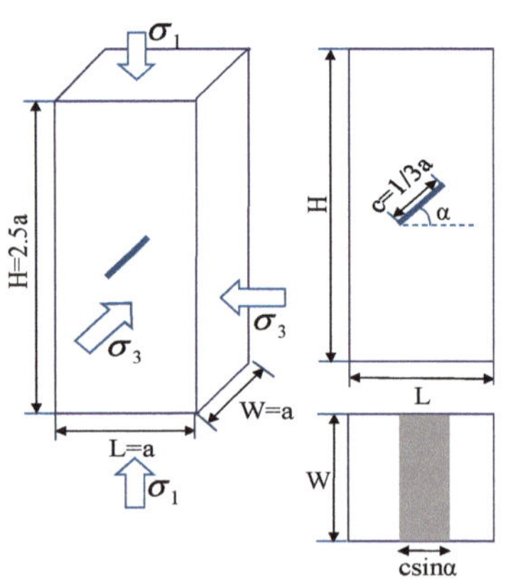

Fig. 2. Geometry of pre-existing fissure and condition of the sample

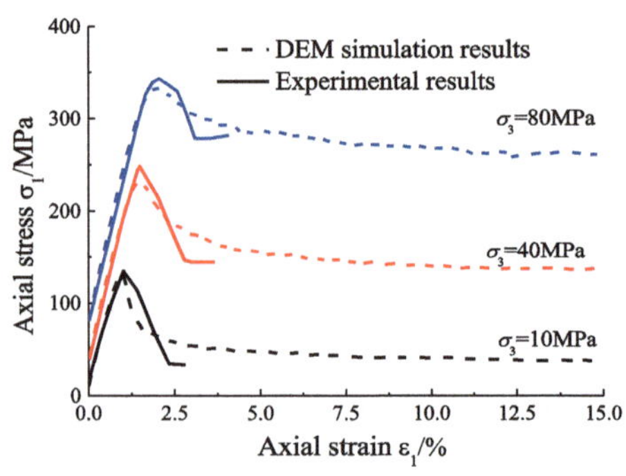

Fig. 3. Comparison between numerical loading and experimental results of intact samples [15]

2.3 Uniaxial and Triaxial Test Simulation

In uniaxial test simulations, each pre-fissured sample with a certain fissure angle was isotropically unloaded close to 0 kPa, then the side walls were deleted. The samples were loaded vertically in a displacement-control manner at a strain rate of 2.5×10^{-4}/s, which is sufficiently low to ensure quasi-static equilibrium. In triaxial simulations, samples was first isotropically consolidated under confining pressure (10 MPa, 30 MPa), and then vertically loaded at strain rate of 2.5×10^{-4}/s under a constant confining pressure. Finally, four uniaxial tests were performed at four fissure angles of 30°, 45°, 60°, 90°. And eight triaxial tests were performed at four angles the same as uniaxial tests and two confining pressures of 10 MPa, 30 MPa.

3 Numerical Simulation Results and Analysis

3.1 Stress-Strain Curves

Triaxial stress-strain curves for pre-fissured samples ($\alpha = 45^{o}$) under different confining pressures are shown in Fig. 4. The stress-strain behavior in a uniaxial test on sandstone with a single fissure can be described into four typical stages, i.e. fissure closure, elastic deformation, crack growth and propagation, and strain-softening, according to Yang and Jing [7]. In the numerical uniaxial tests in this study, the stress-strain curve shows initial nonlinear deformation at the beginning and then changes to elastic deformation stage with a constant elastic modulus. Then, at the stage of crack growth and propagation, the stress-strain shows obvious nonlinear deformation. After that, bonds broke rapidly and the curve shows an abrupt drop. The residual strength is extremely lower than the corresponding peak one.

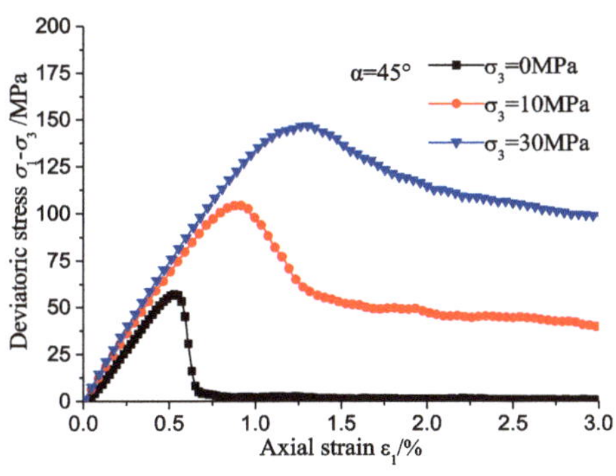

Fig. 4. Stress-strain curves of pre-fissured samples ($\alpha = 45°$)

The stress-strain behavior of triaxial tests share some similarities, while some differences are obvious. Initial nonlinear deformation cannot be seen in the stress-strain curves. The drop after peak stress is slower and the residual strength is much higher compared to that of uniaxial tests. The post-peak deformation changes gradually from softening to hardening when confining pressure rises, which means that the samples change from brittleness to plasticity and ductility. And the peak strength has a prominent enhancement with confining pressure increasing. Similar characteristics can also be observed in the results of samples containing single fissure with $\alpha = 30°$, 60° and 90°.

3.2 Strength and Deformation Properties

Fissures apparently have a significant effect on the strength and deformation behavior. Figure 5(a) presents the effect of the fissure angle on the triaxial peak strength of samples containing a single fissure. It can be seen that the triaxial peak strength has a slightly increase with increasing fissure angle under the same confining pressure. But note that the peak strength of pre-fissured samples is definitely lower than that of intact samples under the same confining pressure. Furthermore, compared to uniaxial results, the fissure angle has less effect on the peak strength when confining pressure is high. The peak strength has a nearly 24.7% increase (from $\alpha = 30°$ to $\alpha = 90°$) in uniaxial tests, while the increase rate is 5.6% when $\sigma_3 = 30$ MPa. The reason may be that the fissure angle plays a major role in the peak strength under low confining pressures while the confining pressure determines triaxial strength when σ_3 is high.

The effect of the fissure angle on the elastic modulus is shown in Fig. 5(b). The elastic modulus is defined as modulus at the stage of elastic deformation. As illustrate, the elastic modulus also increases with the fissure angle increasing. Moreover, a higher confining pressure leads to a higher elastic modulus, which is consistent with the experimental results on pre-fissured granite obtained by Yang and Huang [18].

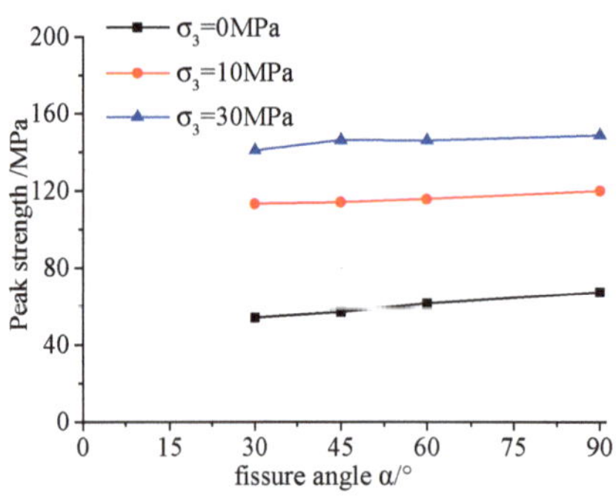

(a)

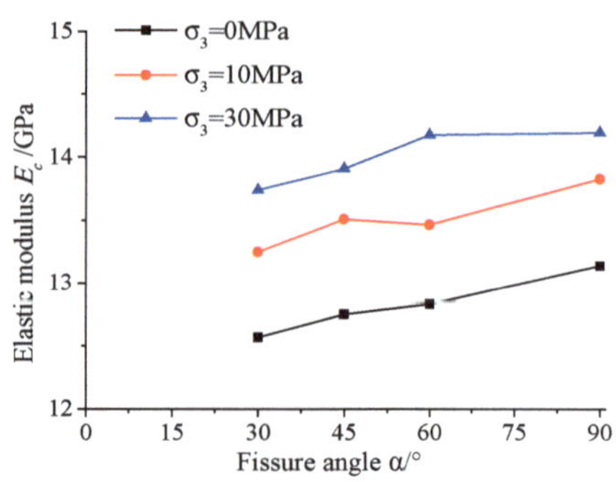

(b)

Fig. 5. The effects of fissure angles on (a) the peak strength (b) the elastic modulus

3.3 Bond Breakage in the Sample

Figure 6 presents the stress-strain curves and the corresponding accumulated broken-bond numbers of the samples ($\alpha = 45°$) under uniaxial test and triaxial test with $\sigma_3 = 30$ MPa. In 3D simulations, shear force, rolling moment and torque all contribute to the breakage of bonds and the type of broken bonds depends on which part playing a primary role. According to the bond failure criterion in Sect. 2.1, the normal force is either positive or negative when bonds break. Therefore, the bond failure mode can be classified into two kinds, i.e. compressive failure if the normal force is positive and tensile failure if the normal force is negative. And each part contains tensile, rolling and twisting components.

The macroscopic failure of the sample is caused by the growth of broken bonds, which explains the consistency between the curve of accumulated broken-bond number and the stress-strain curve. Bonds rarely break at the elastic deformation stage and then break gradually and combine to macro cracks initiating from the fissure tips. After that, the breakage rate has a fast rise. When the stress approaches the peak stress, the breakage rate also reaches the highest value. An obvious difference between uniaxial and triaxial tests can also be observed. The increase rate of broken-bonds number curves is higher under uniaxial condition than that under high confining pressure, e.g. 30 MPa. Furthermore, the bonds rarely break after failure in uniaxial tests but they still gradually break after failure under a high confining pressure. The stronger the brittle response of the stress-strain curve is, the faster the accumulated broken-bond number increases and the shorter the breakage lasts.

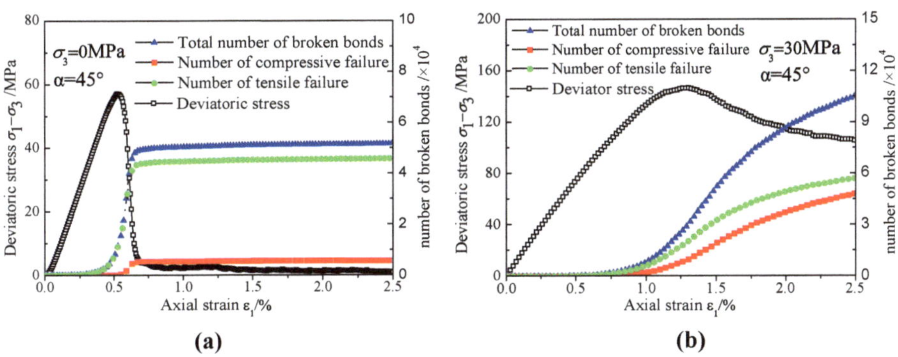

Fig. 6. Stress-strain curves and development of bond-breakage under different confining pressures (a) $\sigma_3 = 0$ MPa (b) $\sigma_3 = 30$ MPa

4 Conclusion

Simulations of uniaxial and triaxial tests on sandstone samples containing a pre-existing single fissure with different angles are performed by 3D DEM. The following conclusions are reached.

1. The confining pressure has a remarkable effect on peak strength, deformation properties and failure mode of pre-fissured sandstone samples. With confining pressure increasing, the peak strength and the deformation modulus increase. Stress-strain curves in uniaxial and triaxial tests are similar in stage division but the behavior at post-peak stages changes from brittle to ductile when confining pressure increases.

2. The single fissure reduces the strength compared to intact rock samples and the reduction extent of peak strength decreases with increasing fissure angle (from 30° to 90°). Also, the elastic modulus has a slight rise with increasing fissure angle.
3. Growth of broken-bonds causes the failure of rock samples so the breakage rate and the number of broken-bonds have great influence on the strength and failure mode. The faster bonds break the stronger brittle response of the stress-strain curve is.

Acknowledgement. The research was funded by the Major Project of Chinese National Programs for Fundamental Research and Development (973 Program) (No. 2014CB046901), which support is greatly appreciated.

References

1. Goodman, R.E.: Introduction to Rock Mechanics. Wiley, New York (1989)
2. Wong, R.H.C., Law, C.M., Chau, K.T., et al.: Crack propagation from 3-D surface fractures in PMMA and marble specimens under uniaxial compression. Int. J. Rock Mech. & Min. Sci. **41**(3), 37–42 (2004)
3. Sahouryeh, E., Dyskin, A.V., Germanovich, L.N.: Crack growth under biaxial compression. Eng. Fract. Mech. **69**(18), 2187–2198 (2002)
4. Baud, P., Wong, T.F., Zhu, W.: Effects of porosity and crack density on the compressive strength of rocks. Int. J. Rock Mech. Min. Sci. **67**(4), 202–211 (2014)
5. Dyskin, A.V., Germanovich, L.N., Ustinov, K.B.: A 3-D model of wing crack growth and interaction. Eng. Fract. Mech. **63**(1), 81–110 (1999)
6. Lee, H., Jeon, S.: An experimental and numerical study of fracture coalescence in pre-cracked specimens under uniaxial compression. Int. J. Solids Struct. **48**(6), 979–999 (2011)
7. Yang, S.Q., Jing, H.W.: Strength failure and crack coalescence behavior of brittle sandstone samples containing a single fissure under uniaxial compression. Int. J. Fract. **168**(2), 227–250 (2011)
8. Yang, S.Q., Jiang, Y.Z., Xu, W.Y., et al.: Experimental investigation on strength and failure behavior of pre-cracked marble under conventional triaxial compression. Int. J. Solids Struct. **45**(17), 4796–4819 (2008)
9. Wang, S.Y., Sloan, S.W., Sheng, D.C., et al.: Numerical study of failure behavior of pre-cracked rock specimens under conventional triaxial compression. Int. J. Solids Struct. **51**(5), 1132–1148 (2014)
10. Duan, K., Kwok, C.Y., Tham, L.G.: Micromechanical analysis of the failure process of brittle rock. Int. J. Numer. Anal. Meth. Geomech. **39**(6), 618–634 (2015)
11. Jiang, M.J., Chen, H., Crosta, G.B.: Numerical modeling of rock mechanical behavior and fracture propagation by a new bond contact model. Int. J. Rock Mech. Min. Sci. **78**, 175–189 (2015)
12. Jiang, M.J., Shen, Z.F., Wang, J.: A novel three-dimensional contact model for granulates incorporating rolling and twisting resistances. Comput. Geotech. **65**, 147–163 (2015)
13. Li, L., Jiang, M.J., Zhang, F.G.: Quantitative simulation of the triaxial test considering residual strength on the deep rock using DEM and parameters analysis. Rock Soil Mech. (accepted for publication). (in Chinese)
14. Shen, Z.F., Jiang, M.J., Wan, R.: Numerical study of interparticle bond failure by 3D discrete element method. Int. J. Numer. Anal. Meth. Geomech. **40**(4), 523–545 (2016)

15. Rutter, E.H., Glover, C.T.: The deformation of porous sandstones: are Byerlee friction and the critical state line equivalent? J. Struct. Geol. **44**, 129–140 (2012)
16. Bera, B., Mitra, S.K., Vick, D.: Understanding the micro structure of Berea Sandstone by the simultaneous use of micro-computed tomography (micro-CT) and focused ion beam-scanning electron microscopy (FIB-SEM). Micron **42**(5), 412–418 (2011)
17. Jiang, M.J., Konrad, J.M., Leroueil, S.: An efficient technique for generating homogeneous specimens for DEM studies. Comput. Geotech. **30**(7), 579–597 (2003)
18. Yang, S.Q., Huang, Y.H.: An experimental study on deformation and failure mechanical behavior of granite containing a single fissure under different confining pressures. Environ. Earth Sci. **76**(10), 364 (2017)

An Investigation of the Pseudo-steady State Approach to Modelling Inter-porosity Flow in Fractured Geomaterials

Lee J. Hosking(✉) and Hywel R. Thomas

Geoenvironmental Research Centre, Cardiff School of Engineering, Cardiff University, The Queen's Buildings, Newport Road, Cardiff CF24 3AA, UK
hoskingL@cardiff.ac.uk

Abstract. This paper examines the assumption of pseudo-steady state inter-porosity mass exchange in dual porosity models of fractured rock. Models of this type rely on the assumption that a pseudo-steady pore pressure distribution prevails in the porous matrix at all times, thereby neglecting transient pressure gradients. The rate of inter-porosity mass exchange is then conveniently expressed as a linear function of the difference between the average pore pressures in the fracture and matrix domains. Whilst providing a relatively simple description of mass exchange, the accuracy of this approach has been debated and it is strictly only valid once the pressure front due to changing conditions in the fracture network reaches the centre of the matrix. The aim of this paper is to compare the pseudo-steady state model of mass exchange with an explicit model of diffusive flow into a rock matrix with parallel-plate geometry. Since the mass exchange coefficient is sometimes described as a function of matrix block geometry and effective diffusivity, an attempt is made to adopt this approach before curve fitting is used. The results indicate that the adopted function underestimates the mass exchange rate compared to the benchmark, although the pseudo-steady state model can provide close agreement if curve fitting is used. It is concluded that the assumption of pseudo-steady state mass exchange may be valid only for cases where calibration of the linear coefficient is possible. Constitutive relationships describing the coefficient should be approached with care, with the possible exception of those considering some level of transiency.

Keywords: Fractured geomaterials · Inter-porosity flow · Mass exchange Dual porosity

1 Introduction

Fractures and discontinuities are commonly important features in geological formations and can have a significant bearing on the water and gas flows and reactive chemical transport. They effectively divide a geomaterial into two distinct porosities, namely, the fracture network and the porous rock matrix [1]. A number of modelling techniques are available to allow the heterogeneous pore structure of a dual porosity geomaterial in a form more amenable to numerical treatment. In broad terms, these are discrete fracture

L. Zhang et al. (Eds.): GSIC 2018, *Proceedings of GeoShanghai 2018 International Conference: Rock Mechanics and Rock Engineering*, pp. 30–38, 2018.
https://doi.org/10.1007/978-981-13-0113-1_4

network (DFN) models, equivalent continuum models, and dual (or higher) porosity models [2]. The most appropriate type of model for a given application depends largely on the problem scale/conditions, the available input data, the type of output data required, and the available computational resources [1, 3]. A discussion of the relative merits of each approach can be found in Hosking et al. [4], whilst the work presented here focuses on the dual porosity model and specifically the mass exchange term used to couple the fracture and matrix flows.

Dual porosity models consider the fracture network and porous matrix as distinct continua over the domain. This approach intends to overcome the loss in accuracy of homogenised, equivalent continuum models as the partition between the fracture and matrix flows becomes more apparent. To reflect the material properties of most fractured rocks, it is generally true that the fracture continuum provides the majority of the flow capacity and the matrix continuum provides the majority of the storage capacity. The fracture continuum is then more highly conductive with a lower porosity and the matrix is poorly (or non-) conductive with a higher porosity [5]. Provided representative properties can be assigned to the respective continua, the capability of the model to predict the salient transport behaviour of fractured rock requires an accurate description of the inter-porosity mass exchange.

Mass exchange models relying on the assumption of a pseudo-steady pore pressure distribution in the matrix are common in the study of dual porosity systems, including water and solute transport in structured porous media [6, 7] and gas flow in coal [8, 9]. This assumption neglects transient pressure gradients in the matrix and allows the rate of inter-porosity mass exchange to be conveniently expressed as a linear function of the difference between the average pore pressures in the fracture and matrix domains. Whilst providing a relatively simple description of mass exchange, the accuracy of this approach has been challenged [10, 11] and it is strictly only valid once the pressure front due to changing conditions in the fracture network reaches the centre of the matrix [6].

This paper examine the soundness of assuming pseudo-steady state mass exchange in dual porosity models. The test cases considered deal with the mass exchange of an ideal, inert gas under isothermal conditions in rigid coal with a parallel-plate geometry, with benchmarks provided by an explicit model of diffusive flow into the coal matrix slabs. Since the mass exchange coefficient is sometimes described as a function of matrix block geometry and effective diffusivity, an attempt is made to adopt this approach in preference to curve fitting. The results presented in this work have been obtained using the coupled thermal, hydraulic, chemical and mechanical (THCM) model, COMPASS, developed incrementally at the Geoenvironmental Research Centre [12, 13].

2 Theoretical Models for Inter-porosity Mass Exchange

Based on a theoretical formulation that can be described as a mechanistic approach, the coupled THCM model used in this work has a background of high performance simulations of three-dimensional multiphase, multicomponent reactive transport in single porosity geomaterials. Recent developments to this platform, presented by

Hosking et al. [4], have extended the capabilities by introducing a dual porosity framework and non-ideal gas behaviour. These new capabilities are employed here alongside the benchmark simulations, which have been performed using the single porosity model assuming purely diffusive gas flow in the coal matrix. A summary of both formulations is provided in the following sections.

2.1 Single Porosity Model for Diffusive Mass Exchange (Benchmark Case)

Based on the principle of conservation of mass, the temporal derivative of the gas chemical accumulation is equal to the spatial gradient of the relevant fluxes. Gas flow from the fracture network into the fully dry, non-deformable coal matrix is assumed to be driven purely by diffusion, which for an ideal, inert gas yields:

$$n\frac{\partial c_g}{\partial t} = \nabla \cdot \left[D_{e,g}\nabla c_g\right] \tag{1}$$

where n is the matrix porosity, c_g is the gas concentration, and $D_{e,g}$ is the gas effective diffusion coefficient, given by [14]:

$$D_{e,g} = n\tau_g D_g \tag{2}$$

where the tortuosity factor, τ_g, is calculated using the relationship by Millington and Quirk [15]:

$$\tau_g = n^{4/3} \tag{3}$$

The gas pressure, u_g, is calculated from c_g, the temperature, T, and the universal gas constant, R, using the ideal gas law, i.e. $u_g = RTc_g$.

2.2 Dual Porosity Model with Pseudo-steady State Mass Exchange

Mass exchange for the dual porosity case is handled through the addition of a sink/source term to the mass balance expression given in Eq. (1). Since this work is concerned with mass exchange and not the bulk flow of gas in the coal, the diffusive flux term is removed, yielding:

$$n_\beta \frac{\partial c_{g,\beta}}{\partial t} = \lambda \Gamma_g \tag{4}$$

where the subscript β is the continuum identifier and becomes F to denote the fracture continuum and M to denote the matrix continuum, Γ_g is the sink/source term for mass exchange, and $\lambda = -1$ if $\beta = F$ or $\lambda = 1$ if $\beta = M$.

The fracture continuum porosity, n_F, is the fraction of the total porosity associated with the fracture network, given by:

$$n_\beta = w_f n_F^L \tag{5}$$

where n_F^L is the local fracture porosity, expressed mathematically as the volume of pores in the fractured network divided by the total volume of the fracture network, i.e. V_F^P / V_F^T. This becomes 1.0 in a clean fracture, but may be less due to mineral infillings and the presence of altered matrix surrounding the fracture. w_f is the volumetric weighting factor, defined as the total volume of the fracture network divided by the bulk rock volume, i.e. $w_f = V_F^T / V_T$, analogous to the following expression for a parallel-plate geometry:

$$w_f = \frac{a_F}{b_F} \tag{6}$$

where a_F is the fracture aperture and b_F is the fracture spacing.

Equation (5) allows the matrix continuum porosity, n_M, to be expressed in terms of the total porosity, n_T, w_F, and n_F^L, giving:

$$n_M = n_T - w_F n_F^L \tag{7}$$

Assuming a pseudo-steady state gas pressure distribution in the matrix slabs at all times, Γ_g becomes a first-order mass exchange term expressed in general as [16]:

$$\Gamma_g = \sigma_D (c_{g,F} - c_{g,M}) \tag{8}$$

where σ_D is the first-order exchange rate, which may be expanded to consider geometrical and material properties, including the coal matrix shape and dimensions and the effective diffusivity of the gas, giving an expression of the form [17]:

$$\sigma_D = \frac{4\psi}{b_M^2} D_{e,g,M} \tag{9}$$

where ψ is a dimensionless shape factor related to the geometry of the matrix blocks, ranging from 3 for rectangular slabs (as in the parallel-plate model in this work) to 15 for spherical aggregates [6].

3 Problem Conditions

Three coal slab geometries are considered in this work, as shown in Fig. 1, taken from Laubach et al. [18] for coal from the San Juan and Black Warrior Basins, USA. Specifically, fracture apertures of 5, 7 and 9 μm have been selected for Tests I, II and III, respectively. The adopted values of fracture spacing correspond to a maximum

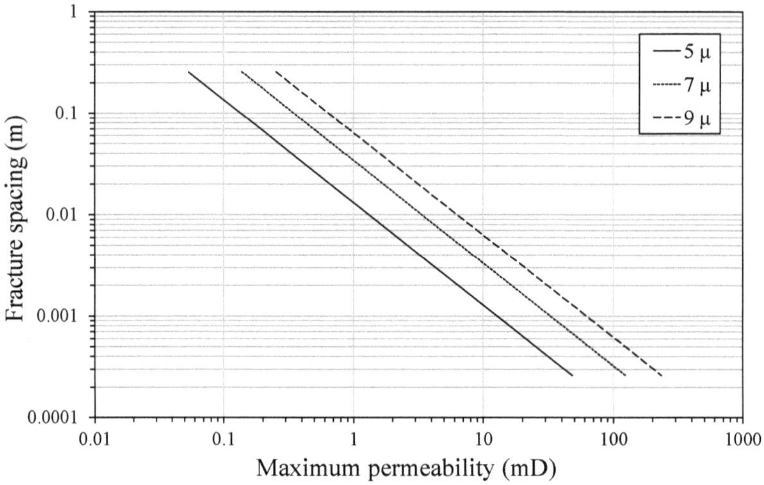

Fig. 1. Relationship between coal permeability and fracture spacing for the three fracture apertures considered in this work (adapted from Laubach et al. [18]). The values of fracture spacing used in the test cases correspond to a maximum permeability of 1 mD.

permeability of 1 mD, giving a_F as 0.013 m, 0.035 m and 0.065 m, respectively. A total porosity of 0.01 was prescribed in each test, with $n_F^L = 1.0$.

The geometrical parameters described above were used to define the terms of the dual porosity Eqs. (4)–(9). The domain for the dual porosity simulations was arbitrary since no bulk flow occurs (only mass exchange via the sink/source term). Owing to the symmetry of the parallel-plate model, the 2-dimensional domains for the benchmark diffusive mass exchange simulations were formed using 0.001 m wide sections of coal slab with thicknesses of $0.5(b_F - a_F)$, i.e. the half width of a single slab. The coal matrix initially contains no gas with the fracture containing gas at a pressure of 100 kPa. The simulation period in each Test corresponded to the time taken for the coal matrix pressure to equilibrate with the prescribed fracture pressure.

4 Results and Discussion

The results for Tests I, II and II are shown in Figs. 2, 3 and 4, respectively. Markers are used to denote the results of the benchmark tests in which diffusive mass exchange into the matrix slabs was considered explicitly. These results show the evolution of gas pressure at the mid-point of the domain, i.e. a quarter into the matrix slab thickness. The dual porosity results for case A used Eq. (9) to define the mass exchange rate, σ_D, whereas for case B the best-fit values were used.

As expected, the benchmark results follow the same trend in each test, reflecting the transient gas pressure gradient in the matrix slab as it tends towards equilibrium with

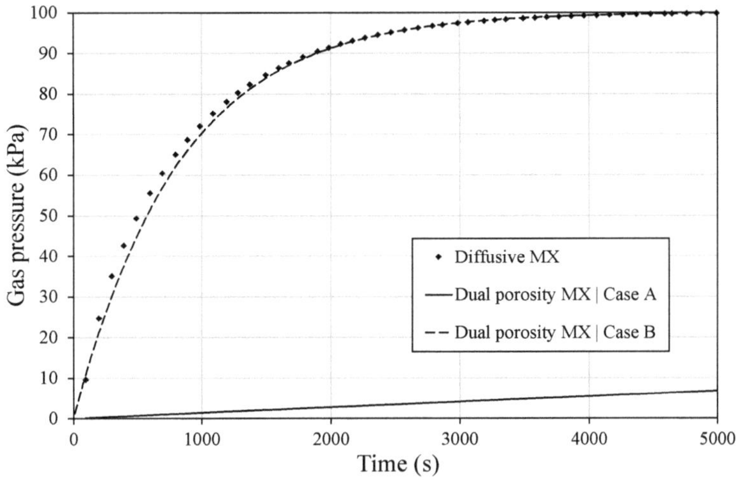

Fig. 2. Results for Test I with $a_F = 5$ µm and $b_F = 0.013$ m, giving $w_F = 3.84 \times 10^{-4}$. Case A used Eq. (9) to give $\sigma_D = 1.4 \times 10^{-5}\,\mathrm{s}^{-1}$, with case B using the best fit value of $\sigma_D = 1.5 \times 10^{-3}\,\mathrm{s}^{-1}$.

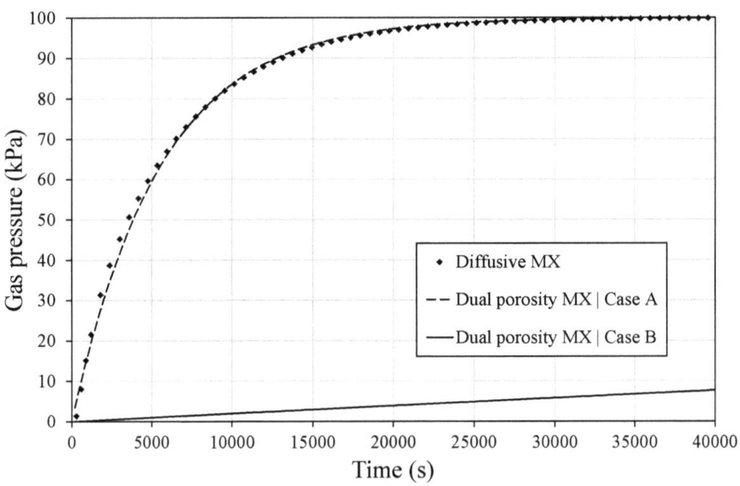

Fig. 3. Results for Test II with $a_F = 7$ µm and $b_F = 0.035$ m, giving $w_f = 2.00 \times 10^{-4}$. Case A used Eq. (9) to give $\sigma_D = 2.0 \times 10^{-6}\,\mathrm{s}^{-1}$, with case B using the best fit value of $\sigma_D = 1.8 \times 10^{-4}\,\mathrm{s}^{-1}$.

the fracture pressure. The only notable difference is in the time scale, which naturally increases with the thickness of the matrix slabs. It can be seen that in all three tests case A underestimated the rate of mass exchange by a considerable margin relative to the benchmark. The results for case B show that the pseudo-steady state model yielded a

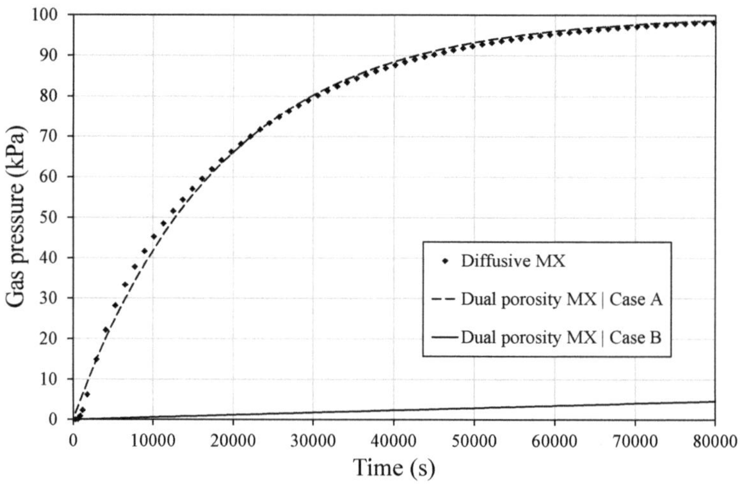

Fig. 4. Results for Test III with $a_F = 9\,\mu\text{m}$ and $b_F = 0.065$, giving $w_f = 1.38 \times 10^{-4}$. Case A used Eq. (9) to give $\sigma_D = 5.9 \times 10^{-7}\,\text{s}^{-1}$, with case B using the best fit value of $\sigma_D = 5.4 \times 10^{-5}\,\text{s}^{-1}$.

close agreement with the benchmark only when the values of σ_D from case A were increased by in the region of 2 orders of magnitude.

These findings imply that the assumption of pseudo-steady state mass exchange may be valid only for cases where calibration of the linear coefficient is possible. In other words, σ_D appears to be a largely empirical value. The use of constitutive relationships describing the coefficient should be approached with care, with the possible exception of those considering some level of transiency, such as those reviewed by Hassanzadeh et al. [10]. Transiency is typically included by taking ψ in Eq. (9) as a function of time. In effect, the rate of mass exchange is increased substantially at early times and decays towards the pseudo-steady state at late times. Referring to the findings in this work, the use of such a transient function would lead to an improved agreement between case A and the benchmark. However, whether this approach improves the theoretical robustness of σ_D beyond being a largely empirical factor is debatable, depending on the exact definition used. Suitable functions may consider different boundary conditions at the fracture-matrix interface (pressure- or flux-dependent), different geometries, and the properties of the solid and pore fluid, including multiphase flow.

5 Conclusions

This paper has investigated the soundness of assuming a pseudo-steady rate of inter-porosity (fracture-matrix) mass exchange in dual porosity numerical models. Three test scenarios have been presented for coal covering a range of fracture apertures and matrix slab thicknesses assuming a rigid parallel-plate geometry. Each test considered the inter-porosity (fracture-matrix) flow of an ideal, inert gas under isothermal conditions. The predicted pseudo-steady state mass exchange rates have been compared with benchmarks provided by explicitly modelling diffusive flow into the coal matrix slabs, thereby taking into account the transient pressure gradient neglected in the pseudo-steady state approach. Since the mass exchange coefficient is sometimes described as a function of matrix block geometry and effective diffusivity, an attempt was made to adopt this approach (case A) before curve fitting was used (case B).

It was found that in all three tests case A underestimated the rate of mass exchange by a considerable margin relative to the benchmark. The results for case B showed that the pseudo-steady state model yielded a close agreement with the benchmark only when the values of the mass exchange rate from case A were increased by in the region of 2 orders of magnitude.

It is concluded that the assumption of pseudo-steady state mass exchange may be valid only for cases where calibration of the linear coefficient is possible. Constitutive relationships describing the coefficient should be approached with care, with the possible exception of those considering some level of transiency.

References

1. Bear, J.: Modeling flow and contaminant transport in fractured rocks. In: Bear, J., Tsang, C.-F., de Marsily, G. (eds.) Flow and Contaminant Transport in Fractured Rock, pp. 1–38. Academic Press, Inc., San Diego (1993)
2. Therrien, R., Sudicky, E.A.: Three-dimensional analysis of variably-saturated flow and solute transport in discretely-fractured porous media. J. Contam. Hydrol. **23**(1–2), 1–44 (1996)
3. Samardzioska, T., Popov, V.: Numerical comparison of the equivalent continuum, non-homogeneous and dual porosity models for flow and transport in fractured porous media. Adv. Water Resourc. **28**(3), 235–255 (2005)
4. Hosking, L.J., Thomas, H.R., Sedighi, M.: A dual porosity model of high pressure gas flow for geoenergy applications. Can. Geotech. J. (2017, submitted)
5. Xu, T.F., Pruess, K.: Modeling multiphase non-isothermal fluid flow and reactive geochemical transport in variably saturated fractured rocks: 1. Methodology. Am. J. Sci. **301**(1), 16–33 (2001)
6. Gerke, H.H., van Genuchten, M.T.: A dual-porosity model for simulating the preferential movement of water and solutes in structured porous-media. Water Resourc. Res. **29**(2), 305–319 (1993)
7. Ma, D.H., Shao, M.G.: Simulating infiltration into stony soils with a dual-porosity model. Eur. J. Soil Sci. **59**(5), 950–959 (2008)
8. Wu, Y., et al.: Dual poroelastic response of a coal seam to CO2 injection. Int. J. Greenhouse Gas Control **4**(4), 668–678 (2010)
9. Chen, D., et al.: Modeling and simulation of moisture effect on gas storage and transport in coal seams. Energy Fuels **26**(3), 1695–1706 (2012)
10. Hassanzadeh, H., Pooladi-Darvish, M.: Effects of fracture boundary conditions on matrix-fracture transfer shape factor. Transp. Porous Media **64**(1), 51–71 (2006)
11. Lemonnier, P., Bourbiaux, B.: Simulation of naturally fractured reservoirs. state of the art-Part 2–matrix-fracture transfers and typical features of numerical studies. Oil Gas Sci. Technol.–Revue de l'Institut Français du Pétrole **65**(2), 263–286 (2010)

12. Thomas, H.R., He, Y.: Modelling the behaviour of unsaturated soil using an elasto-plastic constitutive relationship. Géotechnique **48**(5), 589–603 (1998)
13. Sedighi, M., Thomas, H.R., Vardon, P.J.: Reactive transport of chemicals in unsaturated soils: numerical model development and verification. Can. Geotech. J. **53**(1), 162–172 (2016)
14. Cussler, E.L.: Diffusion: Mass Transfer in Fluid Systems, 2nd edn. Cambridge University Press, Cambridge (1997)
15. Millington, R., Quirk, J.P.: Permeability of porous solids. Trans. Faraday Soc. **57**(8), 1200–1207 (1961)
16. Gwo, J., et al.: A multiple-pore-region concept to modeling mass transfer in subsurface media. J. Hydrol. **164**(1–4), 217–237 (1995)
17. Schwartz, R.C., Juo, A.S.R., McInnes, K.J.: Estimating parameters for a dual-porosity model to describe non-equilibrium, reactive transport in a fine-textured soil. J. Hydrol. **229**(3–4), 149–167 (2000)
18. Laubach, S.E., et al.: Characteristics and origins of coal cleat: a review. Int. J. Coal Geol. **35** (1–4), 175–207 (1998)

Analytical Prediction of Stresses Around Non-circular Tunnels Excavated at Shallow Depth

Hua Ning Wang[1(✉)], Guang Shang Zeng[1], and Ming Jing Jiang[2]

[1] School of Aerospace Engineering and Applied Mechanics, Tongji University, Shanghai 200092, China
wanghn@tongji.edu.cn

[2] Department of Geotechnical Engineering, College of Civil Engineering, Tongji University, Shanghai 200092, China

Abstract. The excavation of non-circular tunnels at shallow depths is becoming increasingly common because of the need to efficiently utilize excavation spaces. However, the stress concentration is likely much greater around shallow tunnels than deep ones. This study provides an analytical solution of stresses around non-circular tunnels at shallow depths by combining the Schwartz alternating method with complex variable theory. The analytical solutions are verified by good agreement between the analytical and FEM numerical results. A parametric study is performed to investigate stress concentrations around elliptical and square tunnels and the effect of tunnels' buried depth. The proposed analytical solutions provide great value to the conceptual understanding of the mechanical behaviour of non-circular tunnels, which are shallowly excavated in rock or medium to stiff clay, and to the verification of numerical models.

Keywords: Non-circular tunnel · Shallow tunnel · Analytical solution

1 Introduction

Non-circular tunnels constructed in urban areas are increasingly commonplace in order to reduce the need for backfilling [1, 2]. However, because of the inferior tunnel shape, a higher stress concentration is likely to arise around non-circular tunnels than circular ones. Moreover, the stress concentration around a shallow tunnel is further intensified by the proximity to the ground surface. This study focuses on the prediction of the stresses induced by the excavation of shallow non-circular tunnels that have significant impacts on tunnel stability.

Numerical modelling [3, 4] and limit analysis [5, 6] are usually employed to predict the ground responses around shallow buried non-circular tunnels. The numerical method is mainly used in detailed tunnel designs, with more consideration given to complex geo-material properties and geological conditions. However, they require long running-times, especially in cases where complete parametric analyses must be performed. Furthermore, the stresses at the regions with a high stress concentration (*e.g.*, stress at square corners) are sometimes difficult to accurately predict via numerical approaches.

L. Zhang et al. (Eds.): GSIC 2018, *Proceedings of GeoShanghai 2018 International Conference: Rock Mechanics and Rock Engineering*, pp. 39–47, 2018.
https://doi.org/10.1007/978-981-13-0113-1_5

In contrast, analytical models can provide the exact stresses and displacement field through strict mathematical derivations with respect to all relevant parameters. They also provide an approach for gaining deep insight into the mechanical mechanism of engineering problems [7]. Most of the analytical studies for non-circular tunnels are mainly focused on deeply-buried tunnels in elastic or viscoelastic ground [8–10].

Regarding the literature reports on analytical methods for shallow buried tunnels, Verruijt derived an analytical solutions for a circular tunnel in an elastic half-plane [11, 12]. Because of the difficulties in finding the direct conformal mapping, the exact analytical solutions for non-circular shallow tunnels subjected to arbitrary boundary stresses are still unavailable in the existing literature. Although the analytical stress expressions for elliptical holes in an anisotropic half-plane were proposed by Dai et al. [13], their solution is only valid for cases where the hole boundary is loaded by particular tension stresses, since these cases can be successfully solved mathematically. In this study, the stresses induced by the excavation of a non-circular tunnel at semi-infinite ground will be analytically derived to provide an alternative approach for preliminary designs.

2 Derivation and Verification of Analytical Solutions

2.1 Assumptions and Problem Definition

A non-circular tunnel contained in semi-infinite ground is investigated where the ground is in a state of small deformation and can be approximately regarded as a homogeneous, isotropic, and linearly elastic medium. Plain-strain conditions are assumed, neglecting the three-dimensional effect of longitudinal tunnel advancement. The additional stresses due to tunnelling can be addressed by the model of a tunnel whose boundary is loaded by the traction $-S_0^1$, where S_0^1 (S_{0x}^1 and S_{0y}^1 are horizontal and vertical components) is the initial stresses along anticipated tunnel boundary. To simplify the derivation, the variation of gravity-induced vertical initial stress across the height of excavation is not considered since it has been proven that the accuracy is sufficient in the prediction of the plastic zone of a shallow tunnel [14] when the buried depth is greater than two times the tunnel radius. Therefore, the uniform initial stress state, which equals the stresses at the tunnel centre, can be assumed. Figure 1 illustrates the tunnel geometry, boundary conditions, and reference coordinate systems in the derivation. The sign conventions are defined as follows: compressive stresses and displacements along the direction of the axes of the coordinate are positive.

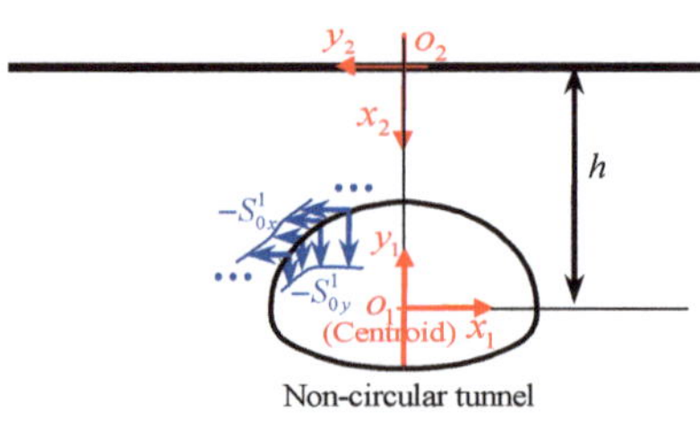

Fig. 1. Geometry, boundary conditions of the problem, and reference coordinate systems.

2.2 Basic Iterative Procedure

A hole in a half-plane belongs to the set of problems with a doubly connected domain. Schwartz [15] proposed an efficient iteration approach for finding the solution of boundary value problems on doubly connected domains by solving the problem on each of the two simply connected domains in turn, taking always the previous values of the approximate solutions as the next boundary conditions. This method has been employed in solving the problem of twin circular tunnels in viscoelastic media [16].

Figure 2 presents the iterative procedure for solving the problem of a half-plane with a non-circular hole subjected to gravity-induced tractions along the hole boundary, where the two following problems are repeatedly solved: Problem (1) is a non-circular hole in an infinite medium subjected to tractions along the hole boundary, and Problem (2) is a half-plane without any hole loaded by tractions along ground surface.

Note from Fig. 2 that the redundant tractions are induced along the removed boundaries during each iterative step. Therefore, tractions with reverse directions and the same magnitude as the redundant tractions (the reverse tractions) are then applied to the corresponding boundary to balance the redundant tractions in the next iterative steps. Through repeated iterations, the redundant tractions along the hole boundary and the ground surface will approach zero with the increase of the iteration number. The solutions are finally attained by superimposing the stresses calculated in all the iterations.

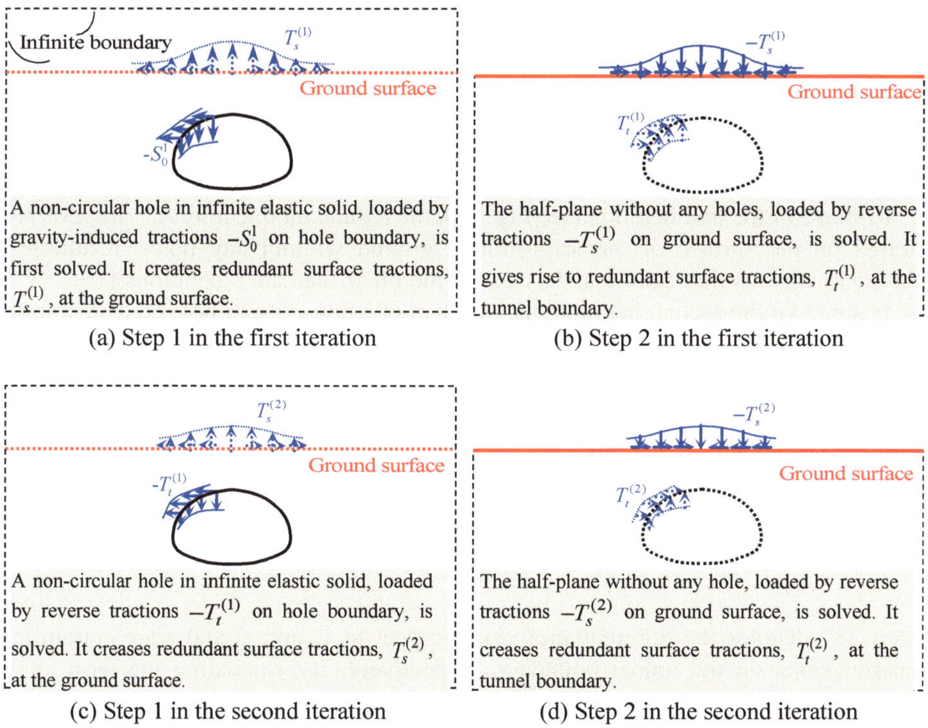

(a) Step 1 in the first iteration (b) Step 2 in the first iteration

(c) Step 1 in the second iteration (d) Step 2 in the second iteration

Fig. 2. Boundary conditions in alternating iterations.

2.3 Potential Determination in Iterative Steps

The derivation is provided for shallow square tunnels as an example. In step 1 of the first iteration, the specific approximate mapping function for square holes is in the following equation, in which the hole boundary and the external infinite region is mapped into a unit circle in the mapped plane [17]:

$$z_1 = \omega(\zeta) = R_0\left(\zeta - \frac{1}{6\zeta^3}\right), \text{ with } \begin{pmatrix} z_1 = x_1 + y_1 i \\ \zeta = \rho e^{i\theta} = \xi + \eta i \end{pmatrix} \tag{1}$$

where z_1 is the complex coordinates of the systems $x_1 o_1 y_1$, ζ is the complex coordinates in the corresponding mapped planes, and R_0 is the coefficient related to tunnel size. The closed-form expressions of the potentials in complex variable theory are obtained as follows [18]:

$$\begin{aligned} \varphi_1^{(1)}(\zeta) &= -\frac{3(1-K_0)p_0R_0}{7\zeta} + \frac{(1+K_0)p_0R_0}{12\zeta^3} \\ \psi_1^{(1)}(\zeta) &= -\frac{39(1-K_0)p_0R_0\zeta}{84\zeta^4+42} - \frac{13(1+K_0)p_0R_0\zeta^3}{24\zeta^4+12} + \frac{(1-K_0)p_0R_0}{12\zeta^3} \end{aligned} \tag{2}$$

where $p_0 = \gamma h$ - with γ being the unit weigh of geo-material- is the vertical initial stress due to gravity at tunnel centre, and K_0 is the ratio of horizontal over vertical initial stresses (*i.e.*, the coefficient of lateral pressure). By substituting the inverse function of Eq. (1) into (2), and using the stress expressions with respect to potentials in complex variable theory, the stress with respect to the variables x_1 and y_1 can be achieved. Accordingly, the redundant tractions along the ground surface, $T_s^{(1)}$, are obtained by substituting $z_1 = x_1 + hi$ into the stress expressions.

In step 2 of the first iteration (see Fig. 2b), the redundant tractions $T_s^{(1)}$ are reversely exerted on the surface of the semi-infinite ground without any holes, resulting in additional stresses that can be addressed by modified Flamant's solutions [19].

In step 1 of the second iteration, the problem of a non-circular hole contained in the infinite medium is tackled, where the redundant tractions $T_{t-x_1}^{(1)}$ and $T_{t-y_1}^{(1)}$, yielded in step 2 of the first iteration, are reversely exerted on the hole boundary. Suppose that $\varphi_1^{(2)}(\zeta)$ and $\psi_1^{(2)}(\zeta)$ are the two potentials in this step. Thus, the potentials should satisfy the stress boundary conditions (hole boundary) as follows:

$$\left[\varphi_1^{(2)}(\zeta) + \frac{\omega(\zeta)}{\overline{\omega'(\zeta)}}\overline{\varphi_1^{(2)'}(\zeta)} + \overline{\psi_1^{(2)}(\zeta)}\right]_{\zeta=t=e^{i\theta}} = -i\int_{t_0}^{t}\left(T_{t-x_1}^{(1)}(\theta) + iT_{t-y_1}^{(1)}(\theta)\right)ds \tag{3}$$

where $(\ )'$ denotes the differentiation with respect to ζ, and t_0 and t are certain and generic points on the tunnel boundary, respectively. By expanding the term at the right side of Eq. (3) as a series, *i.e.*, $-i\int_{t_0}^{t}\left(T^{(1)}_{t-x_1}(\theta)+iT^{(1)}_{t-y_1}(\theta)\right)ds \approx \sum_{j=-L}^{L} C_j t^j$ with $t=e^{i\theta}$, we obtain the potentials in the following [18]:

$$\varphi_1^{(2)}(\zeta)=[(6/7)\mathrm{Re}[C_{-1}]+(6/5)\mathrm{Im}[C_{-1}]i]\zeta^{-1}+\sum_{j=-L}^{-2}C_j\zeta^j$$

$$\psi_1^{(2)}(\zeta)=\frac{13[(6/7)\mathrm{Re}[C_{-1}]+(6/5)\mathrm{Im}[C_{-1}]i]\zeta}{12\zeta^4+6}-\frac{6\zeta^3-\zeta^7}{6\zeta^4+3}\sum_{j=-L}^{-2}jC_j\zeta^{j-1}+\sum_{j=1}^{L}\frac{\overline{C_j}}{\zeta^j} \quad (4)$$

Through the subsequent repeated iterations, the redundant tractions along square tunnel boundary and ground surface will tend towards zero with the increase in the number of iterations. Hence, the solutions of the non-circular hole in the half-plane are finally obtained via the superposition of the stresses calculated in all the iterations.

2.4 Verification of Analytical Solution

To verify the analytical solution presented in this study, the derived solutions were compared with the results from numerical simulations carried out using the code ANSYS.

Consider a square tunnel that has side lengths of 5 m and was excavated in semi-infinite ground at depth h = 15 m. Because of the symmetry of the problem along the y_1 axis, the right half of the structure was analysed in numerical simulation (geometry and the constraint conditions shown in Fig. 3). The parameters employed in the analysis, *i.e.*, the elastic modulus E, Poisson's ratio v and unit weight γ of the geo-material, are 50 MPa, 0.25 and 18.5 kN/m^3, respectively. The coefficient of lateral pressure, K_0, is adopted as 0.35.

Let us consider the points P_1–P_6 shown in Fig. 3. The comparisons of total stresses at these points between the analytical solutions and FEM results are plotted in Fig. 4, where the analytical results are superposed by those in the first and second iterations, as

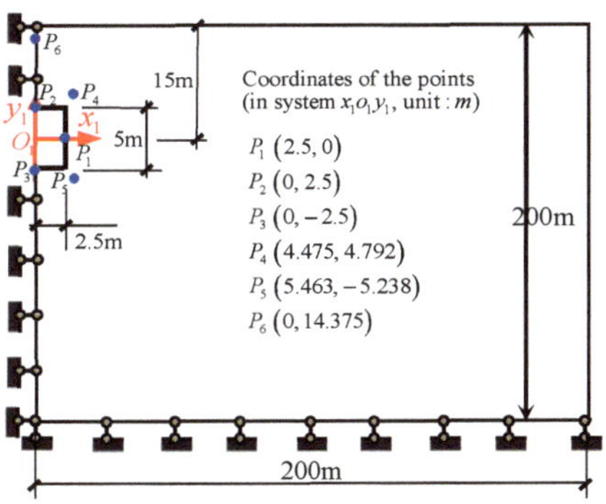

Fig. 3. Geometry of the numerical domain together with the constraint conditions.

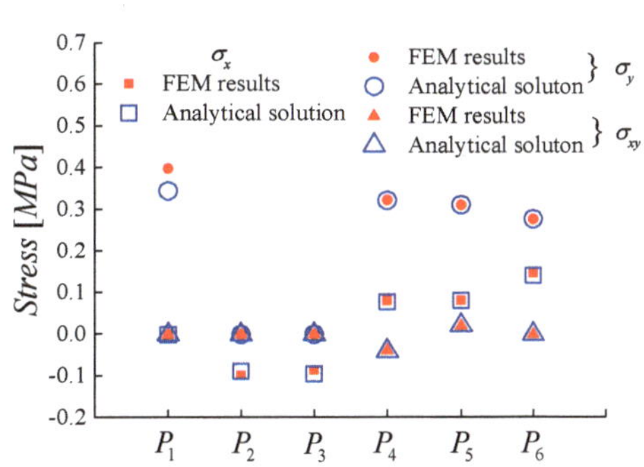

Fig. 4. Comparison of the elastic stresses between analytical and FEM results.

well as step 1 of the third iteration. It is shown that the analytical solutions exhibit close agreement with the numerical results.

3 Parametric Investigation

The stress concentration, as well as the effect of a tunnel's buried depth on the stresses around elliptical and square tunnels, are investigated in this section. For the sake of generality, the stresses are normalized by $p_0 = \gamma h$. All parameters are the same as those in Sub-sect. 2.4. The shapes of the tunnels are shown in Fig. 5.

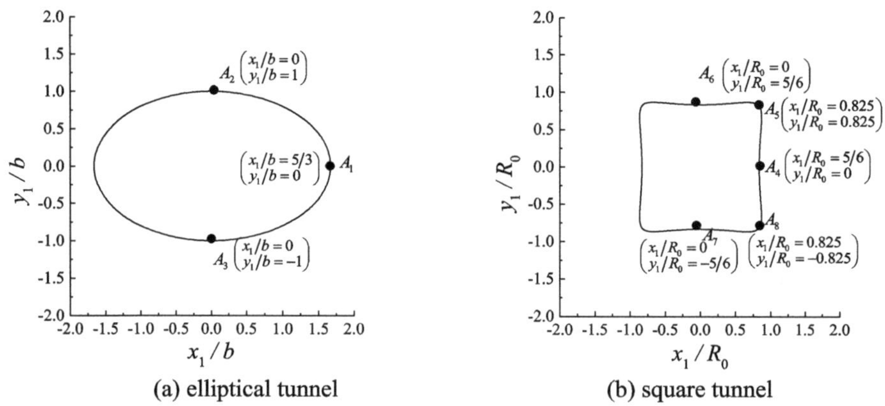

Fig. 5. Geometry of tunnels.

Assume that tunnels with elliptical and square cross-sections are located at the depth $h = 3b$ ($h = 3R_0$), where b is the half minor axis of the elliptical tunnel and R_0 is defined in Eq. (1). The tunnel boundaries, determined by the mapping functions, are illustrated in Fig. 5. The normalized hoop stresses along the tunnel boundary versus θ are plotted in Fig. 6. Owing to the proximity to the ground surface, the tensile stress along the upper boundary of the elliptical tunnel (around $\theta = 90°$) is larger than that along the lower boundary (around $\theta = 270°$) (see Fig. 5a); and the normalized hoop stress at the top two corners of the square tunnel is 5. However, the one at the bottom two corners is only 4.5 (see Fig. 6b).

It can be noted from Fig. 6(a) that the highest compressive stress concentrations around the elliptical tunnel appear at the boundary points $\theta = 0^{\circ}$ and 180°, where the concentration factor $\sigma_\theta/p_0 = 4.9$ is larger than the one around the deep tunnels ($\sigma_\theta/p_0 = 3.98$). The maximum normalized tensile stress of the elliptical tunnel is approximately 0.77, and appears at a point on the tunnel's crown at $\theta = 90^{\circ}$. It can be noted from Fig. 6(b) that the highest compressive stress concentrations occur around the square's corners, while the hoop stresses are mainly tensile along the top and bottom. It can also be observed that the maximum normalized tensile stress is 0.47 (0.29 for the deep square tunnel), and appears at the midpoint of the tunnel's top boundary.

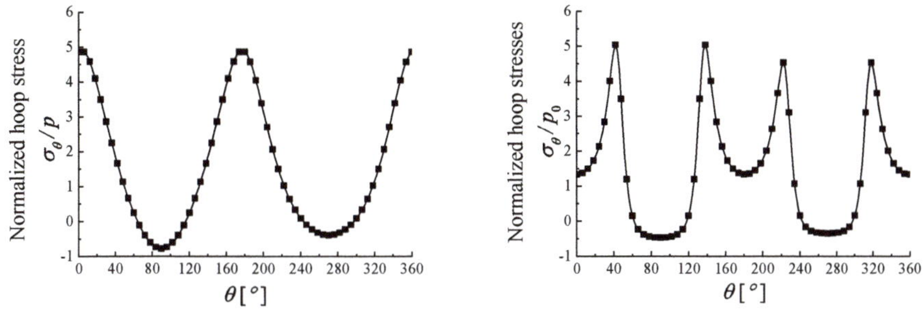

(a) elliptical tunnel (b) square tunnel

Fig. 6. Normalized hoop stresses along tunnel boundary versus θ.

Let us consider points A_1–A_8 around the elliptical/square tunnels shown in Fig. 6. The normalized hoop stresses of these points versus the buried depth *(h)* are plotted in Fig. 7. In general, the hoop stresses all decrease gradually as buried depth increases, and finally tend to become steady values. When the buried depth is greater than $5b$ for the elliptical tunnels ($5R_0$ for square tunnels), the stresses are almost unchanged against h, which indicates that the tunnels can be regarded as deep. The stresses at points A_2 and A_3 (or A_5 and A_8, A_6 and A_7), which are symmetrically located at the tunnel's top and bottom, are very different when the tunnel is shallowly buried; they tend to be equal when h is greater than $5b$ (or $5R_0$) due to the lesser influence of the ground surface. It can also be observed that the stresses at point A_1, A_2, A_5 and A_6, are influenced more significantly by the buried depth than are stresses at other points, as shown in Fig. 7(b).

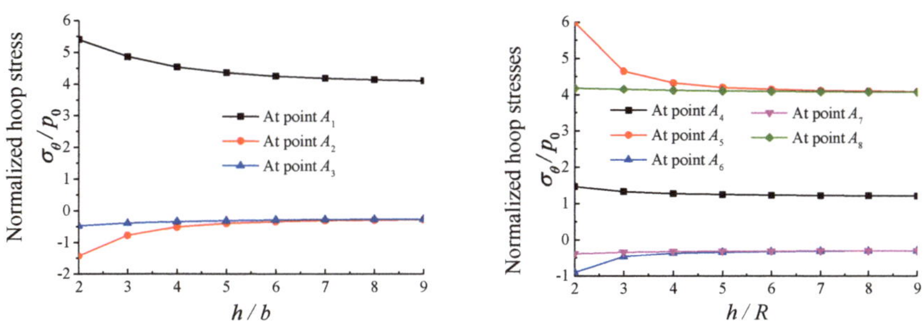

(a) elliptical tunnel (b) square tunnel

Fig. 7. The normalized hoop stresses versus h.

4 Conclusions

In this study, the stress solutions for shallow, non-circular tunnels were analytically proposed, and a good level of agreement was exhibited between the analytical solutions and FEM results. By conducting a parametric investigation, it was shown that the stress concentration of shallow tunnels is larger than that of corresponding deeply buried tunnels with the same shape. It was also shown that when the buried depth is greater than $5b$ for the elliptical tunnels ($5R_0$ for square tunnels), the tunnels can be regarded as deep. The proposed analytical solutions should be of great value for the conceptual understanding of the mechanical behaviour of shallow buried non-circular tunnels and for verifying numerical models.

Acknowledgements. This study was financially supported by the National Natural Science Foundation of China (Grant Nos. 11572228 & 51639008); and the Fundamental Research Funds for the Central Universities. Their support is greatly appreciated.

References

1. Mori, K., Abe, Y.: Large rectangular cross-section tunneling by the multi-micro shield tunneling (MMST) method. Tunn. Undergr. Space Technol. **20**(2), 129–141 (2005)
2. Fang, Y.S., Kao, C.C., Shiu, Y.F.: Double-O-Tube shield tunneling for Taoyuan international airport access MRT. Tunn. Undergr. Space Technol. **30**(4), 233–245 (2012)
3. Abbo, A.J., Wilson, D.W., Sloan, S.W., Lyamin, A.V.: Undrained stability of wide rectangular tunnels. Comput. Geotech. **53**(3), 46–59 (2013)
4. Wilson, D.W., Abbo, A.J., Sloan, S.W., Yamamoto, K.: Undrained stability of rectangular tunnels where shear strength increases linearly with depth. Can. Geotech. J. **54**, 469–480 (2017)
5. Yamamoto, K., Lyamin, A.V., Sloan, S.W., Wilson, D.W., Abbo, A.J.: Stability of a single tunnel in cohesive–frictional soil subjected to surcharge loading. Can. Geotech. J. **48**, 1841–1854 (2011)
6. Pan, Q., Dias, D.: Upper-bound analysis on the face stability of a non-circular tunnel. Tunn. Undergr. Space Technol. **62**, 96–102 (2017)
7. Carranza-Torres, C., Fairhurst, C.: The elastoplastic response of underground excavations in rock masses that satisfy the Hoek-Brown failure criterion. Int. J. Rock Mech. Min. Sci. **36**, 777–809 (1999)
8. Kargar, A.R., Rahmannejad, R., Hajabasi, M.A.: A semi-analytical elastic solution for stress field of lined non-circular tunnels at great depth using complex variable method. Int. J. Solids Struct. **51**, 1475–1482 (2014)
9. Exadaktylos, G.E., Stavropoulou, M.C.: A closed-form elastic solution for stresses and displacements around tunnels. Int. J. Rock Mech. Min. Sci. **39**(7), 905–916 (2001)
10. Wang, H.N., Utili, S., Jiang, M.J., He, P.: Analytical solutions for tunnels of elliptical cross-section in rheological rock accounting for sequential excavation. Rock Mech. Rock Eng. **48**(5), 1997–2029 (2015)
11. Verruijt, A.: A complex variable solution for a deforming circular tunnel in an elastic half-plane. Int. J. Numer. Anal. Meth. Geomech. **21**, 77–89 (1997)
12. Verruijt, A.: Deformations of an elastic half plane with a circular cavity. Int. J. Solids Struct. **35**(21), 2795–2804 (1998)
13. Dai, M., Schiavone, P., Gao, C.F.: Surface tension-induced stress concentration around an elliptical hole in an anisotropic half-plane. Mech. Res. Commun. **73**, 58–62 (2016)
14. Massinas, S., Sakellariou, M.G.: Closed form solution for plastic zone formation around a circular tunnel in half-space obeying Mohr-Coulomb criterion. Géotechnique **60**(7), 691–701 (2009)
15. Sokolnikoff, I.S.: Mathematical Theory of Elasticity. Publisher, McGraw-Hill (1956)

16. Wang, H.N., Zeng, G.S., Utili, S., Jiang, M.J., Wu, L.: Analytical solution of stresses and displacements for deeply buried twin tunnels in viscoelastic rock. Int. J. Rock Mech. Min. Sci. **93**, 13–29 (2017)
17. Savin, G.N.: Stress Concentration Around Holes. Pergamon Press, London (1961)
18. Muskhelishvili, N.I.: Some Basic Problems of the Mathematical Theory of Elasticity. Noordhoff, Groningen (1963)
19. Wang, H.N., Chen, X.P., Jiang, M.J., Song, F., Wu, L.: The analytical predictions on displacement and stress around shallow tunnels subjected to surcharge loadings. Tunn. Undergr. Space Technol. **71**, 403–427 (2018)

Calibration of a Distinct Element Model for Rock Considering the Residual Strength

Lei Li[1,2,3], Mingjing Jiang[1,2,3(✉)], and Fang Liu[1,2,3]

[1] State Key Laboratory for Disaster Reduction in Civil Engineering, Tongji University, Shanghai, China
mingjing.jiang@tongji.edu.cn
[2] Key Laboratory of Geotechnical and Underground Engineering of Ministry of Education, Tongji University, Shanghai, China
[3] Department of Geotechnical Engineering, Tongji University, Shanghai, China

Abstract. The mechanical behavior of rocks at great depth is a research focus in the field of deep underground engineering in recent years. To simulate the mechanical behavior of such rocks, this study implemented an improved three-dimensional (3D) bond model incorporating rolling and twisting resistances into the distinct element method (DEM). The parameters of the distinct element model were calibrated according to the results from the triaxial tests on Berea sandstone by matching both the peak strengths and the residual strengths obtained from the tests under the confining pressure ranging from 10 MPa to 335 MPa. The method and technology roadmap of parameter calibration was proposed to capture the mechanical behavior of deep rock subjected to high stress and large deformation. The results show that the DEM simulation with the well-calibrated model quantitatively reproduced the stress-strain and volumetric strain curves through the whole process of the triaxial tests.

Keywords: Deep rock · Parameter calibration · Triaxial test
Distinct element method

1 Introduction

The distinct element method (DEM) is originally developed by Cundall and Strack [1] for granular materials treated as an assembly of bonded or un-bonded particles. It shows great potential in bridging macroscopic and microscopic responses of the assembly under a wide range of loading conditions. Recently, DEM has been widely applied to analyses of mechanical behavior of rocks such as crack extension [2, 3]. A rational and well-calibrated model is vital to enable a reliable DEM simulation of rocks.

In DEM simulations, model calibration is always a challenge due to a lack of testing techniques at the microscopic scale. In order to obtain reasonable model parameters, many efforts were devoted to calibrate parameters for various rock models [4, 5]. However, rare studies have successfully reached a model calibration that matches every major aspects of laboratory observations (i.e. the DEM simulation cannot quantitatively reproduce the stress-strain and the volumetric strain curves through the

L. Zhang et al. (Eds.): GSIC 2018, *Proceedings of GeoShanghai 2018 International Conference: Rock Mechanics and Rock Engineering*, pp. 48–56, 2018.
https://doi.org/10.1007/978-981-13-0113-1_6

whole process of triaxial tests). In particular, sacrifice in the residual strength is generally to reach a satisfactory fitting in the peak strength of rock samples, and the fitting of the volumetric strain even is usually given up without attempting.

In this paper, an improved 3D bond model incorporating rolling and twisting resistances was employed to calibrate the specimens of Berea sandstone under a wide range of confining pressures. The procedure is introduced to calibrate the model based on the triaxial tests of the rock considering the residual strength. The calibration is intended to quantitatively reproduce the complete stress-strain and volumetric strain curves observed in the laboratory.

2 Contact Model

An improved three-dimensional (3D) bond model incorporating rolling and twisting resistances [6–8] was employed in this study. This model consists of two particles contacting via a bond. When the particles overlap, the contact stress is borne by both the particles and bond, and the particles and bond are connected in parallel. When the particles do no directly contact but interact through the bond, the contact stress is borne by the bond, and the particles and the bond are connected in series. For the mechanical response of particles, this paper only presents a brief introduction, and the details can be referenced in the literature [6]. For the bond strength criterion, the details can be referenced in the literature [7, 8].

2.1 Mechanics Response of the Particles

Figure 1 illustrates the mechanics response of the particles in normal/tangential/rolling/twisting direction, where the rolling and twisting resistance reflects the effect of the shape of the actual particles.

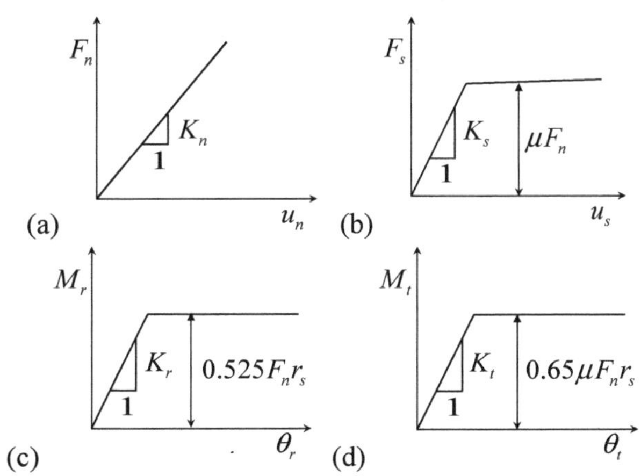

Fig. 1. Mechanical responses of the particle contact [6] in: (a) normal direction, (b) tangential direction, (c) rolling direction, and (d) twisting direction.

In the figure, $K_n = \pi r^2 E/(r_1 + r_2)$ is the normal stiffness, r is the equivalent radius of the contact (i.e. the radius of the smaller particle at the contact), $r_1(r_2)$ is the radius of the particle at the contact, E is the elastic modulus, $K_s = K_n/\kappa$ is the tangential stiffness, κ is the normal-to-shear stiffness ratio, $K_r = 0.25\,K_n r_s^2$ is the rolling stiffness, $r_s = \beta r$ is the shape radius, β is the shape parameter, $K_t = 0.5\,K_s r_s^2$ is the twisting stiffness, μ is the friction coefficient of particle, μ_n is the overlap of the particles, μ_s is the relative tangential displacement, θ_r is the relative rolling angle, θ_t is the relative twisting angle, F_n is normal force, F_s is tangential force, M_r is rolling moment, M_t is the twisting moment.

2.2 Bond Strength Criterion

In the DEM simulation, the bond between the particles is approximated as a cylinder, with a height equal to the sum of the radii of the two contacting spheres. The bond is assumed to be ideal elastic brittleness. When the load exceeds the failure envelope in the shear-rolling-twisting space incorporating the normal force, the bond will break and completely lose strength.

The original formulation of the bond strength criterion is given in [8]. This study employed the simplified version (given below) to ease the model application:

$$\left(\frac{R_s}{mA_b(\sigma_c + \sigma_t)}\right)^2 + \left(\frac{F_n + 0.5A_b(\sigma_t - \sigma_c)}{0.5A_b(\sigma_c + \sigma_t)}\right)^2 = 1, \tag{1}$$

$$R_r = r_b A_b(\sigma_c + \sigma_t)\left[(-\frac{1}{2})(\frac{F_n + A_b\sigma_t}{A_b\sigma_t + A_b\sigma_c} - \frac{1}{2})^2 + n\right], \tag{2}$$

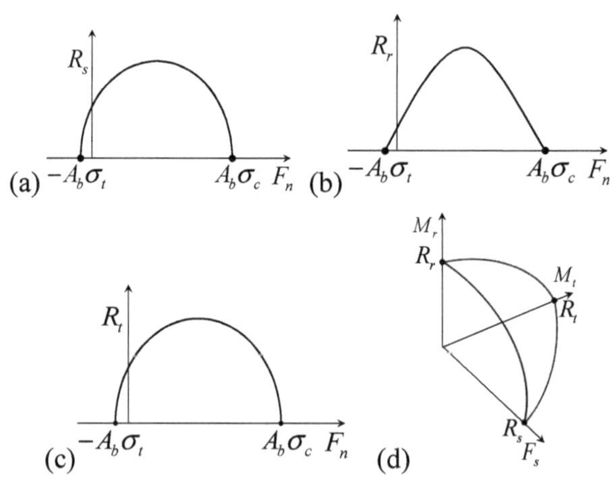

Fig. 2. Bond failure criterion: (a) shear plane, (b) rolling plane, (c) twisting plane, (d) failure envelop in the shear-rolling-twisting space.

$$\left(\frac{R_t}{l r_b A_b(\sigma_t+\sigma_c)}\right)^2+\left(\frac{F_n+0.5A_b(\sigma_t-\sigma_c)}{0.5A_b(\sigma_t+\sigma_c)}\right)^2=1, \tag{3}$$

$$\left[\left(\frac{F_s}{R_s}\right)^2+\left(\frac{M_r}{R_r}\right)^2+\left(\frac{M_t}{R_t}\right)^2\right]=1, \tag{4}$$

where the three fitting parameters are m = 0.193, n = 0.125 and l = 0.138; R_s, R_r, R_t are the absolute value of shear, rolling and twisting resistance, respectively; F_s is the shear force; M_r is the rolling moment; M_t is the twisting moment, $F_n = A_b\sigma_n$ is the normal force; $A_b = \pi r_b^2$ is the area of the bond; r_b is the radius of the bond; σ_t and σ_c are the tensile and compressive strength of the bond. These mathematical expressions are illustrated in Fig. 2.

3 Model Calibration Based on Triaxial Tests

3.1 Specimen Preparation

Berea Sandstone [9] is selected as the target sample for model calibration. Figure 3 shows the distribution of grain size obtained from the CT scanning image [10]. Figure 4 shows a DEM specimen used in this study consisting of 12700 particles, which was generated by the multi-layer with under-compaction method [11] with the ratio of length, width and height as 1:1:2.5. Subsequently, the preloading pressure as vertical pressure was applied on the specimen under the K_0 state to simulate the natural deposition of particles. When the stress of the specimen reached balance, the bonds were activated where the distance between two particles is less than a critical bond thickness to simulate the cement in the strata, and then the specimen was unloading to simulate the sampling process in situ. Finally, the specimen was unloaded to an isotropic stress of 1.0 kPa. When these steps had been completed, the specimen can be used to simulate laboratory triaxial test.

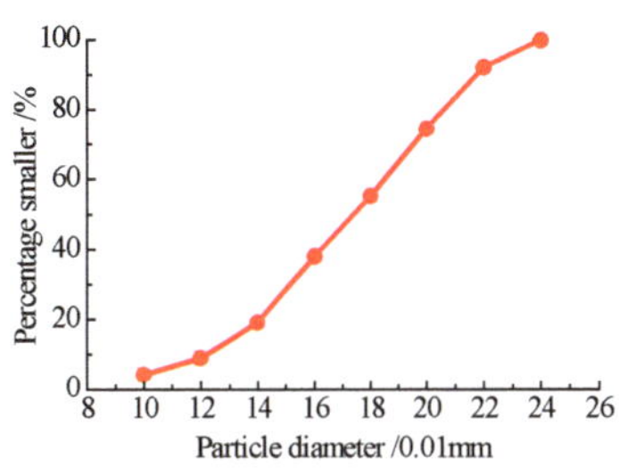

Fig. 3. Distribution of grain size

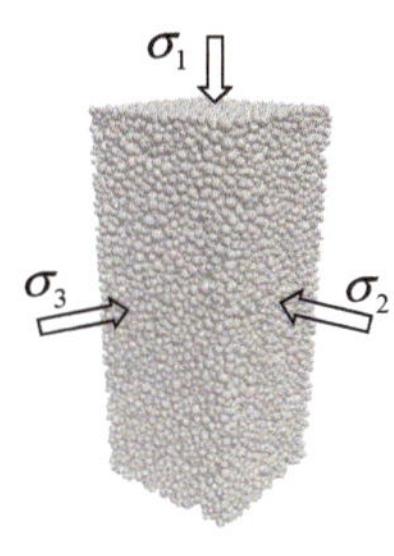

Fig. 4. A DEM specimen

3.2 Procedure of Parameter Calibration

The Technology Roadmap. For the technology roadmap, two issues need special attentions: (1) the parameters incorporating close relation with the stiffness of specimen should be calibrated before the other parameters which have close relation with the strength of specimen, and (2) the parameters incorporating close relation with the residual strength should be calibrated before the other parameters which have close relation with the peak strength. A microscopic parameter may have many effects on the macroscopic level, so a large number of trial calculations are required to determine the value of a microscopic parameter. Figure 5 shows the flow chart of parameter calibration.

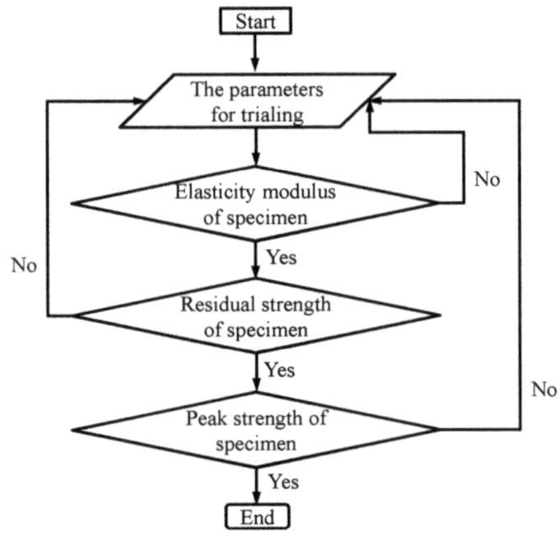

Fig. 5. The technology roadmap of parameter calibration

Getting the Typical Values and Ranges of Parameters for Trialing. During calibration, the physical meaning of microscopic parameters should be considered combining the actual material properties of the simulated object. According to the physical meaning of the parameters, both the typical values and ranges of some parameters can be roughly obtained.

According to several trials, the possible range of the void ratio of the specimen is from 0.6 to 0.9. Considering that sandstone is rather porous, the target void ratio of the specimen is set 0.8. For the preloading consolidation pressure, the vertical pressure of 30 MPa is selected to simulate a rock from deep strata subjected to an extremely high confining pressure. For the elastic modulus of particles, the twice times elastic modulus of specimen of laboratory triaxial test can be taken as the reference value. The normal-to-shear stiffness ratio of particle as well as bond is set to 1.5 to be appropriate according to the experience [6, 8]. The range of friction coefficient of particle is generally between 0 to 1, and the reference value can be set to 0.5. The range of shape parameter of particle is generally between 0.0–2.0, and the reference value can be set to 0.5 [6]. In order to reduce the effect of boundary effect, both the friction coefficient and the shape coefficient between wall and particle are set to 0.0 directly. The elastic modulus between wall and particle is not the key parameter affecting the strength, and can be the same as the elastic modulus of particles. Under quasi-static condition, the local damping coefficient is not the key parameter affecting the strength, and can directly take the common value of 0.7. The elastic modulus of bond can be the same as that of particle, and can also be selected according to the properties of the actual cements and mineral particles of the simulated object. The compressive-to-tensile strength ratio of bond can take 10 as the reference value referring to that of brittle materials such as concrete and rock. The bond thickness is generally much smaller than the particle size, so we can take 1/10 of the medium diameter of the specimen as the reference value, and can also refer to the microscopic scanning image of the object.

The DEM Simulations of Triaxial Tests. In this study, the stress-strain curves through the whole process of the laboratory triaxial tests under two confining pressures of 10 MPa and 40 MPa are selected as the approximating object of DEM simulation to calibrate model parameters. In accordance with the quasi-static conditions, the axial strain rate of loading is set to 2.5×10^{-4}/s. Both the values of parameters for trialing and the finial calibrated values are listed in Table 1.

Table 1. Parameters used in DEM simulation

Parameters	Preloading without bond		Bond activated	
	For trialing	Calibrated	For trialing	Calibrated
Void ratio e	0.6, 0.8, 0.9	0.8	0.6, 0.78, 0.82	0.78
Preloading consolidation pressure (MPa)	1, 10, 30	30	–	–
Particle density ρ(kg/m^3)	2650	2650	2650	2650
Density scaling coefficient of particle	1.0×10^9	1.0×10^9	1.0×10^9	1.0×10^9
Elastic modulus of particle E_p (GPa)	20, 30, 40	30	20, 30, 40	30
Normal-to-shear stiffness ratio of particle κ	1.5	1.5	1.5	1.5
Friction coefficient of particle μ	0.25, 0.3, 0.5	0.3	0.25, 0.3, 0.5	0.3
Shape parameter of particle β	0.5, 1.0, 1.5	1.5	0.5, 1.0, 1.5	1.5
Friction coefficient between wall and particle μ_w	0.0	0.0	0.0	0.0
Particle shape parameter between wall and particle β_w	0.0	0.0	0.0	0.0
Elastic modulus between wall and particle E_w (GPa) (always equal to E_p)	20, 30, 40	30	20, 30, 40	30
Local damping coefficient	0.7	0.7	0.7	0.7
Elastic modulus of bond E_b (GPa)	–	–	12, 13.5, 20, 30	13.5
Normal-to-shear stiffness ratio of bond κ^*	–	–	1.5	1.5
Tensile strength of bond σ_t (MPa)	–	–	20, 30, 35, 40	35
Compressive strength of bond σ_c (MPa)	–	–	350, 700, 1400, 2100	1400
Reduction coefficient of bond diameter λ	–	–	0.5, 0.75, 1.0	1.0
Thickness of bond g_c (mm)	–	–	0.005, 0.01, 0.02	0.01

Determining the Final Values of Model Parameters. In order to select the reasonable values of model parameters, the effects of model parameters on macroscopic mechanical behavior should be analyzed. Based on the final calibrated values listed in Table 1, the effects of some key parameters are shown in Fig. 6 as examples.

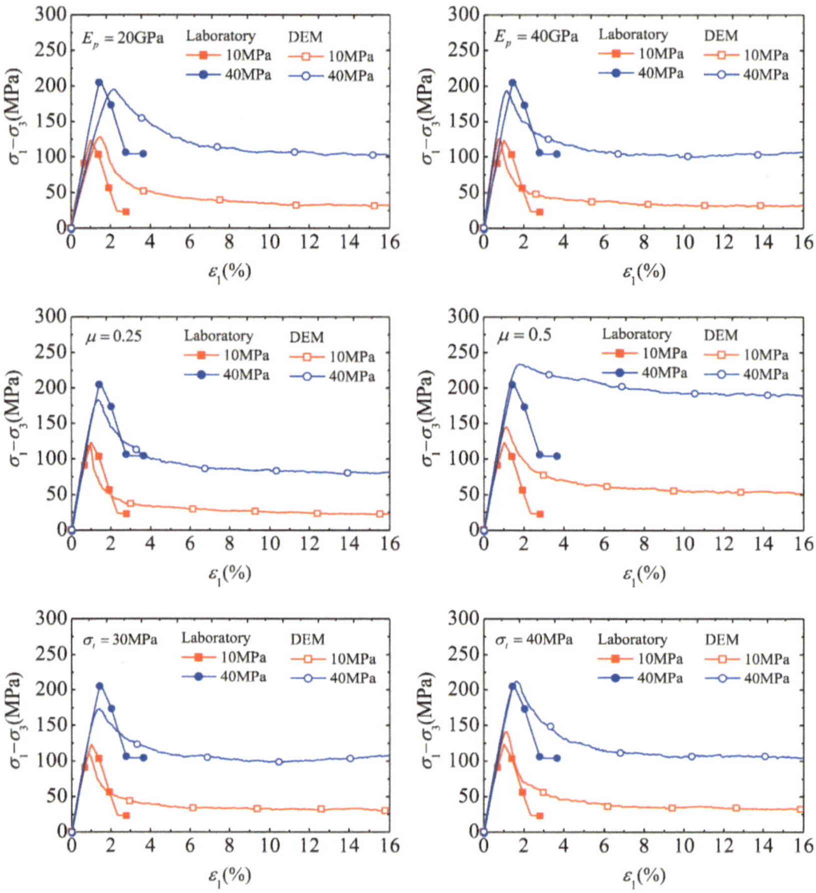

Fig. 6. Relationships of deviatoric stress vs axial strain

The parameters to be analyzed are to be presented in the graphs. We look for the relationships between model parameters and mechanical behavior in this way, and then calibrate the parameter values.

4 Calibration Results

Figure 7 compares the test data with the simulations using the calibrated model. The range of confining pressure of the triaxial tests is from 10 MPa to 335 MPa. As shown in the figure, the curves obtained from DEM are generally consistent with the results of laboratory tests under most confining pressures, although relatively more scatter is noted at confining pressures of 200 MPa and 240 MPa. For the volumetric strain curves, the simulations match the test data better at larger axial strains than that at smaller one. This may be due to larger measurement error of the laboratory tests at

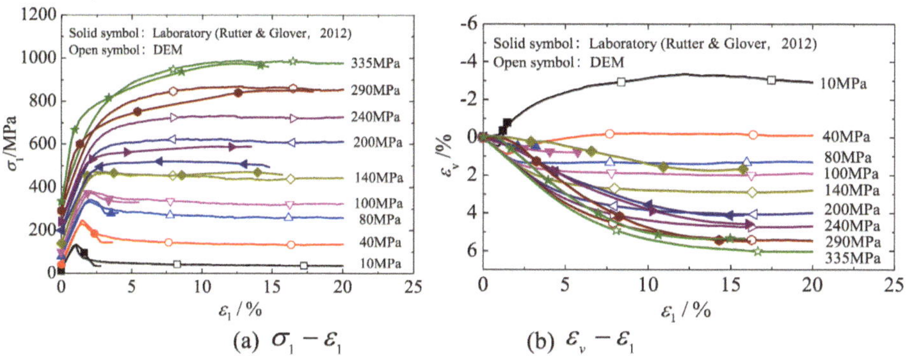

Fig. 7. Comparison between DEM simulations and the triaxial test data

small axial strains. Nevertheless, the DEM simulations quantitatively reproduced the stress-strain and volumetric strain curves to a acceptably satisfactory level.

5 Conclusions

This study calibrates an improved 3D bond model incorporating rolling and twisting resistances for rocks. The calibration procedure is proposed to obtain a set of model parameters based on available triaxial test data from the specimens of Berea Sandstone. The results show that the calibrated model is able to quantitatively reproduce the complete stress-strain and volumetric strain curves including the tails where the specimens reach the residual strengths. The lessons learnt from this study for an efficient calibration are as follows:

(1) The parameters affecting the bulk stiffness of the specimen should be calibrated before the parameters affecting the bulk strength of the specimen.
(2) The parameters affecting the residual strength should be calibrated before the ones affecting the peak strength.
(3) The typical values and ranges of some parameters should be roughly obtained according to their physical meanings to reduce trial calculation and improve efficiency.
(4) In order to determine the final values of model parameters, the relationships between model parameters and mechanical behavior should be found out by comparison between DEM simulations and the triaxial tests.

Acknowledgments. This work is funded by the National Basic Research Program of China (with grant No. 2014CB046901).

References

1. Cundall, P.A., Strack, O.D.L.: A discrete numerical model for granular assemblies. Geotechnique **29**(1), 47–65 (1979)
2. Hoek, E., Martin, C.: Fracture initiation and propagation in intact rock–a review. J. Rock Mech. Geotech. Eng. **6**(4), 287–300 (2014)
3. Lisjak, A., Grasselli, G.: A review of discrete modeling techniques for fracturing processes in discontinuous rock masses. J. Rock Mech. Geotech. Eng. **6**(4), 301–314 (2014)
4. Jiang, M.J., He, J., Shen, Z.F.: Preliminary investigation on parameter inversion for three-dimensional distinct element modeling of methane hydrate. Chin. J. Geotech. Eng. **36**(4), 736–744 (2014)
5. Yang, B.D., Jiao, Y., Lei, S.T.: A study on the effects of microparameters on macroproperties for specimens created by bonded particles. Eng. Comput. **23**(6), 607–631 (2006)
6. Jiang, M.J., Shen, Z.F., Wang, J.F.: A novel three-dimensional contact model for granulates incorporating rolling and twisting resistances. Comput. Geotech. **65**, 147–163 (2015)
7. Li, L., Jiang, M.J., Zhang, F.G.: Quantitative simulation of the triaxial test considering residual strength on the deep rock using DEM and parameters analysis. Rock Soil Mech. (2018, in press)
8. Shen, Z.F., Jiang, M.J., Wan, R.: Numerical study of inter-particle bond failure by 3D discrete element method. Int. J. Numer. Anal. Meth. Geomech. **40**(4), 523–545 (2016)
9. Rutter, E.H., Glover, C.T.: The deformation of porous sandstones; are Byerlee friction and the critical state line equivalent? J. Struct. Geol. **44**, 129–140 (2012)
10. Bera, B., Mitra, S.K., Vick, D.: Understanding the micro structure of Berea Sandstone by the simultaneous use of micro-computed tomography (micro-CT) and focused ion beam-scanning electron microscopy (FIB-SEM). Micron **42**(5), 412–418 (2011)
11. Jiang, M.J., Konrad, J.M., Leroueil, S.: An efficient technique for generating homogeneous specimens for DEM studies. Comput. Geotech. **30**(7), 579–597 (2003)

Characteristics of Hydraulic Fracture in Heterogeneous Rock with Distributed Hard Inclusions

Ming Li(✉), Li Liang, and Yunyun Fan

Northeastern University, Shenyang 110819, China
liming@mail.neu.edu.cn

Abstract. Hydraulic fracture is an important technic for oil/gas production. A smeared cracked method is adopt for simulating hydraulic fracturing propagation within the framework of a fully coupled, pore pressure-solid phase interaction formulation. The stress-strain behavior of the solid phase is modeled as a porous media by using elasto-plastic, Drucker-Prager description. The inherent assumption of the fracture model is that the permeability can be expressed as a function of the difference between the mean effective stress and the tensile strength of the rock. For two-phase flow, the relative permeability is used for including the effect of fluid mixture between the original fluid and injection fluid. The rock sample with randomly distributed irregular hard inclusions is studied with the vertical stress, horizontal stress and pore pressure being 29.13 MPa, 19.77 MPa and 10.34 MPa, respectively. By running over 30 numerical cases, the characteristics of injection pore pressure and equivalent opening versus injection time are well simulated with different relevant dynamic viscosity, injection rate, intrinsic conductivity and tensile strength. Finally, the paper ends with some useful observations on the characteristics of hydraulic fracture.

Keywords: Hydraulic fracture · Smeared crack method
Heterogeneous rock material · Hard inclusion

1 Introduction

Hydraulic fracturing is an important technique for oil/gas production. The first industrial use is in 1949 (Clark 1949). Hereafter, various kinds of mechanical models and numerical models have been developed for solving two-dimensional and three-dimensional problems. KGD fracture model (Geertsma and de Klerk 1969), PKN fracture model (Perkins and Kern 1961) and P-3D-C model (Adachi et al. 2007; Rahman and Rahman 2010) are three of the traditional mechanical models for hydraulic fracture solutions. To conquer the restriction of the mechanical models and considering the complex heterogeneity of rock, numerous numerical approaches for simulating hydraulic fracture are introduced, e.g., finite element method (FEM) (Wangen 2011); cohesive zone model (Sarris and Papanastasiou 2011); distinct element method (DEM) (Fatahi et al. 2016); extended finite element method (XFEM)

L. Zhang et al. (Eds.): GSIC 2018, *Proceedings of GeoShanghai 2018 International Conference: Rock Mechanics and Rock Engineering*, pp. 57–64, 2018.
https://doi.org/10.1007/978-981-13-0113-1_7

(Lecampion 2009; Mohammadnejad and Khoei 2013); meshfree (meshless) method (Nguyen et al. 2016) and smeared crack method (Hu et al. 2014; Li et al. 2016).

The smeared crack approach adopted in this paper is based on the assumption that the permeability can be expressed as a function of the difference between the mean effective stress and the tensile strength of the rock. This method has been implemented in ABAQUS software by Li et al. (2016).

The major work in the present paper is to study the hydraulic fracture characteristics of a rock sample with hard inclusions randomly distributed using the smeared crack method. The behavior of the matrix of a rock sample is considered to be an elasto-plastic porous media following Drucker-Prager description. And for the hard inclusions, it is simplified to be elastic. More than 30 numerical cases are run for studying the influences of ratio of fluid viscosity, injection rate, permeability and tensile strength.

The next section provides a brief introduction to the smeared crack mode for hydraulic fracture simulations. The modeling methods for a heterogeneous rock are introduced in Sect. 3. Section 4 describes the characteristics of hydraulic fracture under different parameters. The paper ends with some conclusions in Sect. 5.

2 A Smeared Crack Model for Hydraulic Fracture Simulation

The smeared crack model used in the present paper is introduced by Li et al. (2016, 2017), which is based on the framework of a fully coupled, pore pressure-solid phase interaction formulation. The major equations for the intrinsic permeability modification relating to effective mean stress (σ'_{m}) and tensile strength (P_{t}) are listed as belows.

$$K = \frac{1+h}{2} K^{UB} + \frac{1-h}{2} K_0 \tag{1}$$

$$h = \tanh(\xi(\sigma'_m - P_t)) \tag{2}$$

The variable K is the hydraulic conductivity in the FEM software ABAQUS. The parameter K_0 and K^{UB} are the initial hydraulic conductivity of rock and the upper bound value after fracturing, respectively. K^{UB} = 0.1 m/s is adopted in the paper. The damage coefficient ξ influences the change rate of hydraulic conductivity that is first small and then becomes large, this feature has been studied by Li et al. (2016).

The effects of fluid mixture between the original fluid and injection fluid in the crack zone is considered by using Corey correlation (Corey 1954; Brooks and Corey 1964) which has been adopted by Li et al. (2016, 2017). Equations for the relative permeability (k_{r}) calculation are

$$k_r = k_{r0} \left(\frac{S_w - S_{wi}}{1 - S_{wi} - S_{orw}} \right)^{N_w} \tag{3}$$

$$S_w = \frac{\Phi_0(1-\Phi)S_{w0} + \Phi - \Phi_0}{\Phi(1-\Phi)} \tag{4}$$

where S_{wi} is the irreducible (minimal) degree of water saturation, and S_{orw} is the residual oil saturation after flooding by water. The physical property k_{r0} is called the end point of the water relative permeability. S_{w} is the degree of saturation. The variable Φ is porosity of a rock sample with Φ_0 being its initial value. In order to simplify the problem, we assume $N_{\mathrm{w}} = 3.66$, $S_{\mathrm{wi}} = S_{\mathrm{orw}} = 0.15$, $k_{\mathrm{r0}} = 0.5$ for the case study in the paper.

3 Heterogeneous Rock Sample Modeling

The rock sample used in the paper is modelled according a section from underground near –1054 m. The outline of inclusions is enhanced by using MATLAB script (programmed by Yaqian Liu, McMaster University). The sketch of the model was finally generated by AutoCAD, and then the mechanical model was imported to ABAQUS software. The hard inclusions extracted from the section is randomly positioned. To reduce the influence of horizontal boundary condition, the matrix part of the model is extended to be 890 mm in width. The height of mechanical model is 300 mm. The finite element mesh of matrix and inclusions are shown in Fig. 1. The plane strain 8-node biquadratic displacement, bilinear pore pressure element (CPE8P) is used in the present paper. There are totally 31430 nodes and 10437 elements in the mesh.

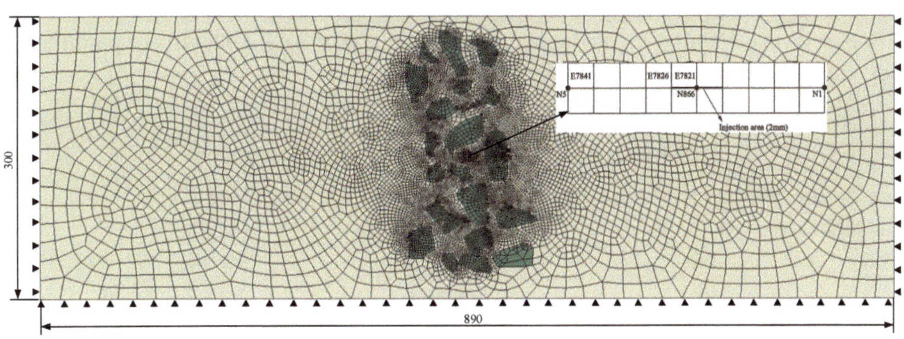

Fig. 1. Finite element mesh of a rock sample with hard inclusions randomly distributed and element and node number near injection.

Comparing with a field problem, the mechanical model adopted in the paper can be considered to be a material point, so the gravity influence is omitted and the total stress state are set to be constant through the whole model with vertical stress, horizontal stress and pore pressure being 29.13 MPa, 19.77 MPa and 10.34 MPa, respectively.

The flux rate (for applying injection pore pressure) is applied on a 2 mm length line located at the center of the finite element model, shown in Fig. 1. The displacement boundaries are applied on horizontal direction on both side of the model and vertical direction on the bottom of it. The pore pressure on the four edge of model is equal to 10.34 MPa. Material properties of the rock matrix from field tests are listed in Table 1. The hard inclusions are assumed to be elastic during the simulations.

Table 1. Material properties of rock matrix and hard inclusion.

Property Name	Rock matrix	Hard inclusion
Logarithmic bulk modulus	0.001	–
Permeability (mm/s)	5.0E–7	–
Tensile strength (MPa)	13.8	–
Friction angle (°)	30	–
Dilation angle (°)	10	–
Elastic modulus (GPa)	13.4	41.4
Poisson's ratio	0.33	0.33

4 Characteristics of Hydraulic Fracture

The injection pore pressure and equivalent opening of hydraulic fracture versus injection time are studied in the section. Influence of material properties of rock matrix, tensile strength and permeability, are first carried out. The ratio of fluid viscosity between injection fluid (μ_{inf}) and original fluid (μ_{pf}) is the second parameter studied with two levels. Finally, the influence of flux rate is performed. Details of parameters of rock sample for case study are listed in Table 2.

Table 2. Parameters considered in the case study.

Tensile strength (P_t, MPa)	Hydraulic conductivity (k, mm/s)	Ratio of fluid viscosity ($m = \mu_{inf}/\mu_{pf}$)	Flux rate (V, mm/s)
13.8	5.0E–5	1, 100	96, 144, 192, 240, 300
13.8	5.0E–7	1, 100	3, 9, 24, 48, 96
15.3	5.0E–7	1, 100	3, 9
12.3	5.0E–7	1, 100	3, 9

When the tensile stress and hydraulic conductivity are 13.8 MPa and 5.0E–7 mm/s, take the cases with flux rate being 3 and 9 mm/s as an example. The effective mean stress distribution is shown in Fig. 2. The fractured zone is where the effective mean stress becomes tension.

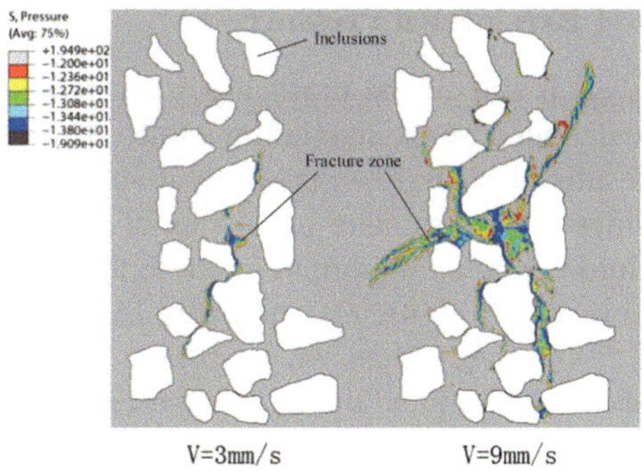

Fig. 2. Effective mean stress distributions with flux rate being 3 and 9 mm/s, respectively.

4.1 Stress Path Characteristics

The stress path results are plotted according to the cases with tensile strength, hydraulic conductivity and ratio of fluid viscosity being 13.8 MPa, 5E–7 mm/s and 1, respectively, as shown in Fig. 3. Comparison under different flux rate from 3 mm/s to 96 mm/s can be observed from this figure. When the flux rate is lower, e.g. 3 mm/s, the shear stress initially increases and then decreases to zero. For the high flux rate, e.g. 96 mm/s, the shear stresses is directly decreasing to zero. The effective mean stress (σ'_{m}) changes from compressing to tension no matter what the condition of flux rate is. When σ'_{m} becomes tension, the hydraulic fracture appears and the hydraulic conductivity (K) of the fractured element will be modified according to equations form (1) to (4).

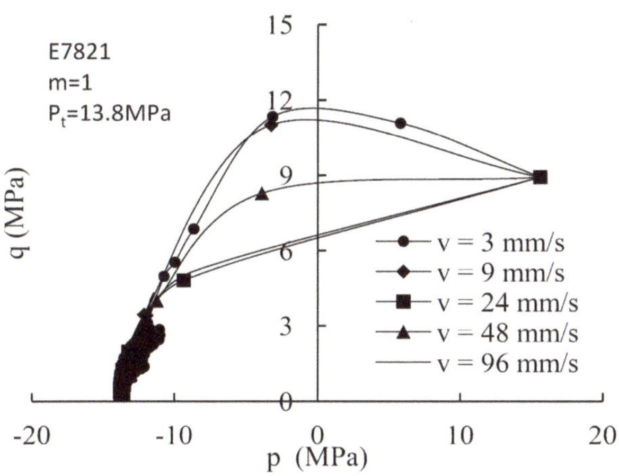

Fig. 3. Stress path of injection element 7821 under different flux rate with tensile strength, hydraulic conductivity and ratio of fluid viscosity being 13.8 MPa, 5E–7 mm/s and 1, respectively.

4.2 Characteristics of Breakdown Pressure

Breakdown pressure (P_{bp}) is one of the most important parameters in hydraulic fracturing technique. It is calculated by the maximum value of the pore pressure during the injection procedure. Results of breakdown pressure are plotted in Fig. 4 for case study listed in Table 2. Besides the stress state and material properties near injection, the viscosity of injection fluid is also an important factor to the P_{bp}. When the ratio of fluid viscosity m changes from 1 to 100, the P_{bp} increases 15 ~ 35 MPa when the flux rate increases from 100 to 300 mm/s. When we keep m = 1, the hydraulic conductivity K decreases from 5E–5 to 5E–7 mm/s, the P_{bp} increases a level about 25 MPa.

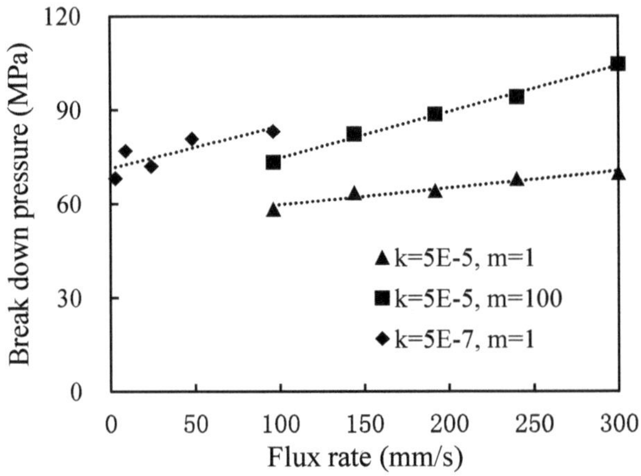

Fig. 4. Breakdown pressure of injection element 7821 under different flux rate with tensile strength being 13.8 MPa, considering two level of ratio of fluid viscosity (1 and 100) and hydraulic conductivity (5E–5 mm/s and 5E–7 mm/s).

4.3 Injection Pore Pressure Under Different Flux Rate

The initial part of injection pore pressure versus injection duration are plotted in Fig. 5 with flux rate changing from 3 to 96 mm/s. The whole propagation could be divided into two part: cracking and propagation. The cracking starts at the very beginning of the fluid injection. When the flux rate increases, the propagation pressure increases too.

4.4 Equivalent Opening Under Different Flux Rate

Since the smeared crack model is adopted in the paper, there is no explicit fracture in the simulation results. The equivalent opening is used, and its definition in the case study is the displacement difference in horizontal direction between node 1 and node 5 as shown in Fig. 1. We take the case (parameters of tensile strength, hydraulic conductivity and ratio of fluid viscosity are 13.8 MPa, 5E–7 mm/s and 100, respectively) as the example, the increasing of flux rate influences the equivalent opening, as shown in Fig. 6.

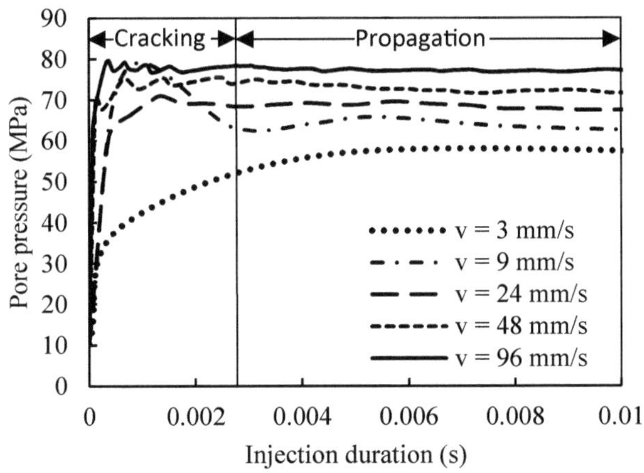

Fig. 5. Injection pore pressure at the beginning period from 0 to 0.01 s under different flux rate with tensile strength, hydraulic conductivity and ratio of fluid viscosity being 13.8 MPa, 5E–7 mm/s and 1, respectively.

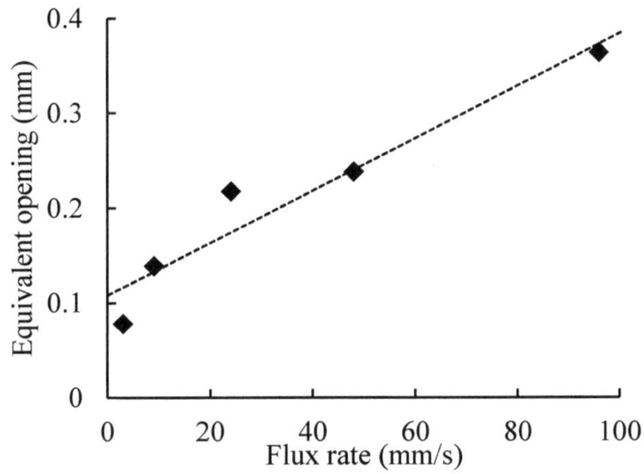

Fig. 6. Equivalent opening under different flux rate with tensile strength, hydraulic conductivity and ratio of fluid viscosity being 13.8 MPa, 5E–7 mm/s and 100, respectively.

5 Conclusion

A smeared crack model is adopt in the present paper for studying the characteristics of hydraulic fracturing considering a heterogeneous rock matrix with hard inclusions randomly distributed. The stress path, breakdown pressure, injection pore pressure and equivalent opening are well simulated in the paper. Results shows that (1) the smeared crack model is useful for simulating the part of hydraulic fracture procedure before crack closing; (2) the viscosity of injection fluid is also an important factor to the breakdown pressure; (3) The equivalent opening could be used for checking the opening state in the selected smeared crack model, and the increasing of flux rate influences the equivalent opening.

Acknowledgement. Thanks for the support of the General Program of National Natural Science Foundation of China (51474048).

References

Adachi, J., Siebrits, E., Peirce, A., Desroches, J.: Computer simulation of hydraulic fractures. Int. J. Rock Mech. Min. Sci. **44**(5), 739–754 (2007)

Brooks, R.H., Corey, A.T.: Hydraulic Properties of Porous Media. Hydrological Papers (1964)

Clark, J.: A hydraulic process for increasing the productivity of wells. J. Pet. Technol. **1**(1), 1–8 (1949)

Corey, A.T.: The interrelation between gas and oil relative permeabilities. Producers Monthly **19** (2), 38–41 (1954)

Fatahi, H., Hossain, M.M., Fallahzadeh, S.H., Mostofi, M.: Numerical simulation for the determination of hydraulic fracture initiation and breakdown pressure using distinct element method. J. Nat. Gas Sci. Eng. **33**(1), 1219–1232 (2016)

Geertsma, J., de Klerk, F.: A rapid method of predicting width and extent of hydraulically induced fractures. J. Petrol. Technol. **21**, 1571–1581 (1969)

Hu, Y., Chen, G., Cheng, W., Yang, Z.: Simulation of hydraulic fracturing in rock mass using a smeared crack model. Comput. Struct. **137**(6), 72–77 (2014)

Lecampion, B.: An extended finite element method for hydraulic fracture problems. Int. J. Numer. Methods Biomed. Eng. **25**(2), 121–133 (2009)

Li, M., Guo, P., Stolle, D., Liang, L.: Development of hydraulic fracture zone in heterogeneous material based on smeared crack method. J. Nat. Gas Sci. Eng. **35**(1), 761–774 (2016)

Li, M., Guo, P., Stolle, D., Liang, L.: Modeling method for a rock matrix with inclusions distributed and hydraulic fracturing characteristics. J. Pet. Sci. Eng. **157**(1), 409–421 (2017)

Mohammadnejad, T., Khoei, A.R.: An extended finite element method for hydraulic fracture propagation in deformable porous media with the cohesive crack model. Finite Elem. Anal. Des. **73**(15), 77–95 (2013)

Nguyen, T., Quoc, N., Nguyen, V.: Mesh free approach for hydraulic fracture. Eng. Geol. (2016). https://doi.org/10.1016/j.enggeo.2016.11.009

Perkins, T.K., Kern, L.R.: Widths of hydraulic fractures. J. Petrol. Technol. **13**(1), 937–949 (1961)

Rahman, M.M., Rahman, M.K.: A review of hydraulic fracture models and development of an improved pseudo-3D model for stimulating tight oil/gas sand. Energy Sources Part A Recovery Utilization Environ. Eff. **32**(15), 1416–1436 (2010)

Sarris, E., Papanastasiou, P.: The influence of the cohesive process zone in hydraulic fracturing modelling. Int. J. Fract. **167**(1), 33–45 (2011)

Wangen, M.: Finite element modeling of hydraulic fracturing on a reservoir scale in 2D. J. Pet. Sci. Eng. **77**(3), 274–285 (2011)

Damage-Based Cohesive Modelling for Fractures Initiation and Propagation Around a Deep Tunnel in Quasi-brittle Rocks

Edoardo Trivellato[1,2(✉)], Amade Pouya[1], Minh-Ngoc Vu[2], and Darius Seyedi[2]

[1] Laboratoire Navier, UMR 8205, École des Ponts ParisTech, Champs-sur-Marne, France
edoardo.trivellato@enpc.fr
[2] Andra R&D, 1/7, rue Jean Monnet, 92298 Châtenay-Malabry Cedex, France

Abstract. This work presents numerical simulations of the excavation-induced fractures around galleries in Andra Underground Research Laboratory (URL-MHM) in France. An instrumented gallery with geological surveys of the surrounding fractures network was taken as case study. Fractures form a system of distinct, almost parallel discontinuities surrounding the excavation. Finite Element simulations with joint elements (JFEM) constituting the potential fractures were performed to reproduce their spacing and length in the rock. A damage cohesive fracture model representing the stress-displacements behaviour and failure of discontinuities was used. Two configurations were analysed. The first one consists in a JFEM model with a single joints series to analyse the fracture evolution. The second one considers the occurrence of multiple fractures around the gallery. Here, the maximal cracks length approximates correctly the extension of the failure zone observed in-situ. The average radial displacements at the tunnel wall are well estimated. The constitutive modelling and numerical tools introduced in this work contribute to improve rock fracturing simulations in deep rocks tunneling.

Keywords: Fracturing · Cohesive model · Damage-softening · Galleries
Joint-enriched FEM

1 Introduction

This work proposes to model the fractures occurrences around galleries excavated in the Andra Underground Research Laboratory, located in the Meuse-Haute-Marne department, France (URL-MHM). An anisotropic initial stress state is observed in the Callovo-Oxfordian claystone at the main level of the URL, at −490 m: vertically, $\sigma_v \approx 12.5$ MPa, and, horizontally, $\sigma_h \approx \sigma_v$ and $\sigma_H \approx 16$ MPa [1, 2, 11]. The GCS gallery, excavated parallel to σ_H and with no support, was considered.

The damaged zone mechanically induced around galleries by the drifting action was studied. In particular, it is possible to distinguish between an Excavation *Damaged* Zone (EDZ), in which the hydro-mechanical properties in the gallery wall proximity are modified and, farther, an Excavation *disturbed* Zone (EdZ) [2]. Mostly in the EdZ,

L. Zhang et al. (Eds.): GSIC 2018, *Proceedings of GeoShanghai 2018 International Conference: Rock Mechanics and Rock Engineering*, pp. 65–73, 2018.
https://doi.org/10.1007/978-981-13-0113-1_8

analyses of core drillings unveil a homogeneous cracks pattern, consisting of shear fractures. They expand deeper in the rock and form a system of distinct, almost parallel fractures known as *Chevron* or *herringbone* fractures. Their propagation length in the host rock should be limited to the area of EdZ. In the URL-MHM, it extends from 2 m up to one diameter from the gallery wall. A reconstruction of their orientation for GCS drift is reported in Fig. 1a. Figure 1b provides a 3D conceptual model deduced from in-situ observations [2].

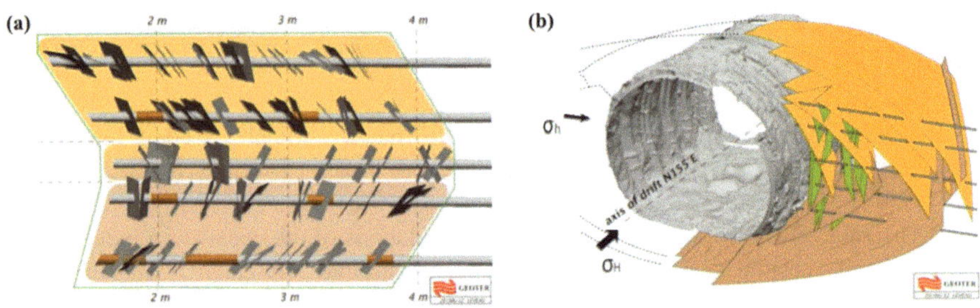

Fig. 1. Setup of the excavation-induced fractures in the GCS drift. (a) Zoom on the shear zone. (b) 3D conceptual model. Modified from [2].

A model for the axisymmetric numerical simulations is proposed in Fig. 2. In particular, Fig. 2a reports the bi-dimensional scheme already presented by Armand et al. [2] while Fig. 2b reports the simplified model proposed in numerical simulations. Chevron fractures are assumed to be mutually parallel with a certain inclination to the drift wall, characterized by a spacing $\overline{S}$ and a propagation length $\overline{L}$. An initial single-fracture problem is presented to analyse the fracture propagation and estimate the model parameters. Then, multi-fractures simulations are presented, modelling the fractures occurrence while the excavation moves forward.

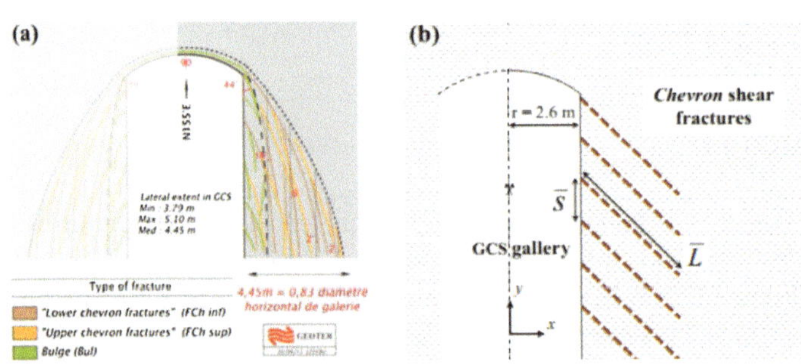

Fig. 2. Axisymmetric schemes of the excavation-induced fractures. (a) 2D proposed model for the case study along its longitudinal and horizontal plane, modified from [2]. (b) Conceptual model for the numerical simulations.

2 Constitutive Modelling

According to theory and observations, a progressive degradation of strength and stiffness in the rock may induce fractures. Damage mechanics could describe these phenomena. In the constitutive model adopted, material damage as yielding process was integrated in the *cohesive fracture* framework proposed by Pouya and Bemani Yazdi [9]. A cohesive fracture assumes the existence of a process zone ahead of the crack tip with infinitesimal thickness [7]. Here, a cohesive constitutive law relates the stress state $\boldsymbol{\sigma}$ with the discontinuities displacements $\mathbf{U}$. In a bi-dimensional modelling, both $\boldsymbol{\sigma}$ and $\mathbf{U}$ have two components, normal and tangential.

The fracture propagates when the failure criterion is equalized at the cohesive zone tail (i.e. the physical fracture tip).

In case of compressive behaviour (fracture closing), the damage cohesive model respects the condition of non-penetration of the cohesive surfaces, meaning $u_n \geq -e$ [3]. u_n is the normal component of the displacements vector $\mathbf{U}$ a e is the initial joint aperture. The damage stress-displacements law for cohesive joints writes:

$$\boldsymbol{\sigma} = (1 - D)\mathbf{KU} + \frac{\partial \Phi}{\partial \mathbf{U}} + \boldsymbol{\sigma}_0, \ \ \Phi \neq 0 \ \text{ if } \ u_n < 0 \tag{1}$$

D is the damage variable, $\mathbf{K}$ the joints stiffness tensor and $\boldsymbol{\sigma}_0$ the initial stress state. The potential function Φ accounts for the differences between compression and extension behaviours (respectively, $u_n < 0$ and $u_n > 0$).

At failure, fractures develop in mode I, II or mixed, according to the criterion already proposed by Carol et al. [4]. The failure criterion, expressed as function of tangential and normal stresses, τ and $\sigma_{\rm n}$, and damage, D, writes:

$$F(\boldsymbol{\sigma}, D) = \tau^2 - \sigma_n^2 \tan^2 \varphi + 2g(D)\sigma_n \sigma_c - g^2(D)C^2 \tag{2}$$

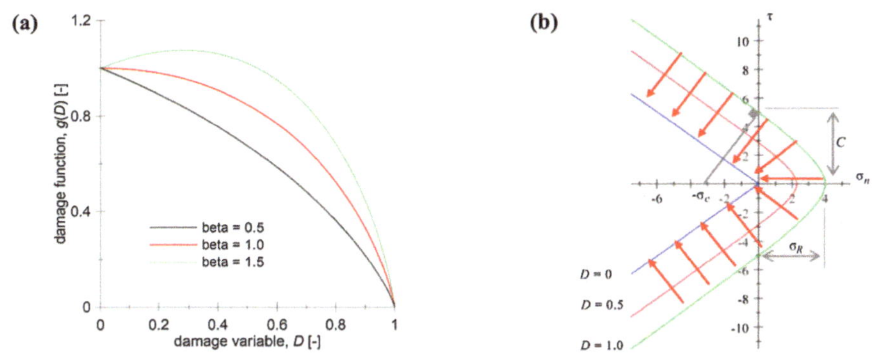

Fig. 3. (a) Damage law for different values of the parameter β. To be noticed that, with $\beta > 1$, $g(D)$ is not monotonic (hardening yielding). (b) Evolution of the failure criterion from the initial surface at D = 0 to the residual frictional surface for $D = 1$; modified from [9].

where σ_c relates the failure parameters: the cohesion C, the friction angle φ and the traction resistance. $g(D)$ is the yielding law of the criterion, a function of damage D and a parameter β describing the hardening/softening yielding process (Fig. 3a). If $D = 1$, f turns into a Mohr-Coulomb failure criterion (purely frictional joint), as shown in Fig. 3b [9].

3 Numerical Modelling

Axisymmetric simulations were performed with the Finite Element code *POROFIS* [7], developed for hydro-mechanical coupling in porous and fractured media. It enables to create FEM meshes containing built-in discontinuities modelled with *joint elements* [5] (JFEM). Discontinuities in the displacement field may exist with the relative stress values that verify the constitutive laws at the integration points. Fruitful applications of *POROFIS* in cracking modelling may be found in the works of Nguyen [8] and Vo [10].

The stress tensor $\boldsymbol{\sigma}_0$ (Table 1) was used as initial geostatic stress state [1, 2, 11] for every simulation presented, before the excavation. Potential fractures are located in correspondence of consecutive joint elements and the damage cohesive model was assigned as constitutive law. The rock matrix was modelled by linear elasticity with $E = 4000$ MPa and $\nu = 0.25$ (e.g. [1, 2]).

Table 1. Components of the initial stress tensor imposed in each simulation.

Principal stress component	Value (MPa)
$\sigma_{xx,0}$	−12.5
$\sigma_{yy,0}$	−16.0
$\sigma_{zz,0} = \sigma_{xx,0}$ (axisymmetric conditions)	−12.5

Firstly, a single-fracture simulation was performed to analyse the crack generation and evolution. The model is reported in Fig. 4. The gallery excavation was simulated through a steady state unloading σ_T applied up to the fracture tip, along the wall. The axis system (x; y) corresponding to the principal stress directions is reported.

For the multi-fractures simulations, the numerical method simulating the excavation is following introduced. The excavation was applied to the model in Fig. 5, including 44 potential fractures, as sequences of joint elements, parallel and equally distant.

To simulate the advancement of a front excavating the gallery, the function $d(y, t)$ in Eq. 3 below was defined along the drift axis, y, and in time, t. It varies asymptotically between (0; 1) from the initial excavation time t_0 to the final, t_{fin}:

$$d(y,t) = \frac{1}{2}\left[1 - \tanh\left(\frac{y - y_0 - V \cdot t}{\tilde{y}_{ELEM}}\right)\right] \quad (3)$$

y_0 is the (optional) excavation offset, V the excavation velocity and $\tilde{y}_{ELEM}$ a parameter related to the mesh size. It results $d(y, t = t_0) \rightarrow 0$ and $d(y, t = t_{fin}) \rightarrow 1$ for every y within the excavation length, equal to 60 m for the model in Fig. 5.

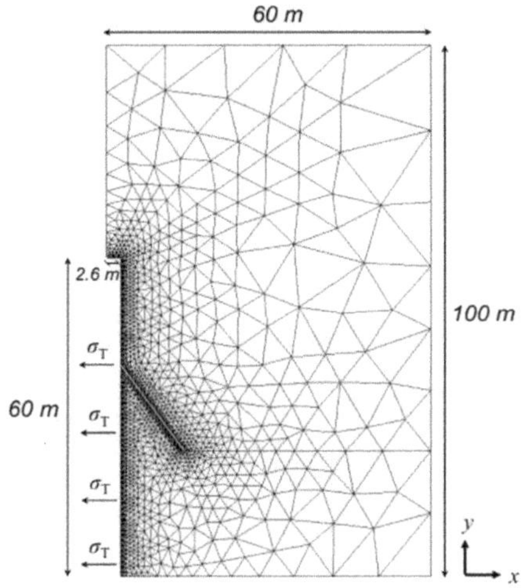

Fig. 4. *Single-fracture* model with the unloading boundary conditions σ_T, simulating a 40 m excavation. The potential fracture is located in correspondence of the finest mesh inside the domain.

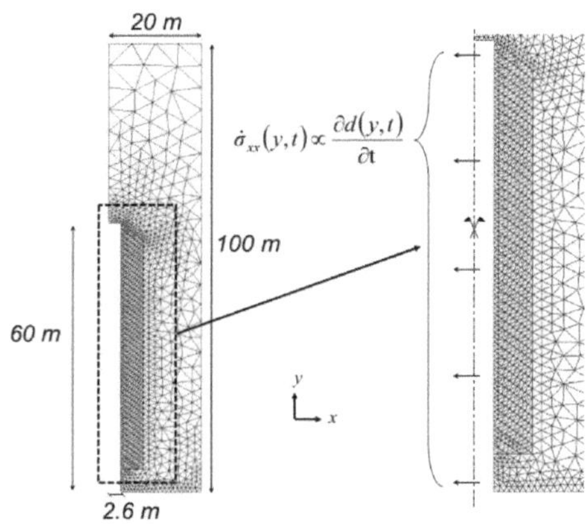

Fig. 5. *Multi-fractures* model. On the right, boundary conditions for the time-dependent excavation are shown. Potential fractures are highlighted in green and are surrounded by a finer mesh to locally enhance the stress-strain field computation precision.

An initial normal stress, $\sigma_{n,0}$, acts on the gallery wall (pre-drifting conditions). According to the reference system, $\sigma_{n,0} = \sigma_{xx,0}$ (Table 1). The following relation was introduced as a time-dependent boundary condition at the gallery wall, for the excavation:

$$\sigma_{xx}(y,t) = -\sigma_{n,0} \cdot d(y,t) \tag{4}$$

Defining *d*(*y*, *t*) according to Eq. 3, is possible to verify that σ_{xx} (y, $t = t_0$) $\rightarrow$ 0 and σ_{xx} (y, $t = t_{\mathrm{fin}}$) $\rightarrow$ $-\ \sigma_{n,0}$, for every y within the excavation length. In transient conditions, the implementation of $\dot{\sigma}_{xx}$(the time-derivative of σ_{xx}) was required:

$$\int_{t_0}^{t_{fin}} \dot{\sigma}_{xx} d\tau = -\sigma_{n,0} \rightarrow \dot{\sigma}_{xx} = \frac{\partial}{\partial \mathrm{t}}\left[-\sigma_{n,0} \cdot d(y,t)\right] \tag{5}$$

From Eq. 5, the time-dependent boundary condition writes:

$$\dot{\sigma}_{xx}(y,t) = \frac{\sigma_{n,0}}{2} \frac{V}{\tilde{y}_{ELEM}} \left[1 - \tanh^2\left(\frac{y - y_0 - V \cdot t}{\tilde{y}_{ELEM}}\right)\right] \tag{6}$$

4 Results

4.1 *Single Fracture* Analyses

The result, shown in Fig. 6, was obtained by the steady-state unloading σ_{T} on the model in Fig. 4. An intermediate and the final steps are reported in Fig. 6a. A *brittle* fracture opening occurs at 0.18 • σ_{T} meaning that several consecutive joint elements reach $D = 1$ (complete failure). This corresponds, physically, to an instantaneous dissipation of the elastic energy in the rock around the structure. Then, the fracture propagates in a *stable* evolution, with only one joint element opening for a certain unloading (Fig. 6b). The failure criterion parameters [1, 2, 6], the distance from the tunnel wall and the choice of a monotonic damage law, *g*(*D*) ($\beta \leq 1$, according to Fig. 3a), influence the combination of these mechanisms.

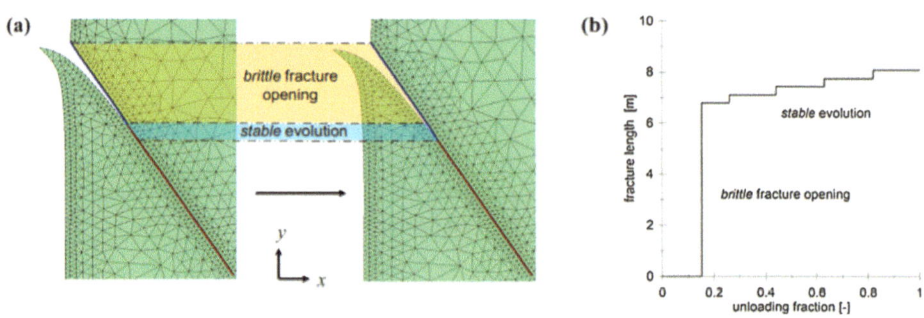

Fig. 6. (a) Initial unstable fracture opening (unloading = 0.18 • σ_{T}), followed by a progressive joint-by-joint cracking. (b) Length of the fracture progression with the unloading fraction.

4.2 *Multi-fracture* Analyses

The validity of the Eq. 3 is tested and results are reported in Fig. 7 on an elastic model. The same type of boundary conditions shown in Fig. 5 are applied to simulate the excavation. The function $d(y, t)$ is plotted along 2 m, for five time steps, t_n, such that $t_n - t_{n-1} = n \cdot \Delta t$. Each t_n corresponds to the moment when the excavation front crosses the vertical coordinate $y_n = V \cdot t_n$. In Fig. 7, it is verified, for every $t_n, d(y, t = t_n) \rightarrow 1$ for $y \leq y_n$ (excavation occurred) and $d(y, t = t_n) \rightarrow 0$ for $y > y_n$ (excavation not yet occurred).

Figure 8 shows the results of the unloading process, according to Eqs. 3–6, applied to the model in Fig. 5, at the end of the excavation ($t = t_{fin}$). A unique velocity $V = 2.3$ m per week was assumed, corresponding to approximately 6 months to complete the excavation of 60 m. In Fig. 8a and b, the opened fractures and the radial displacements, u_r, are shown, respectively to the left and right of each figure. A sensitivity analyses was performed on the tensile strength of the potential fractures, with a variation from 1.5 MPa [6] (Fig. 8a), to 1 MPa (Fig. 8b). In Fig. 8a, a coupled-fractures

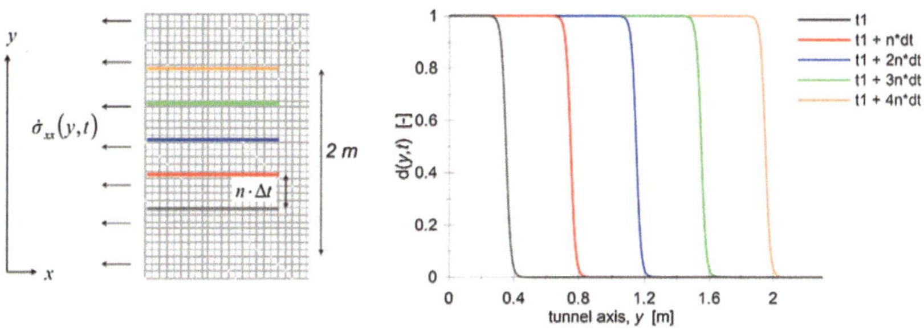

Fig. 7. Test for the function $d(y, t)$ in Eq. 3. Each line in colour defines the excavation front at t_n.

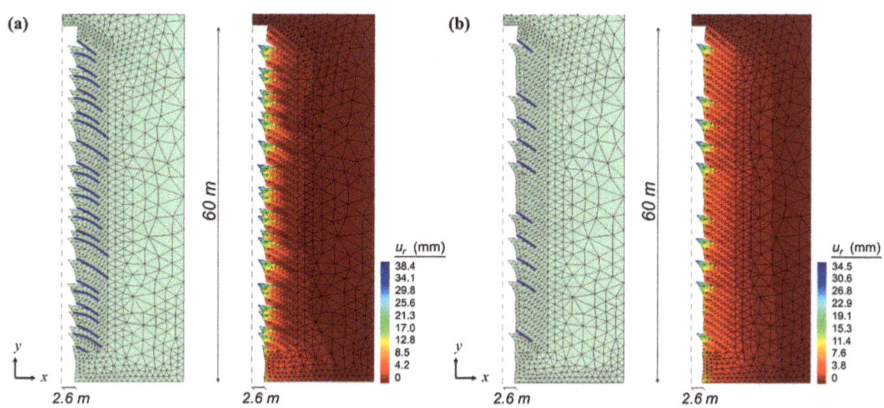

Fig. 8. (a) Damage on joints and displacements (m) for the model in Fig. 5, applying the transient excavation method in Eqs. 3–6. (b) Analogue results reducing the joints tensile strength from 1.5 to 1 MPa.

Table 2. Measured displacements at the gallery wall (u_H, horizontal and u_V, vertical) [1, 2], average radial displacement, u_r, and correspondent values numerically estimated, $\bar{u}_r$, according to Fig. 8a and b.

u_H (mm)	u_V (mm)	Average u_r (mm)	$\bar{u}_r$ (mm)	
~20	~10	~15	19.2	Fig. 8a
			17.2	Fig. 8b

pattern was obtained with about half of the potential fractures activated. The fracture length is in the range of the EdZ extension observed in-situ ($\bar{L}$ varies between 4 and 4.5 m) but the average displacement at the tunnel wall results slightly overestimated with respect to in-situ monitoring [1, 2], according to values reported in Table 2. In Fig. 8b, fractures $\bar{L}$ is similar to the same result in Fig. 8a. Moreover, the average displacement estimates better the short term convergences measured along the instrumented gallery wall (Table 2). With respect to Fig. 8a, a smaller fraction of activated (damaged) fractures on the total amount was obtained, corresponding to 20% of the 44 potential discontinuities. The spacing $\overline{S}$ varying from 2.5 to 8 m. Differently from the displacements at the gallery wall and the estimated fractures length, there are no corresponding field observations for $\overline{S}$, according to literature reviews.

5 Conclusion and Perspectives

Excavation-induced fractures around galleries were numerically modelled with a hardening\softening damage cohesive model for discontinuities with a JFEM geometrical discretization. Geological surveys of the gallery wall displacements and damage extension in Andra URL constituted the validation references, for an instrumented case-study gallery. A first single-fracture problem allowed a study of the model functioning for the forthcoming simulations and the evolution of the joint elements opening was analysed. Multi-fractures simulations showed a periodicity in the fractures occurrence together with a correct estimation of their length, related to the extension of the damaged area around the gallery. By a sensitivity analysis on the discontinuities failure criterion, a reduced amount of opened fractures (where $D = 1$) on the totality was obtained, in both multi-fractures simulations (Fig. 8a and b). Hence, it is possible to affirm that the number of the potential discontinuities set in the mesh was sufficiently high. This confirms the validity of the numerical model. The second multi-fractures simulation (Fig. 8b) provides a good estimation of the radial displacements at the gallery wall.

In general, the numerical methods introduced propose, possibly, a contribution in the simulation of the short-term mechanical response of rock fracturing due to underground excavations. This is performed, at the same time, through a constitutive model accounting for physical degradation of the material (i.e. damage) and numerical tools simulating conveniently the discontinuities and the excavation process.

Further developments of this work will account for a better modelling of residual fractures conditions due to damage, together with more sensitivity analyses of the failure parameters related to the fractures spacing, $\overline{S}$. On numerical aspects, changes in the excavation parameters (time, velocity) and mesh size or structure dependencies will be considered.

References

1. Armand, G., Noiret, A., Zghondi, J., Seyedi, D.M.: Short- and long-term behaviors of drifts in the Callovo-Oxfordian claystone at the Meuse/Haute-Marne underground research laboratory. J. Rock Mech. Geotech. Eng. **5**(3), 221–230 (2013)
2. Armand, G., Leveau, F., Nussbaum, C., De La Vaissiere, R., Noiret, A., Jaeggi, D., Landrein, P., Righini, C.: Geometry and Properties of the excavation-induced fractures at the Meuse/Haute-Marne URL drifts. Rock Mech. Rock Eng. **47**(1), 21–41 (2014)
3. Bandis, S.C., Lumsden, A.C., Barton, N.R.: Fundamentals of rock joint deformation. Int. J. Rock Mech. Mining Sci. Geomech. Abs. **20**(6), 249–268 (1983)
4. Carol, I., Prat, P.C., Lopez, C.M.: Normal/shear cracking model: application to discrete crack analysis. J. Eng. Mech. **123**(8), 765–773 (1997)
5. Goodman, R.E.: Methods of Geological Engineering in Discontinuous Rocks, 1st edn. West Publishing Company, San Francisco (1976)
6. Guayacán-Carrillo, L.M., Ghabezloo, S., Sulem, J., Seyedi, D.M., Armand, G.: Effect of anisotropy and hydro-mechanical couplings on pore pressure evolution during tunnel excavation in low-permeability ground. Int. J. Rock Mech. Min. Sci. **97**, 1–14 (2017)
7. Jin, Z.-H., Sun, C.T.: A comparison of cohesive zone modeling and classical fracture mechanics based on near tip stress field. Int. J. Solids Struct. **43**, 1047–1060 (2006)
8. Nguyen, V.N.: Modélisation numérique du comportement hydromécanique des milieux poreux fractures: analyse des conditions de propagation des fractures. Ph.D. thesis. Laboratoire Navier, École des Ponts ParisTech, France (2015)
9. Pouya, A., Bemani, Y.P.: A damage-plasticity model for cohesive fractures. Int. J. Rock Mech. Min. Sci. **73**, 194–202 (2015)
10. Vo, T.D.: Modélisation numérique et analytique de la fissuration de séchage des sols argileux. Ph.D. thesis. Laboratoire Navier, École des Ponts ParisTech, France (2016)
11. Wileveau, Y., Cornet, F.H., Desroches, J., Blumling, P.: Complete in situ stress determination in an argillite sedimentary formation. Phys. Chem. Earth **32**(8–14), 866–878 (2007)

Investigating Effects of Fracture Aperture and Orientation on the Behaviour of Weak Rock Using Discrete Element Method

Xiangyu Zhang(✉), Behzad Fatahi, and Hadi Khabbaz

School of Civil and Environmental Engineering,
University of Technology, Sydney, Australia
Xiangyu.Zhang@uts.edu.au

Abstract. The effects of the fracture aperture and orientation on the behaviour of weak rock were numerically investigated using discrete element method (DEM). In this study, the mechanical behaviour of the intact and fractured rock specimens was simulated by adopting the discontinuum based software PFC3D. The rock specimens with various fracture apertures and orientations were replicated, and the effects of these two fracture characteristics were studied through triaxial tests. The flat-joint model was employed for simulating the stress-strain behaviour of intact rock and had the ability to reproduce the cementation effect. The smooth-joint contact model was utilised to simulate the sliding effect of the fractures. The effects of five different fracture orientations were investigated in the combination of three different fracture aperture categories, namely very tight, open, and moderately wide. It can be concluded that the strength of the fractured weak rock specimens reduces as the fracture aperture width increases. The amount of alternation in strength and deformability that were contributed by fracture apertures differed with the orientations of the fracture. With the fracture orientation that was parallel to the deviatoric loading, the effect of fracture aperture on the strength and deformability of the specimens was less evident.

Keywords: Fracture aperture · Fracture orientation · Weak rock

1 Introduction

The existence of fractures and discontinuities contributes to the complexity of investigating the mechanical behaviour of rock masses significantly [1–5]. The fracture characteristics such as the fracture aperture and orientation can influence the strength and deformability of rock mass to different extent [1, 2, 4]. The fracture aperture largely affects the overall strength of rock mass, and can be classified into different categories based on the opening width [1, 5]. The fracture orientation alters the failure mode and diminishes the strength of the rock mass [1, 2, 5, 6]. Certain laboratory and numerical studies have investigated the effects of fracture orientation on the mechanical behaviour of rock mass subjected to unconfined compressive loading [4, 5]. In this study, the combined effects of the fracture orientation and aperture on the strength and deformability of weak rock specimens have been numerically investigated under

L. Zhang et al. (Eds.): GSIC 2018, *Proceedings of GeoShanghai 2018 International Conference: Rock Mechanics and Rock Engineering*, pp. 74–81, 2018.
https://doi.org/10.1007/978-981-13-0113-1_9

triaxial confined conditions. The discrete element method based on particle mechanics has been adopted in this study, which allows explicit consideration of the discrete nature of the rock mass, replicating the natural fractures, and reproducing the cementation degradation effect of cemented soils [3, 5, 7, 8, 10, 11]. For a controlled fracture density, various fracture orientations and apertures have been simulated by employing the particle flow code in three-dimension (PFC3D).

2 Discrete Element Simulation

The numerical model of the intact specimen was initially calibrated with the experimental data reported for weak weathered porphyrite rock samples by Sun et al. [9]. The numerical results were compared with the triaxial test measurements for a confining pressure of 392 kPa of the weak rock specimen. In the numerical simulation, two types of contact models were adopted to replicate the mechanical behaviour of the rock mass, namely, the flat-joint model and the smooth joint contact model. The former was adopted for the simulation of the intact rock specimen, which could simulate the particle contacts using rigidly connected surfaces. These surfaces could be further discretised into 4 elements, and each element allowed sole breakage, which could result in the partial damage of the contact bonds. The later was applied to the discrete fracture network to replicate the natural fractures. The discrete fracture network was introduced into the rock specimen as circular shaped plates, then the smooth joint contact model was assigned to the particles which had the contacts lied on both sides of the fractures. The illustrations of the flat joint model between particles, the smooth joint contact model for fractures, the fracture orientation (α), and the fracture aperture (ω) are shown in Fig. 1(b).

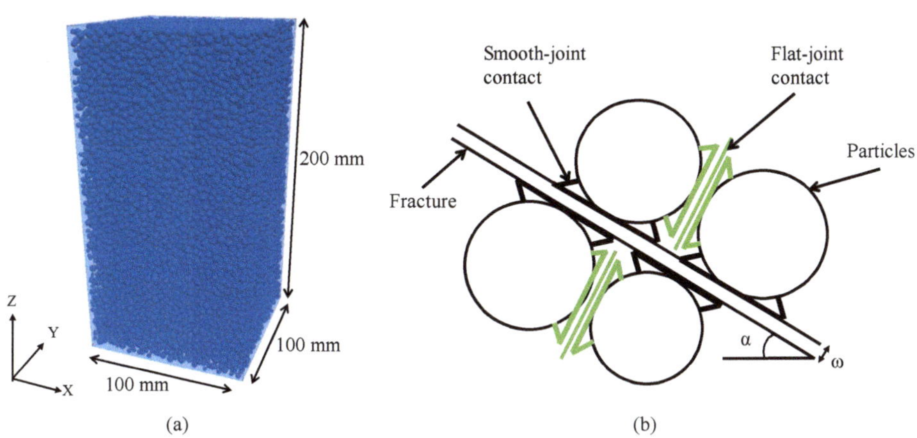

Fig. 1. (a) Numerical model of the intact rock specimen, and (b) numerical representation of the rock grains with the flat-joint contact and the fracture with the smooth-joint contact

2.1 Calibration of Micro-parameters for the Intact Rock Mass

The numerical model of the intact rock specimen is shown in Fig. 1(a), which has the dimensions of 100 × 100 × 200 mm and contains 32,500 particles. Particles were randomly distributed with a median diameter of 3.5 mm. Calibration of micro-parameters for the flat-joint model was carried out by simulating the triaxial test in which the specimen was loaded with the isotopic compression pressure of 392 kPa. To replicate the behaviour of the rock accurately, the void ratio of 1.25 was obtained after the confining pressure of 392 kPa was reached, which had the same value as the experimental measurements. Then the axial loading was applied on the top and bottom walls at a low speed to keep the specimen under quasi-static equilibrium. The side walls were controlled by a wall-servo mechanism to maintain the cell pressure throughout the test.

The strength and deformability of the specimen were mainly controlled by the normal and shear stiffness values of the interface at particle contacts. The shear strength was controlled by cohesion and friction angle at contacts. In order to replicate the macro-mechanical behaviour of the weathered porphyrite, numbers of trial-and-error processes were conducted until the stress-strain behaviour could be properly predicted in the numerical model. In this study, to replicate the mechanical behaviour of the weak rock, both the adopted normal stiffness and shear stiffness at contacts were 0.7 GPa/m, and the adopted cohesion, tensile strength and friction coefficient at contacts were 50 MPa, 10 MPa and 0.5, respectively. As shown in Fig. 2, the numerical prediction of the strength and deformability of the intact rock specimen had a good agreement with the experimental results with the adopted micro-parameters.

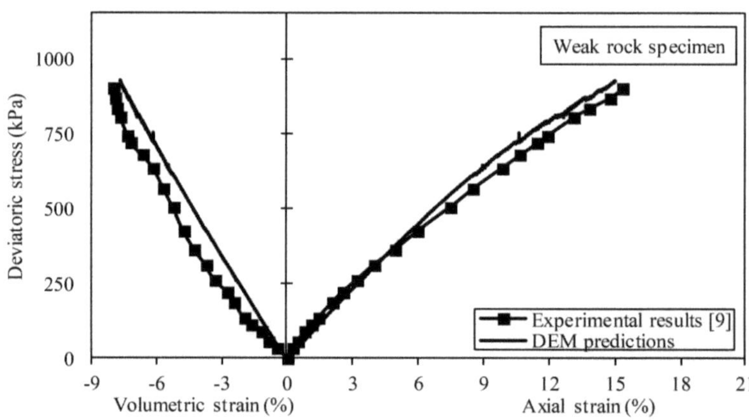

Fig. 2. The stress-strain curve of experimental measurements and model predictions of the intact rock specimen for confining pressure of 392 kPa

2.2 Modelling of Fractured Rock Mass

The discrete fracture network and smooth joint contact model were adopted for simulating the rock fractures. Similar to previous research works [2, 5], reduced contact normal stiffness and shear stiffness were adopted to simulate the fractures since the strength of the fractures should be weaker than the intact grain assembly. Fractures were randomly distributed into the system with orientations (α) of 0°, 30°, 45°, 60°, and 90° to the loading direction of the cell pressure, as illustrated in Fig. 3. A controlled fracture density $\xi = 7\ m^2/m^3$ (known as the fracture area over the volume of the specimen) was adopted to study the effects of fracture apertures and orientations on the mechanical behaviour of the weak rock specimens. The fracture aperture was simulated by controlling the contact gap of the smooth joint contact model. Indeed, by introducing a particular fracture aperture (ω), the two sides of fracture would not contact and exchange forces unless the gap was closed during the analysis. The adopted fracture aperture (ω) values in this paper were 0, 1, and 4 mm, which corresponded to very tight, open, and moderately wide fracture opening, respectively [1]. The fracture characteristics of the tested fracture specimens are summarized in Table 1. In order to analyses the effects of these two fracture characteristics (i.e. orientation and aperture), the fractured specimens were subjected to the same isotropic compression pressure as the intact specimen reported in the calibration exercise.

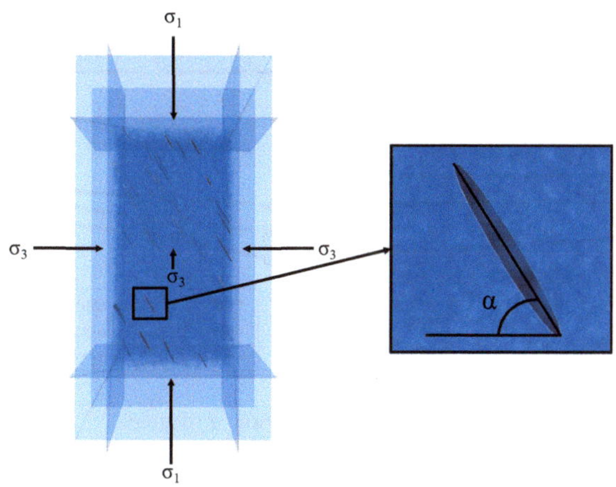

Fig. 3. Representative of fractured rock specimen with the fracture orientation (α) to the direction of cell pressure

Table 1. The fracture apertures and orientations investigated in the study

Fracture aperture description	Width of aperture, ω (mm)	Fracture orientation, α (degrees)
Very tight	0	0, 30, 45, 60, 90
Open	1	0, 30, 45, 60, 90
Moderately wide	4	0, 30, 45, 60, 90

3 Effects of Fracture Orientation on the Mechanical Properties of Fractured Rock

As expected, the fracture orientation influences the strength of the fractured specimen as shown in Fig. 4. It is noted that the lowest value for the ratio of deviatoric stress at 15% axial strain of the fractured rock specimen to the corresponding value for the intact rock specimen, namely the normalized deviatoric strength, was encountered when fracture orientation α = 45° was adopted. With the same fracture aperture width, when α = 45°, the particle assembly was more likely to slide on the fracture plane compared to other fracture orientations. Similar to [2, 5], for these steeper fracture planes (i.e. α = 45° and 60°), the sliding on the fracture plane was more obvious. The fracture plane was weaker than the intact grain assembly, hence, with the failure in the intact material, the sliding effect would result in further reduction in the strength. However, when α = 90°, the existence of fractures had the least effect on the reduction of strength. In this case, since the fracture orientation was parallel to the axial loading direction, the fractures and the slip surfaces between grains were not aligned. Comparing to the intact specimen, the most reduction in normalized deviatoric strength at 15% axial strain was encountered in the specimen with 45° moderately open fracture (i.e. α = 45° and ω = 4 mm).

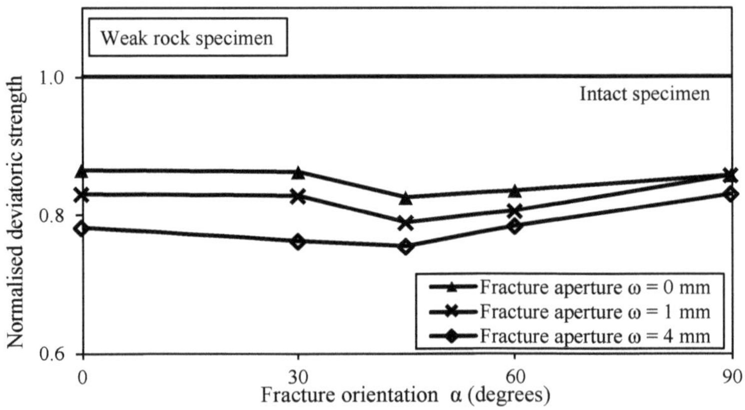

Fig. 4. Normalized deviatoric strength at 15% axial strain of fractured specimen for different fracture orientations and apertures

The deformability of the fractured specimens was varied for different fracture orientations. Since the fractures did not dilate in this study (i.e. dilation angle = 0), no volume change during shearing of fractures were expected. Therefore, as shown in Fig. 5, the fractured specimens had less volume change compared to the intact specimen. It is noted that for a given fracture aperture, for the specimens with α = 45° and 60°, the volumetric strains of specimens with α = 45° and 60° were less than the corresponding values for the specimens with α = 0° and 30°. Furthermore, as the facture has been distributed randomly, the effect of random distribution with different distribution methods on the stress-strain behavior of fractured rock can be studied further in future.

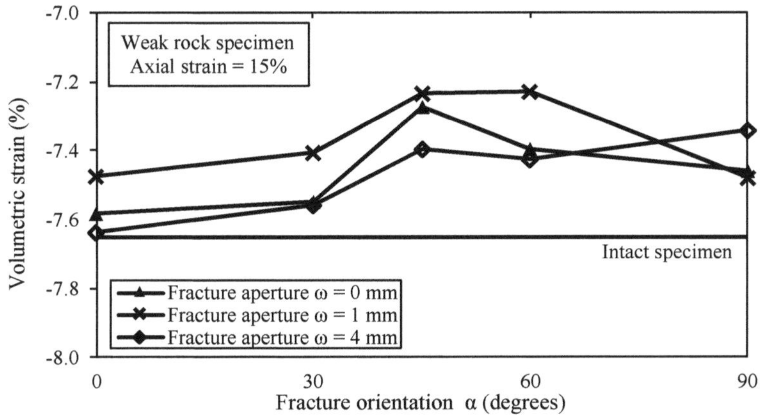

Fig. 5. Volumetric strain at 15% axial strain of fractured specimens with different fracture orientations and apertures

4 Effects of Fracture Aperture on the Mechanical Properties of Fractured Rock

As plotted in Fig. 6, for the fracture orientation $\alpha = 30°$, the strength of the fractured rock specimens decreased with increasing fracture aperture. The specimen with moderately wide fracture aperture (i.e. $\omega = 4$ mm) had a 12% further reduction in strength, comparing to the specimen with very tight fracture aperture (i.e. $\omega = 0$). It is noticed that the specimens with larger fracture apertures experienced more deformation under the same deviatoric stress.

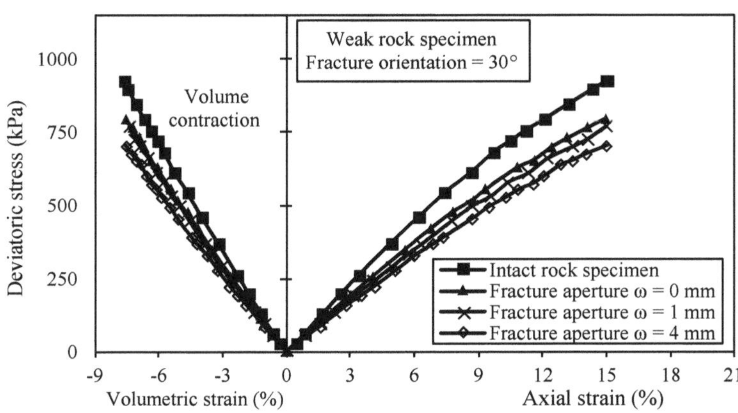

Fig. 6. Comparisons of the intact specimen and fractured specimens with different fracture apertures for the given fracture orientation of 30°

Referring to Fig. 4, the extent in which fracture aperture contribute the reduction in the strength of the rock specimens depends on the orientations of the fractures. With different fracture apertures, the changes in strength were the least for the specimens with α = 90°. In contrast, the most alteration was encountered for the fracture orientation of 30°. When α = 90°, the fracture aperture could close during the isotropic compression loading, while this phenomenon would readily evident for inclined fractures. Therefore, the effects of the fracture aperture were minimal for the specimens with fracture orientation of 90°. Referring to Fig. 6, for a given deviatoric stress, samples with wider aperture experienced larger axial strains. This can be attributed to the fact that more shear forces would be mobilized in the intact parts of the specimen inducing more axial displacement.

As expected, at a given deviatoric stress (e.g. 600 kPa), the fractured specimens had higher volumetric strains (attributed to larger axial displacement of dilative material) comparing to the intact rock specimen, as illustrated in Fig. 7. As the studied weak rock specimen had a high void ratio of 1.25, the specimens only experienced contraction during shearing. As increases the fracture aperture, more volume contraction was observed for fracture orientation of 0°, 30°, 45° and 60°, while for α = 90°, the effect of fracture apertures on volume change was insignificant. Moreover, as the effect of apertures on permeability has not been considered in this paper, hence, further study inspired by the effects of excess pore water pressure can be studied in future [12, 13].

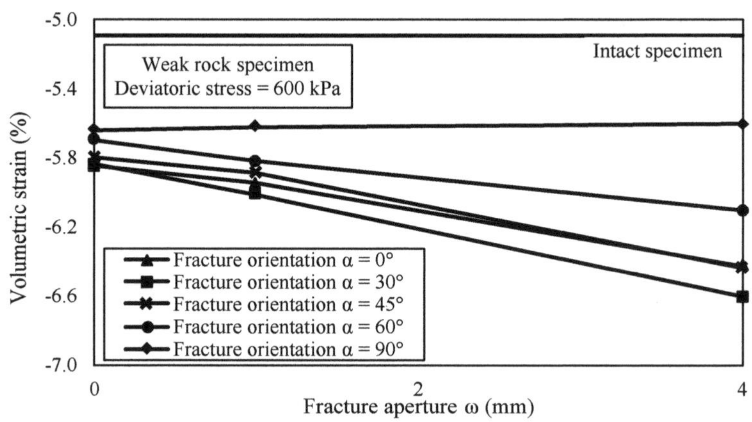

Fig. 7. Volumetric strain of different fracture apertures for fractured rock specimens

5 Conclusions

In this study, the effects of fracture aperture and orientation on the strength and deformability of weak rock specimens have been investigated numerically using discrete element method. Two contact models have been adopted, namely the flat-joint model and smooth-joint contact model, to simulate the cementation effects of the rock mass and sliding effects of fractures, respectively. For a given fracture density, comparisons on the strength and deformability of rock specimens have been conducted through triaxial tests on 15 series of tests with fracture apertures of 0, 1 and 4 mm, and fracture orientations of 0°, 30°, 45°, 60° and 90°. It can be concluded that:

1. The fracture aperture could diminish the strength of the weak rock specimen; as the fracture aperture increased, the strength of the fractured specimens reduced.
2. The extent in which fracture aperture alters the strength and deformability of rock specimens depends on the orientation of the fractures.
3. It is noticed that with the fracture orientation parallel to the axial loading direction, the diminishing effects of fracture aperture on the weak rock strength was minimal.
4. In contrast, the fracture apertures influenced the strength and deformability of the rock specimen with inclined fractures.

References

1. Bell, F.G.: Engineering Properties of Soils and Rocks (2000)
2. Huang, D., Wang, J., Liu, S.: A comprehensive study on the smooth joint model in DEM simulation of jointed rock masses. Granular Matter **17**(6), 775–791 (2015)
3. Ivars, D.M., et al.: The synthetic rock mass approach for jointed rock mass modelling. Int. J. Rock Mech. Min. Sci. **48**(2), 219–244 (2011)
4. Wasantha, P., et al.: Do joint geometrical properties influence the fracturing behaviour of jointed rock? An investigation through joint orientation. Geomech. Geophys. Geo-Energy Geo-Resour. **1**(1–2), 3–14 (2015)
5. Bahaaddini, M., Sharrock, G., Hebblewhite, B.: Numerical investigation of the effect of joint geometrical parameters on the mechanical properties of a non-persistent jointed rock mass under uniaxial compression. Comput. Geotech. **49**, 206–225 (2013)
6. Zhang, X.-P., Wong, L.N.Y.: Loading rate effects on cracking behavior of flaw-contained specimens under uniaxial compression. Int. J. Fract. **180**(1), 93–110 (2013)
7. Chiu, C.-C., et al.: Modeling the anisotropic behavior of jointed rock mass using a modified smooth-joint model. Int. J. Rock Mech. Min. Sci. **62**, 14–22 (2013)
8. Lambert, C., Coll, C.: Discrete modeling of rock joints with a smooth-joint contact model. J. Rock Mech. Geotech. Eng. **6**(1), 1–12 (2014)
9. Sun, D.A., Feng, T., Matsuoka, H.: Stress–strain behaviour of weathered weak rock in middle-sized triaxial tests. Can. Geotech. J. **43**(10), 1096–1104 (2006)
10. Nguyen, L., Fatahi, B.: Behaviour of clay treated with cement & fibre while capturing cementation degradation and fibre failure–C3F model. Int. J. Plast. **81**, 168–195 (2016)
11. Nguyen, L., Fatahi, B., Khabbaz, H.: Development of a constitutive model to predict the behavior of cement-treated clay during cementation degradation: C3 model. Int. J. Geomech. **17**(7), 04017010 (2017)
12. Le, T.M., Fatahi, B.: Trust-region reflective optimisation to obtain soil visco-plastic properties. Eng. Comput. **33**(2), 410–442 (2016)
13. Le, T.M., Fatahi, B., Khabbaz, H., Sun, W.: Numerical optimization applying trust-region reflective least squares algorithm with constraints to optimize the non-linear creep parameters of soft soil. Appl. Math. Model. **41**, 236–256 (2017)

Numerical Investigation of the Dynamic Response of Tunnel Structure and Surrounding Rock Mass to Seismic Loads Based on DEM Simulation

Jie Mei[1], Lei Yang[1(✉)], Weimin Yang[1], Shucai Li[1], Yujing Jiang[2], Bo Zhang[3], and Kongling Guo[1]

[1] Geotechnical and Structural Engineering Research Center, Shandong University, Jinan, China
yanglei@sdu.edu.cn
[2] Faculty of Engineering, Nagasaki University, Nagasaki, Japan
[3] School of Civil Engineering, Shandong University, Jinan, China

Abstract. The dynamic response of underground structures and jointed rock masses remains lack of effective evaluations owing to the high complexity of rock masses and seismic waves. This study investigates numerically the seismic response of a tunnel structure and the surrounding rock mass subjected to a harmonic S-wave by using the DEM (Distinct Element Method) code of UDEC, and the effects of joint spacing, joint stiffness and wave frequency on the seismic response of tunnel lining and rock mass are evaluated. For a given wave frequency, the Peak Particle Velocities (PPVs) of the tunnel lining decrease as the joint spacing increases. The joint stiffness (including the normal and shear stiffness) has opposite effects on the PPVs of tunnel lining and surrounding rock mass. For given properties of joints, the PPVs decrease as a result of an increasing wave frequency. In addition, the study reveals that stress concentration occurs near the top of the tunnel vault. The main results could provide a technical support for the stability analysis and seismic design of underground structures.

Keywords: Distinct Element Method · Seismic response · Jointed rock mass Underground structure · Joint properties · Wave frequency

1 Introduction

The seismic analysis of underground structures has gained a growing emphasis due to the rapid development of underground engineering and the frequent occurrence of earthquakes in recent years [1]. Owing to the brokenness of jointed rock masses and the irregularity of seismic waves, the dynamic problems of underground engineering are extremely complicated, and have not yet been satisfactorily solved so far.

To date, lots of efforts have been made to investigate the seismic response of underground structures using theoretical and experimental methods [2, 3]. However, both of these two methods have some limitations in solving complex problems. E.g., the theoretical method is difficult to deal with the problem with random joint distribution,

L. Zhang et al. (Eds.): GSIC 2018, *Proceedings of GeoShanghai 2018 International Conference: Rock Mechanics and Rock Engineering*, pp. 82–90, 2018.
https://doi.org/10.1007/978-981-13-0113-1_10

and the experimental method has limitations in establishing large-scale models and in setting up reliable boundary conditions.

Currently, many researchers have tended to numerical solutions. Some numerical methods based on the continuum mechanics such as Finite Element Method (FEM) are generally difficult to make an accurate prediction of the mechanical response of fractured rock masses. Thereby, the Distinct Element Method (DEM) has an increasing application in recent years. The common DEM treats the rock masses as a combination of discrete blocks and contacts between them, in which the blocks represent rocks and the contacts represent joints [4]. To date, there have been many attempts to verify the reliability of DEM in dynamic analyses and to acquire new understanding of seismic response of jointed rock masses [5–7]. Zhao et al. [6] investigated the normal transmission of S-waves across parallel fractures with Coulomb slip behavior, and a set of recurrence equations were developed. Zhu et al. [7] studied numerically the wave transmission in jointed rock masses, and multiple parameters were considered and analyzed.

The previous studies have confirmed the reliability of the DEM code of UDEC in dynamic analysis and conducted many parameter studies on wave transmission in jointed rock masses. However, the seismic behavior of tunnel structure in the jointed rock masses is still lack of sufficient and systematic study. This paper investigates the seismic response of a lined tunnel structure in jointed rock masses under a harmonic S-wave by using the code of UDEC. Due to that the joint spacing and stiffness affect the strength and deformation properties of rock masses, and the frequency is an important parameter of seismic wave, the effects of the above parameters on seismic response were evaluated. The Peak Particle Velocities (PPVs) and shear stress distribution in the tunnel lining and rock masses were analyzed.

2 Establishment of Numerical Model

The numerical model has a rectangle shape with the size of 60 m × 80 m (H × L), and a U-shaped tunnel is located in the center of the model (see Fig. 1). The buried depth of the tunnel is set to 200 m. The cross section of the tunnel has a radius of $R = 3.6$ m, a height of $h = 4.0$ m, and a width of $w = 8.0$ m. The lining structure with a uniform thickness of $t = 0.4$ m is set inside the tunnel. Note that the tunnel section in the dashed frame is enlarged for highlighting. Commonly, the joints existing in rock masses appear in sets with almost the same dip angles. Therefore, a set of parallel joints with a specific dip angle of $\theta = 45°$ is set in the model. The joint spacing s is defined as the normal distance between two adjacent joints.

Table 1 shows the physico-mechanical properties of the numerical model. In this study, the mechanical behavior of rocks was simulated using the Mohr-Coulomb model, and the physico-mechanical properties referred to the limestone commonly encountered in engineering practices [8]. Considering that the Mohr-Coulomb joint model can simulate effectively the deformation behavior of large-scale joints, the Mohr-Coulomb joint model was adopted to characterize joints, and the joint properties also referred to literature [8]. The tunnel lining was modelled by using the elastic model, and the mechanical properties of C25 concrete were utilized [9].

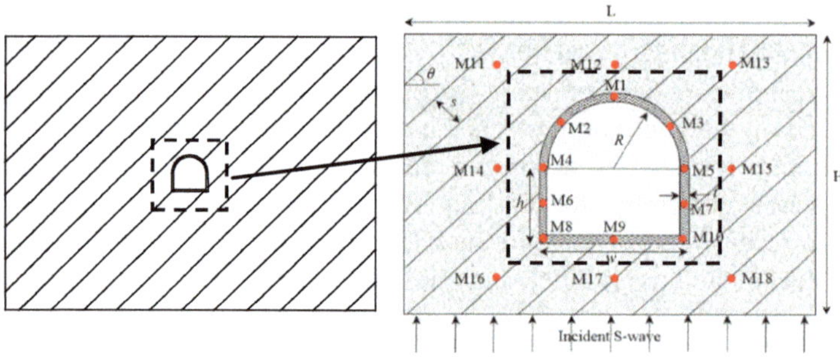

Fig. 1. The numerical model and the tunnel and rock mass structures.

Table 1. Physico-mechanical parameters of the numerical model [8, 9].

Parameters	Index (Unit)	Joint	Lining	Rock
Density	ρ (kg/m^3)	-	2500	2200
Young's modulus	E (GPa)	-	28	28.5
Poisson's ratio	v (-)	-	0.2	0.29
Friction angle	φ (°)	25	-	42.0
Cohesion	c (MPa)	0.84	-	6.27
Uniaxial compressive strength	σ_c (MPa)	-	16.7	72
Tensile strength	t (MPa)	0.1	1.78	1.58

Generally, the S-wave causes larger damage to structures than the P-wave, and the complex seismic wave can be decomposed into a series of simple harmonic waves with different frequencies and amplitudes. To simplify the calculation, a harmonic incident S-wave was considered, which propagated in the rock masses from the bottom of model and passed across the lined tunnel (see Fig. 1). It was assumed that the wave amplitude A is 1.95 cm/s (the corresponding acceleration is 0.1 g, equating to a 7-magnitude earthquake), and the wave duration t is 10 s.

The viscous boundaries were applied to the top and the bottom surfaces of numerical model to remove reflected wave. Due to that it is hard to apply velocity wave to the viscous boundary, a stress wave boundary condition was used at the model bottom. Besides, the free-field boundaries were applied to the both lateral boundaries of model, and thus the fixed-size model could be regarded as infinite in the horizontal direction.

Seismic behavior was measured at 18 points in the tunnel lining and rock masses. The monitoring points, as denoted by red dots in Fig. 1, were numbered orderly, in which M1–M10 corresponded to the points in the tunnel lining, and M11–M18 corresponded to the points located in the rock masses.

3 Numerical Results and Analysis

The Peak Particle Velocities (PPVs) can directly reflect the vibration behavior of rock masses and lining, and the shear stress is a key factor to characterize the dynamic response. In order to study the effects of joint spacing *s*, joint stiffness (normal stiffness k_n and shear stiffness k_s) and wave frequency *f* on the seismic response of tunnel lining and surrounding rock masses, the PPVs and shear stress were monitored during numerical simulation.

3.1 Influence of Joint Spacing on Seismic Response

Figure 2 shows the variation tendencies of PPVs at all monitoring points with the joint spacing, and the related shear stress contours of the model are shown in Fig. 3.

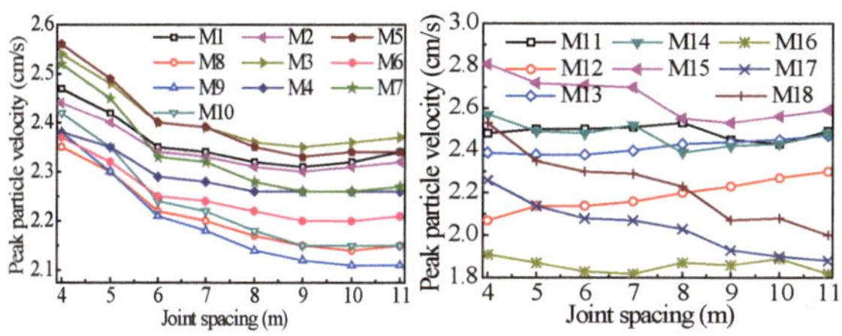

(a) The PPVs of the tunnel lining (b) The PPVs of the rock masses

Fig. 2. Variation tendency of PPVs with joint spacing.

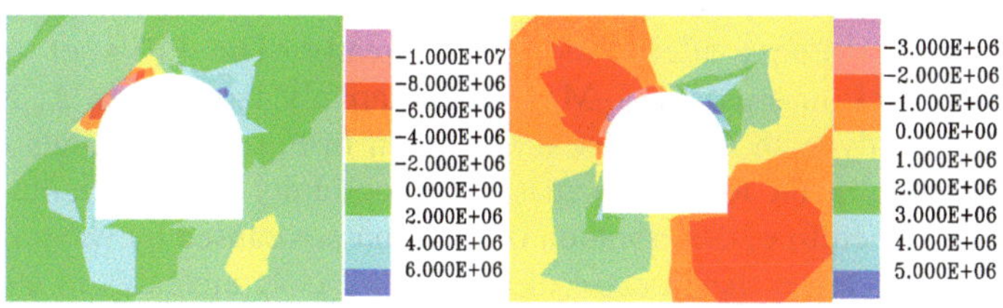

(a) $s = 4$ m (b) $s = 11$ m

Fig. 3. Shear stress contours of the tunnel (unit: Pa).

From Fig. 2, it can be confirmed that the PPVs of all monitoring points are generally larger than the amplitude of incident wave (A = 1.95 cm/s) due to the superposition of reflected and transmitted waves among joints. The joint spacing has an obvious influence on the seismic response of tunnel structure and rock mass. As shown in Fig. 2(a), the PPVs of tunnel lining first decrease remarkably and then tend to a relatively stable value, as *s* increases from 4 to 11 m. This vibration tendency may be due to the decrease of multiple wave superposition [4]. In the region around the tunnel, the reflected, transmitted and incident waves that arrive at the same time superimpose with each other to control the overall vibration. The change of *s* affects the superposition of waves in this region. In the spacing range of this study, an increasing s enhances the possibility of wave superposition of reverse-direction vibration, thus leading to the decrease of PPVs of the lining. It can also be speculated that the larger the s, the more intact the rock masses, which leads a relatively stable lining vibration under seismic loading.

Variation tendencies of the PPVs in rock masses at different height are different with the increase of s (see Fig. 2(b)). First, the PPVs of the rock masses above the tunnel (M11–M13) increase with s, due to that the integrity of rock mass is improved and the attenuation effect of joints becomes weak as s increases. Besides, the increase of PPVs in M11–M13 may also be caused by the participation of reflected waves at the tunnel surface. Second, the PPVs of the rock masses below the tunnel (M16–M18) decrease continuously with s due to the wave superposition of reverse-direction.

The joint spacing affects the stress distribution greatly. It can be concluded from Fig. 3(a) that when s is small, the stress concentration of the tunnel vault is remarkable, especially the left side vault has a maximum shear stress of around 10 MPa. The result shows that, the dynamic loading can cause greater stress concentration in the weak area with many joints, which leads to damage easily. However the values of the shear stresses decrease as s increasing to 11 m. The maximum value is around 5 MPa. The probable cause is that the relatively intact rock mass has a better ability to propagate stress, and shear stress is reduced and the tunnel is safer.

3.2 Influence of Joint Stiffness on Seismic Response

The angle between the incident wave and the joints is 45°, therefore, both the normal and shear stiffness make contributions to the deformation of joints, and thus they affect the seismic response of the tunnel lining and rock masses.

Influence of Joint Normal Stiffness

The joint normal stiffness affects the PPVs greatly. The PPVs of tunnel lining (M1–M10) and rock masses (M13–M18) are relatively large when k_n is small, and decrease significantly with k_n increases, as shown in Fig. 4. The joints compression deformation under seismic loading increases the continuity of the rock masses when k_n is at a low level. Therefore, the vibration of rock masses caused by wave superposition is violent and thus causes the increase of the PPVs. As k_n increases to a certain value (k_n = 5.0 GPa/m), all the PPVs are insensitive to the change of k_n, and reach relatively stable values. However, the PPVs of points below the tunnel (M16–M18) are obviously smaller than the above ones (M11–M15) as shown in Fig. 4(b) and (c). This may be due to that the reflected waves caused by the joints near the tunnel structure reversely superimpose with the incident waves at the lower points, resulting in smaller PPVs.

The PPVs of some individual points (M11–M12) above the tunnel increase as k_n increases (see Fig. 4(b)). The reason might be that the barrier effect of the tunnel space and the absorption effects of the joints on the seismic waves are better when k_n is small. And the waves can propagate with k_n increases, which leads to the vibration of the rock.

Effects of k_{n} on stress distribution is not obvious. The shear stress of the right arch reduce slightly when k_{n} increase to 20 GPa/m as shown in Fig. 5. Because the compressive deformation of joints decreases with the increase of k_{n}.

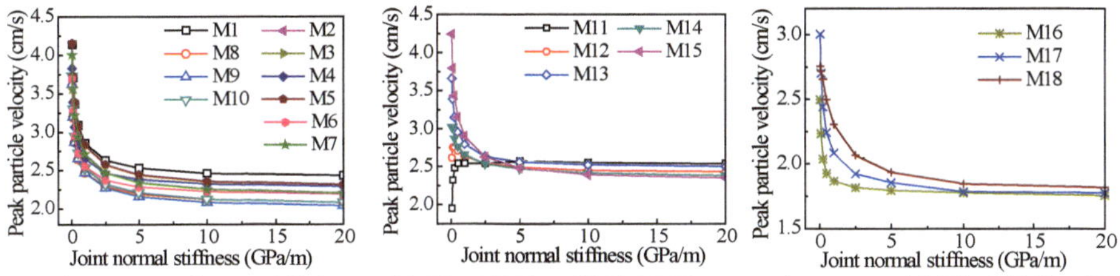

(a) The PPVs of tunnel lining (b) The PPVs of M11-M15 at rock masses (c) The PPVs of M16-M18 at rock masses

Fig. 4. Variation tendency of PPVs with different joint normal stiffness.

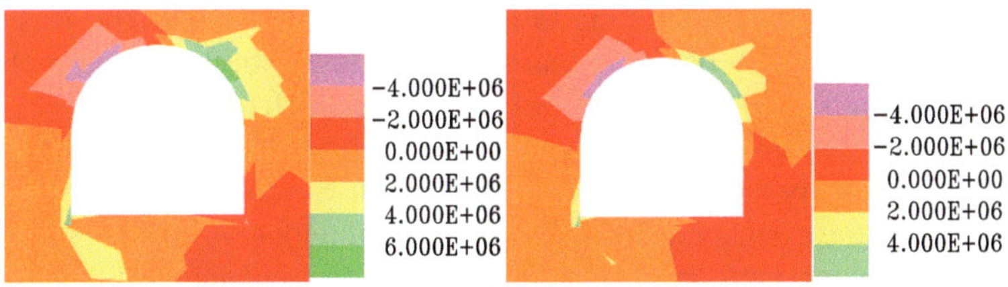

(a) k_n = 0.625 GPa/m (b) k_n =20 GPa/m

Fig. 5. Shear stress contours of the tunnel (unit: Pa).

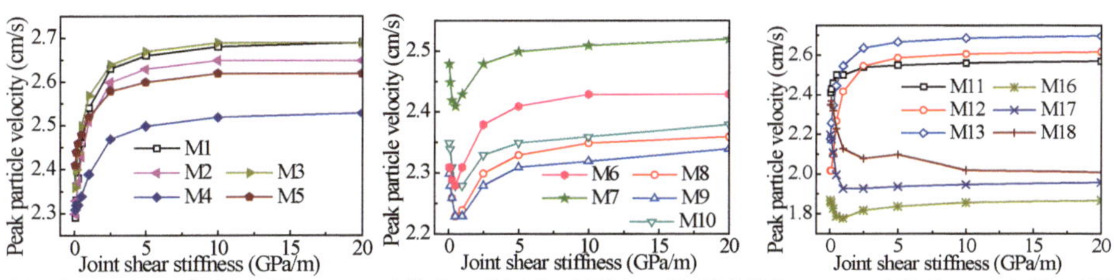

(a) The PPVs of M1-M5 at tunnel lining (b) The PPVs of M6-M10 at tunnel lining (c) The PPVs of rock masses

Fig. 6. Variation tendency of the PPVs with joint shear stiffness.

Influence of Joint Shear Stiffness

The joint shear stiffness has complex effects on PPVs. As shown in Fig. 6(a) and (c), the PPVs of M1–M5 and M11–M13 increase remarkably as k_s increasing from 0.0625-5.0 GPa/m. The S-wave will cause movements and slipping of joints, which results in the energy consumption and attenuation of the seismic wave. Owing to that the energy of the seismic wave is less absorbed by the joints with a large k_s, the vibration can propagate upward, leading to the increase of PPVs. Additionally, Fig. 6(b) and (c) indicate that, as k_s increasing from 0.0625 to 0.5 GPa/m, the PPVs of M6–M10 and M16–M18 at the lower parts of the model show a certain declining tendencies. The reflected waves caused by the free surface of the tunnel superposes the incident waves,

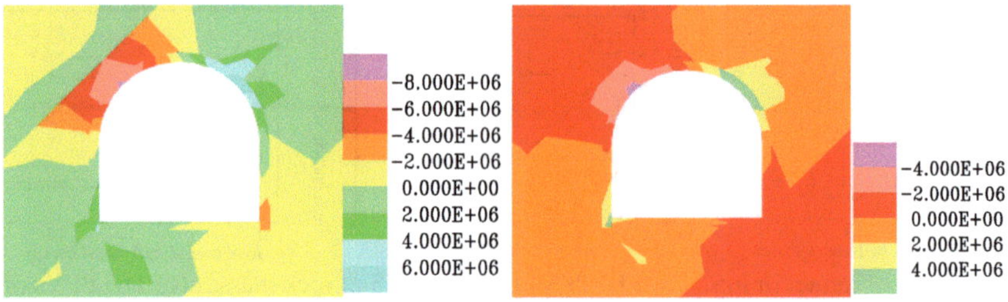

(a) $k_s = 0.625$ GPa/m (b) $k_s = 20$ GPa/m

Fig. 7. Shear stress contours of the tunnel (unit: Pa).

leading to the violent vibration of the lower parts with a small k_s. The reason for this decline is that the vibration of rock masses is weakened with the increase of k_s.

The representative stress contours of the tunnel with different k_s are shown in Fig. 7. Compared with the tensile and compressive deformations, the slip deformation of joints is more easily to occur under the action of S-wave. Therefore, a smaller k_s results in greater stress concentration in the tunnel lining than k_n. The joints are difficult to slide and rub along surfaces when k_s is large under seismic loading, and thus the stress concentration area and shear stress of the tunnel lining decrease significantly.

k_n and k_s have different effects on the dynamic response of rock masses and tunnel lining. The fundamental reason is that the slip deformation caused by S-wave increases the discontinuity of rock masses, while compression deformation enhances the continuity. From the stress distribution characteristics, it can be concluded that s and k_s are two important properties of joints to focus on in seismic design of the tunnel structure.

3.3 Influence of Wave Frequency on Seismic Response

The wave frequency affects the PPVs greatly. As shown in Fig. 8, the PPVs of the tunnel lining (M1–M10) and rock masses (M11–M18) decrease continuously with the increase of f, and the lower the location, the greater the curves slope. This is ascribed to that the jointed rock masses can be regarded as a low-pass filter [10], and seismic waves are easily absorbed by joints with the increase of frequency, resulting in greater attenuation of the energy and decrease of the PPVs.

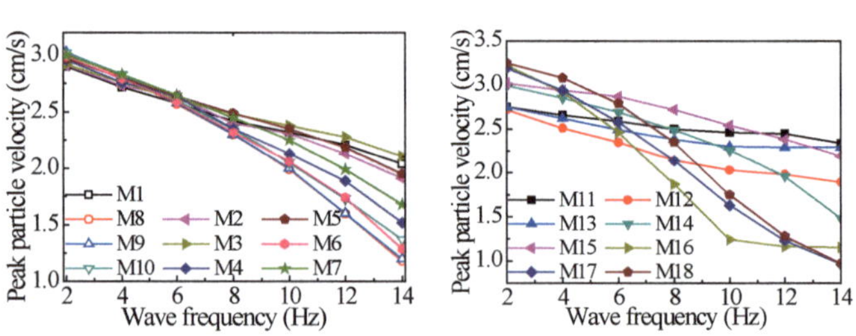

(a) The PPVs of tunnel lining (b) The PPVs of rock masses

Fig. 8. Tendency of the PPVs with varying wave frequency.

(a) $f = 2$ Hz

(b) $f = 14$ Hz

Fig. 9. Shear stress contours of the tunnel (unit: Pa).

The stress contours of model with different f are almost the same (see Fig. 9), which means that stress distribution of the tunnel lining is less affected when wave frequency increases from 2 to 14 Hz. This might be due to that the study range of frequency variation is not large enough to have a significant impact on the stress distribution.

4 Conclusions

(1) The PPVs of all monitoring points are enlarged due to the wave superposition. An increasing s enhances the possibility of wave superposition of reverse-direction vibration, thus leading to the decrease of PPVs of tunnel lining and rock masses below the tunnel. And the PPVs in rock masses above the tunnel increase due to the decrease of wave attenuation. Besides, the shear stress of tunnel with a small s is larger, due to that the broken jointed rock masses are more easily to produce stress concentration.

(2) Most of the PPVs are relatively large when k_n is small, and decrease significantly as increase of k_n. Due to that the wave superposition and rock vibration are easily to occur when k_n is small. The PPVs of tunnel lining and rock masses in upper parts increase as the increase of k_s, while the PPVs of rock masses in lower parts decrease. k_s has a greater influence on the stress distribution of rock masses than k_n owing to the significant joint shear deformation caused by S-wave.

(3) The PPVs of the tunnel lining and rock masses decrease continuously as a result of the increasing of wave frequency due to the low-pass filtering effects of joints. A larger frequency range should be adopted in the future studies to investigate its effects on stress distribution.

Acknowledgement. This research work was supported by the National Natural Science Foundation of China (Grant No. 51509146, 51379114, 51479107) and the Fundamental Research Funds of Shandong University (2014HW012).

References

1. Hashash, Y.M.A., Hook, J.J., Schmidt, B., et al.: Seismic design and analysis of underground structures. Tunn. Undergr. Space Technol. **16**(4), 247–293 (2001)
2. Yi, C., Zhang, P., Johansson, D., et al.: Dynamic response of a circular lined tunnel with an imperfect interface subjected to cylindrical P-waves. Comput. Geotech. **55**(1), 165–171 (2016)
3. Li, J.C., Ma, G.W.: Experimental study of stress wave propagation across a filled rock joint. Int. J. Rock Mech. Min. Sci. **46**(3), 471–478 (2009)
4. Chen, S.: Discrete Element Modeling in Rock Tunnels. Southwest Jiaotong University Press, Chengdu (2015)
5. Souley, M., Homand, F.: Stability of jointed rock masses evaluated by UDEC with an extended Saeb-Amadei constitutive law. Int. J. Rock Mech. Min. Sci. Geomech. Abstr. **33** (3), 233–244 (1996)
6. Zhao, X.B., Zhu, J.B., Zhao, J., et al.: Study of wave attenuation across parallel fractures using propagator matrix method. Int. J. Numer. Anal. Meth. Geomech. **36**(10), 1264–1279 (2012)
7. Zhu, J.B., Deng, X.F., Zhao, X.B., et al.: A numerical study on wave transmission across multiple intersecting joint sets in rock masses with UDEC. Rock Mech. Rock Eng. **46**(6), 1429–1442 (2013)
8. Lan, H.X., Fa-Quan, W.U.: The decision of mechanical parameters for numerical simulation of rock and soil mechanics. World Geol. (2001)
9. Code for Design on Tunnel of Railway (TB 10003-2016). China Railway Publishing House (2017)
10. Pyrak-Nolte, L.J., Myer, L.R., Cook, N.G.W.: Transmission of seismic waves across single natural fractures. Geophys. Res. **95**(B6), 8617–8638 (1990)

Numerical Simulation on Acoustic Emission of Composite Stratum

Jianxiu Wang[1,2,3(✉)], Yansheng Deng[1], and Linbo Wu[1]

[1] College of Civil Engineering, Tongji University, Shanghai 200092, China
wang_jianxiu@163.com
[2] State Key Laboratory for Geo Mechanics and Deep Underground Engineering, China University of Mining and Technology, Xuzhou 221008, China
[3] Key Laboratory of Geotechnical and Underground Engineering of Ministry of Education, Tongji University, Shanghai 200092, China

Abstract. Composite stratum is composed by soft and hard alternated composite stratum with different lithological character. So it exists interfaces between soft and hard layers, which have influence to the deformation failure of composite stratum. Based on particle flow discrete element theory, the different acoustic emission samples of composite stratum were established. By recording the number of contact bond breaks between particles at same time step, the failure characteristics and AE distribution of different composite stratum samples are analyzed under uniaxial compression numerical tests.

Keywords: Acoustic emission · Composite stratum · Failure characteristics

1 Introduction

Composite stratum is a common geological conditions in all types of engineering construction. The transverse isotropy is a typical characteristic of composite stratum. Layered rock mass is a typical formation of composite stratum, and has been studied widely. The bedding plane dip angle has great influence in strength and failure mode of layered rock [9, 10, 19, 20]. Many researchers investigated mechanical response and failure mode of layered rock mass by improving continuum modeling [9–11, 22, 24]. However, traditional laboratory experiments have widely been used as a simple and effective method to study the failure process in rock, especially the uniaxial compression test [21, 25, 26]. Combined with acoustic emission (AE) technology, the microcracks in failure process can be recorded and investigated under uniaxial compression test [5–8, 16–18, 23]. Numerical methods have been developed to investigated extensively the microcracks initiation and growth in rocks under compression, like RFPA model based on FEM [13–15], PFC model [2, 3], and other model [1]. So the failure process of composite stratum is rarely studied.

The fracture process of composite stratum with different combination is investigated under uniaxial compression numerical test by using particle flow code (PFC^{2D}) software. By recording the number of bond break between particles in same time step represents AE event in the test process. Finally, failure characteristics and AE distribution of different composite stratum samples are analyzed.

L. Zhang et al. (Eds.): GSIC 2018, *Proceedings of GeoShanghai 2018 International Conference: Rock Mechanics and Rock Engineering*, pp. 91–98, 2018.
https://doi.org/10.1007/978-981-13-0113-1_11

2 Simulation Mechanism of PFC

PFC^{2D} provides two basic contact models for bond between particles: contact bond model and parallel bond model. The contact bond can be considered as elastic springs with constant normal and shear stiffnesses acting at the contact point between particles, which can only transmit force between particles. The parallel bond provides the mechanical behavior of a finite-sized piece of cement-like material deposited between the two contacting pieces, and can transmit both force and moment between the particles [4, 12]. Therefore, in this work, the parallel bond model was used to simulate the behavior of the complex rock mass. The bond breakage in PFC represents the formation of cracks, the constitutive behavior in shear and tension of parallel bond model in PFC is shown in Fig. 1.

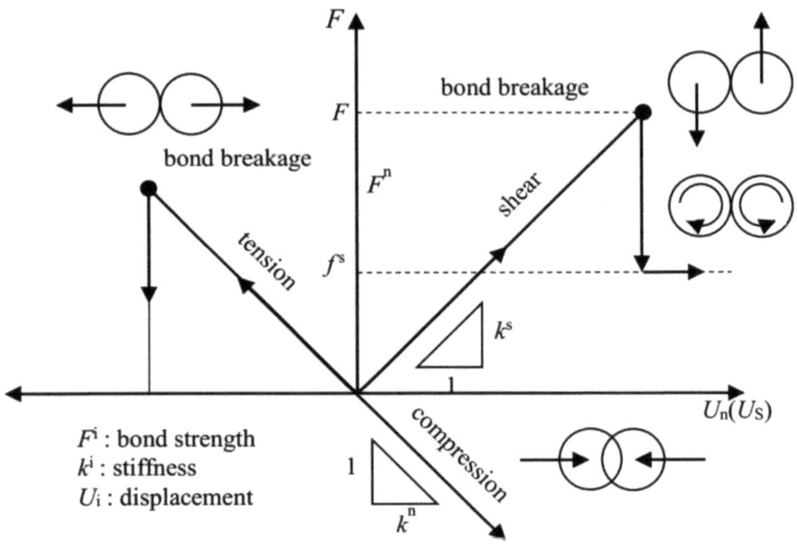

Fig. 1. Constitutive behavior in shear and tension of parallel bond model in PFC

The formation of each tensile crack or shear crack represents the increasing damage, and release the strain energy. In PFC^{2D}, the fracture number in the same interval is approximately equivalent to an acoustic emission event in the calculation process. In this way, the numerical method can be used to study the acoustic emission law of composite stratum.

3 Numerical Simulation Results and Discussions

3.1 Numerical Model

This paper aims to discuss the failure law of composite stratum, the micro-parameters were assumed to be a certain proportion to simulate the soft and hard stratum, like

Fig. 2. The two dimensional numerical samples of uniaxial compression (The green part represents soft stratum, and blue part represents hard stratum)

Table 1. Parameters of soft and hard stratum

Parameters	Value	
	Soft stratum	Hard stratum
Effective modulus (GPa)	6	12
Stiffness ratio	1	1
Bond effective modulus (GPa)	6	12
Bond stiffness ratio	1	1
Tensile strength (MPa)	40	100
Cohesion (MPa)	80	500
Normal critical damping ratio	0.5	0.5

effective modulus, bond effective modulus, tensile strength and cohesion. According to the different combinations of soft stratum and hard stratum, this paper took soft stratum (SS), hard stratum (HS), soft-hard stratum (SHS), soft-hard-soft stratum (SHSS) and hard-soft-hard stratum (HSHS) numerical samples as research object to investigate deformation failure and AE characteristics of composite stratum. The two dimensional numerical samples of uniaxial compression are shown in Fig. 2. The SHS model is composed of the same two parts (soft stratum and hard stratum). The SHSS and HSHS model is composed of the same three parts (soft stratum, hard stratum, soft stratum and hard stratum, soft stratum, hard stratum.) All of the models are 0.05 m in width and 0.1 m in height, conforming to typical laboratory core samples. The porosity is 0.1, the density is 2500 kg/m^3 and the total amount of particles is 3602. The particles of all of samples radiuses have a uniform distribution between the maximum and minimum radii. The particles are randomly distributed in the model. The minimum and maximum radius is 0.5 m and 0.75 m, respectively. The parameters of soft stratum and hard stratum are shown in Table 1.

The samples were loaded until the axial stress falls below 70% of the peak stress. In order to reduce the influence of loading rate, the loading rate was 7.5e−8 mm/step in the calculation process. In this paper, the number of bond break in 40 steps was regarded as an AE event.

3.2 Numerical Results

Failure Characteristics of Different Composite stratum Samples. By recording the number of bond fracture of models during loading, the results of AE and stress-strain under uniaxial compression are shown in Fig. 3. According to Fig. 3, during the initial loading period, all of the models had no AE event due to less stress. With increasing load, bond break began to appear. Due to the existence of defects in the natural stratum, the initial is not a complete entity, but there is a complex network of stress structure system. The bond break can lead to stress adjustment of structure system, which caused stress concentration of adjacent elements. When stress of adjacent elements was up to bond strength, the bond broke and AE event occurred. When the stress was up to peak value, the number of AE event leaped at a rapid speed, and the number of crack increased sharply, formed macroscopic failure.

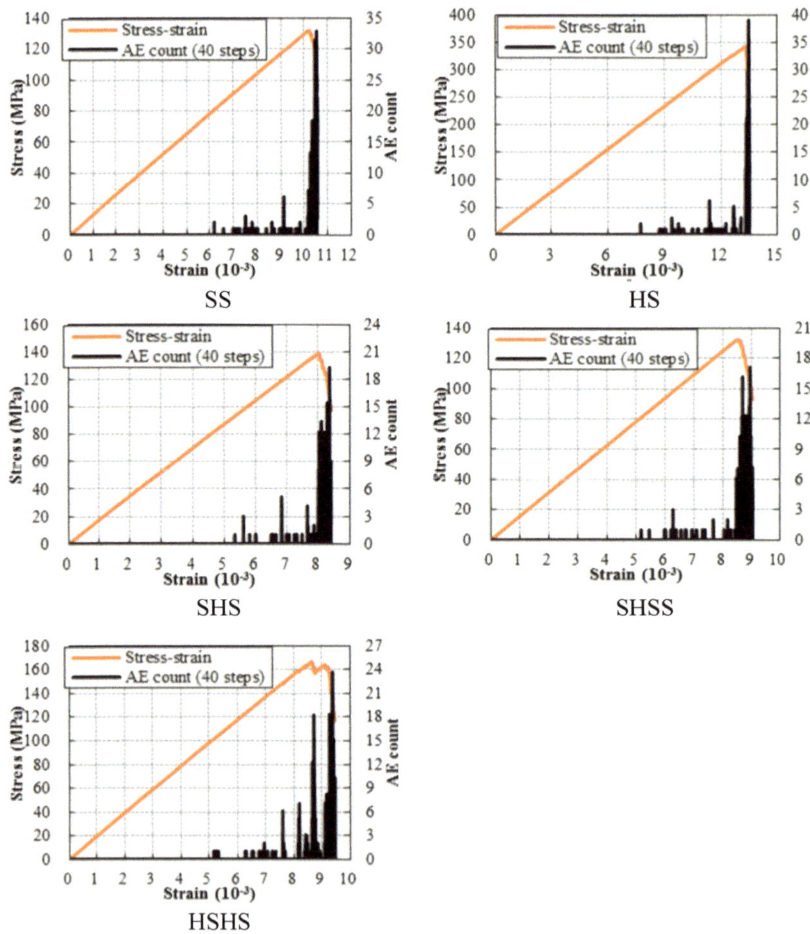

Fig. 3. AE and stress-strain results for models under uniaxial compression

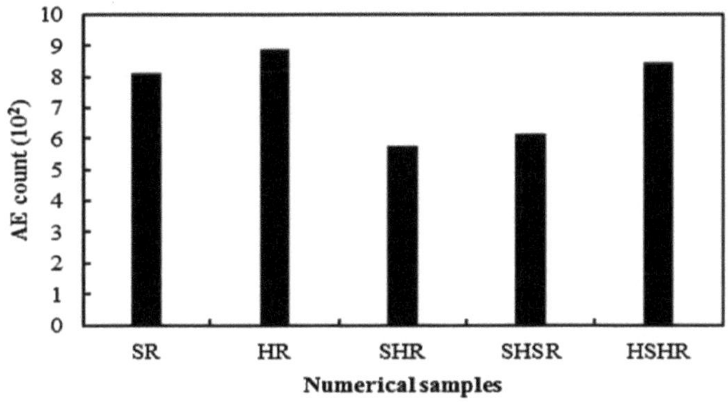

Fig. 4. Cumulative AE count of different numerical samples

In addition, the results of SHS, SHSS and HSHS samples in Fig. 3 showed the AE event still had high intensity when the stress exceeded the peak strength. It demonstrated that compared with the peak stress, the AE event existed hysteresis phenomenon caused by plastic deformation of soft stratum parts. The more soft stratum part was, the more obvious hysteresis phenomenon.

From the result of HSHS samples in Fig. 3, the stress didn't decreased rapid when the stress was up to peak stress. The stress increased gradually to peak stress again after a small drop, and then decreased sharply like the other samples. At the same time, the AE event was sharply increasing when the stress was equal to peak stress. This phenomenon illustrated the hard stratum parts has some restriction effect on the soft stratum part. The restriction effect can improve the toughness of sample to a certain extent.

The cumulative AE count of different numerical samples is shown in Fig. 4. The HS sample had the maximum AE count, and SHS sample showed the minimum AE count. It means that the hard stratum had the most bond break, and the different combinations of soft stratum and hard stratum had great influence on bond break.

Distribution Characteristics of AE Event. PFC^{2D} can record deformation and distribution of AE event in the loading process. The AE event spatial distribution of different numerical samples is shown in Fig. 5. From Fig. 5, the SS and HS samples had same spatial distribution of AE event. Their AE event distributed mainly at the failure path. The failure face was formed when the stress was up to peak stress. According to the results of SHS, SHSS and HSHS samples in Fig. 5, the AE event occurred mainly in soft stratum parts, and hard stratum parts had few AE events due to higher bond strength. In addition, the distribution of AE event in soft stratum was more dispersed, and the AE event in hard stratum part was mainly occurred in very few major cracks.

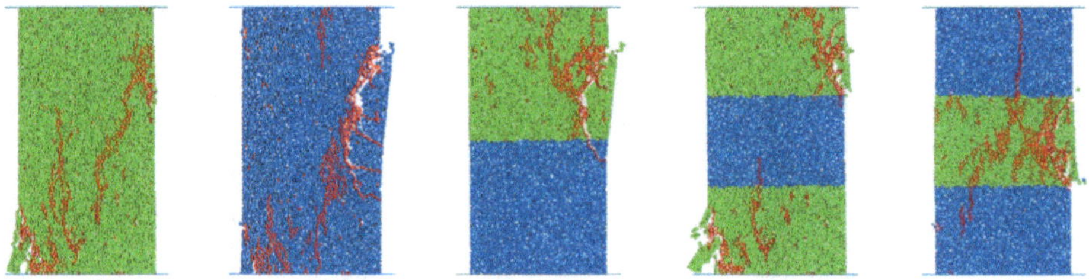

Fig. 5. AE event spatial distribution of numerical samples

4 Conclusions

Based on PFC^{2D}, the deformation failure and AE characteristics of five composite stratum samples (SS, HS, SHS, SHSS and HSHS) were investigated using uniaxial compression numerical tests. By comparison, the AE peak of combination samples with soft stratum part had hysteresis phenomenon, and the hysteresis was more obvious with increasing soft stratum parts. The hard stratum parts of HSHS sample showed restriction effect to soft stratum, which can improve toughness of samples to some extent. By compared with the cumulative AE count, the different combination had great influence on the bond break, and the SHS and SHSS showed less bond break. In addition, to combination samples, the AE event mainly concentrated in soft stratum parts due lower strength and the distribution of AE event was more dispersed. However, the AE event in hard stratum concentrated on few major cracks. Because it is only a discussion on failure process of composite stratum, the micro-parameters are not calibrated based on laboratory tests. The numerical samples are not absolutely uniform, there may be failure to both sides, once the damage on either side of the left and right side occurs.

References

1. Ann, B., Juha, A., Tobias, B., et al.: Numerical modeling of uniaxial compressive failure of granite with and without saline porewater. Int. J. Rock Mech. Min. Sci. **45**(7), 1126–1142 (2008)
2. Cho, N., Martin, C.D., Sego, D.C., et al.: Modelling Dilation in Brittle Rocks. ARMA/NARMS 04–483 (2004)
3. Hajiabdolmajid, V., Kaiser, P.K., Martin, C.D.: Modelling brittle failure of rock. Int. J. Rock Mech. Min. Sci. **39**, 731–741 (2002)
4. Itasca Consulting Group Inc.: PFC3D-Particle Flow Code in 3 Dimensions. Version 4.0, Minneapolis (2008)
5. Lei, X., Kusunose, K., Nishizawa, O., Cho, A., Satoh, T.: On the spatio-temporal distribution of acoustic emissions in two granitic rocks under triaxial compression: the role of pre-existing cracks. Geophys. Res. Lett. **27**(13), 1997–2000 (2000)

6. Lei, X., Kusunose, K., Rao, M.V.M.S., Nishizawa, O., Satoh, T.: Quasi-static fault growth and cracking in homogeneous brittle rock under triaxial compression using acoustic emission monitoring. J. Geophys. Res. Solid Earth **105**(B3), 6127–6139 (2000)
7. Lei, X., Masuda, K., Nishizawa, O., Jouniaux, L., Liu, L., Ma, W., et al.: Detailed analysis of acoustic emission activity during catastrophic fracture of faults in rock. J. Struct. Geol. **26**(2), 247–258 (2004)
8. Lei, X., Satoh, T.: Indicators of critical point behavior prior to rock failure inferred from pre-failure damage. Tectonophysics **431**(1–4), 97–111 (2007)
9. Nasseri, M., Rao, K.S., Ramamurthy, T.: Anisotropic strength and deformational behavior of Himalayan schists. Int. J. Rock Mech. Min. **40**(1), 3–23 (2003)
10. Niandou, H., Shao, J.F., Henry, J.P., Fourmaintraux, D.: Laboratory investigation of the mechanical behaviour of Tournemire shale. Int. J. Rock Mech. Min. **34**(1), 3–16 (1997)
11. Nova, R.: An extended Cam Clay model for soft anisotropic rocks. Comput. Geotech. **2**(2), 69–88 (1986)
12. Potyondy, D.O., Cundall, P.A.: A bonded-particle model for rock. Int. J. Rock Mech. Min. Sci. **41**(8), 1329–1364 (2004)
13. Tang, C.A., Kou, S.Q.: Crack propagation and coalescence in brittle materials under compression. Eng. Fract. Mech. **61**(3–4), 311–324 (1998)
14. Tang, C.A., Liu, H., Lee, P.K.K., et al.: Numerical studies of the influence of microstructure on rock failure in uniaxial compression. Part I: effect of heterogeneity. Int. J. Rock Mech. Min. Sci. **37**, 555–569 (2000)
15. Tang, C.A., Liu, H., Lee, P.K.K., et al.: Numerical studies of the influence of microstructure on rock failure in uniaxial compression. Part II: constraint, slenderness and size effect. Int. J. Rock Mech. Min. Sci. **37**, 571–583 (2000)
16. Thompson, B.D.: Observations of premonitory acoustic emission and slip nucleation during a stick slip experiment in smooth faulted Westerly granite. Geophys. Res. Lett. **32**(10) (2005)
17. Thompson, B.D., Young, R.P., Lockner, D.A.: Fracture in westerly granite under AE feedback and constant strain rate loading: nucleation, quasi-static propagation, and the transition to unstable fracture propagation. Pure Appl. Geophys. **163**(5–6), 995–1019 (2006)
18. Thompson, B.D., Young, R.P., Lockner, D.A.: Premonitory acoustic emissions and stick-slip in natural and smooth-faulted Westerly granite. J. Geophys. Res. Solid Earth **114** (B2) (2009)
19. Tien, Y.M., Kuo, M.C.: A failure criterion for transversely isotropic rocks. Int. J. Rock Mech. Min. Sci. **38**(3), 399–412 (2001)
20. Tien, Y.M., Kuo, M.C., Juang, C.H.: An experimental investigation of the failure mechanism of simulated transversely isotropic rocks. Int. J. Rock Mech. Min. **43**(8), 1163–1181 (2006)
21. Wong, R.H.C., Lin, P., Tang, C.A.: Experimental and numerical study on splitting failure of brittle solids containing single pore under uniaxial compression. Mech. Mater. **38**(1–2), 142–159 (2006)
22. Xu, D.P., Feng, X.T., Chen, D.F., et al.: Constitutive representation and damage degree index for the layered rock mass excavation response in underground openings. Tunn. Undergr. Space Technol. **64**(2017), 133–145 (2017)
23. Zang, A., Wagner, F.C., Stanchits, S., Janssen, C., Dresen, G.: Fracture process zone in granite. J. Geophys. Res. Solid Earth **105**(B10), 23651–23661 (2000)
24. Zhang, X.M., Feng, Y., Yang, J.S.: Experimental study on anisotropic strength properties of sandstone. Electron. J. Geotech. Eng. **15**, 1325–1335 (2010)

25. Zhou, H.M., Yang, Y., Zhang, Y.H., et al.: Fine test on progressive fracturing process of multi-crack rock samples under uniaxial compression. Chin. J. Rock Mech. Eng. **29**(3), 465–470 (2010). (in Chinese)
26. Zhou, J.W., Xu, W.Y., Yang, X.G.: A microcrack damage model for brittle rocks under uniaxial compression. Mech. Res. Commun. **37**, 399–405 (2010)

Research on Construction Simulation and Early Warning Method of High Steep Rock Slope

De-yong Wang[1,2,3(✉)], Yong-ping Wang[2,3], Hai-hong Mo[1], and Ping-shan Chen[2,3]

[1] Civil and Transportation Institute, South China University of Technology, Guangzhou 510641, China
de_yong_wang@163.com

[2] CCCC Fourth Harbor Engineering Institute Co., Ltd., Guangzhou 510230, China

[3] Key Laboratory of Environmental Protection and Safety of Communication Foundation Engineering, CCCC, Guangzhou 510230, China

Abstract. The slopes in southwest of China are high and steep. Moreover, their geological conditions and form history are complicated. Stability of slope was one the most important for the expressway constructions in these areas. Based on the excavation and reinforcement of rock slopes in DaoAn highway project, combining with the strata occurrence, calculation models of slope excavation are established. Then the comprehensive process of slope excavation was simulated. Site monitoring and testing were implemented, monitoring data such as surface displacement, vertical displacement, deep horizontal displacement and anchor cable force of three typical slopes during construction are analyzed. Then dynamic warning method and index of slope monitoring during the slope construction are studied. Based on designed safety factor and failure mode of slopes, combined with field monitoring and numerical simulation, a new method to determine the monitoring & early warning index was proposed. They could give references to evaluate the safety state of rock slopes during excavation.

Keywords: High steep rock slope · Excavation and support
Construction simulation · Monitoring warning

1 Introduction

The slopes in southwest of China are high and steep, their geological conditions are complicated, and the forming history are special. For meeting the requirement of construction of Highway engineering in this area, slope stability and early-warning during excavation was the most important topic to safety evaluation of slopes along the proposed highway.

Landslide prediction is the key problem of landslide research [1]. Slope monitoring is the most direct and reliable method, which is widely used in prediction of slope failure. The main purpose of slope monitoring is to investigate the slope evolution, and

L. Zhang et al. (Eds.): GSIC 2018, *Proceedings of GeoShanghai 2018 International Conference: Rock Mechanics and Rock Engineering*, pp. 99–108, 2018.
https://doi.org/10.1007/978-981-13-0113-1_12

to provide reliable information and scientific basis for the safety construction. While in fact, most important displacement information lose effectiveness due to later monitoring reaction than the excavation activity, moreover the damage evolution range during excavation is very difficult to monitor.

In recent years, people pay more attention to the prediction of slope instability; constantly enrich the early-warning theory and method research [2, 3]. Most of the researches were focused on the landslide prediction by means of nonlinear scientific theory.

Based on the studies of slope safety evaluation during excavation engineering in DaoAn highway, combined with the site monitoring of the high rock slope, the slope construction process was simulated numerically, the dynamic warning method & index of monitoring and forecast during excavation are proposed.

2 Early Warning Method of During Excavation

For the slope deformation monitoring and early warning of rock slope during excavation, in order to overcome the shortcoming of single warning index from analogy method, early warning method base on safety factor and failure mode was proposed, and the procedure was shown on Fig. 1.

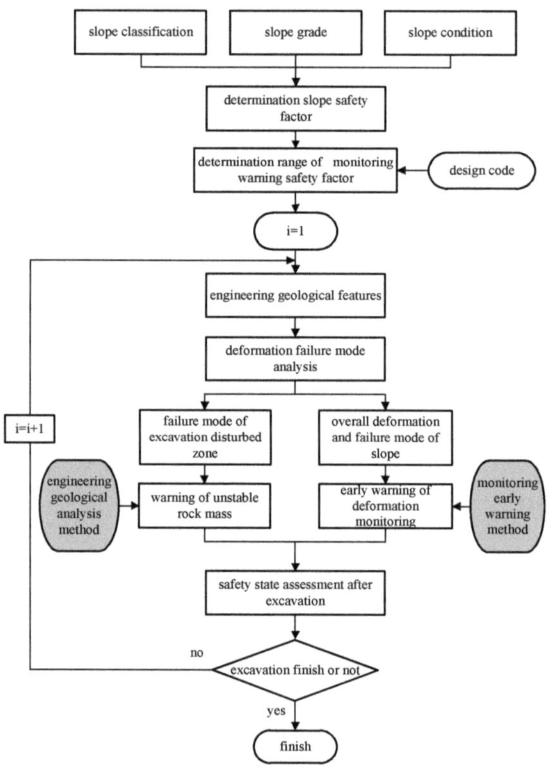

Fig. 1. Flow chart of slope early warning during excavation

The specific steps are as follows:

(a) Firstly determine the type and grade of research slope as per design code, then determine the construction condition and design safety factor Fs. (b) Secondly, in order to determine whether the state of slope is safe, analyze the failure mode of slope by geological analysis method and monitoring early warning method. (c) According to the actual safety state of the slope after excavation. If not finished, gain the deformation modulus with intelligent displacement back analysis method, and then repeat preceding steps; if finished, the deformation monitoring and early warning for slope excavation close.

3 Case Study of Slope Excavation

3.1 Project Introduction

The research object is a rock slope besides highway which located in Daoan, Guizhou province, and the height is 56.93 m. Exploration revealed that, the 0–2 m overly clay gravel, bedrock is argillaceous siltstone from thin to medium, and the rock mass weather strongly which easily induced slope collapse during excavation. The preliminary design of slope support was shown in Fig. 2, three kinds of gradability as 1:0.75, 1:1.1 and 1:1.25 to slow down the slope, the anchor frames were set to fix the foot, anchor frame beam were used as shallow support.

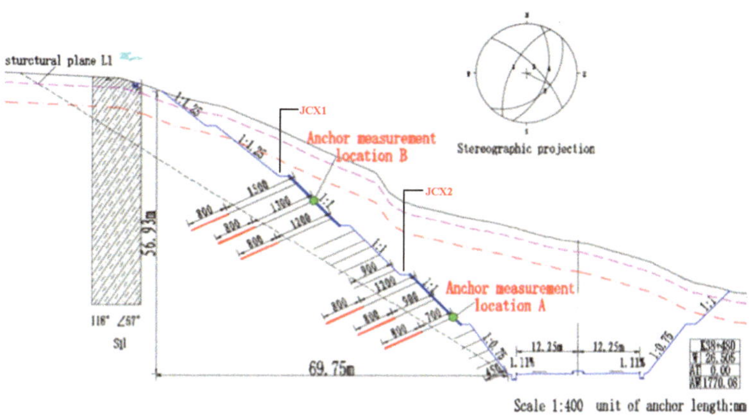

Fig. 2. Typical profile of the slope

The overburden soil was mainly gravel silty clay, which consists of strongly weathered shaly siltstone and middle weathered shaly siltstone. Combined with the test results and engineering geological analogy, the physical and mechanical parameters used in slope stability analysis are shown on Table 1.

Table 1. The physical and mechanical parameters of overburden soil and rock mass

Classification	Unit weight (KN/m^3)	Rock shear strength		Substrate friction coefficient	E (GPa)	Poisson's ratio
		C/MPa	ϕ/°			
Pebbly silt clay	18.0	22	15		0.016	0.43
Strongly weathered argillaceous siltstone	22.0	0.10	24	0.40	0.1	0.35
Moderate weathered argillaceous siltstone	24.0	0.35	37	0.50	1.7	0.24
Moderate weathered argillaceous siltstone	23.0	0.20	10	0.45	0.8	0.30

3.2 Construction Simulation

The slope was excavated from top to bottom layer by layer, and support measurement was taken after excavation for each layer. Combined with the attitude of stratum, the simulated model and mesh generation was shown on Fig. 3, then the slope excavation process was simulated.

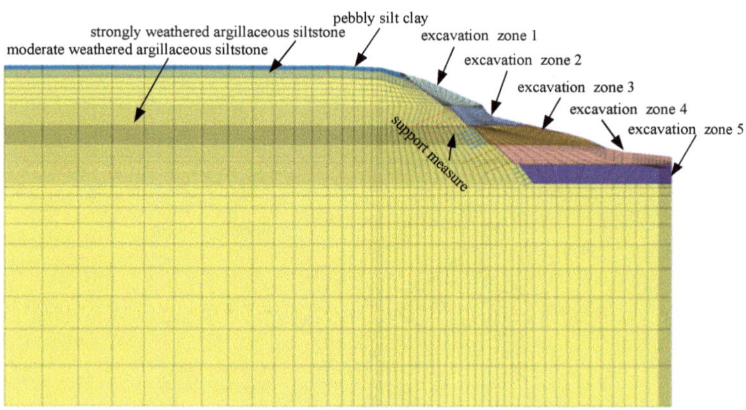

Fig. 3. Simulated model and mesh generation

Strength reduction method is widely used in the slope stability analysis considering excavation and unloading effect, and could find the most probable sliding surface or collapse boundary, which also provide basis for the whole failure mode assessment of slope. By continually reducing the strength of rock mass during disturbing to reaches the ultimate state, gets the destruction sliding surface and the corresponding safety factor, which was the strength reduction method for slope limit analysis. So strength reduction method using FEM are applied for excavation simulation.

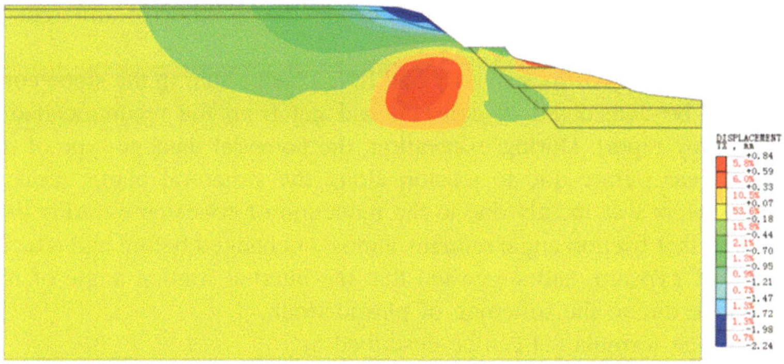

Fig. 4. Horizontal displacement nephogramag after excavation stage 1

Fig. 5. Horizontal displacement nephogramag after excavation stage 5

Figures 4 and 5 are the horizontal displacement variations of slopes after first and fifth excavation stage respectively. This can reflect the change of slope displacement during excavation, and could determine dangerous area for layout of monitoring point.

4 Warning Index Determination During Slope Excavation

4.1 Determination Warning Index

As for slope warning, the important first step to determine an early warning index to help deformation monitoring. Take the slope mentioned above as an example, to state the warning index determination method and steps.

The slope state and safety coefficient Fs were mainly affected by the rock cohesion c and internal friction angle φ and density γ, so Fs can be expressed as a function as:

$$F_s = f(c_1, \varphi_1, \gamma_1, c_2, \varphi_2, \gamma_2, c_3, \varphi_3, \gamma_3, \ldots, c_n, \varphi_n, \gamma_n) \tag{1}$$

Where the subscript 1, 2, 3,…, n indicate the number of different materials respectively.

Because the density of rock mass γ had little change during the slope construction process, it can be considered as constant and get from the recommended value of geological survey report. During excavation, the potential damage type of rock mass was tension shear failure due to tension along the structural plane; weaken of the strength parameters was mainly due to the reduction of cohesion c during excavation. Sun [4] proved that friction angle remains almost unchanged before and after its failure by shear test of gypsum, and suggested that the internal friction angle of rock value does not change due to the softening of plastic strain.

Therefore, the formula (1) can be described as:

$$F_s = f(c_1, c_2, c_3, \ldots, c_n) \quad (2)$$

In order to determine the non-linear relationship safety coefficient and cohesion of rock mass, the uniform design method [5] was adopted to design the combinations of different cohesion c in each stratum. Then the strength reduction method was used to calculate each combination. According to the obtained samples, aiming at the randomness of parameter selection, a PSO-SVM model [6] was used to determine the highly nonlinear mapping relationship between the cohesion of rock mass and safety coefficient Fs.

According to geological data, combined with the excavation disturbed range, sensitivity analysis was considered for the cohesion of rock slope and related safety factor Fs, the range of cohesion value to weathering sandstone was shown on Table 2.

Table 2. Range value of rock mass cohesion in disturbed zone (kPa)

Strongly weathered argillaceous siltstone	Moderate weathered argillaceous siltstone
25.0–50.0	50.0–100.0

According 25 kinds of combination, the safety factor of each group was calculated by strength reduction finite element method, and the results were shown on Table 3.

Table 3. Caculated safety factor with different combination of rock mass cohesion c (kPa)

Scheme	Strongly weathered *c* (kPa)	Moderate weathered *c* (kPa)	Safety factor
1	25	50	0.989
2	26	52	0.996
3	27	54	1.021
4	28	56	1.042
5	29	58	**1.051**
6	30	60	**1.057**
7	31	62	**1.061**
8	32	64	**1.064**

(*continued*)

Table 3. *(continued)*

Scheme	Strongly weathered c (kPa)	Moderate weathered c (kPa)	Safety factor
9	33	66	**1.082**
10	34	68	**1.097**
11	35	70	**1.104**
12	36	72	**1.127**
13	37	74	**1.149**
14	38	76	**1.186**
15	39	78	**1.212**
16	40	80	1.23
17	41	82	1.25
18	42	84	1.32
19	43	86	1.37
20	44	88	1.40
21	45	90	1.42
22	46	92	1.45
23	47	94	1.47
24	48	96	1.49
25	49	98	1.58
26	50	100	1.65

When analyzed the failure mode of the object rock slope, the possible type was mainly manifested arc sliding during excavation, and the deformation instability mainly embodied in the horizontal displacement increasing, so the horizontal displacement increment index was considered as the warning index of monitoring.

As per slope code [7], the grade of highway slope belongs to I, and design safety coefficient ranged from 1.20 to 1.30, in construction the temporary safety coefficient of slope should not be less than 1.05, therefore safety coefficient between 1.05 and 1.20 was considered as the early warning range.

Among schemes 5 to 15 cohesion c was selected as parameters, displacement caused by excavation of above combinations were calculated, and then the relation of the horizontal displacement increment and safety coefficient were fitted by polynomial, the fitting curves of JCX1 and JCX2 were shown on Figs. 6 and 7 respectively.

From 1.05 to 1.2, the catastrophe point of displacement increment was considered as the key points for early warning. According to the fitting curve, firstly the key point A was determined as per max change trend, we called it unstable start point; then the key point C was determined while the displacement change rate increased gradually to stable, we called it blue warning start point; thirdly the key point B was determined where the displacement increment change accelerating, we called it yellow warning start point. And corresponding safe factor were shown together on Table 4.

According to the determined blue warning index and the red warning index, the grade of slope warning as per the monitoring horizontal displacement increment during excavation was recommended on Table 5.

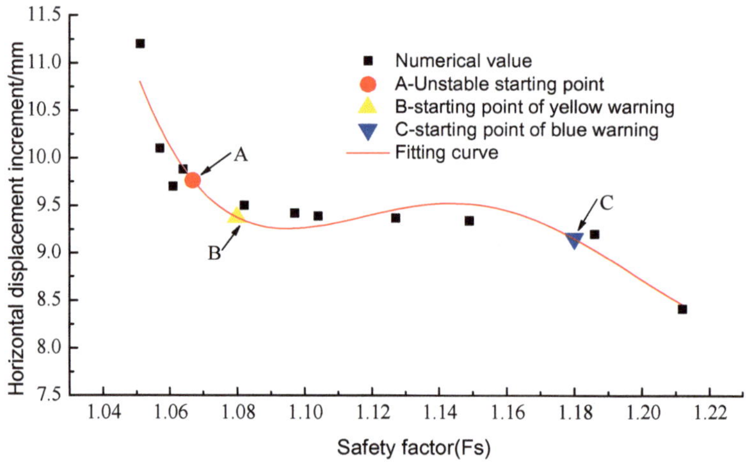

Fig. 6. Relation of horizontal displacement increment vs. safety factor curve for JCX1

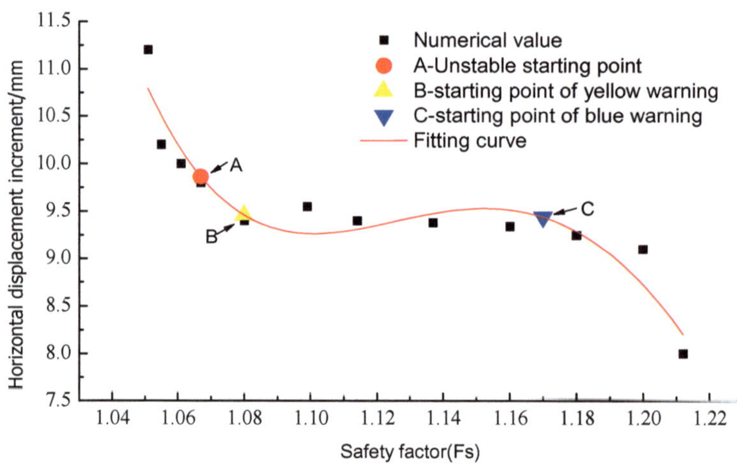

Fig. 7. Relation of horizontal displacement increment vs. safety factor curve for JCX2

Table 4. Relation of limit safety factor and corresponding horizontal displacement increment

Monitoring point No	Horizontal displacement increment(mm)			Limit safety factor		
	A	B	C	A	B	C
JCX1	9.1	9.4	9.8	1.067	1.08	1.18
JCX2	9.0	9.4	9.8	1.067	1.08	1.17

Table 5. warning grade recommendation

Monitoring point No	JCX1	JCX2	Warning grade
Horizontal displacement increment range(mm)	<9.1	<9.0	Safe
	9.1–9.4	9.0–9.4	Blue warning
	9.4–9.8	9.4–9.8	Yellow warning
	>9.8	>9.8	No safe

4.2 Comparison with Site Monitoring

The site monitoring of the horizontal displacement increment were shown on Fig. 8, compared with the warning value of Table 6, It can see that the actual monitor horizontal displacement increment was little less than the warning value, which indicated that the slope was in a stable state during excavation process. This is consistent with actual engineering condition.

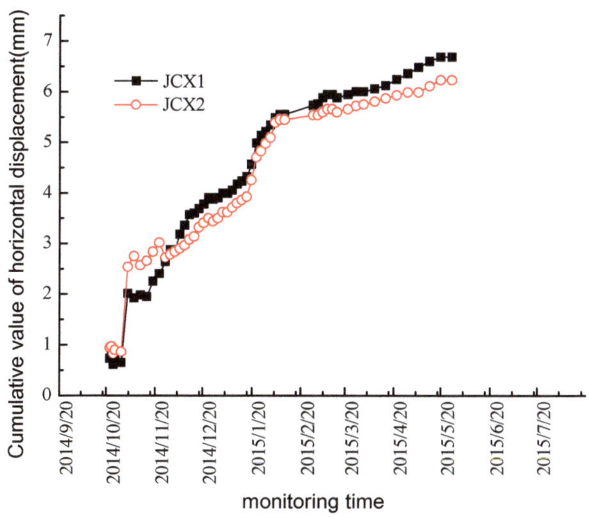

Fig. 8. Relation of horizontal displacement increment vs. monitoring time

Table 6. Comparison of monitoring horizontal displacement and early warning value

Monitoring point No	Horizontal displacement increment (mm)	Range of warning value (mm)
JCX1	7.17	9.1–9.8
JCX2	6.45	9.0–9.8

5 Conclusions

High steep rock slopes early warning has been studied by specified project case from DAOAN highway. Combined with the survey and design data, numerical model of slope excavation are established. Then the slope excavation and support were simulated step by step. As per numerical results and site displacement monitoring data, a new study route and corresponding method based on designed factor of safety and failure model was proposed, which gives scientific route and method for dynamic monitoring warning of deformation and failure of slopes during excavation.

What's more, the dynamic warning index of slope monitoring during excavation were confirmed, comparison with the monitoring data showed that, the warning method was valid and the index was reasonable, which could gives references to evaluate the safety state of the slope excavation in similar engineering.

References

1. Wu, X., Shen, S., Niu, R.: Landslide susceptibility prediction using GIS and PSO-SVM. Geomatics Inf. Sci. Wuhan Univ. (2016)
2. Osanai, N., Shimizu, T., Kuramoto, K., et al.: Japanese early-warning for debris flows and slope failures using rainfall indices with Radial Basis Function Network. Landslides **7**(3), 325–338 (2010)
3. Atzeni, C., Barla, M., Pieraccini, M., et al.: Early warning monitoring of natural and engineered slopes with ground-based synthetic-aperture radar. Rock Mech. Rock Eng. **48**(1), 235–246 (2015)
4. Sun, J.: Rheology of geotechnical materials and engineering applications. China Architecture & Building Press, Beijing (1999)
5. Li, R., Lin, D.K.J., Chen, Y.: Uniform design: design, analysis and applications. Int. J. Mater. Prod. Technol. **20**(1–3), 101–114 (2004)
6. Zhang, Z., Feng, X., Zhou, H., Zhang, C., Cui, Q.: Research on dynamic early warning method of slope deformation monitoring during excavation based on designed safety factor and failure mode. Rock Soil Mechan. **30**(3), 604–613 (2009)
7. Ministry of Transport of the People's Republic of China, Specifications for Design of Highway Subgrades (JTG D30-2015). China Communications Press, Beijing (2015)

Theoretical Analysis of Gas Hydrate Dissociation in Sediment

X. B. Lu[1,3(✉)], L. Lu[2], X. H. Zhang[1,3], and S. Y. Wang[1]

[1] Institute of Mechanics, Chinese Academy of Sciences, Bei Si Huan Xi Road 15, Beijing 100190, China
xblu@imech.ac.cn
[2] School of Science, China University of Mining and Technology, Xuzhou, China
[3] Department of Engineering Sciences, University of Chinese Academy of Sciences, Beijing, China

Abstract. Theoretical analysis was carried out to investigate the dissociation of gas hydrate in stiff sediment. First the mathematical model for gas hydrate dissociation was decoupled by asymptotic expansion method considering the order differences of time scales among seepage, dissociation of gas hydrate and heat conduction. The multi-scale perturbation method was used to solve the problem. It is shown that seepage is the fastest process. The heat conduction is the slowest process. With the pressure decreases at the boundary, pressure changes first while no hydrate dissociation and heat conduction occur. Gradually, dissociation causes the decrease of temperature. After a long time, heat can conduct to cause the increase of temperature. Otherwise, the decreased temperature will cause the stop of dissociation if no heat is supplied in time.

Keywords: Gas hydrate · Dissociation · Mathematical model

1 Introduction

Natural gas hydrate (NGH) is treated as a potential energy resource for the 21st century because a large amount of NGH is trapped in hydrate reservoirs though it can induce the large deformation [1] and even kinds of hazards in some cases such as marine landslide, crack extensions [2]. In the past 30 years, considerable effort has been made for commercial production of NGH from hydrate reservoirs. Until now, all methods are still limited to experimental scale, except for one gas-hydrate field in western Siberia, which was exploited successfully [3].

To recover NGH from hydrate reservoir, depressurization, temperature falling and replacing are three presented methods. Extensive reviews of NGH reported by Englezos [4] and Sloan [5]. Selim and Sloan [6], Tsypkin [7] and Lu et al. [8] presented different methods for analysis of the dissociation process of NGH by depressurization. In those models, various assumptions were adopted in order to obtain the analytical solution, such as the water in the reservoir remained stationary and the well temperature was kept constant. Moridis et al. [9] added a module for NGH dissociation into the TOUGH2 (general-purpose reservoir simulator). Swinkels and Drenth [10], Zhang et al. [11] and

L. Zhang et al. (Eds.): GSIC 2018, *Proceedings of GeoShanghai 2018 International Conference: Rock Mechanics and Rock Engineering*, pp. 109–116, 2018.
https://doi.org/10.1007/978-981-13-0113-1_13

Ji et al. [12] used thermal stimulation for NGH dissociation and studied the enthalpy of dissociation for hydrates formed by different gases.

Either the formation or the recovery of NGH is related with the thermal conduction/convection, migration of gas and water, dissociation of gas hydrate, stress/deformation variation of stratum. These processes are very complex because of the multi-components and complex sediment's conditions such as fracture system [13, 14]. To analyze the recovery of gas hydrate, these four processes must be clarified.

In this paper, theoretical analysis on NGH dissociation was studied. The governing equations were first decoupled into three courses by asymptotic expansion method: seepage, dissociation of NGH and heat conduction, considering the order differences of the time scales. Then solutions were given under two-dimensional conditions.

2 Formation of Problem

We consider the hydrate-bearing sediment constituted of free methane (gas), methane hydrate (solid), water (liquid) and soil/rock skeleton (stiff solid). The four components are in thermodynamic equilibrium. The formation is assumed to be uniform and the flow of free methane and water can be regarded as flow through a porous media with porosity $\varepsilon_g + \varepsilon_w$, which are fractional volume of gas and water. Soil/rock skeleton is assumed to be stationary. The flow of gas and water in pores obeys the Darcy law. Though the permeability of gas and water changes with the properties of media such as porosity, components, they are assumed as constants for simplicity of analysis.

Dissociation will happen once the pressure and temperature changes. Seepage and temperature change inside the sediment will occur also once the pressure changes at the boundary.

The fractional volume of each component is ε_g, ε_h, ε_w and ε_m respectively and

$$\varepsilon_g + \varepsilon_w + \varepsilon_h + \varepsilon_m = 1 \tag{1}$$

in which ε_h, ε_g, ε_w, ε_m are fractions of hydrate, gas, water and skeleton, respectively. For a unit volume, ε_m is a constant, ε_g, ε_h, ε_w will change with the dissociation of gas hydrate.

Methane (gas) satisfies the ideal gas law, i.e.

$$pV = NRT \tag{2}$$

in which p is the pressure, V is the volume, N is the Mole number of gas, R is the ideal gas constants, T is the temperature.

Mass conservation equations are as follows by neglecting diffusion term:

$$\begin{aligned} \frac{\partial \varepsilon_g \rho_g}{\partial t} + \nabla \cdot \varepsilon_g \rho_g \bar{u}_g &= -\chi \rho_h \frac{\partial \varepsilon_h}{\partial t} \\ \frac{\partial \varepsilon_w \rho_w}{\partial t} + \nabla \cdot \varepsilon_w \rho_w \bar{u}_w &= -(1-\chi) \rho_h \frac{\partial \varepsilon_h}{\partial t} \end{aligned} \tag{3}$$

in which ρ_g, u_g are density and velocity of gas, ρ_w and u_w are density and velocity of water, $\chi = M_g/M_h, M_h$ and M_g are Mole numbers of gas hydrate and methane gas, respectively.

Momentum conservation equations are as follows by neglecting inertia,

$$\begin{aligned} \bar{u}_g &= -\frac{K_g}{\varepsilon_g \mu_g} \nabla p \\ \bar{u}_w &= -\frac{K_w}{\varepsilon_w \mu_w} \nabla p \end{aligned} \tag{4}$$

in which K_g and K_w are relative permeabilities of gas and water, respectively, μ_g and μ_w are cohesions of gas and water, respectively.

The energy conservation equation is

$$\begin{aligned} &\varepsilon_g \rho_g \frac{\partial C_g T}{\partial t} + (\varepsilon_w \rho_w C_w + \varepsilon_h \rho_h C_h + \varepsilon_m \rho_m C_m) \frac{\partial T}{\partial t} + \varepsilon_g \rho_g u_g \bar{\cdot} \nabla C_g T + \varepsilon_w \rho_w C_w \bar{u}_w \cdot \nabla T \\ &= [\Delta H + \chi C_g T + (1 - \chi) C_w T] \rho_h \frac{\partial \varepsilon_h}{\partial t} + K \nabla^2 T \end{aligned} \tag{5}$$

in which C_g, C_w, C_h, C_m are thermal capacities of gas, water, hydrate and skeleton respectively, ΔH is the latent heat, K is the heat conduction coefficient which is assumed to be a constant.

The dissociation rate of NGH is

$$\frac{\partial \varepsilon_h}{\partial t} = -k_d M_g A_s (f_e - f) \tag{6}$$

in which A_s is the area, k_d are coefficients, f_e and f are the three-phase equilibrium fugacity and methane fugacity, respectively. In the following analysis, they are substituted by the phase equilibrium pressure of methane hydrate and pore pressure in the hydrate-bearing sediment. Generally, k_d changes with temperature or pore pressure. For simplicity, it is assumed as a constant here. The above equations can be normalized by a characteristic specific heat C, T_0 and ρ_h and p_0. T_0 and p_0 can be chosen as the initial temperature and pressure of the media.

$$\begin{aligned} &\frac{\partial \varepsilon_g \bar{\rho}_g}{\partial \tau} + \nabla (\bar{\rho}_g \cdot \nabla \bar{p}) = -\chi \bar{\rho}_h \frac{\partial \varepsilon_h}{\partial \tau} \\ &\frac{\partial \varepsilon_w}{\partial \tau} + \nabla^2 \bar{p} = -(1 - \chi) \bar{\rho}_h \frac{\partial \varepsilon_h}{\partial \tau} \\ &\frac{\partial \varepsilon_h}{\partial \tau} = \bar{k}_d M_g \bar{A}_s (\bar{f}_e - \bar{f}) \end{aligned} \tag{7}$$

$$\begin{aligned}&\varepsilon_g\bar{\rho}_g\frac{\partial\bar{C}_p\bar{T}}{\partial t}+(\varepsilon_w\bar{\rho}_w\bar{C}_w+\varepsilon_h\bar{\rho}_h\bar{C}_h+\varepsilon_m\bar{\rho}_m\bar{C}_m)\frac{\partial\bar{T}}{\partial t}\\&-\frac{p_0k_g}{\mu_g}\bar{\rho}_g\nabla\bar{p}\cdot\nabla\bar{C}_p\bar{T}-\frac{p_0k_w}{\mu_w}\bar{\rho}_w\bar{C}_w\nabla\bar{p}\cdot\nabla\bar{T}\\&=[\frac{\Delta H}{CT_0}+\chi\bar{C}_p\bar{T}+(1-\chi)\bar{C}_w\bar{T}]\frac{\partial\varepsilon_h}{\partial t}+\frac{K}{C\rho_h}\nabla^2\bar{T}\end{aligned}\tag{8}$$

For simplicity, the "-" in the equations are omitted. In this problem, let $\tau=p_0k_gt/(\mu_g r_0^2)$, $\zeta=x/r_0$. Since $\eta_1=C\rho_h\mu_g/(Kp_0k_g)$ and $\eta_2=K\alpha A_s\mu_g/(p_0k_g)$ are two small parameters, we can give the asymptotic expansions by using of multi-scale method:

$$\begin{cases}f=f^{(0)}(x_i,\tau_0,\tau_1,\tau_2)+\sum\limits_{n=1}^{\infty}\eta_2^nf^{(n)}(x_i,\tau_0,\tau_1,\tau_2)\\f^{(0)}=f^{(0)(0)}(x_i,\tau_0,\tau_1,\tau_2)+\sum\limits_{n=1}^{\infty}\eta_1^nf^{(0)(n)}(x_i,\tau_0,\tau_1,\tau_2)\end{cases}\tag{9}$$

in which $\tau_0=\tau,\tau_1=\eta_1\tau,\tau_2=\eta_2\tau$. Instituting these expressions into Eqs. (7) and (8), we can decouple the problem. It is interesting to note that the equations of the first three orders indicate the three physical courses: 0th order, the development of seepage, the fast course; 1st order, the dissociation of gas hydrate, the second fast course; 2nd, heat conduction, the third fast course. The 0th order equations are as follows:

$$\begin{cases}\frac{\partial\varepsilon_g^{(0)(0)}\rho_g^{(0)(0)}}{\partial\tau_0}-\nabla(\rho_g^{(0)(0)}\nabla P^{(0)(0)})=0\\\frac{\partial\varepsilon_w^{(0)(0)}}{\partial\tau_0}-\nabla^2p^{(0)(0)}=0\\\frac{\partial\varepsilon_h^{(0)(0)}}{\partial\tau_0}=0\\\varepsilon_g^{(0)(0)}\rho_g^{(0)(0)}\frac{\partial C_gT^{(0)(0)}}{\partial\tau_0}+\varepsilon\rho C\frac{\partial T^{(0)(0)}}{\partial\tau_0}+\left[\frac{k_w\mu_g}{k_g\mu_w}\rho_wC_w+\rho_g^{(0)(0)}C_g\right]\nabla p^{(0)(0)}\nabla T^{(0)(0)}=0\end{cases}\tag{10}$$

Summing the last three equations together and neglecting the smallness, the controlling equation of $P^{(0)(0)}$ can be obtained

$$\varepsilon_{g0}\frac{\partial p^{(0)(0)}}{\partial\tau_0}-2p^{(0)(0)}\nabla^2p^{(0)(0)}=0\tag{11}$$

We can obtain the following simplified equation when the pore pressure difference is small [15].

$$\varepsilon_{g0}\frac{\partial p^{(0)(0)}}{\partial\tau_0}-2p_0\nabla^2p^{(0)(0)}=0\tag{12}$$

The 0th order solution can be obtained as follows:

$$\begin{cases} T^{(0)(0)} = T_0 \\ p^{(0)(0)} = p_1 + \frac{p_0 - p_l}{l} x + \sum_{n=0}^{\infty} \frac{2(p_0 - p_l)}{n\pi} e^{-\frac{n^2\pi^2 a^2}{l^2}\tau_0} \sin\frac{n\pi}{l} \\ \varepsilon_h^{(0)(0)} = \varepsilon_{h0} \\ \varepsilon_w^{(0)(0)} = \sum_{n=1}^{\infty} \frac{2(p_0 - p_l)}{n\pi} \frac{1}{a^2} e^{-\frac{n^2\pi^2 a^2}{l^2}\tau_0} \sin\frac{n\pi}{l} x + \varepsilon_{w0} - \frac{p_0 - p_l}{a^2 l}(l - x) \end{cases} \tag{13}$$

in which p_0 is the initial pore pressure. The pore pressure at the end l is kept as p_0, p_1 is the pore pressure at the end $x = 0, a^2 = 2p_0/\varepsilon_{g0}$, initially $\varepsilon_w = \varepsilon_{w0}$ at $t = 0$.

1st order equations are as follows:

$$\begin{cases} \left(\varepsilon_g^{(0)(0)} \rho_g^{(0)(0)} \frac{\partial C_g T^{(0)(1)}}{\partial \tau_0} + \varepsilon\rho C \frac{\partial T^{(0)(1)}}{\partial \tau_0}\right) + \left[\frac{k_w \mu_g}{k_g \mu_w} \rho_w C_w + \rho_g^{(0)(0)} C_g\right] \nabla p^{(0)(0)} \nabla T^{(0)(1)} \\ = \left[\frac{\Delta H}{C T_0} + \chi C_g + (1-\chi) C_w T^{(0)(0)}\right] A\left(p^{(0)(0)} - p_e\right) \\ \frac{\partial\left(\varepsilon_g^{(0)(1)} \rho_g^{(0)(0)} + \varepsilon_g^{(0)(0)} \rho_g^{(0)(1)}\right)}{\partial \tau_0} - \nabla(\rho_g^{(0)(1)} \nabla P^{(0)(0)} + \rho_g^{(0)(0)} \nabla P^{(0)(1)}) = -\chi A\left(p^{(0)(0)} - p_e\right) \\ \frac{\partial \varepsilon_w^{(0)(1)}}{\partial \tau_0} - \nabla^2 p^{(0)(1)} = -(1-\chi) A_1 \left(p^{(0)(0)} - p_e\right) \\ \frac{\partial \varepsilon_h^{(0)(1)}}{\partial \tau_0} = A_2 \left(p^{(0)(0)} - p_e\right) \end{cases} \tag{14}$$

in which $A = k_d M_g A_s \rho_h, A_1 = k_d M_g A_s \rho_h/\rho_w, A_2 = k_d M_g A_s$. Neglecting the effects of convection on the variation of temperature, then the variation of pore pressure $p^{(0)(0)}$ is mainly due to dissociation of GH, the 1st order solution solution can be obtained as

$$\begin{cases} T^{(0)(1)} = \left[\frac{\Delta H}{C T_0} + \chi C_g + (1-\chi) C_w T_0\right] \frac{A}{\varepsilon\rho C + \varepsilon_g^{(0)(0)} \rho_g^{(0)(0)}} \\ \left(T_1 \tau_0 + \frac{T_0 - T_l}{l} x \tau_0 - \sum_{n=0}^{\infty} \frac{2(T_0 - T_l)}{n\pi} \frac{l^2}{n^2\pi^2 a^2} e^{-\frac{n^2\pi^2 a^2}{l^2}\tau_0} \sin\frac{n\pi}{l} x - T_e \tau_0\right) \\ p^{(0)(1)} = [A_2 - \chi A - (1-\chi) A_1]\left(p_1 x^2 + \frac{p_0 - p_l}{3l} x^3 - \sum_{n=0}^{\infty} \frac{2(p_0 - p_l) l^2}{n^3\pi^3} e^{-\frac{n^2\pi^2 a^2}{l^2}\tau_0} \sin\frac{n\pi}{l} - p_e x^2\right) \\ \frac{\partial \varepsilon_w^{(0)(1)}}{\partial \tau_0} = -(1-\chi) A_1 \left(p^{(0)(0)} - p_e\right) + \nabla^2 p^{(0)(1)} \\ \varepsilon_h^{(0)(1)} = A_2 \left(p_1 \tau_0 + \frac{p_0 - p_l}{l} x \tau_0 - \sum_{n=0}^{\infty} \frac{2(p_0 - p_l)}{n\pi} \frac{l^2}{n^2\pi^2 a^2} e^{-\frac{n^2\pi^2 a^2}{l^2}\tau_0} \sin\frac{n\pi}{l} - p_e \tau_0\right) \end{cases} \tag{15}$$

$T^{(1)}$ and $p^{(1)}$ are small parameters. Neglecting the product term of these two smallness and considering $\varepsilon_g^{(1)}\rho_g^{(0)} \gg \varepsilon_g^{(0)}\rho_g^{(1)}$, $\rho_g^{(1)}\nabla P^{(0)} \gg \rho_g^{(0)}\nabla P^{(1)}$, 2nd order solution can be simplified as

$$\begin{cases} \left(\varepsilon_g^{(0)}\rho_g^{(0)} + \varepsilon\rho C\right)\frac{\partial T^{(1)}}{\partial \tau_1} = \nabla^2 T^{(1)} \\ \frac{\partial \varepsilon_g^{(1)}\rho_g^{(0)}}{\partial \tau_0} - \nabla\rho_g^{(0)}\nabla P^{(1)} = -\chi A p^{(1)} \\ \frac{\partial \varepsilon_w^{(1)}}{\partial \tau_0} - \nabla^2 p^{(1)} = -(1-\chi)A_1 p^{(1)} \\ \frac{\partial \varepsilon_h^{(1)}}{\partial \tau_0} = -(1-\chi)A_2 p^{(1)} \end{cases} \tag{16}$$

Then the solution for $T^{(1)}$ can be obtained as follows

$$T^{(1)} = T_1 + \frac{T_0 - T_l}{l}x + \sum_{n=1}^{\infty}\frac{2(T_0 - T_l)}{n\pi}e^{-\frac{n^2\pi^2 b^2}{l^2}\tau_1}\sin\frac{n\pi}{l}x \tag{17}$$

T_0 is the initial temperature, T_l is the temperature at the end x = 0, the temperature at the end x = l is kept as T_0, $b^2 = 1\Big/\left(\varepsilon_g^{(0)}\rho_g^{(0)}C_g + \varepsilon\rho C\right)$. In the same way, $p^{(1)}$, $\varepsilon_w^{(1)}$ and $\varepsilon_s^{(1)}$ can be solved.

It can be seen clearly from the above analysis that the dissociation process can be divided into three decoupled courses: seepage, phase change and thermal conduct in the sequence of the development speed of each course in a GH reservoir with low permeability. Thermal conduction is the slowest course while seepage is the fastest. The analytical solution can be obtained easily by the decoupled courses relative to obtain the solution directly from the original equation. The solution shows that no matter what kind of method to induce the dissociation of GH, thermal conduction is the key in dissociation of GH because it is the slowest course and phase change needs heat consumption. In other words, without enough heat supply phase change will undoubtedly stop no matter what the method we adopted so the fast heat supply is very important in exploitation of GH.

For certification, we compute the development of pore pressure and temperature with time by the above analytical solutions first and, compare them with experimental results in literature [16]. The parameters adopted here are shown in Table 1. Initial pore pressure is 4.0 MPa. At one end the pore pressure is 0.1 MPa and at the other end the pressure is kept as the initial value. Initial temperature is 16.5 °C. At one end temperature increases to 40 °C while at the other end it keeps the initial value. It can be seen from Figs. 1 and 2 that the analytical solution is close to the experiments'. The measurement of pore pressure is relatively accurate. The temperature is measured by thermocouple whose accuracy is 1°, thus the measured value of temperature is not as precise as pore pressure. That is why the differences between theoretical and experimental temperature is a little large. The fact that the theoretical value is only approximation leads to error.

Table 1. Parameters adopted in computation

Parameter	Value	Parameter	Value	Parameter	Value	Parameter	Value
C_a/kJ/kg/K	−1.05	N	6	$\rho_w/10^3$ kg/m^3	1.00	a_0/m	0.007
P_{gc}/MPa	3.5	λ_g/W/m/K	0.06	$\rho_h/10^3$ kg/m^3	0.92	$k_0/10^4$mol/m^2/Pa/s	3.6
P_{g0}/MPa	0.1	λ_w/W/m/K	0.50	$\rho_s/10^3$ kg/m^3	2.60	n_w	4
T_0/K	289.5	λ_h/W/m/K	0.46	$\mu_w/10^{-3}$ Pa·s	1.00	n_g	2
T_e/K	413.0	λ_s/W/m/K	2.90	λ_a/W/m^2/K	0.26	n_c	0.65
S_{h0}	0.0	C_g/kJ/kg/K	2.18	D/m	0.038	ϕ	0.27
S_{g0}	0.32	C_w/kJ/kg/K	4.20	L/m	0.78		
S_{w0}	0.0	C_h/kJ/kg/K	2.22	$H_0/10^6$J/kg	3.53		
K_0/mD	86.1	C_s/kJ/kg/K	0.89	ΔE/kJ/mol	81		

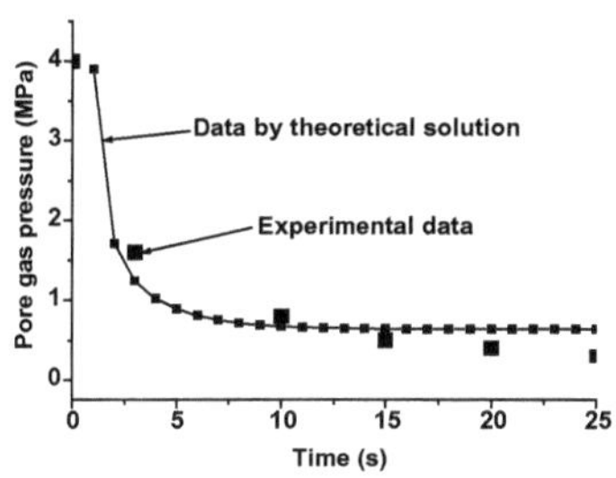

Fig. 1. Comparison of theoretical and experimental results in literature (Liu, 2013)

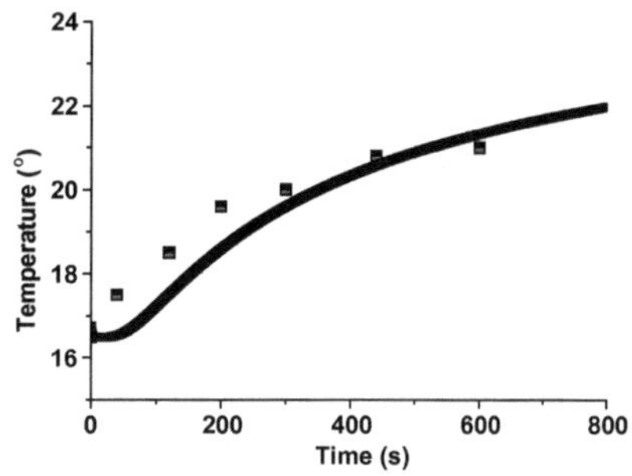

Fig. 2. Comparison theoretical and experimental results in literature (Zhang, et al., 2010)

3 Conclusions

A new method to obtain the analytical solution of equations describing the dissociation of NGH in stiff sediment was presented. By this method it can be seen clearly that the dissociation process can be divided into three decoupled courses which are thermal conduct, phase change and seepage by development speed in a NGH reservoir with low permeability. It is easily to obtain the analytical solution by the decoupled courses. The solution shows that no matter what kind of method to induce the dissociation of GH, thermal conduction is the key because it is the slowest course while phase change needs heat consumption. In other words, without enough heat supply phase change will undoubtedly stop no matter what the method we adopted. The analytical solution is close to that of experimental results by comparison.

Acknowledgement. This project is supported by the Open Research Fund of Shanghai Key Laboratory of Mechanics in Energy Engineering and China natural Science Fund (No.11272314 and No.51239010).

References

1. Jin, Z.H., Johnson, S.E., Cook, A.E.: Crack extension induced by dissociation of fracture-hosted methane gas hydrate. Geophys. Res. Lett. **42**(20), 8522–8529 (2015)
2. Lu, X.B., Wang, L., Wang, S.Y., Li, Q.P.: Instability of seabed and pipes induced by NGH dissociation. In: Proceedings of the Nineteenth International Offshore and Polar Engineering Conference, Beijing, pp. 110–114 (2010)
3. Makogon, Y.F.: Hydrates of Hydrocarbons. Penn Well, Tulsa (1997)
4. Englezos, P.: Reviews: clathrate hydrates. Ind. Eng. Chem. **32**, 1251–1274 (1993)
5. Sloan, E.D.: Clathrate hydrates of natural gases. Marcel Dekker, New York (1998)
6. Selim, M.S., Sloan, E.D.: Heat and mass transfer during the dissociation of hydrates in porous media. AIChE J. **35**, 1049–1052 (1989)
7. Tsypkin, G.: Mathematical models of gas hydrates dissociation in porous media. In: Holder, G.D., Bishnoi, P.R. (eds.) New York Academy of Sciences, New York, pp. 428–436 (2000)
8. Lu, X.B., Wang, S., Zhang, X., Li, Q.P., Yao, H.Y.: A mathematical model for dissociation of gas hydrate. In: Proceedings of the Nineteenth International Offshore and Polar Engineering Conference, Osaka, Japan (2009)
9. Moridis, G., Apps, J., Pruess, K., Myer, L.: EOSHYDR: a TOUGH2 module for ch4 – hydrate release and flow in the subsurface. Lawrence Berkeley National Laboratory, Berkeley, CA, LBNL-42386 (1998)
10. Swinkels, W.J.A.M., Drenth, R.J.J.: Thermal reservoir simulation model of production from naturally occurring gas hydrate accumulations. In: SPE 56550, Annual Technical Conference, Houston, TX October, pp. 465–477 (1999)
11. Zhang, X.H., Lu, X.B., Li, Q.P.: Thermally induced evolution of phase transformations in gas hydrate sediment. Sci. China Phys. Mech. Astron. **53**(8), 1530–1535 (2010)
12. Ji, C., Ahmadi, G., Smith, D.H.: Natural gas production from hydrate decomposition by depressurization. Chem. Eng. Sci. **56**, 5801–5814 (2001)
13. Hillman, J.I.T., Cook, A.E., Daigle, H., Nole, M., Malinverno, A., Meazell, K., Flemings, P. B.: Gas hydrate reservoirs and gas migration mechanisms in the Terrebonne Basin, Gulf of Mexico. Mar. Pet. Geol. **2017**(86), 1357–1373 (2017)
14. You, K.H., Flemings, P.B.: Methane hydrate formation in in thick sand reservoirs: 1. Short-range diffusion. Mar. Pet. Geol. **89**, 428–442 (2018)
15. Yan, K.X.: Advanced Mechanics of Fluids in Porous Media, University of Science and Technology of China Press, Beijing (1999)
16. Liu, L.L., Lu, X.B., Zhang, X.H.: A theoretical model for predicting the spatial distribution of gas hydrate dissociation under the combination of depressurization and heating without the discontinuous interface assumption. J. Pet. Sci. Eng. **133**, 589–601 (2015)

Time-Dependent Failure of Marble Under Long-Term Loading

Ning Liu(✉), Chunsheng Zhang, Weijiang Chu, and Pingzhi Chen

PowerChina Huadong Engineering Corporation Limited,
Hangzhou 311122, Zhejiang, China
liu_n@ecidi.com

Abstract. Observations of marble brittle failure at the laboratory scale indicates that the failure process involves the initiation, growth and accumulation of micro-cracks. With the micro-cracks increasing, marble subjected to long-term loading will be damaged. The rate of cracking is sensitive to the magnitude of long-term loading. The static-failure tests were conducted and the test results indicate that the time-dependent is being. Process simulation tests have been conducted to investigate the time-dependent failure in JPII marble. The static-failure data were used to match the parallel-bonded stress corrosion (PSC) model parameters which based on PFC2D (Particle Flow Code in 2 Dimension). The time-dependent failure was simulated. The response was characterized by a creep curve. The damage mechanisms and deformation behavior of marble were discussed. The model exhibits damage mechanisms and deformation behavior that are similar to those of the rock.

Keywords: Time-dependent failure · Static-failure tests · PSC model
Brittle rock

1 Introduction

Underground rock engineering design needs to know the long-term strength of a rock. It is also important for minimize potential pathways for the transport of water in the rock as demonstrated in [1]. But the mechanical behavior of rock is controlled by its microstructures. Complex macroscopic behaviors like fracture and failure attribute to the micro-cracks initiation, accumulation and interaction. When the loading stress below the peak strength, the rocks is also be damaged and fractured after some time. This phenomenon is called time-dependent failure. After found the effect, the failure behavior of brittle rock is researched. Because the problem is so difficult, the study was progressed systematically in Canada's Underground Research Laboratory (URL) until 1990s and obtained some main reorganization in references [2–4].

(1) When the stress reaches certain level, the micro-cracks will appear in brittle rock. The stress level is about 40% of UCS. This stress is called crack initiation strength σ_{ci};
(2) When the stress reaches 80% of UCS, the micro-cracks is at large. The space and length is approximately equal. The macro nonlinear characteristic became appear. The stress level is called crack damage strength σ_{cd} or long-term strength.

L. Zhang et al. (Eds.): GSIC 2018, *Proceedings of GeoShanghai 2018 International Conference: Rock Mechanics and Rock Engineering*, pp. 117–125, 2018.
https://doi.org/10.1007/978-981-13-0113-1_14

Fig. 1. Delay fracture of intact marble.

The initiation strength and damage strength are all below peak strength, so not satisfy the traditional strength criterion. The traditional mechanical concept and description method can not be used to describe the failure. Developing with the monitoring technique, the acoustic emission was put into monitor brittle failure. PFC (Particle Flow Code) also satisfies the need for researching the failure problem as demonstrated in [5]. The PSC (parallel-bonded stress corrosion) model in reference [6] describes the time effect of crack by stress and environment. When some part stress satisfies one condition, the crack will grow and the environment factors including moisture, temperature and water will change the crack growth velocity as shown in [7]. In order to use the model to simulate the time effect, must confirm some parameters which can not be obtained by original tests.

This paper aims to study the time-dependent failure of marble and strength degradation under long-term constant loading. The static-fatigue tests were put up to help fix the parameters of PSC model. The PSC model parameters can be chosen to match both the static-fatigue curve (time-to-failure versus applied load) and the damage mechanisms and deformation behavior (a creep curve showing primary, secondary and tertiary creep) of marble.

2 Laboratory Test

For this study the test rock specimens used in the test were all prepared from Jinping II hydropower station tunnel, China, and they are referred to as JPII marble. The JPII marble is classified as classical brittle rock. The time-dependent behavior of rock can be characterized by performing a series of static-fatigue tests to produce a time-to-failure curve in reference [8]. The static-fatigue test consisted of two stages: initial loading and constant loading. In the initial loading stage, the specimen was axially loaded at a constant lateral deformation rate of 0.05 mm/min as demonstrated in

[9, 10]. The specimens are about 50 mm in diameter and 100 mm in height. When the predetermined stress level was achieved above crack initiation strength σ_{ci}(about 40% of UCS), the constant loading stage was then started [6]. The load was maintained until the rock specimen failed. If the specimen did not fail after a long period of time, up to several days, the constant loading test was terminated in reference [11].

2.1 Uniaxial Compression Test

The uniaxial compression tests were carried out on MTS. The uniaxial compression strength is 95–107 MPa. Because the marble is typical brittle rock, the sudden failure immediately occurred after peak strength and the axial strain and lateral strain are small (see Fig. 2). Under uniaxial compression state, rock is priority to shear failure and the angle between failure face and loading is 45° (see Fig. 3).

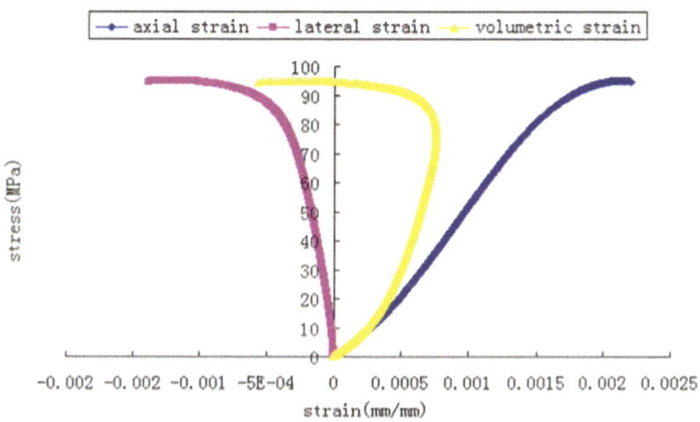

Fig. 2. Typical marble stress-strain curve.

Fig. 3. Some damaged rock sample.

2.2 Long-Term Loading Test

Multi-stage creep test is a common tool for the investigation of the time-dependent prosperities of materials such as rock and soil as demonstrated in [12]. During tests on marbles, recorded the data including stress, strain, time, temperature to study the deformation characters, damage accumulation and strength degradation of JPII marbles with time. When a brittle rock is held at constant loading, its creep strain is also typically composed of three phases: a very short duration primary by an almost linearly increasing transient, or secondary, creep phase and a tertiary creep phase leading to the failure of the specimen according to reference [13]. Figure 1 presents a typical constant loading test on JPII marble. The initial deformation rate is high and then decreases continuously (transient creep phase). When the load holds constant, a steady state creep occurs. It is observed in the tertiary phase, and the deformation rate keeps increasing and ends in creep fracturing (Fig. 4).

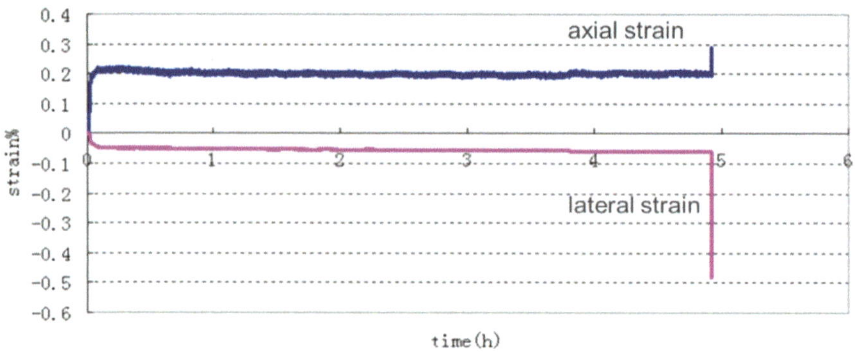

Fig. 4. A typical static-fatigue test and its three creep phases.

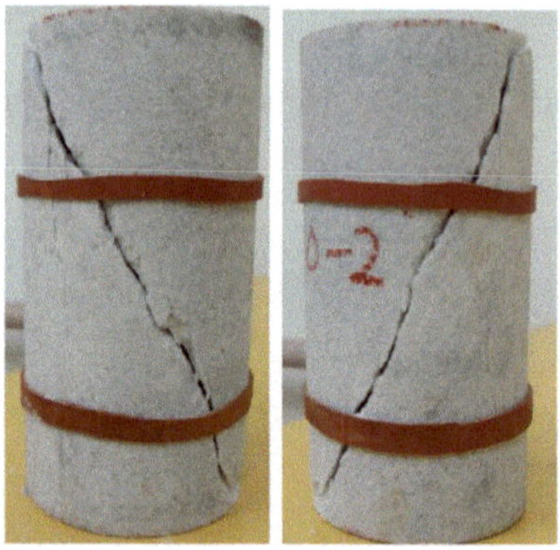

Fig. 5. Failure rock sample including macro sliding surface and axial crack.

2.3 Failure Characteristics

The constant loading in fatigue damage test tend to produce non-elastic and unrecoverable, in which the increase volume strain means the axial crack growth, and increase of shear strain means shear slip. The two failure modes draw in each other. In the test, both the micro crack along axial and shear sliding show obvious (see Fig. 5).

Because the test can not be stopped in process, the whole shear failure took place. However there are lots of approximate parallel axial cracks close to surface.

3 Simulating Time-Dependent Failure

3.1 Model Formulation

The PSC model extends the formulation of the BPM in reference [14] (bonded-particle model) to include time-dependent behavior by adding a damage-rate law to the parallel-bond formulation. The PSC model is implemented in the two- and three-dimensional distinct-element programs PFC2D and PFC3D. In this paper, the term PFC denotes information that applies to PFC2D.

Static-fatigue tests are performed biaxial testing environment whereby the top and bottom walls act as loading platens, and the velocities of the lateral walls are controlled by a servo-mechanism that maintains a constant confining stress. Use the material-genesis to produce the numerical model in Fig. 6.

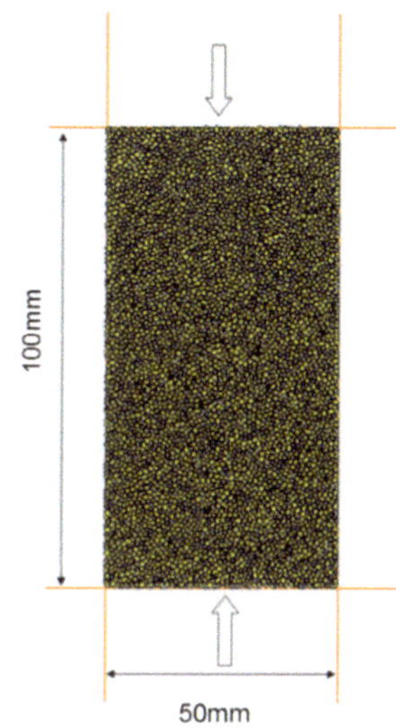

Fig. 6. Specimens model by PFC2D.

3.2 Numerical Experiment by PSC Model

Static-fatigue tests were performed upon the model to produce the static-fatigue data shown in Figs. 7 and 8, and the micro physico-mechanical parameters were listed in Table 1. The static-fatigue curve produced by the PSC model is approximately a straight line for $(\sigma/\sigma_c)_{th} \leq \sigma/\sigma_c \leq 0.85$. For high driving-stress ratios, the model curves approach minus infinity (indicating a zero time-to-failure); for low driving-stress ratios, the model curves approach plus infinity (indicating a static-fatigue limit).

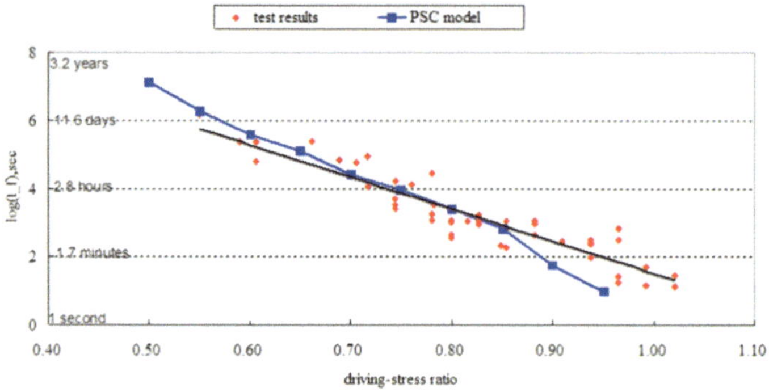

Fig. 7. PSC simulation results contrast with liner fits.

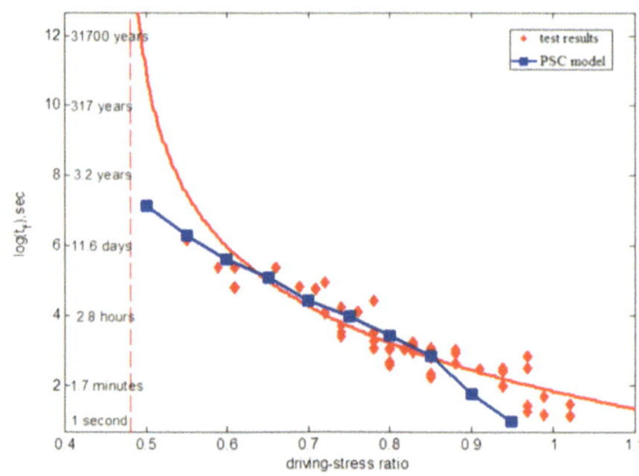

Fig. 8. PSC simulation results contrast with exponential fits.

Table 1. Micro physico-mechanical parameters.

Parameters	Value
Particle number	5173
Density/kg·m^{-3}	2600
The minimum size/mm	0.4
The ratio of particle size	1.66
Particle modulus/GPa	30
Parallel bond modulus/GPa	50
The ratio of stiffness	1.5
Friction factor	1
Normal strength/MPa	100 ± 20
Tangential strength/MPa	80 ± 15

4 Time-Dependent Failure Behavior by PSC Model

4.1 Deformation Behavior

Because the axial strain is larger in initial loading stage and failure stage, the constant loading stage is magnified. Take the driving-stress ratio is 0.6 as an example, the constant loading stage can also be divided in three stage, transient creep, steady state creep and tertiary creep. In the transient creep stage, the deformation velocity is rather fast. When turn into steady creep stage, the velocity is gradually slow. Only if there is macro fracture formed, the crack increase will be interrupted and then come silent. When the damage accumulates to some degree, the strain will increase rapidly and enter the tertiary creep stage. The rock specimen will be damaged in small time (Fig. 9).

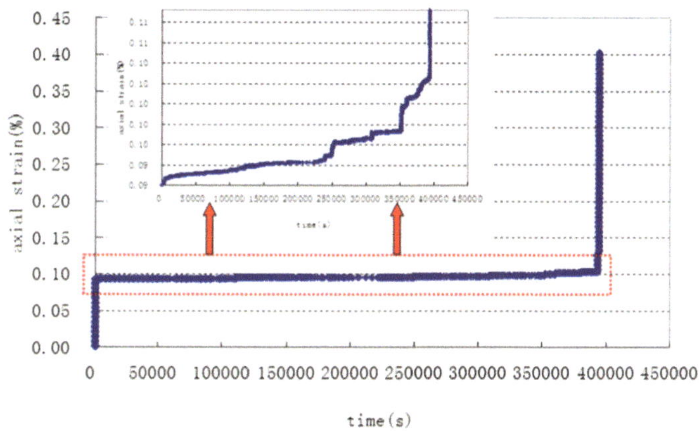

Fig. 9. Representative creep curve by PFC2D at driving-stress ratio of 0.6.

4.2 Failure Behavior

The critical and total crack numbers under different driving-stress ratio are plotted in Figs. 10 and 11, in which c is total number of microcracks, and c^* is c when $t/t_f = 0.99$. The crack number increases greatly in final failure stage, and reflect the brittle failure character. With the driving-stress ratio decrease, the crack number is also put up the creep behavior with time. When under the high driving-stress ratio (0.85–0.95), the crack number keeps the line in the constant loading stage. Only enter into the failure stage, the number increase rapidly. When under the low driving-stress ratio (0.5–0.8), the crack number is increasing. The crack number of final failure is one time than the constant loading stage start.

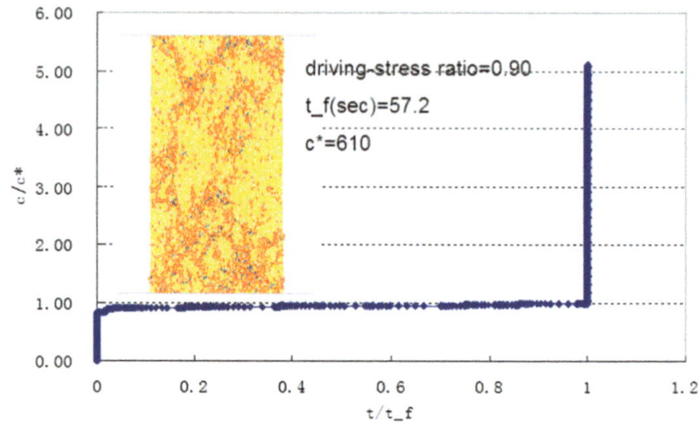

Fig. 10. Crack number curve at a driving-stress ratio of 0.9.

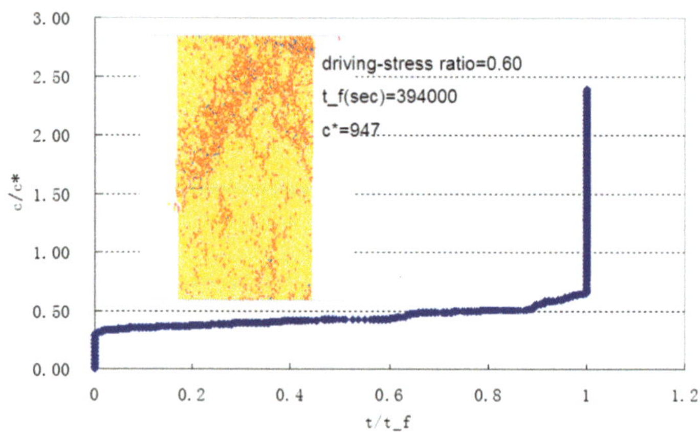

Fig. 11. Crack number curve at a driving-stress ratio of 0.55.

5 Conclusions

The time-dependent failure of JPII marble is simulated by PSC model. This model can be used to better understand the time effect and the weakening process in rock under long-term loading. The PSC model is a comprehensive rock model. In order to match the micro parameters of PSC model, a series of static-fatigue tests need to be done. The PSC model parameters can be chosen to match the static-fatigue curve (time-to-failure versus applied load) for JPII marble. The failure process and crack behavior simulated by PSC model is accordance with the laboratory test. But the PSC model is a comprehensive rock model that combines spatial and temporal scales. How to apply the model into an excavation and predict the excavation stability and long-term strength is a difficult research topic.

References

1. Martin, C.D.: The strength of massive Lac du Bonnet granite around underground openings. University of Manitoba, Winnipeg (1993)
2. Bieniawski, Z.T.: Mechanism of brittle fracture of rock, Parts I, II and III. Int. J. Rock Mech. Min. Sci. Geomech. Abstr. **4**(4), 395–430 (1967)
3. Wawersik, W.R., Fairhurst, C.: A study of brittle rock fracture in laboratory compression experiments. Int. J. Rock Mech. Min. Sci. Geomech. Abstr. **7**(7), 561–575 (1979)
4. Kranz, R.L.: Crack growth and development during creep of Barre granite. Int. J. Rock Mech. Min. Sci. Geomech. Abstr. **16**(1), 23–35 (1979)
5. Itasca Consulting Group, Inc.: PFC2D/3D (Particle Flow Code in 2/3 Dimensions), Version 4.0. ICG, Minneapolis (2008)
6. Potyondy, D.O.: Simulating stress corrosion with a bonded-particle model for rock. Int. J. Rock Mech. Min. Sci. **44**(5), 677–691 (2007)
7. Okui, Y., Horii, H.: Stress and time-dependent failure of brittle rocks under compression: a theoretical prediction. J. Geophys. Res. **102**(B7), 14869–14881 (1997)
8. Martin, R.J., Noel, J.S., Boyd, P.J., Price, R.H.: Creep and static fatigue of welded tuff from Yucca Mountain, Nevada. Int. J. Rock Mech. Min. Sci. **34**(3), 382–412 (1997)
9. Tapponier, P., Brace, W.F.: Development of stress-induced microcracks in Westerly granite. Int. J. Rock Mech. Min. Sci. **13**(2), 103–112 (1976)
10. Cruden, D.M.: A theory of brittle creep in rock under uniaxial compression. J. Geophys. Res. **75**(17), 3431–3442 (1970)
11. Martin, C.D., Chandler, N.A.: The progressive fracture of Lac du Bonnet granite. Int. J. Rock Mech. Min. Sci. Geomech. Abstr. **31**(4), 643–659 (1994)
12. Kemeny, J.M.: Time-dependent drift degradation due to the progressive failure of rock bridges along discontinuities. Int. J. Rock Mech. Min. Sci. Geomech. Abstr. **42**(1), 35–46 (2005)
13. Lockner, D.: A generalized law for brittle deformation of westerly granite. J. Geophys. Res. **103**(B3), 5107–5123 (1998)
14. Potyondy, D.O., Cundall, P.A.: A bonded-particle model for rock. Int. J. Rock Mech. Min. Sci. Geomech. Abstr. **41**(8), 1329–1364 (2004)

Experimental Study of Rock Behavior

A Comparative Study of Grout Placement in a Rock Joint

Jiaxing Zhang[1], Xiangjun Pei[1], Yixuan Sun[2], and Haiying Huang[2(✉)]

[1] State Key Laboratory of Geohazard Prevention and Geoenvironment Protection, Chengdu University of Technology, Chengdu, Sichuan, China
[2] School of Civil and Environmental Engineering, Georgia Institute of Technology, Atlanta, GA, USA
haiying.huang@ce.gatech.edu

Abstract. A comparative study is conducted to investigate the grouting process in a rock joint. The aim of the present work is to establish a baseline to aid the development of theoretical models that can be applied to design the grouting treatment in a high angle large aperture rock joint. An experimental apparatus, consisting of a joint simulation system, a grouting control system and a data acquisition system, has been constructed with the capability to vary the dip angle and the joint aperture. A transparent viewing window also allows real-time imaging of the grout placement. A series of experiments is performed under pressure control with a dip angle of 0°. Variations of the grout placement radius with time obtained from the experiments are compared with the theoretical predictions based on radial flow of a Bingham fluid in between two parallel plates. The experimental and theoretical results agree in the overall trend. However, the maximum placement radius is overestimated by the theoretical solution. Further development in the theoretical model is needed in order to take into account the complexities such as the interface condition between the grout and the walls and the time-dependent nature in the grout rheology.

Keywords: Grout placement · Rock joint · Radial flow

1 Introduction

Grouting is a ground improvement technique that has been commonly employed to improve rock mass quality. By filling the void spaces in rock joints with materials such as cement, grouting can not only increase the strength and stability of the jointed rock mass, but also help control groundwater flow.

Rock joint characteristics, grout properties and grouting operation conditions all affect the placement of the grout in the rock mass. Grouting treatment is particularly challenging if a rock joint has a large aperture, high dip angle and long extension. Gravity tends to drive the grout to flow downwards along the steep slope instead of spreading out laterally to fully seal the joint. For a grout with a long pumpable time, e.g., the ordinary Portland cement grout, excessive downward flow could result in grout leakage and poor sealing effect, as was observed in the Jinping-I hydropower station on

L. Zhang et al. (Eds.): GSIC 2018, *Proceedings of GeoShanghai 2018 International Conference: Rock Mechanics and Rock Engineering*, pp. 129–136, 2018.
https://doi.org/10.1007/978-981-13-0113-1_15

Yalong River where unloading joints subjected to high in situ stresses are present in the left bank slope [1]. The issue in this particular project was remediated primarily by the improvement in the grout mix design [1, 2]. Additives were included to adjust the cement hydration process, which essentially affected the time-dependent rheological behaviors of the grout.

Grout placement in a rock joint has often been modeled based on steady state radial flow of a non-Newtonian fluid of power law or Bingham rheology [3–5]. Transient flow of a Bingham fluid in a fracture with parallel walls was investigated in [6]. Spread of grout in a fracture with varying aperture as well as in a three-dimensional network was discussed in [7]. Similar problems have also been investigated in the drilling industry for the issue of drilling mud loss [8–10]. Previously in the literature, most experimental and theoretical analyses on grouting dealt with the case when the fracture is horizontally oriented.

In order to better understand how the complex grout rheology such as time-dependence and the joint configurations including the aperture and the dip angle affect the grout placement, a large-scale flow cell was constructed at the Chengdu University of Technology (CDUT) to model the grouting process in a single rock joint. In this work, we report a preliminary study where the injection experiments in this laboratory grouting simulator are compared with a theoretical analysis based on radial flow of a Bingham fluid in between two parallel plates. The aim of the present work is to establish a baseline to aid further development of the theoretical model. Experimental setup and theoretical background are first described. The experimental and theoretical results are then presented and compared.

2 Experimental Setup

2.1 Grouting Simulator

The grouting simulator at CDUT consists of a joint simulation system, a grouting control system and a data acquisition system, see Fig. 1. A distinct feature of this simulator is that the dip angle of the joint model can be adjusted in between 0 and 90°. Aperture of the joint can also be varied from 0 to 25 mm. The largest dimensions of the joint in the longitudinal and lateral directions are 6.4 m and 3.8 m, respectively. There are five injection ports with a diameter of 20 mm on the joint surface. Different injection ports can be selected to allow grout flow for as long as possible at a given dip angle.

The experiments can be conducted under either injection pressure-controlled or rate-controlled with the joint being initially empty or filled with water to simulate the condition inside a natural fracture. A grid laid out on top of the viewing windows on the joint surface allows recording of the grout placement profile in real-time.

2.2 Testing Materials

Grouts made of ordinary Portland cement are tested. The cement is produced by the Southwest cement plant in Chengdu, Sichuan, China. The chemical composition of the

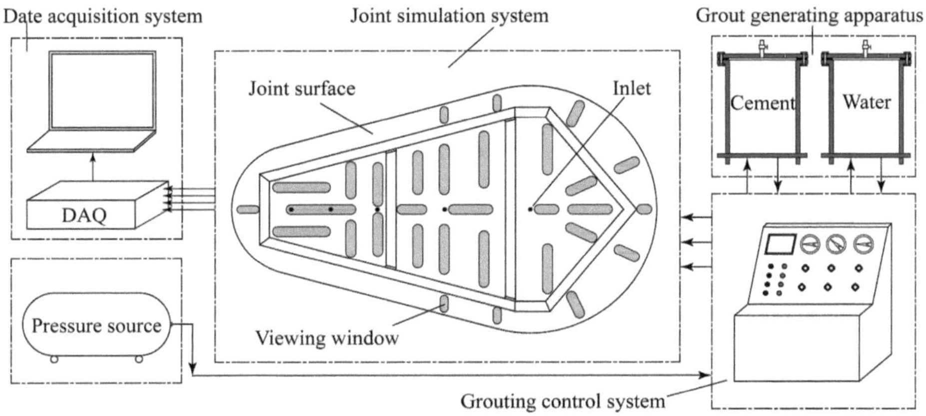

Fig. 1. Schematic of the laboratory grouting simulator.

cement as deduced from X-ray fluorescence (XRF) and X-ray diffraction (XRD) analyses. The mass content of SiO_2 and CaO is 20.91% and 63.12%, respectively.

The grout is made with a water-cement ratio of 0.5. Initial density of the grout is 1.82 g/cm^3 and the pumpable time is about 6 h. The initial fluidity obtained from a cone model is 210 mm.

The grout is assumed to be of Bingham rheology, i.e., $\tau = \tau_0 + \mu\dot{\gamma}$, where τ and $\dot{\gamma}$ are the shear stress and strain rate. In our experiments, the yield stress τ_0 and the plastic viscosity μ are determined from the measurements taken at shear rates of 300 rpm and 600 rpm in a six-speed rotational viscometer.

2.3 Testing Parameters

A series of four injection tests is performed under pressure control with the joint oriented horizontally (dip angle = 0°). The joint is initially empty and dry. Both the injection pressure and the grout rheology are varied. Variation of the grout rheology is achieved by starting the tests at different times after the cement is mixed. The testing parameters are listed in Table 1, where P_0 is the pressure in the grout generating apparatus; P_1 is the line pressure loss and P_{in} is the injection pressure at the injection port. History of the grout front radius is recorded during the tests.

Table 1. Parameters in the grouting tests.

Test No.	P_0 (kPa)	P_1 (kPa)	P_{in} (kPa)	μ (mPa.s)	τ_0 (Pa)
1	60	50	10	50	25.55
2	70	60	10	50	30.66
3	80	60	20	50	30.66
4	90	60	30	62	45.0

3 Theoretical Model

Though the grouting process in the aforementioned experiments is transient in nature, the process can be approximated as quasi-steady state since there is no back pressure beyond the fluid front. In other words, the boundary condition at the grout front can be treated as drained. Effect of pressure diffusion may be regarded as secondary. A theoretical model can be constructed based on radial flow of a Bingham fluid in between two parallel plates.

The grout may be assumed to be an incompressible single phase fluid. No slip condition may be assumed for the interfaces between the grout and the two walls. After combining the rheological equation, the equilibrium equation, the geometrical equation and the continuity equation with the no slip condition at the walls, the injection rate at the inlet can be expressed as a function of the local pressure gradient through [8],

$$Q_o = 2\pi r\left(\frac{\tau_0 b^2}{3\mu\psi}\right)\left(\psi^3 - 3\psi + 2\right) \tag{1}$$

where r is the radial coordinate and b is half of the joint aperture. Symbol ψ denotes the ratio between the yield stress and the shear stress at the walls, τ_w,

$$\psi = \frac{\tau_o}{\tau_w} = \frac{\tau_o}{|dp/dr|b} \tag{2}$$

where $|dp/dr|$ is the magnitude of the local pressure gradient. For a yield stress fluid, flow occurs only if the magnitude of the wall shear stress exceeds the yield stress, $\tau_w \geq \tau_o$ or $\psi \leq 1$. Based on global mass balance, the growth rate of the grout front radius can be written as,

$$\frac{dr_f}{dt} = \frac{Q_o}{4\pi b r_f} \tag{3}$$

Boundary conditions at the inlet $(r = r_w)$ and the fluid front can be expressed as $p = P_{\text{in}}$ at $r = r_w$ and $p = 0$ at $r = r_f$.

An explicit moving mesh algorithm based on Eqs. (1)–(3) is developed to solve for the position of the fluid front r_f as a function of time. An initial guess is needed to start the numerical calculation. Such a guess can be obtained from the early time asymptotic of Eq. (1).

At early time when the flow rate is relatively large, $\tau_w \gg \tau_o$ or $\psi \to 0$. Discarding the higher order term in Eq. (1), the pressure gradient can be expressed as,

$$-\frac{dp}{dr} = \frac{Q_o}{2\pi r}\left(\frac{3\mu}{2b^3}\right) + \frac{3\tau_0}{2b} \tag{4}$$

Integrating Eq. (4) and substituting the injection rate Q_o with Eq. (3) yield a first order ordinary differential equation for the growth rate of the grout placement radius,

$$\frac{dr_f}{dt} = \frac{P_{\text{in}} - \frac{3\tau_0}{2b}\left(r_f - r_w\right)}{\left(\frac{3\mu}{2b^2}\right) r_f \ln\left(\frac{r_f}{r_w}\right)} \tag{5}$$

While Eq. (5) can be directly integrated with the result expressed in complex functions, it is also straightforward to integrate it numerically.

Meanwhile, at late time when the wall shear stress at the fluid front reaches the yield stress, i.e., $\tau_w \to \tau_o$ or $\psi \to 1$, fluid flow stops. The threshold pressure gradient for flow cessation to occur is therefore $|dp/dr| = \tau_0/b$. Consequently, the maximum grout placement radius can be determined from,

$$r_{f\max} = r_w + \frac{P_{\text{in}} b}{\tau_0} \tag{6}$$

If the inlet radius r_w is relatively small, the maximum front radius and the maximum grout volume that can be placed into a horizontally oriented joint can be determined from,

$$r_{f\max} \cong \frac{P_{\text{in}} b}{\tau_0} \qquad V_{\max} \cong 2\pi b^3 \left(\frac{P_{\text{in}}}{\tau_0}\right)^2 \tag{7}$$

It can be shown that Eqs. (6) and (7) are valid not just for a Bingham fluid, but for any complex fluids with yield stress.

4 Experimental and Theoretical Results

Grout front position as a function of time from Tests 1–4 is summarized in Fig. 2(a). As expected, nearly circular grout fronts are observed in this series of tests, see Fig. 2(b). In general, the grout front radius r_f increases rapidly with time initially, but the growth rate decreases as the grouting process continues. Eventually, the grout front becomes almost stagnant.

Between Tests 1 and 2, only the yield stress is slightly different, $\tau_o = 25.55$ Pa in Test 1 and $\tau_o = 30.66$ Pa in Test 2. The yield stress seems to have little impact at the early time, but the effect gradually manifests itself at late time, see Fig. 2. As the overpressure increases from $P_{\text{in}} = 0.01$ MPa in Test 2 to $P_{\text{in}} = 0.02$ MPa in Test 3, the placement radius is nearly doubled; at $t = 300$ s, $r_f = 195$ mm in Test 2, and $r_f = 352$ mm in Test 3. Increases in the overpressure, the plastic viscosity and the yield stress in Test 4 seem to have resulted in a relatively similar flow behavior to that in Test 2.

Theoretical predictions for the growth of the placement radius in these four tests are presented in Fig. 3. Compared with the experiments, the theoretical model yields much fast growth rates for the front radius at early time. The grout front radius has almost reached its maximum around $t = 50$ s for all the tests. In addition, the maximum placement radius $r_{f\max}$ as predicted by the theoretical model, $r_{f\max} = 597, 499, 988, 1001$ mm for Tests 1–4, is also larger than what is observed at the last measurement point in the experiments, $r_{f\max} = 255, 255, 383$ and 335 mm for Tests 1–4. In addition,

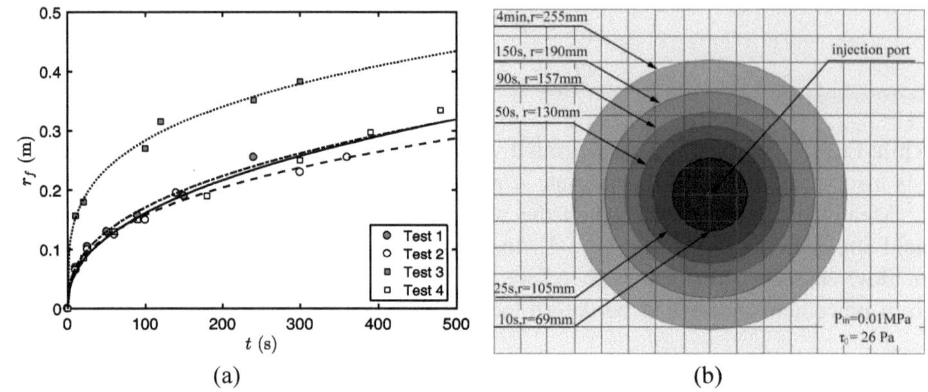

Fig. 2. Growth of the grout front radius with time; (a) comparison for Tests 1–4, (b) plan view from Test 1.

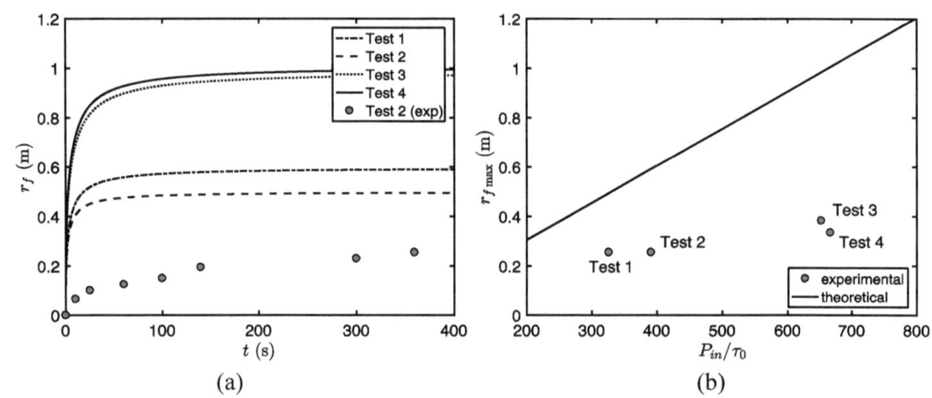

Fig. 3. Comparison between the theoretical predictions and the experimental results; (a) growth of the grout front radius with time, (b) dependence of the maximum grout placement radius on P_{in}/τ_0; comparison with the last experimental measurement points.

it is expected from the theoretical model that variations of r_f with time in Tests 3 and 4 should be relatively close since P_{in}/τ_0 is similar. However, the experimental results in these two tests show marked differences.

Several factors may have contributed to the differences between the experiments and the theoretical model. There is certainly room for improvement in the quantification of the inlet pressure as well as the yield stress of the grout. We assume that the pressure loss along the line is constant during the test, which is debatable since the pressure drop also depends on the injection rate. In the experiments, the yield stress is obtained by extrapolating the measurements at rates of 300 rpm and 600 rpm to the zero shear rate. Rheology in the near zero shear rate range could be rather complex. Such an extrapolation may not have yielded the true yield stress of the grout. On the other hand, the theoretical prediction is relatively sensitive to the yield stress,

$r_{f\max} \propto 1/\tau_0$. Slight change in the yield stress magnitude could have a relatively large impact in the maximum placement radius. For example, $P_{in} = 10$ kPa and $\tau_0 = 55$ MPa yield $r_{f\max} = 283$ mm, which is indeed rather close to the results in Tests 1 and 2. Other factors such as the interface condition between the grout and the walls and the time-dependent nature of the cement grout rheology may also be the sources that have caused the differences between the experiments and the theoretical model.

The comparison between the experimental results and the theoretical predictions is nevertheless encouraging since the theoretical model is able to capture the overall trend of the grout front growth. The differences between the two could help guide us in further improving the experimental design and material characterization as well as in adding complexities in the theoretical model in future work.

5 Concluding Remarks

Experimental and theoretical analyses have been conducted to investigate the grouting process in a rock joint. The grouting experiments are conducted in a newly constructed large-scale laboratory grouting simulator with unique features that allow adjustments in the joint aperture and inclination. A series of four tests are performed with the dip angle equal to zero under a pressure-controlled condition. The experiments are then modeled theoretically based on radial flow of a Bingham fluid in between two parallel plates. Results from this comparative study can be summarized as follows.

- Under the pressure-controlled boundary condition, both the experimental observations and the theoretical predictions show that the grout front extends rather rapidly at the early time (within ~50 s during the experiments and ~10 s from the theoretical solution). Growth of the grout front gradually slows down and flow of the grout eventually stops due to the yield stress of the grout. As expected, the shape of the grout front is observed to be nearly circular in the experiments.
- Both the overpressure and the grout yield stress affect the grout placement. The maximum placement radius from the theoretical solution and from the last measurements in the experiments increases with the overpressure, but decreases with the grout yield stress.
- The maximum placement radius is overestimated by the theoretical solution. While further improvement is needed in better control of the experimental conditions and in incorporating more complexities in the theoretical model, the theoretical model provides a relatively simple and first order estimate for the design of the grouting treatment, which could be quite beneficial from an engineering practice stand point.

Acknowledgements. Financial support to JZ and XP is provided through the Major Program of the National Natural Science Foundation of China (Grant No. 41790445) and Team Key Program of the State Key Laboratory of Geohazard Prevention and Geoenvironment Protection (No. SKLGP2014Z001). The support to YS and HH is provided from the National Science Foundation through grant NSF/CMMI-1055882.

References

1. Pei, X.J., Huang, R.Q., Li, Z.B., Luo, J.L.: Research on grouting reinforcement of unloading fractured loose rock mass in left bank of Jinping-I hydropower station. Chin. J. Rock Mech. Eng. **30**(2), 285–288 (2011)
2. Zhang, J.X., Pei, X.J., Wang, W.C., He, Z.H.: Hydration process and rheological properties of cementitious grouting material. Constr. Build. Mater. **139**, 221–231 (2017)
3. Bird, R.B., Stewart, W.E., Lightfoot, E.N.: Transport Phenomena, 2nd edn. Wiley, New York (2007)
4. Dai, G., Bird, R.B.: Radial flow of a Bingham fluid between two fixed circular disks. J. Non-newton. Fluid Mech. **8**, 349–355 (1981)
5. Hässler, L.: Grouting of rock—simulation and classification. Ph.D. thesis, p. 159. Royal Institute of Technology, Sweden (1991)
6. Amadei, B., Savage, W.Z.: An analytical solution for transient flow of Bingham viscoplastic materials in rock fractures. Int. J. Rock Mech. Min. Sci. **38**(2), 285–296 (2001)
7. Gustafson, G., Stille, H.: Prediction of groutability from grout properties and hydrogeological data. Tunn. Underground Space Technol. **11**(3), 325–332 (1996)
8. Majidi, R., Stefan, M.Z., Ahmed, R., Yu, M., Thompson, L.G.: Radial flow of yield-power-law fluids: numerical analysis, experimental study and the application for drilling fluid losses in fractured formations. J. Petrol. Sci. Eng. **70**(3), 334–343 (2010)
9. Sun, Y., Huang, H.: Drilling mud loss in a natural fracture - a PKN fracture geometry model based analysis. J. Energy Challenges Mech. **1**(1), 32–36 (2014)
10. Guillot, D.: Rheology of well cement slurries. In: Nelson, E.B. (ed.) Well Cementing. Elsevier, New York (1990)

Effect of Injection Rate on Micro-seismicity Due to the Hydraulic Fracturing of Granite

Inderdeep Grewal[1], Bing Q. Li[2], and Bruno Gonçalves da Silva[1(✉)]

[1] New Jersey Institute of Technology, Newark, NJ 07102, USA
bmgsilva@njit.edu
[2] Massachusetts Institute of Technology, Cambridge, MA 02139, USA

Abstract. Enhanced Geothermal Systems (EGS) utilize the heat of the earth to produce electricity or to heat buildings. In EGS, a crystalline rock, usually granite, is hydraulically-fractured to produce a network of fractures through which the water is circulated. While several pilot projects have proven that this technology can be successful, the strong seismicity caused during the hydraulic fracturing stage is limiting its widespread use. Therefore, the causes of this micro-seismicity need to be better understood. This research evaluated the effect of the rate of water injected in granite specimens on the magnitude and number of micro-seismic events produced during recently-conducted hydraulic fracturing tests. In these tests, granite specimens subjected to two different vertical loads (0 and 5 MPa) were hydraulically-fractured while visual and micro-seismic monitoring was performed. The objectives of this paper were to examine the relationship between the number and amplitudes of the first P-wave arrivals of micro-seismic hits, and the rates of water injected used in the tests. It was found that in pressure increments closer to failure, more AE (Acoustic Emission) hits occur after the highest injection rates. This delayed seismic response may be due to the release of stresses accumulated during the high injection-rate stages of the tests. The study also showed that considerable energy is released throughout the tests as hits with relatively high P-wave amplitudes are observed before the visible fractures propagate

Keywords: Micro-seismic events · Hydraulic fracturing · Granite
Injection rate

1 Introduction

The continuous development of alternative renewable energy sources is necessary to bring about an environmentally sensible era. One area of potential takes the form of Enhanced Geothermal Systems (EGS) which, according to a recent report, may be capable of providing more than 100,000 MWe for the next 50 years in the United States [1] and have been gradually adapting and becoming more efficient as time progresses. In EGS, hydraulic fracturing of rock is primarily used to induce fractures in rock formations. The network of fractures creates a permeable track in which the injected water can flow and exchange heat with the surrounding rock in an efficient manner. While some pilot projects have proven the feasibility of EGS [2, 3], there are many

L. Zhang et al. (Eds.): GSIC 2018, *Proceedings of GeoShanghai 2018 International Conference: Rock Mechanics and Rock Engineering*, pp. 137–144, 2018.
https://doi.org/10.1007/978-981-13-0113-1_16

issues that may affect the productivity and feasibility of this process, such as water escaping the production loop, short-circuiting of the water path, and induced seismicity [4, 5].

In this particular research, the issue of understanding the interaction between the hydraulic fracturing processes and seismic activity induced from such processes was investigated. Such interactions are difficult to analyze at the field scale as they occur deep within the ground. Therefore, laboratory experimentation plays a crucial role in the understanding of the induced seismicity produced during the hydraulic fracturing of rocks. This paper analyzes the effects of the rate at which water is injected in granite specimens on the resulting seismic activity, particularly on the number of micro-seismic events and on the amplitude of the first P-wave arrivals.

2 Methodology

The experiments that were the basis of this study were performed by Gonçalves da Silva [6] and Gonçalves da Silva et al. [7]. In these experiments, granite specimens were placed into a pressure enclosure, which held the specimen such that water could be injected into precut fractures, or flaws, and viewed through a polycarbonate viewport on the opposite side (see Fig. 1). To simulate the overburden pressure, a vertical stress was applied to the granite specimens. Both sides of the specimen had four acoustic emission sensors attached. In this particular research, only the P-waves were analyzed, since the sensors used were not able to record S-waves. The P-wave sensors picked up the micro-seismic P-waveforms that propagated away from the location of the seismic events which are primarily caused by the development of micro- and macro-cracks. Based on these waveforms, the amplitudes of the first P-wave arrivals of micro-seismic "hits" and events, as well as their corresponding location, could be determined. In the context of this paper, an acoustic emission "hit" occurs when an acoustic emission sensor picks up a P-wave with an amplitude higher than a user-defined threshold which, for the current research, was 36 dB.

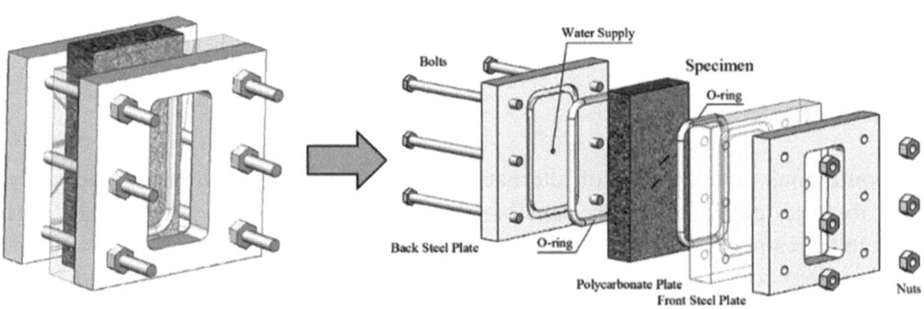

Fig. 1. Testing apparatus.

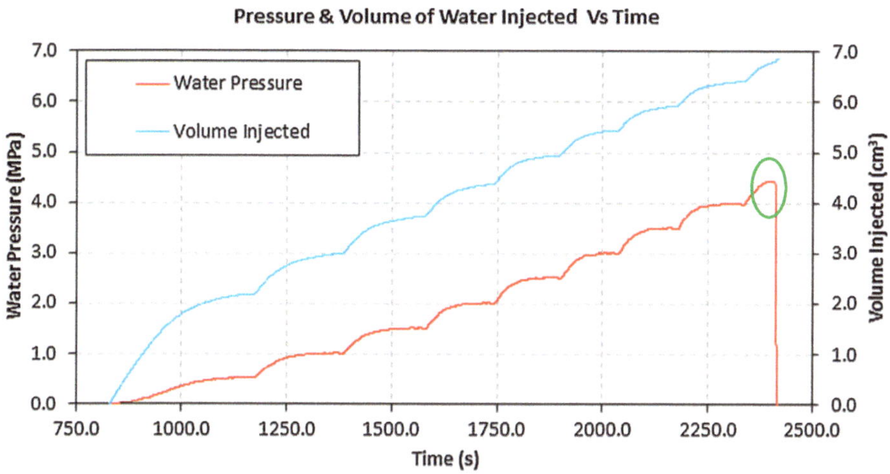

Fig. 2. Pressure and water injected over time. The green ellipse highlights the pressure drop when visible fractures begin to propagate.

The specimen was incrementally subjected to water pressures. As shown in Fig. 2, the total volume injected as well as the water pressure were incrementally increased until cracks propagated.

As illustrated in Fig. 3, the tests conducted in granite specimens were named by flaw geometry spacing (L), flaw angle (β) bridging angle between the flaws (α) and vertical loading (VL) applied on the specimens, which can be 0 or 5 MPa. Although most tests consist of a double-flaw configuration, tests with a single-flaw were also investigated, in which the angle β and vertical load VL were used to identify the specimens tested. The specimens studied in this paper are shown in Table 1, which is a subset of the specimens tested by Gonçalves da Silva [6].

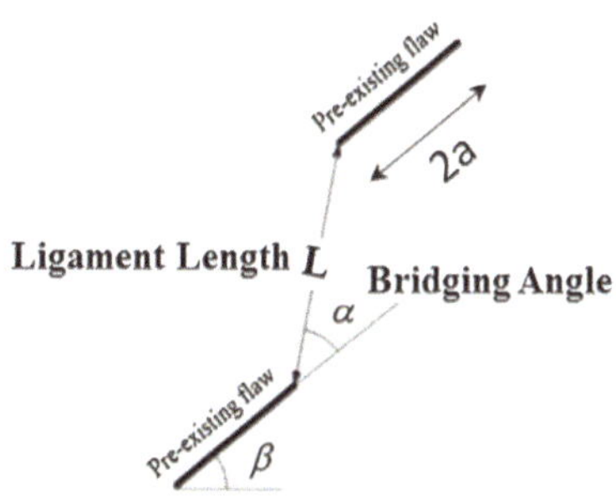

Fig. 3. Nomenclature for double-flaw geometries: $L\text{–}\beta\text{–}\alpha\text{–}$VL.

The plot in Fig. 4 denotes "I1" to "I6", or "Increment 1" to "Increment 6". These "Increments" were defined for each analyzed test in order to determine the injection rates for each increment leading to failure. Figure 4 shows the breakup of these

Table 1. Specimens analyzed in this study.

Specimens	
2a-30-0-VL0-B	2a-30-0-VL5-C
2a-30-30-VL5-B	2a-30-60-VL0-C
2a-30-120-VL5-B	30-VL5-C
2a-30-30-VL0-C	30-VL5-B

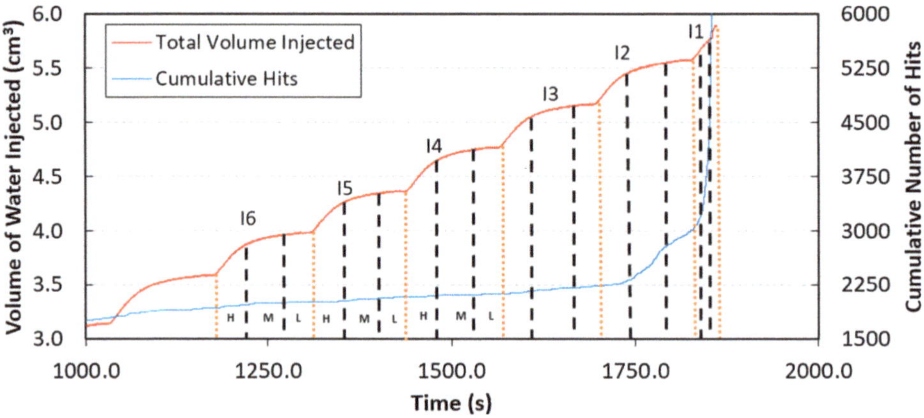

Fig. 4. Volume of water injected and cumulative hits vs time for test 2a-30-0-VL0-B. The "H", "M", and "L" correspond to the High, Middle, and Low injection rates per increment (only shown in I6, I5, and I4 above).

increments for test 2a-30-0-VL0-B. Each increment was subdivided into three sections of different slopes: "High" (H), "Middle" (M), and "Low" (L). The slopes denote volume of water injected over time; in other words, the rate of injection of water. The "Increment" numbering starts from the failure of each test and ends at the 6th increment which corresponds to approximately 50% of the total volume of water injected.

3 Results

The results presented in this section will focus primarily on one of the tests analyzed, namely 2a-30-0-VL0-B, since the results of this test are representative of all the tests analyzed. Given the number of figures that would have been necessary, it would not be feasible to show and separately analyze the results for all the tests investigated.

3.1 Number of Hits Versus Injection Rate

As noted in Fig. 4, each increment was broken down into three separate intervals whose slope represents the rate of injection. "High" (initial), "middle" (intermediate), and "low" (end) injection rates were considered. Figure 5 shows the injection rates used in this study for the tests analyzed. The "high" injection rate corresponds to

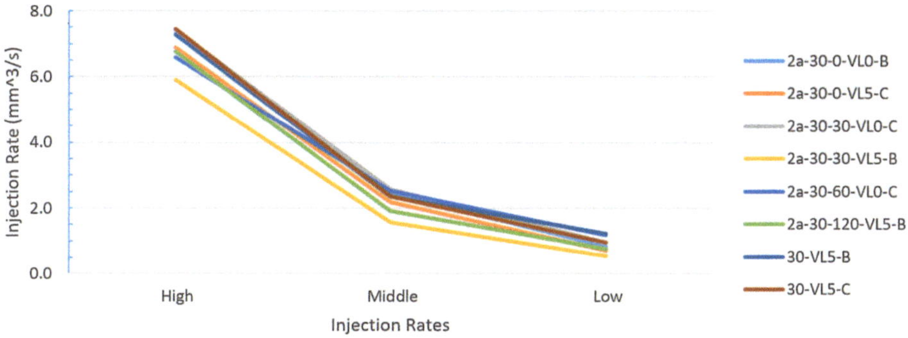

Fig. 5. Injection rates used in this study for each specimen analyzed.

approximately 6–8 mm^3/sec, the "medium" to 2 mm^3/sec and the "low" to 1 mm^3/sec. Figure 6 shows the number of AE hits increasing as the volume injected increases from about 4 cm^3 to 4.35 cm^3 for a particular pressure increment. Sudden increases in the cumulative hit count can be observed in the three injection rates investigated; these sudden increases are interpreted as microscopic cracking.

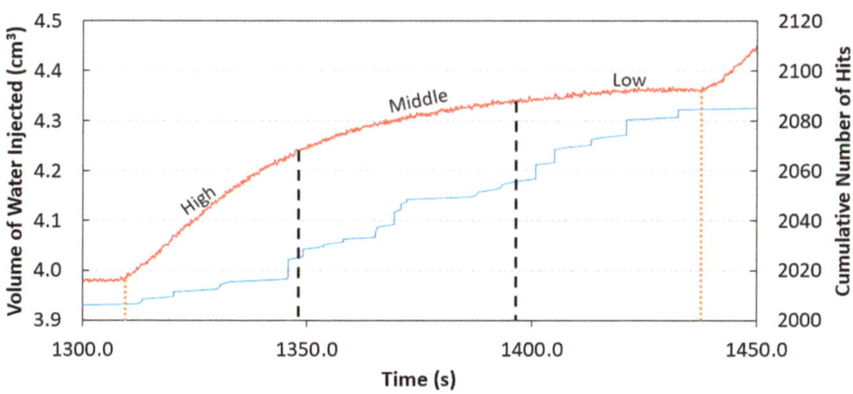

Fig. 6. Volume of water injected and cumulative AE hits vs time for increment I5 in test 2a-30-0-VL0-B.

Figure 7 shows the number of hits and the average of the first P-wave amplitudes versus injection rates for Increments I6 to I1 for the test 2a-30-0-VL0-B. Focusing on the number of hits, it is observed that there are more hits in the "middle" injection rates for the initial Increments I6 to I4. For the increments closer to failure – I3 to I1 – there are significantly less hits in the "high" injection rate periods than in the "medium" and "low" injection rates. This relatively counter-intuitive observation may be caused by a delayed release of strain energy accumulated in the "high" injection rate periods. This behavior may be material-related; however, other rocks would need to be tested in a similar way to confirm this hypothesis.

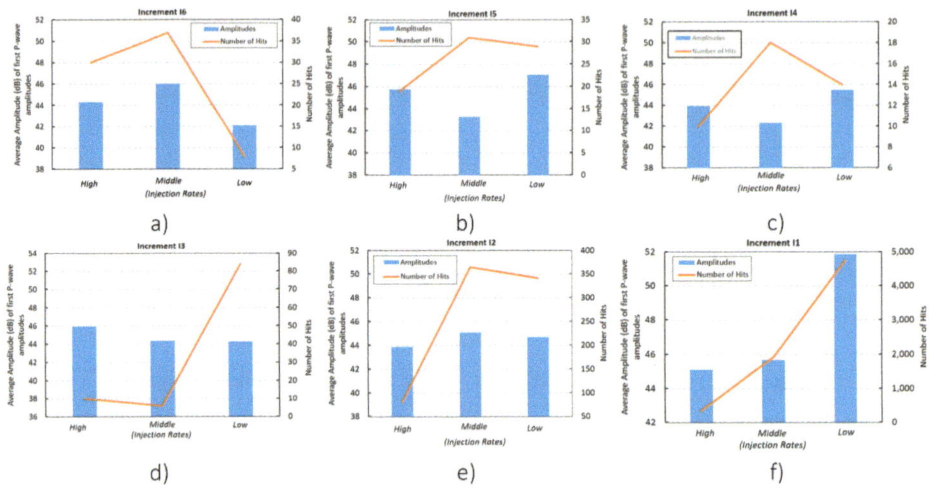

Fig. 7. Increments I-6 to I-1 for specimen 2a-30-0-VL0-B.

It is also interesting to note that significant microseismic activity can be observed from early stages of the test. In Increment I6, for example, 30, 36 and 7 AE hits were recorded for the "high", "middle" and "low" injection rate periods, respectively. These hits may be caused not only by the development of micro-cracks, but also by the closure of existing micro-cracks, crushing of grains and sliding between particles. More research would be necessary to determine the mechanisms responsible for this early microseismic activity. Furthermore, the material heterogeneity could have an effect in the results (such as grain size, shape and distribution), but separate testing of different rocks would need to be conducted to observe possible effects.

3.2 Amplitude of the First P-Wave Arrivals Versus Injection Rates

Figure 7 shows that there is not a clear trend in the variation of the average P-wave amplitudes with the injection rates. In general, the average P-wave amplitudes range between 42 dB and 48 dB regardless of the injection rate. The exception is the "low" injection rate of Increment I1 (52 dB), which occurs immediately before the development of visible cracks, therefore releasing more energy than the microscopic cracks that developed previously.

In order to better assess the effect of the injection rates in the first P-wave amplitudes of the micro-seismic hits recorded, the ten largest seismic hits at each period of injection were isolated and their average amplitude calculated. By doing this, one can better analyze the effect of the injection rates in the amplitudes of the largest micro-seismic events. In fact, when the average of all the hits is used, the amplitudes of the largest events are "diluted" in the amplitudes of the many lower-amplitude events. Figure 8 shows the average amplitude of the ten largest micro-seismic hits during the increments and injection rates studied for test 2a-30-0-VL0-B.

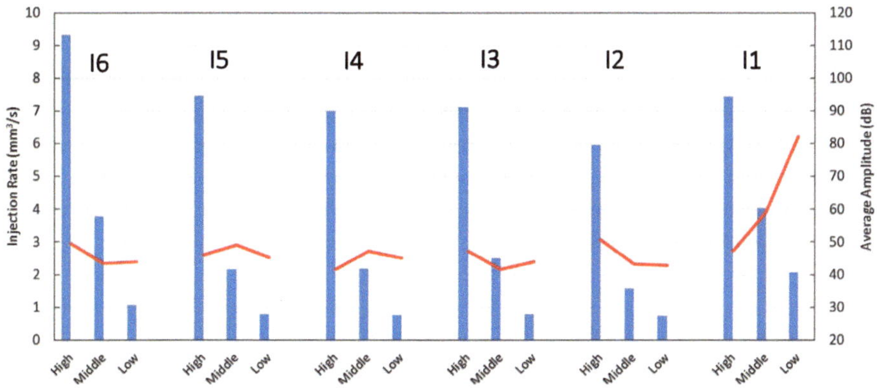

Fig. 8. Injection rates (blue bars) vs average of the ten largest first P-wave amplitudes (red lines) for test 2a-30-0-VL0-B.

In general, the average P-wave amplitudes of the top-ten AE hits range between 45 dB and 50 dB from Increments I6 to I2. In Increment I1, the top-ten P-wave amplitude increase to 60 dB and 80 dB for the "middle" and "low" injection rates, respectively. The injection rates do not appear to have a strong effect on the average top-ten amplitudes; for Increments I5 and I4, the maximum amplitudes occur in the "middle" injection rate, while in Increments I6, I3 and I2, the maximum amplitudes occur in the "high" injection rate period. In Increment I1, closer to failure, the maximum amplitudes occur in the "low" injection rate; this is related to the fact that the visible cracks developed at this injection rate, releasing more energy and therefore creating large-amplitude events.

The fact that the average of the amplitudes of the P-wave hits (all and top-ten) does not vary substantially, except for Increment I1, indicates that the mechanisms responsible for this microseismicity may be very similar throughout the tests regardless of the injection rate used. If one assumes that the microseismicity generated between Increments I6 and I2 is entirely caused by micro-cracks, the relatively constant average amplitudes of the first P-waves may indicate that these micro-cracks have comparable size and relative motion. This does not apply to Increment I1, in which higher-amplitude AE hits are related to larger, usually visible, cracks.

4 Conclusion

The effect of the injection rates on the number and amplitude of AE hits was investigated using data obtained from hydraulic fracturing tests conducted in granite specimens with pre-fabricated flaws.

From the analyses presented, it was observed that while injection rates do not seem to affect the number of hits in the initial pressure/volume increments, it is clear that there are less hits in the "high" than in the "middle" and "low" injection rates for increments closer to the failure of the specimens. This response may be due to a delayed release of the stresses accumulated during the high injection periods of the tests.

Furthermore, the amplitudes of the microseismic AE hits do not appear to be significantly affected by the injection rates used, except in the increment before failure. This may indicate that the mechanisms responsible for the microseismicity are, in general, very similar regardless of the injection rate used.

Finally, significant microseismic activity does not occur solely when visible fracture propagation takes place. Instead, considerable energy is also released throughout the tests, since many hits with relatively high P-wave amplitudes are observed before the visible fractures propagate.

References

1. The Future of Geothermal Energy: Impact of Enhanced Geothermal Systems (EGS) on the United States in the 21st Century, Massachusetts Institute of Technology (2006). https://www1.eere.energy.gov/geothermal/pdfs/future_geo_energy.pdf
2. Hodson-Clarke, A., Rudolf, R., Bour, D., Russell, P.: Key factors to successful drilling and completion of EGS well in Cooper Basin. In: Proceedings of the 41st Workshop on Geothermal Reservoir Engineering Stanford University, Stanford, California, 22–24 February 2016, SGP-TR-209 (2016)
3. Bendall, B., Huddlestone-Holmes, C., Goldstein, B.: The current status of geothermal projects in Australia – a national review. In: Proceedings of the Thirty-Eighth Workshop on Geothermal Reservoir Engineering Stanford University, Stanford, California, 11–13 February 2013, SGP-TR-198 (2013)
4. Sausse, J., Dezayes, C., Genter, A., Bisset, A.: Characterization of fracture connectivity and fluid flow pathways derived from geological interpretation and modelling of the deep seated EGS reservoir of Soultz (France). In: Proceedings of the Thirty-Third Workshop on Geothermal Reservoir Engineering Stanford University, Stanford, California, 28–30 January 2008, SGP-TR-18 (2008)
5. Ghasemi, A., Alexis D.A., Vardcharagossad, P., Krishnamoorthy, V.: Feasibility and design of an engineered geothermal system using dry and abandoned holes as a prospective location. In: EME 580: Integrative Design of Energy and Mineral Engineering Systems, Pennsylvania State University (2010)
6. Gonçalves da Silva, B.: Fracturing processes and induced seismicity due to the hydraulic fracturing of rocks. Ph.D thesis, Massachusetts Institute of Technology (2016)
7. Gonçalves da Silva, B., Einstein, H.: Development of a test setup capable of producing hydraulic fracturing in the laboratory with image and acoustic emission monitoring. In: Proceedings of the 49th U.S. Rock Mechanics and Geomechanics Symposium, San Francisco. American Rock Mechanics Association, ARMA 15-141 (2015)

Experiment and Numerical Simulation on Contact and Breakage of Marble Particles

Jian Zhou[1,2], Shicai Yu[1,2], Jin Zhang[1,2], and Cheng Zhao[1,2](✉)

[1] College of Civil Engineering, Tongji University, Shanghai, China
09184@tongji.edu.cn
[2] Key Laboratory of Geotechnical and Underground Engineering of the Ministry of Education, Tongji University, Shanghai, China

Abstract. Current research on coarse-grained soils focuses mostly on the physical and mechanical properties of particle aggregates. For the study of the coarse-grained soil's breaking behavior, the particle contact experiments involving marble particles are carried out by employing Rock Rheological Testing System, through which the mechanical properties and microscopic mechanism of marble particles' contact breakage are analyzed. Moreover, the numerical simulation with parameters for strength reduction is performed to reveal the breakage process. The study shows that, the local breakage is occurred firstly in the contact area and then the Elastic Core is formed under normal contact force. The existing of the Elastic Core changes the path of force transfer that results in overall breakage of the particle after crack coalescence. Considering the stress concentration in contact breakage, parameters calibrated through uniaxial compression test are unsuitable for numerical simulation. The introduction of strength reduction parameters contribute to the simulation on the morphology after breakage.

Keywords: Contact test · Particle breakage · Elastic core
Numerical simulation

1 Introduction

Coarse-grained soil is with high strength, small deformation, good permeability and other characteristics, which is widely used in earth rock dams, highways, railways, airports, dams, bridge piers, artificial islands, and ground treatment engineering applications. In recent years, super high dam construction has developed rapidly in China, so that the research on the mechanics mechanism and engineering characteristics of coarse-grained soil under high stress has caught the attention of many scholars.

The breakage of soil particles refers to the phenomenon that the coarse-grained component in soil crack and slit into multiple particles under the influence of external factors [1]. Particle breakage may result in a change of particle gradation that may influence the structure and has a noticeable impact on the mechanical behaviors of soil.

Study of soil particle's breakage was first started in the beginning of the last century, but there was no notable successes limited to the experimental conditions. In 1948, Terzaghi, the founders of modern soil mechanics, found that the phenomenon of

L. Zhang et al. (Eds.): GSIC 2018, *Proceedings of GeoShanghai 2018 International Conference: Rock Mechanics and Rock Engineering*, pp. 145–157, 2018.
https://doi.org/10.1007/978-981-13-0113-1_17

particle breakage is obvious in the compression test of sand [2]. Since that a growing number of researchers began to study on particle breakage [3–7]. Through contact test between two particles, large-scale direct shear test and centrifuge modeling test, Wang [8] studied the mechanical characteristics between coarse grained soil particles under contact conditions. Moreover, a force-deformation fitting formula was proposed based on the tests. The rise of Discrete Element Method (DEM) breaks through the limitations of conventional indoor geotechnical tests in the study of meso-mechanism [9], so that a new direction has been opened for the study of mechanical behavior and particle fragmentation of granular systems. DEM has been used by many scholars in the study of particle contact mechanics [10]. Using PFC^{3D} to simulate the contact test of Gypsum granules, Wang [7] found that the numerical parameters calibrated by uniaxial compression test are not suitable for numerical model of particle contact. In the past, the researches predominantly focused on the physical and mechanical properties of particle aggregations, research on coarse-grained soils and particle contact characteristics is rare. At present, there is little research on the process, morphology and mechanism of particle breakage from the angle of particle contact, and the experimental data cannot directly get the physical and mechanical parameters related to particle contact when the particle is broken, it is difficult to reveal the crushing process and mechanism of individual particles.

From the angle of particle contact, this study investigate the crushing mechanism and morphology of marble particles through ball-surface and ball-ball contact test by employing Rock Rheological Testing System. Moreover, this study also introduces a contact model with strength reduction parameter in PFC^{3D} to simulate the process of marble particles' breakage. And the result of breakage morphology of single particle is similar to that of the experiment. The numerical simulation provide a good supplement to the experimental results by observing the development of the local breaking zone and analyzing the basic particle's displacement, to further reveals the meso mechanism of the contact crushing process of marble particles.

2 Experimental Study

2.1 Experimental Material

It is difficult to carry out duplicate test considering all factors (particle size, shape and etc.). As a preliminary research, this study chooses the spherical marble particle with a diameter of 5 cm. If the particle size is too large, point to point contact cannot be guaranteed; on the other hand, if the particle size is too small, the particle can easily slip below the experimental limits. Basic mechanical parameters measured by standard test for marble materials used in the test are shown in Table 1.

Table 1. Physical parameters of marble

Compressive strength	Elastic modulus	Poisson's ratio	Tensile strength	Specific gravity
34.9 MPa	57.94 GPa	0.210	2.43 MPa	2.71

Considering the large discreteness of the rock material itself, the crushing mechanism of spherical specimen under ball-ball contact and ball-surface contact is studied by repeated experiment.

2.2 Experimental Instrument

As shown in Fig. 1, the test adopts Rock Rheological Testing System, which can provide stable displacement control and servo control biaxial loading conditions. The suitable loading system and spheric joint support are specially customized. The measurement and control precision for specimen stress and deformation were 5 N and 0.001 mm, respectively. The data acquisition frequency was approximately 1 Hz.

(a) Rock Rheological Testing System (b) ball-surface (c) ball-ball

Fig. 1. Experimental facilities and experimental conditions

2.3 Test and Results

The test process can be divided into the main steps of test preparation, loading and data acquisition.

(1) The fixture is lubricated before the test starts to eliminate the effect of friction on the final results.
(2) The error in placing specimen is permissible because of the ball hinge support. Start and reset the measuring system when the specimen is ready. To ensure fully

contact between test facilities, a normal force of about 200 N is applied in advance.

(3) The loading process is controlled by the displacement to observe the change of contact force-displacement relation the test. The loading rate of ball-surface and ball-ball contact test are both 0.002 mm/s. The maximum displacement of ball-surface contact is set to 1 mm and that of ball-ball test is set to 0.7 mm, because a large displacement may let the specimen drop from the loading platform.

It is necessary to note that for the standard rock specimens, the methods recommended by International Society of Rock Mechanics and China's Water Conservancy Regulations require the loading rate of 0.5 MPa/s. However, there is no unified standard for contact test between particles. A loading rate of 5 N/s was performed in the prophase research, which may stop the measuring system because the protection limit of system would be triggered by the time specimens crush. Therefore, the method of displacement controlled loading is employed. It is generally believed that a low strain rate or astatic procedure refers to the strain rate below $10^{-4}s^{-1}$, and 10^{-4}–10^{-2} s^{-1} for a quasi-static procedure [11]. A contact test does not have the concept of strain. But, as a reference, the loading rate divided by the diameter of the specimen belongs to a static or quasi-static procedure.

A total of 40 groups of ball-surface contact tests and 20 groups of ball-ball contact tests are conducted to explore the morphology after breakage. The experimental results show that the morphology after breakage under two kinds of contact conditions are similar.

When the test start, the local breakage first occurs at the contact point due to the stress concentration (see Figs. 2(b) and 3(b)). And the local crushing area will form a circular platform gradually with debris generated around it. With the loading going on, a cone area is formed under the contact platform, which is tentatively called Elastic Core [12]. In the vicinity of the elastic core, a number of cracks are gradually developed, and the particle is crushed into 2–4 blocks after crack coalescence. Only one particle is broken in every test under ball-ball contact condition, and local breakage is observed on another.

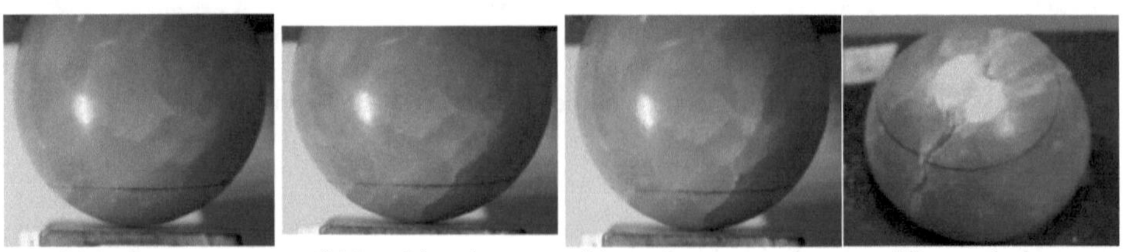

(a) before test (b) local breakage (c) crack propagation (d) morphology

Fig. 2. 5 cm marble breakage process under ball-surface contact condition

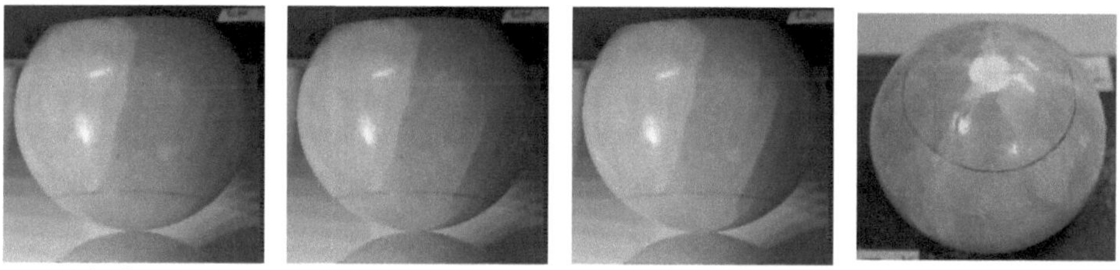

(a) before test (b) local breakage (c) crack propagation (d) morphology

Fig. 3. 5 cm marble breakage process under ball-ball contact condition

The morphologies after breakage are shown in Figs. 4 and 5, and the detail of enlarged scale of the elastic core is shown in Figs. 4(b) and 5(b). A penetrated crack can be observed after breakage that divides the spherical particle into blocks.

(a) (b)

Fig. 4. Marble particle breakage morphology under ball-surface contact condition

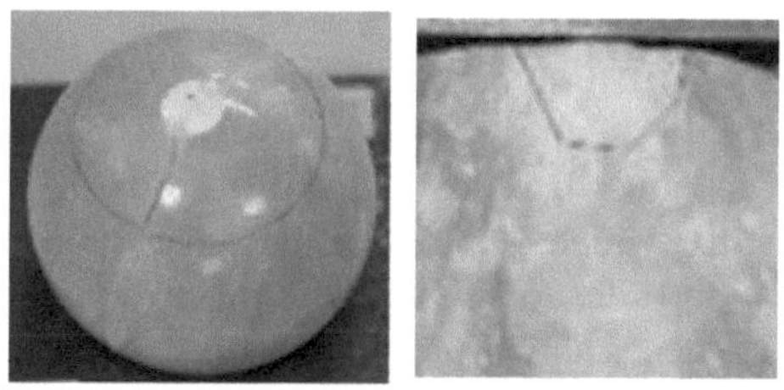

(a) (b)

Fig. 5. Marble particle breakage morphology under ball-ball contact condition

The force-displacement statistical relationships obtained by the experiment above are shown in Fig. 6. At the beginning, when the loading force reaches a certain value, the contact point of marble particles is partially broken, and the first ascending segment of the curve is the stage for the local breakage, along with the elastic core's generating and developing. The end point of the first ascending segment is the dividing point between local breakage and overall breakage.

According to the results of the ball-surface contact test (see Fig. 6(a)), the average contact force on overall breakage's occurring is 5.156 *kN*, and the average slope of first stage is 26.559 *kN/mm*. After the overall breakage, the strength and the slope of the curve are related to the stress state after the crushing. However, the uncertainty of the number of blocks and the width of the crack make the force state of the specimen after crushing of lots of randomness.

Besides, According to the results of the ball-ball contact test (see Fig. 6(b)), the average contact force on overall breakage's occurring is 4.001 *kN*, and the average slope of first stage is 16.885 *kN/mm*.

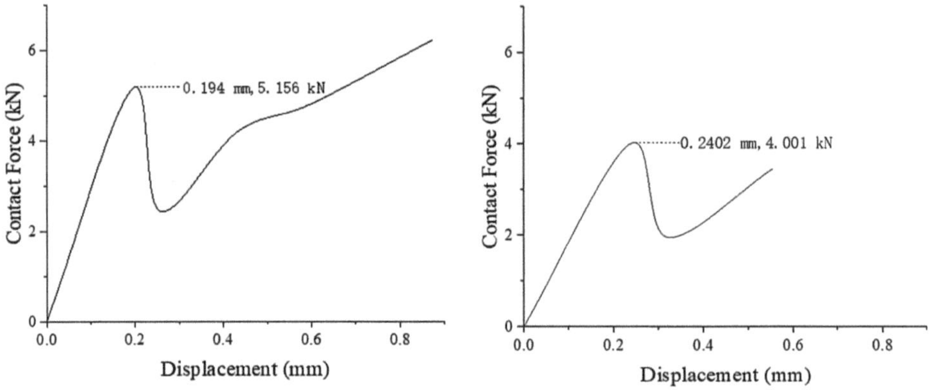

(a) ball-surface contact (b) ball-ball contact

Fig. 6. Force-displacement statistical relationship

3 Numerical Simulation

This study uses Particle Flow Code 3D (PFC^{3D}) for numerical modeling and simulating, which is based on Discrete Element Method (DEM).

In order to simulate the macroscopic phenomena of contact breakage, it is necessary to establish the relationships between the macroscopic mechanical properties of the material and the meso-mechanical parameters of the model. Since there is no direct mathematical relations between them, calibration and interactive debugging is necessary to match multiple sets of parameters. To get the results similar to the experiment, repeated calculation and adjustment are required.

3.1 Numerical Model

The hemispherical model is used to simulate the marble spherical particles in this study. The shape and size of the numerical model are consistent with the specimen in the laboratory test. The diameter of the hemispherical model is 50 mm, and the specific particle number and size are adjusted according to the situation. The wall unit in PFC^{3D} is used to simulate the contact surface.

Besides, a strength reduction parameter is introduced in the parallel bonding model to take the influence of the stress concentration and local fragmentation to strength into consideration. By using this parameter, the bond strength of the surrounding particles is reduced when the bonding fracture occurs. The influence of this parameters on the result is mainly reflected in the morphology after breakage. Through numerical calculation, it is found that the affected area of the reduction parameter is related to the particle size distribution of the hemispherical model, which needs to be determined according to the specific numerical model.

In the numerical model, a subroutine is programmed to show local breakage by changing the color of bond-broken particles to red.

The following 5 groups of meso parameters need to be calibrated for numerical simulation.

(1) Contact modulus E_c and parallel bond modulus $\bar{E}_c$
(2) Ratio of normal stiffness to tangential stiffness (Particle) k_n/k_s
(3) Ratio of normal stiffness to tangential stiffness (Parallel bond) $\bar{k}_n/\bar{k}_s$
(4) Average normal strength $\bar{\sigma}_c$ and tangential strength $\bar{\tau}_c$ and the strength after reduction (Parallel bond)
(5) Value of isotropic stress

According to the previous study [7, 8, 13], the parameters in (1) mainly affect the slop in the force-displacement curve. The parameters in (4) mainly affect the crushing strength. And the parameter in (5) mainly affect the degree of breakage. The specific numerical parameters should be obtained from the calibration and trial calculation.

The hemispherical model is composed of basic particles of varying sizes, as shown in Figs. 7 and 8. According to previous studies [7, 13], the minimum particle size of the numerical particle is 1.5 mm, with the ratio of the maximum diameter to the minimum size of 1.8. The total number of numerical particles is about 30,000 for ball-surface contact test and 60,000 for ball-ball contact test. In this situation, the range of the strength reduction effect is 1.5 times the sum of the two particle radius of the bonding fracture.

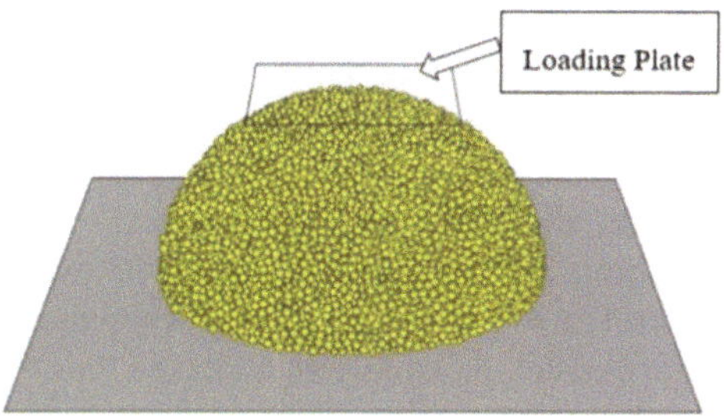

Fig. 7. Model for ball-surface contact test

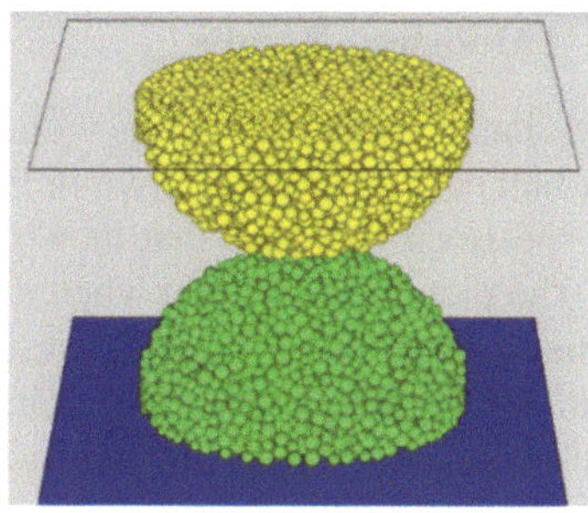

Fig. 8. Model for ball-ball contact test

Previous research indicates that, parameters calibrated through uniaxial compression test are unsuitable for numerical simulation. The parameter calibration process of contact test simulation is as follows.

(1) To calibrate E_c and $\bar{E}_c$, we take the secant modulus (average) of contact test as the target value. We fix the material's strength to 60 Mpa that higher than its compressive strength at first, and then adjust E_c and $\bar{E}_c$ to make the secant modulus of the numerical model similar to that of contact test's result.
(2) To calibrate k_n/k_s and $\bar{k}_n/\bar{k}_s$, we take the Poisson's ratio corresponding to the secant modulus (average) of contact test as the target value.
(3) To calibrate $\bar{\sigma}_c$ and $\bar{\tau}_c$, we take the value of compressive strength of specimen as target value.

Parameter debugging methods are referred to the PFC official handbook. The strength reduction parameters introduced in this model are obtained in actual simulation and calibrate with reference to the crushing process and morphology. The value of isotropic stress is according to the experience of rock material simulation.

The parameters for the contact model are shown in Table 2

Table 2. Modeling parameters of ball-surface contact simulation

Parameters	Value
E_c/Pa	0.5e9
$\bar{E}_c/Pa$	0.95e9
k_n/k_s	4.78
$\bar{k}_n/\bar{k}_s$	4.78
$\bar{\sigma}_c/Pa$	13e6
$\bar{\tau}_c/Pa$	13e6
Average normal strength after reduction/*Pa*	10e6
Average tangential strength after reduction/*Pa*	10e6
Functional range of strength reduction parameter(multiple of particle diameter)	1.5
Value of isotropic stress/*N*	−5e3

3.2 Modeling Results

The shape changing along with the increasing of loading under ball-surface and ball-ball conditions are shown in Figs. 9 and 10 respectively (top view). The red part of the specimen indicates that in which place the local breakage occurs.

The picture shows that when the loading start, the top of the hemispherical model is partially broken due to the stress concentration. And the fractured region expands continuously with the loading going on. The broken area begin to join up when the load reaches a certain value, and the fractured surface develops rapidly as well (see Figs. 9(c) and 10(c)). Finally, the breaking surface has been connect into one piece and overall breakage take place (see Figs. 9(d) and 10(d)). The morphology after breakage is basically consistent with the contact test.

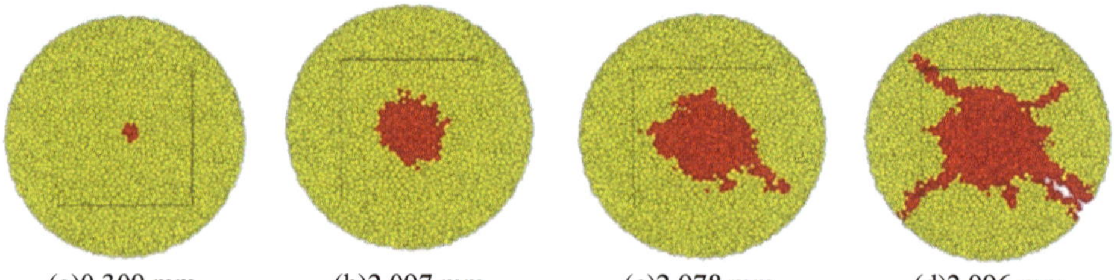

(a)0.309 mm (b)2.097 mm (c)2.978 mm (d)2.996 mm

Fig. 9. The breaking process of ball-surface contact simulation

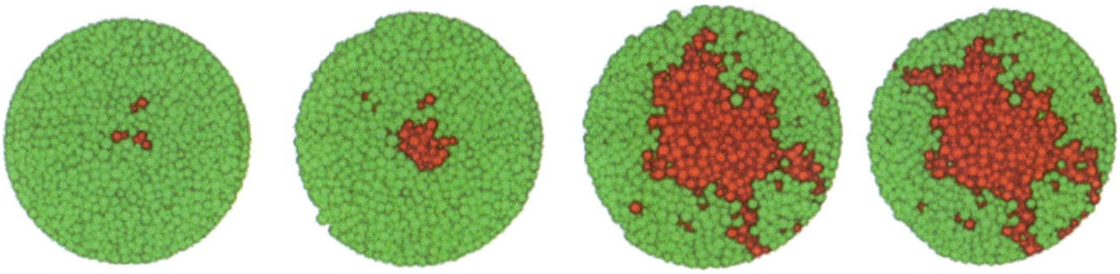

(a)0.180 mm (b)1.200 mm (c)2.760 mm (d)2.880 mm

Fig. 10. The breaking process of ball-ball contact simulation

In the vertical direction of the breaking surface, the development of the internal crushing area can be observed by capturing the cross section of the contact surface (see Figs. 11 and 12). It shows that the crushing area is developing in both horizontal and vertical directions with the loading going on. The shape of the crushing area is basically unchanged but the range of which is enlarged. The front edge of the local fragmentation area gradually formed the crushing surface, which is almost the same time with that in the top view. Then the crushing surface is rapidly developed to the bottom of the model, and the overall breakage happened to the specimen.

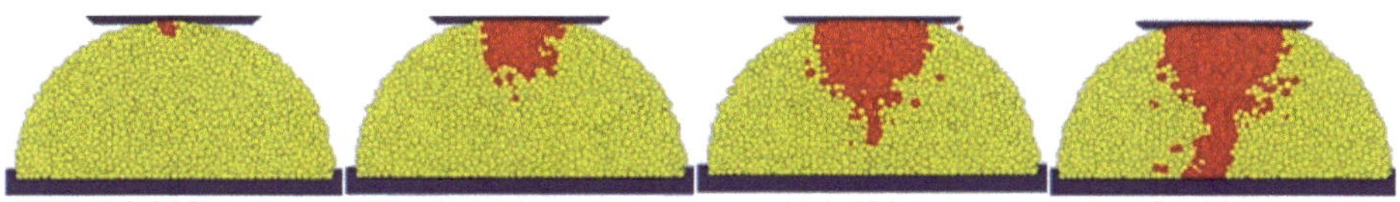

(a)0.309 mm (b)2.097 mm (c)2.789 mm (d)2.990 mm

Fig. 11. The breaking area's development of ball-surface contact simulation

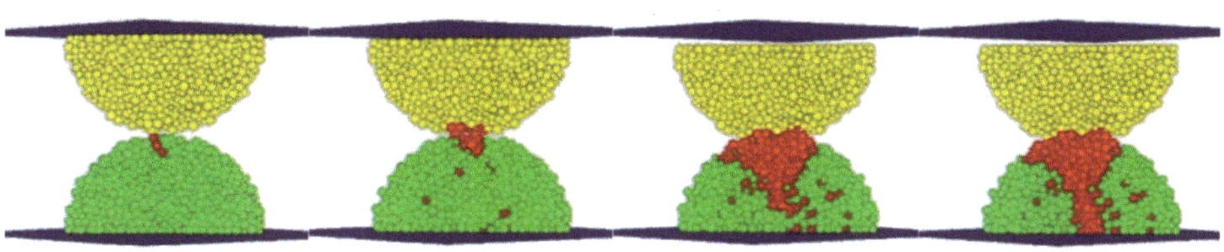

(a)0.180 mm (b)1.200 mm (c)2.760 mm (d)2.880 mm

Fig. 12. The breaking area's development of ball-ball contact simulation

The displacement of the internal particles can reflect the development trend of local breakage in different regions of the specimen. Figures 13 and 14 are displacement vector diagrams of the layer of particles perpendicular to the breaking surface, the direction of the arrow represents the direction of displacement, and the length of the arrow represents the size of displacement.

It can be found that the local breakage area is cone-shaped and develop gradually. The displacement vector between particles in the cone area is basically the same, that pushes the particles in the front move downward. After the cone area develops to a

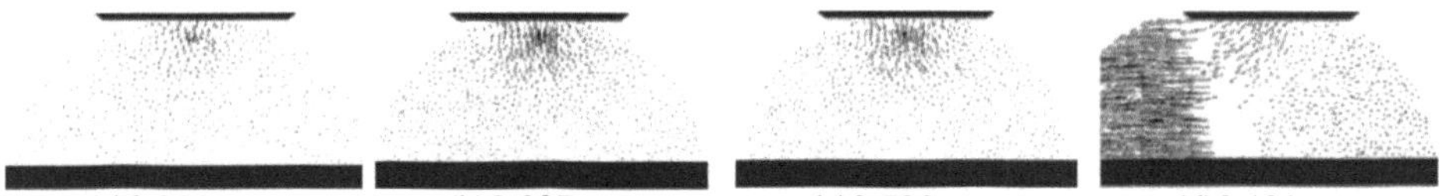

(a)0.309 mm (b)2.097 mm (c)2.789 mm (d)3.120 mm

Fig. 13. The diagram of particle displacement for ball-surface contact simulation

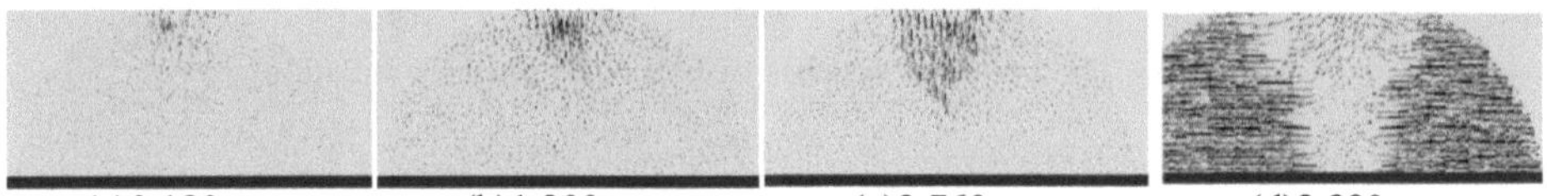

(a)0.180 mm (b)1.200 mm (c)2.760 mm (d)2.880 mm

Fig. 14. The diagram of particle displacement for ball-ball contact simulation

certain extent, it squeezes the surrounding particles to the sides and causes the specimen to crush.

The force-displacement curve obtained by numerical simulation of ball-surface contact test is linearly changed before overall breakage. The slope of first stage is 11.962 *(kN/mm)*. And the model crush under the contact force of 33.78 *kN* (see Fig. 15)

And for the ball-ball contact simulation, there is a big fluctuation in the curve before the overall breakage occurs (see Fig. 16). By observing the loading process when simulating, we find that the fluctuation of the curve is caused by the whole dislocation of the specimen. The contact position is not a smooth surface because of the limitation of modeling method. With the loading going on, the ball-ball model has a certain dislocation to overcome the friction.

The fact that the numerical particles cannot bond with the wall result in the difference between force-displacement curve and the calibration result. Because the boundary conditions in simulation are different from those required by symmetric

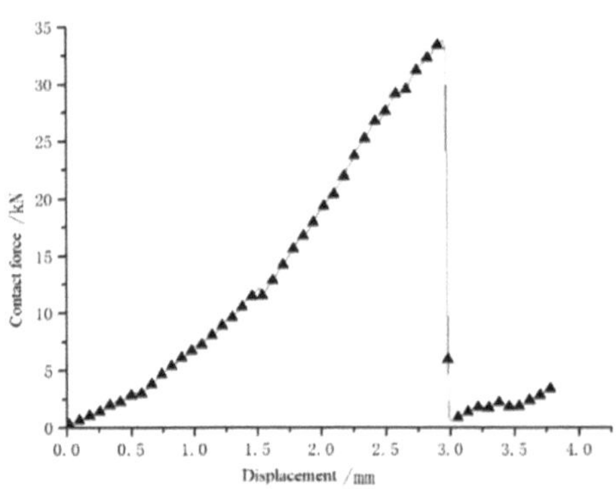

Fig. 15. The force-displacement curve of ball-surface contact simulation

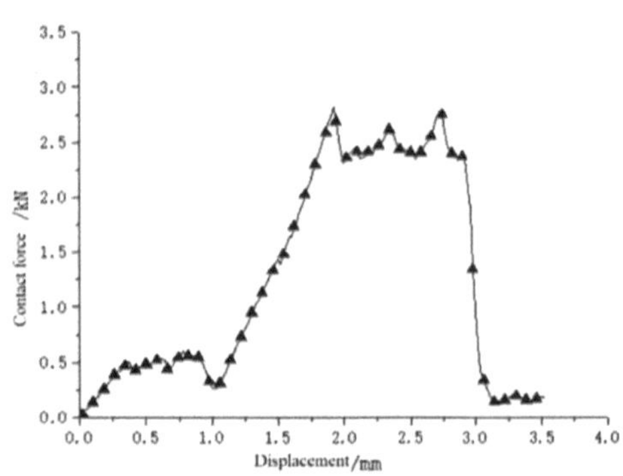

Fig. 16. The force-displacement curve of ball-ball contact simulation

modeling. But the form and development law of crushing are basically consistent with the experimental results.

4 Comparative Analysis

In the numerical simulation, the cone-shape crushing zone is similar to the elastic core in the results of experiment in shape and development trend (see Fig. 17).

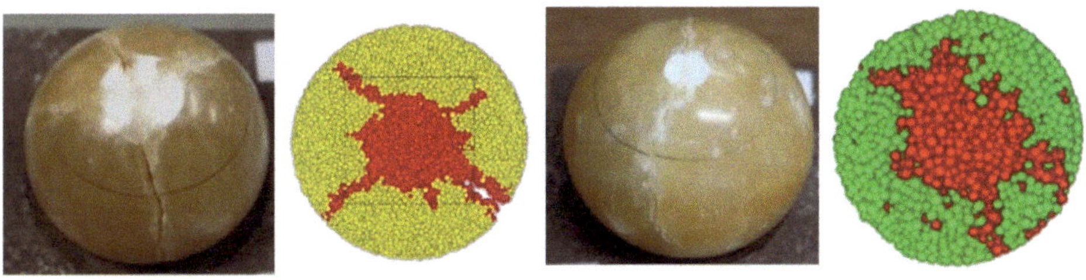

(a)ball-surface contact test (b) ball-ball contact test

Fig. 17. The comparison of shape between experiment and simulation after breakage

As shown in Fig. 17 the contact area of numerical simulation is much larger than that of the laboratory test when the overall breakage occurs. The possible reason might be that the overall breakage only take place at a large displacement in numerical simulation, which is clearly reflected in the contact force-displacement curve. Comparing the results shown in Fig. 15 with that shown in Fig. 6(a), we can find that the slope in laboratory test is about twice as much as that in numerical simulation. Besides, the contact force and displacement are different between experiment and numerical simulation on crushing.

In addition to the differences in boundary conditions, another reason is that the PFC^{3D} numerical model is composed of small particles that need for overlap between them to transmit the force, which will make the sample perform like granular mixtures. However, the strength reduction parameters introduced above only take actions when local breakage takes place. Thus, the overall breakage occurs at a large displacement.

Though the result carried out by numerical simulation with strength reduction parameters cannot be used in quantitative analysis, the morphology after breakage is similar to the laboratory test. The numerical method in this study can be used to analyze the process and mechanism of crushing development. Through numerical simulation, we can study the formation, development process and meso-mechanism of local breakage, as well as the overall breakage of particles. The result of numerical simulation is a good supplement to laboratory experiment.

5 Conclusions

By employing Rock Rheological Testing System, the particle contact experiments are carried out along with the numerical simulation using PFC^{3D} to study the marble particles' breakage process, morphology after breakage. The main results are as follows.

(1) In marble particles' contact test, the local breakage is occurred firstly in the contact area and then the Elastic Core is formed under normal contact force. The existing of the Elastic Core changes the path of force transfer, and results in overall breakage of the particle after crack coalescence
(2) This study introduce a strength reduction parameter in numerical model in consideration of the stress concentration's influence on the strength of the adjacent area in the crushing zone, which can simulate the overall breakage phenomenon of the particles quite well.
(3) The result of numerical simulation is a good supplement to laboratory experiment, which further clarified the meso-mechanism of the contact breakage process of marble particles.
(4) This study only involves marble particles with particle diameter of 5 cm. Further research should be carried out considering the influence of different size, shape and material.

References

1. Zhao, G., Zhou, G., Zhu, F.: Experimental research on the influence of particle crushing on direct shear strength of sand. J. Chin. Univ. Min. Technol. **37**(3), 291–294 (2008)
2. Terzaghi, K., Peck, R.B.: Soil Mechanics in Engineering Practice. Wiley, New York (1948)
3. Lee, K.L., Farhoomand, I.: Compressibility and crushing of granular soil in anisotropic triaxial. Can. Geotech. J. **4**(1), 68–86 (2011)
4. Miura, N., Yamamoto, T.: Particle-crushing properties of sands under high stresses. Technology Reports of the Yamaguchi University, vol. 1, pp. 439–447 (1976)
5. Hardin, B.O.: Crushing of soil particles. J. Geotech. Eng. **111**(10), 1177–1192 (1985)
6. Potapov, A.V., Campbell, C.S.: Parametric dependence of particle breakage mechanisms. Powder Technol. **120**(3), 164–174 (2001)
7. Wang, Q.: A Study on Contact Behaviors and Particle Breakage of Coarse-grained Soils. Tongji University, Shanghai (2014)
8. Wang, Z.: A study on contact behaviors and microscopic constitutive relations of coarse-grained soils. Tongji University, Shanghai (2013)
9. Jiang, M., Li, X.: DEM simulation of the shear band of sands in biaxial test. J. Shandong Univ.(Eng. Sci.) **40**(2), 52–58 (2010)
10. Cole, D.M., Peters, J.F.: A physically based approach to granular media mechanics: grain-scale experiments, initial results and implications to numerical modeling. Granular Matter **9**(5), 309–321 (2007)
11. Yin, X., Ge, X., Li, C., et al.: Influences of loading rates on mechanical behaviors of rock materials. Chin. J. Rock Mech. Eng. **29**(Supp. 1), 2610–2615 (2010)

12. Zhou, Y.: Fracture characteristics of coarse spherical particles under normal contact condition. Tongji University, Shanghai (2017)
13. Zhang, J.: A Preliminary Study on Normal Contact Behaviors and Particle Breakage of Marble Samples. Tongji University, Shanghai (2016)

Experiment on the Failure Process of Jointed Rock Specimen Under Compression and Numerical Test

Hao Zhou[1,2](✉), Yang Song[1], Qimin Wang[1], Weishen Zhu[2], and Yanqing Men[1]

[1] Jinan Rail Transit Group Co., Ltd., No. 2000 Shunhua Rd., Jinan 250101, Shandong, People's Republic of China
haozhouchina@163.com
[2] Geotechnical and Structural Engineering Research Center, Shandong University, No. 17923 Jingshi Rd., Jinan 250061 Shandong, People's Republic of China

Abstract. Numerous fissures, joints and faults exist widely in natural rock masses which played an important role in the geotechnical engineering. Experimental and numerical methods are very important to the researches of jointed rock mass. The improvement measures of Discontinuous Deformation Analysis numerical method make it more suitable for the failure process of jointed rock mass. A rock specimen was tested under compression by experimental method and numerical method. The numerical test of failure process that improved Discontinuous Deformation Analysis method matches well with the experiment of rock specimen. The process proved that the improved Discontinuous Deformation Analysis method could effectively simulate the failure process of jointed rock mass. Additionally, detailed initiation and propagation process of secondary cracks were shown in the research. The initiation and propagation of secondary wing cracks were significant behaviors during the failure process of jointed rock mass. Detailed crack extending process had been discussed deeply and it could be useful for the theoretical research.

Keywords: Rock specimen · Experimental test · Numerical method

1 Introduction

Numerous fissures, joints and faults exist widely in natural rock masses which played an important role in geotechnical engineering. Fissures, joints and faults could initiate to propagate facing compressions during the excavation of engineering. As a result, the surrounding rock in tunnels or hydropower stations could be damaged by the propagation of secondary cracks. Griffith [1] established the brittle fracture mechanics in 1920. In 1960s, Cook [2] tested many specimens with precast cracks. Sih [3] created S principle to explain the propagation of cracks. Primeval researches on jointed rock mass lacked efficient research methods which make it difficult to explore the detailed failure process of rock mass.

L. Zhang et al. (Eds.): GSIC 2018, *Proceedings of GeoShanghai 2018 International Conference: Rock Mechanics and Rock Engineering*, pp. 158–165, 2018.
https://doi.org/10.1007/978-981-13-0113-1_18

Experimental and numerical methods are very important to the researches of jointed rock mass. Those methods simplify the complex natural rock mass for further explorations. More and more researches were made by experimental and numerical methods for the explorations of failure process. Gay [4] compared the results of specimens in experiments with the practical engineering. Rossi [5] proposed a statistic numerical method based on finite element. Huang [6] did some researches on the problem of jointed rocked mass by numerical simulation. Zhu [7] illustrated the experimental failure process of fissured cracks. Cracks which widely exist in natural rock mass could obviously influence the mechanical characteristics, but the analysis based on numerical simulation is still quite few, or not enough in depth.

Discontinuous Deformation Analysis method was a new numerical simulation method created by Shi [8]. Joints and cracks which widely exist in rock mass could not be separated completely by other existing numerical methods. To solve the problem, the numerical method of Discontinuous Deformation Analysis was improved by Jiao [9] who perfected this method. Jiao came up with Discontinuous Deformation Analysis for rock failure method to solve the failure problems in jointed rock mass. This method suited the researches of rock mass fracture criterion generation plane of fracture, especially for the failure process of jointed rock mass.

2 Experiment on Failure Process

The experimental method was a kind of traditional research method which could be extensively used in geotechnical engineering. Various variables of rock mass could be effectively controlled by the experimental method. Although rock masses are widely exist in nature, the aeolotropism of natural rock mass makes it very hard to test in laboratory. Numbers of rocklike materials were used to construct rock specimen for better test results. A kind of hybrid material containing cement mortar was selected to simulate jointed rock mass in the following experiment. The failure process of specimen was recorded based on the results of experiments.

A kind of rocklike material which adopts special formulas was concluded after a series of tests to simulated rock mass. The rocklike material mainly contains Portland cement, quartz sand and admixtures. The mix proportion was 1:0.54:3.5:0.01. Specimen making process included stirring, concreting and other requisite steps. The mould of specimen was removed after 24 h. Then the specimen was cured in the thermostatic water tank for 5 days. After that, the specimen was transferred in drying oven for 2 days and then the specimen needed further requisite curing steps. The size of specimen was decided by a special iron mould which was designed especially for the specimen. The length, width and height of specimen were all 100 mm. 2D crack was constructed by polyvinyl chloride film with the width of 1 mm during specimen making process. The angle of the 2D preset crack was 45. The key physical and mechanical parameters of specimen were similar to natural rock mass which was able to simulate the failure process of jointed rock mass. The elasticity modulus of specimen was 10 GPa and the strength of extension was 2.01 MPa.

Uniaxial compression was pressured upon the specimen in the experiment. The loading rate of the uniaxial compression test was 0.002 mm/s. The rough failure

Fig. 1. Experiment on failure process of rock specimen

process could be distinctly revealed in experiments the initiation and propagation of secondary cracks will be detailed described in the following content. At the beginning of the test, secondary cracks did not exist with the increasing uniaxial compression. Wing cracks subsequently initiated from the top and bottom of preset crack almost simultaneously. Compared with the preset crack, the direction of initiation cracks on the endpoints of preset crack initiated almost vertically. Secondary wing cracks continue processing as the increasing of uniaxial compression. The propagation of wing cracks' trended to upright directions and extended toward the edges of specimen. The curve of propagation tracks warped was shown in Fig. 1. Later in the failure process, few tiny secondary cracks initiated which could be revealed on the surface of specimen. Those tiny secondary cracks increased gradually as time went on. The combination of secondary wing cracks and preset crack developed rapidly in the last stage of failure process and finally formed a splitting crack crossing though the specimen. The specimen was destroyed by the splitting crack when the uniaxial compression reached 19.3 GPa.

3 The Improved Discontinuous Deformation Analysis Method

A large number of numerical methods had been adopted to analyze the discontinuous planes since Goodman [10] invented a classic numerical method for analyzing discontinuous planes in 1968. Abundant research achievements had been approached by numerical methods. However, the unsolved problem of contiguous bodies and discontinuous deformations puzzled all the attempts to establish a perfect numerical model. Shi [8] created Discontinuous Deformation Analysis method to solve the problem by using separated block systems. The numerical method was based on Newton's Second Law and Minimum Potential Energy Principle.

3.1 Automatic Generation of Random Joints and Triangular Block

The length, dip angle, tendency and many other physical parameters and attributes of cracks determined the characters of jointed rock mass as some kind of macroscopically statistic law in nature. It is very important to select one statistic law in numerical method. Las-Vegas method [11] and Monte-Carlo [12] method take an important position in statistics. Monte-Carlo method is used in the automatic generation of random joints to improve the Discontinuous Deformation Analysis method for achieving good effects on rock mass. Numbers of random number peek methods could be applied to generating random numbers. The improved Discontinuous Deformation Analysis method introduces Linear congruential method to peek random numbers in the section of [0, 1]. The original formula is shown in formula (1).

$$\begin{cases} x_n = (ax_{n-1} + c)(modM) \\ r_n = \frac{x_n}{M}, n \in N \\ x_0 \end{cases} \tag{1}$$

where r_n denotes the generated random number in the section of [0, 1], a denotes multiplier, c denotes increment, *(mod M)* denotes the remainder of M, and x_0 denotes the original value.

The details of random cracks could be determined by the center situation, length and direction angle of cracks in 2D numerical simulation. The distributions of center situation follow the uniform distribution or Poisson distribution according to the previous researches. The uniform distribution was used in the improved measure by the following formula (2).

$$\xi_i = (b - a)r_i + a \tag{2}$$

where r_n denotes the generated random number in the section of [0, 1], and ζ_i denotes the generated random number in the section of [a, b] following uniform distribution.

The distributions of length and direction angle of cracks follow the Gaussian distribution according to the previous researches. Gaussian distribution was used in the improved measure by the following formula (3).

$$\eta_i = \sigma \sum_{j=1}^{6} (r_{12\times(i-1)+2j} - r_{12\times(i-1)+2j-1}) + \mu \tag{3}$$

where r_n denotes the generated random number in the section of [0, 1], η_i denotes the generated random number in the section of [a, b] following Gaussian distribution, σ denotes the standard deviation of Gaussian distribution, and μ denotes the mathematical expectation of Gaussian distribution.

Triangular blocks are separated by traveling wave methods for constructing meshes in the calculation area. Meanwhile, virtual joints are also generated by the separation of triangular block.

3.2 The Inhomogeneity of Material and Fracture Algorithm of Boundary

The natural rock mass represents special characters of inhomogeneity which is different from many numerical tests. Weibull distribution was used to simulated rock material which could be illuminated by the following formula (4):

$$\Phi(h, \beta) = 1 - e^{-\left(\frac{\beta}{\beta_0}\right)^k} \tag{4}$$

where h denotes the homogeneity coefficient, β denotes the mechanical parameter, and β_0 denotes the average of the mechanical parameter.

Virtual joints are generated by the separation of triangular block. The adhesive algorithm is used to connect those virtual joints. The algorithm could be determined by the indicator of rock strength. The principle could be illuminated by the formula (5):

$$\begin{cases} f_n = -T_0 \cdot l \\ f_\tau = c \cdot l + f_n tan\, \varphi \end{cases} \tag{5}$$

where f_n denotes normal contact force, f_τ denotes tangential contact force, c denotes the cohesion, φ denotes internal friction angle, and l denotes the length of contact.

4 Numerical Modeling Test

The numerical method of improved Discontinuous Deformation Analysis for rock failure would be good at simulating the failure process of crack's propagation according to the aforementioned content. The separated blocks could obviously display the initiation, propagation and combinations of the secondary cracks under additional loads. A numerical model which simulated the aforementioned rock specimen would be constructed by the Discontinuous Deformation Analysis method. Further researches were explored in the following content.

The numerical model was constructed by improved Discontinuous Deformation Analysis method as shown in Fig. 2. The length and width of numerical model were both 100 mm. The preset crack was settled in the center of the model. The angle of preset crack was also 45 as the aforementioned rock specimen in the experiment. The basic physical and mechanical parameters were the same with the rock specimen before. Triangular blocks were produced automatically by the former numerical improvement. Seven monitoring points were settled on the top and bottom edges to record the displacement and strength during the failure process. Uniaxial loading situation was also similar with the specimen in experiment.

Detailed failure process of cracks' initiation and propagation could be described in the following content. The secondary wing cracks initiated from the top and bottom of the preset crack in the second stage after the first stage which have no secondary crack initiated. The wing cracks' initiated angle was close to 90 compared with the preset crack. Secondary wing cracks on the both sides of preset cracks extended to the top and bottom edges as the increasing of uniaxial compression. The track of the wing crack curved to vertical direction during its extending process. The secondary crack hardly

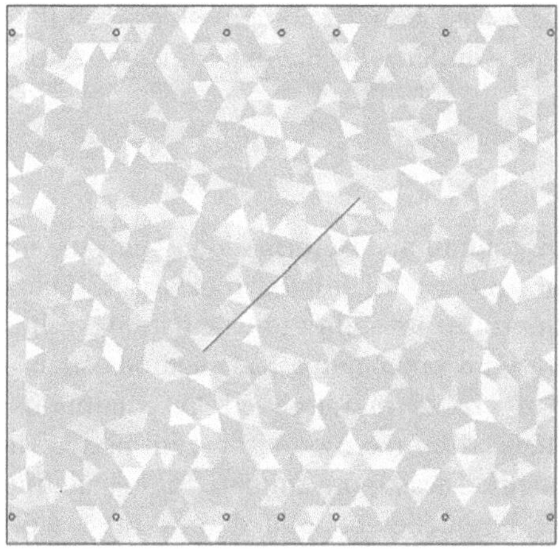

Fig. 2. Construction of rock specimen model in numerical test

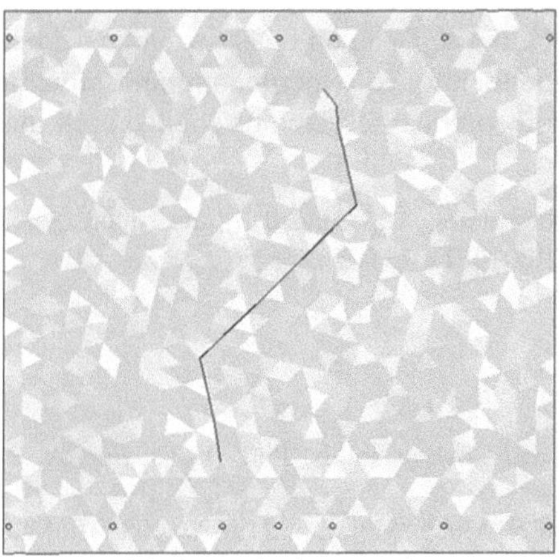

(a) Initiation of wing cracks

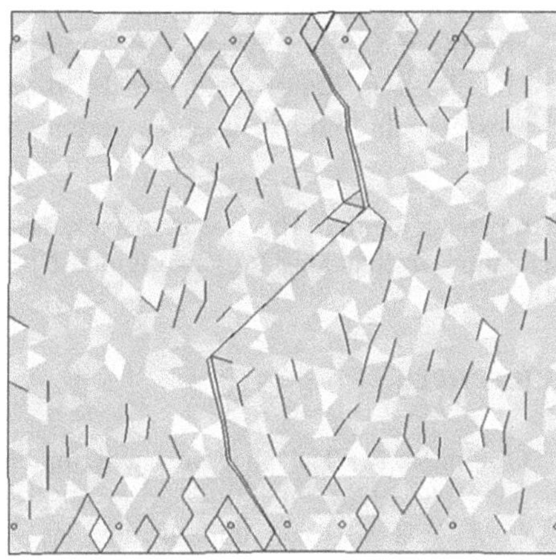

(b) Propagation of secondary cracks

Fig. 3. Detailed failure process in numerical test

initiated except the wing cracks in this stage. The failure process of this stage is shown in Fig. 3(a).

Comparing the aforementioned stage, the root of secondary wing crack became widen besides the increasing of length which indicated stronger strength and larger deformation. Secondary wing cracks continue extending to the edges of specimen with its track curved to vertical direction. More and more micro cracks existed besides the development of secondary wing cracks, especially around the progressive directions of wing cracks as shown in Fig. 3(b). Adjacent secondary cracks combined together and finally constructed a huge separated crack in the center of the specimen along with the direction of uniaxial compression direction. The numerical specimen was damaged by the huge separated crack at last with the compressive strength of the numerical test was 24.2 GPa which also was similar with the experimental test.

5 Conclusions

(1) The improved Discontinuous Deformation Analysis method could effectively simulate the failure process of jointed rock mass. The numerical improvement measure greatly improved its ability to calculate the initiation and propagation of cracks. Comparing with other numerical methods such as XFEM, the initiation and propagation of cracks in improved Discontinuous Deformation Analysis method were influenced by the motion of rigid body. The Discontinuous Deformation Analysis method was more suitable for calculating failure process of rock mass. Microscopic researches of jointed rock mass would be promoted by the improved Discontinuous Deformation Analysis method. However, the Discontinuous Deformation Analysis method was immaturity which could only simulated 2D specimen.

(2) The initiation and propagation of secondary wing cracks are significant behaviors during the failure process of jointed rock mass. Detailed crack extending process has been in depth discussed by the aforementioned experimental and numerical test which could be useful reference for theoretical research on jointed rock mass.

Acknowledgements. The research work described herein was funded by the Nature Science Foundation of Shandong Province (Grant No. ZR2016EEQ13).

References

1. Griffith, A.A.: The phenomena of rupture and flow in solids. Phil. Trans. Ser. A. **221**, 163–198 (1920)
2. Cook, N.G.W.: The failure of rock. Int. J. Rock Mech. Min. Sci. **21**, 389–403 (1965)
3. Sih, G.C.: Energy-density concept in fracture mechanics. Eng. Fract. Mech. **5**, 1037–1040 (1973)
4. Gay, N.C.: Fracture growth around openings in large blocks of rock subjected to uniaxial and biaxial compression. Int. J. Rock Mech. Min. Sci. Geomech. Abstr. **13**, 231–243 (1976)
5. Rossi, P., Richer, S.: Numerical modelling of concrete cracking based on stochastic approach. Mater. Struct. **20**, 334–337 (1987)
6. Li, L.Y., Huang, K.Z., Han, Z.C., et al.: Experimental and theoretical analyses of 3D surface crack propagation. Chin. J. Rock Mech. Eng. **31**, 311–318 (2012)
7. Zhu, W.S., Tang, C.A.: Micromechanical model for simulating the fracture process of rock. Rock Mech. Rock Eng. **37**, 25–56 (2004)
8. Shi, G.H.: Producing joint polygons, cutting joint blocks and finding key blocks for general free surfaces. Chin. J. Rock Mech. Eng. **25**, 2161–2170 (2006)
9. Jiao, Y.Y., Zhang, X.L., Liu, Q.S., et al.: Simulation of rock crack propagation using discontinuous deformation analysis. Chin. J. Rock Mech. Eng. **26**, 682–691 (2007)
10. Goodman, R.E., Taylor, R.L., Brekke, T.L.: A model for the mechanics of joint rock. J. Soil Mech. Found. Div. ASCE **94**(3), 637–659 (1968)

11. Zavalnij, B: Speeding up parallel combinatorial optimization algorithms with Las Vegas method. In: International Conference on Large-scale Scientific Computing, pp. 258–266 (2015)
12. Chen, Y., Li, S., Zhu, W., et al.: New method of DDARF network simulation and its application on highway tunnel. J. Cent. South Univ. **44**, 2494–2499 (2013)

Experimental Investigations of Rock Dynamical Characteristics Under Cyclic Loading

Xiao-zhang Lei[1,2], Jian-feng Liu[1,2(✉)], Lu Zheng[1], Lu Wang[1,3], and Hui-ning Xu[1,2]

[1] State Key Laboratory of Hydraulics and Mountain River Engineering, Sichuan University, Chengdu 610065, China
ljfscu@163.com
[2] Key Laboratory of Energy Engineering, Safety and Mechanics on Disasters, Sichuan University, Ministry of Education, Chengdu 610065, China
[3] School of Civil Engineering and Environment, Xihua University, Chengdu 610039, China

Abstract. Rock dynamic characteristics are important factors influencing the long-term stability of rock masses. In addition, they are important parameters for seismic response analysis and safety evaluation. In this study, a solution was discussed for the evaluation of the damping coefficient of rock masses under cyclic loading. Then, low cyclic loading tests with a frequency of 3 Hz were carried out on sandstone samples. Through these tests, the influences of the amplitude stress and cycle number on the dynamic elastic modulus and Poisson's ratio, damping ratio and coefficient, and their relationships were evaluated. Under the cyclic loading condition, the dynamic elastic modulus and Poisson's ratio increased parabolically and linearly with the increase in the amplitude stress, respectively; however, the damping ratio and coefficient decreased in accord with a power function. The damping ratio and coefficient increased linearly with the increase in the cycle number, and their increments for low-amplitude stresses were larger than those at high stresses. The results suggest that the damping parameters of rock masses could be obtained from the hysteresis loop measured from the intermediate cycle during a 31-cycle cyclic loading test.

Keywords: Cyclic loading test · Dynamic elastic modulus and poisson's ratio Damping ratio · Damping coefficient

1 Introduction

The mechanical characteristics of rock and soil subjected to cyclic loading are important factors influencing their long-term stability (Ge 1987). The dynamic elastic modulus and damping ratio of rock and soil masses are the key factors for the evaluation of their stress-strain characteristics under seismic excitation. The dynamic characteristics of soil masses have been extensively studied (He 1995; Worthington et al. 2001) and abundant outcomes related to the strength, deformation characteristics,

L. Zhang et al. (Eds.): GSIC 2018, *Proceedings of GeoShanghai 2018 International Conference: Rock Mechanics and Rock Engineering*, pp. 166–175, 2018.
https://doi.org/10.1007/978-981-13-0113-1_19

fatigue damage and dynamic response subjected to cyclic loading have been achieved (Xi et al. 2002; Jafari et al. 2004; Bagde and Petroš 2005; Yang et al. 2007; Nie et al. 2011; Liu et al. 2014; Ni 2014; Fathi et al. 2016). However, compared with soil dynamics, the dynamic parameters of rocks remain to be further investigated.

Nishi et al. (1983) and Chen et al. (2004) obtained relationships between the rock dynamic shear modulus, damping ratio, confining pressure and static mechanical parameters. The irreversible deformation and fatigue threshold of rocks subjected to cyclic loading were systematically studied by (Ge 1987; Ge and Lu 1992; Ge et al. 2003). Ge summarized the observed bedrock deformation data and provided a relationship between the irreversible deformation and cycle number. According to the loading-unloading tests of red sandstones and marbles, Mo (1988) found that the relationship between the fatigue life and loading speed, frequency, and stress amplitude under dynamic cyclic loading would be expressed as the S-N curve. The dynamic parameters of rocks were studied by cyclic loading tests (Nie et al. 2011). They continued to be widely used thereafter (Zhu et al. 2009; Xiao et al. 2010; Zhu et al. 2010; Jiang et al. 2011). However, few studies have addressed the influences of the amplitude stress and cycle number on the dynamic elastic modulus and Poisson's ratio, damping ratio and coefficient, and their relationships.

At present, the measurement of rock damping ratio is mainly based on the method of damping ratio in soil mechanics, which is obtained by low-cycle loading. However, there is no appropriate method to measure the damping coefficient of rock. In this study, the analysis methods and theoretical formulae of rock damping parameters were first analyzed based on the structural dynamics to provide a theoretical basis for the tests. The sandstone samples were then tested using cyclic loading tests to discuss the rationality of selecting the data from the intermediate cycle to obtain the damping parameters of the rocks. Based on the study, the influences of the amplitude stress and cycle number on the dynamic elastic modulus and Poisson's ratio were analyzed; the relationships between the amplitude stress, cycle number and the dynamic elastic modulus and between Poisson's ratio, the damping ratio and the coefficient were investigated.

2 Experimental Principle

For a linear single degree of freedom system with a damping force F_d under a sinusoidal compelling force $F(t) = F_0 \sin(\omega t)$ at $t = 0$, its differential equation of motion could be expressed as

$$m\ddot{x} + F_d + kx = F_0 \sin(\omega t) \quad (1)$$

where m is the mass of the rocks, k is the stiffness, and x is the vibration displacement.

The viscous damping is related to the strain rate or deformation rate of the rocks. If it is assumed to be proportional to the motion velocity, the damping force $F_d = C\dot{x}$. C is the damping coefficient. Thus, Eq. (1) could be expressed as

$$m\ddot{x} + C\dot{x} + kx = F_0 \sin(\omega t) \tag{2}$$

Let $\omega_0^2 = \frac{k}{m}$, $\lambda = \frac{C}{2\omega_0 m}$, $f(t) = \frac{F(t)}{m}$, ω_0 is the natural frequency, λ is the damping ratio, and $f(t)$ is the forced vibration force per unit mass. Equation (2) could be expressed in plural form:

$$\ddot{x} + 2\omega_0\lambda\dot{x} + \omega_0^2 x = f(t)e^{i\omega t} \tag{3}$$

Its particular solution is

$$P^* = \frac{\frac{F_0}{m}e^{i\omega t}}{(i\omega)^2 + 2i\omega\lambda\omega_0 + \omega_0^2} \tag{4}$$

Based on the homogeneous differential equation corresponding to the characteristic equation of Eq. (3) and its two latent roots, Eq. (4) could be transformed into

$$P^* = \frac{\frac{F_0}{k}e^{i\omega t}}{1 - \left(\frac{\omega}{\omega_0}\right)^2 + 2i\lambda\frac{\omega}{\omega_0}} \tag{5}$$

Taking the imaginary part of Eq. (5), the particular solution of Eq. (1) could be obtained as

$$x^* = X \sin(\omega t - \alpha) \tag{6}$$

where X is the amplitude of the vibration response, and α is a constant.

The general solution of the corresponding homogeneous equation to Eq. (1) could be presented as

$$x(t) = A_1 e^{r_1 t} + A_2 e^{r_2 t} \tag{7}$$

where A_1 and A_2 are constants. Then Combining Eqs. (6) and (7), the general solution of Eq. (1) is

$$\bar{x} = A_1 e^{r_1 t} + A_2 e^{r_2 t} + X \sin(\omega t - \alpha) \tag{8}$$

Although Eq. (1) provides a mathematical description of the viscous damping vibration, the accurate calculation of the damping force is very difficult because of the complexity of damping itself. Therefore, equivalent viscous damping is used for an approximate calculation. Whatever the damping is, its essence is to cause the energy dissipation of the vibration system. Thus, it could be assumed that the energy dissipation of the vibration system replaced by an equivalent viscous damping is equal to the energy consumed to overcome the actual damping within a cycle. Supposing that the equivalent viscous damping coefficient is $\tilde{C}$, then the damping force could be expressed as $F_d = \tilde{C}\dot{x}$. For one cycle, the energy dissipation of a system is

$$E_{\mathrm{dis}} = \int_0^T F_{\mathrm{d}} \mathrm{d}x = \int_0^T \tilde{C}\dot{x} \mathrm{d}x \tag{9}$$

Under a sinusoidal compelling force, the steady displacement and velocity could be expressed as

$$x = X \sin(\omega t - \alpha) \tag{10}$$

$$\dot{x} = \omega X \cos(\omega t - \alpha) \tag{11}$$

Thus, the energy dissipation induced by the equivalent viscous damping in a vibration cycle could be expressed as

$$E_{\mathrm{dis}} = \int_0^T \tilde{C}\dot{x} \mathrm{d}x = \int_0^{\frac{2\pi}{\omega}} \tilde{C}X^2\omega^2\cos^2(\omega t - \alpha)\mathrm{d}t = \pi\tilde{C}X^2\omega \tag{12}$$

When the actual damping force is R, the energy dissipation for a vibration cycle is

$$E_R = \int_0^T R\mathrm{d}x = \int_0^T R\dot{x}\mathrm{d}t \tag{13}$$

According to the assumption that the energy dissipation of equivalent viscous damping is equal to the energy consumed to overcome the actual damping, we have

$$\tilde{C} = \frac{E_R}{\pi X^2 \omega} \tag{14}$$

Based on Eq. (14), the damping coefficient of rocks could be expressed by the loading frequency of the forced vibration, the dissipation energy and the corresponding amplitude.

As shown in the stress-strain curve of rocks under cyclic loading, the unloading curve does not follow the loading curve (Fig. 1). The area under the loading curve represents the work of the external load on the rock specimen, and the area under the unloading curve represents the elastic energy released from the rocks. The loading and unloading curves form a hysteresis loop. The area enclosed by this hysteresis loop represents the dissipated energy from the rocks and their damping characteristics. In terms of Eq. (14), the energy E_R consumed by the damping force in a vibration cycle could be described using the area A_R of the corresponding hysteresis loop. The damping ratio λ, the damping coefficient C, the dynamic elastic modulus E_{d} and Poisson's ratio could be determined by the equations (The Professional Standards Compilation Group of People's Republic of China SL237– 1999)

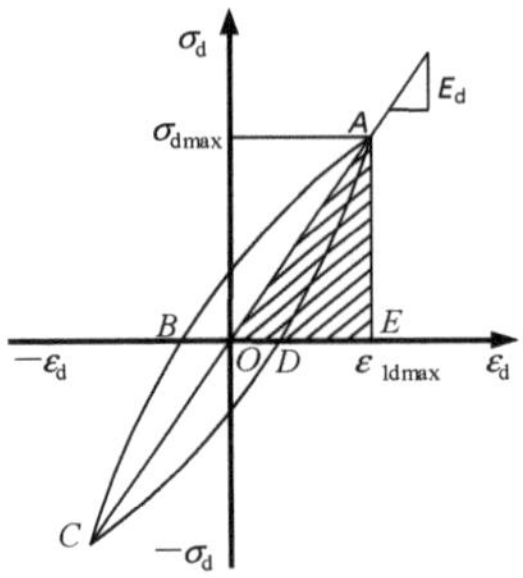

Fig. 1. Hysteresis loop of dynamic strain and stress

$$\lambda = \frac{A_R}{4\pi A_S} \tag{15}$$

$$C = \frac{A_R}{\pi X^2 \omega} \tag{16}$$

$$E_d = \sigma_{dmax} / \varepsilon_{1dmax} \tag{17}$$

$$\mu_d = \Delta\varepsilon_{2d} / \Delta\varepsilon_{1d} \tag{18}$$

where A_R is the area of hysteresis loop *ABCDA*, A_S is the area of triangle *AOE*, *X* is the corresponding axial vibration amplitude to the hysteresis loop, σ_{dmax} is the maximum dynamic stress of the hysteresis loop, ε_{1dmax} is the maximum axial dynamic strain of the axial hysteresis loop, and $\Delta\varepsilon_{1d}$ and $\Delta\varepsilon_{2d}$ are the strain increments corresponding to the loading point of the axial hysteresis loop and the unloading point of the lateral hysteresis loop, respectively.

3 Experimental Test

3.1 Sample Preparation and Test Equipment

The rock sample was sandstone with a natural average density of 2.44 g/cm^3, which was cored from the foundation of an intended site for a nuclear power station at a depth of 51 m from the ground. The main minerals include montmorillonite 12%, illite 10%, quartz 45%, feldspar 24% and others about 9%. The rock cores were immediately sealed to maintain their natural water content in the field. The rock cores were then reprocessed into $\phi 50\,\text{mm} \times 100\,\text{mm}$ cylinders by dry turning according to a standard for engineering rock testing methods in the lab (The Professional Standards Compilation Group of People's Republic of China GB/T 50 266- 2013).

An MTS815 Flex Test GT rock mechanics testing system was used for the cyclic loading test. Its maximum axial load is 4600 kN. The vibration frequency can exceed 5 Hz, and its waveform could be a sine, triangular, square, oblique or random wave. The vibration phase difference could vary from 0 to 2π, and the vibration mode could be controlled by an axial load, confining pressure, or permeability pressure. Combinations of any of the above waveforms, frequencies, phases, and modes are also possible.

3.2 Test Method

The test loading mode was controlled by the axial load and the schematic diagram of the tests was as shown in Fig. 2(a). The loading waveform was a sine wave with a 3Hz frequency. The loading level and its corresponding stress amplitude for each grade dynamic stress were as shown in Fig. 2(b). σ_{max} and σ_{min} are the upper and lower

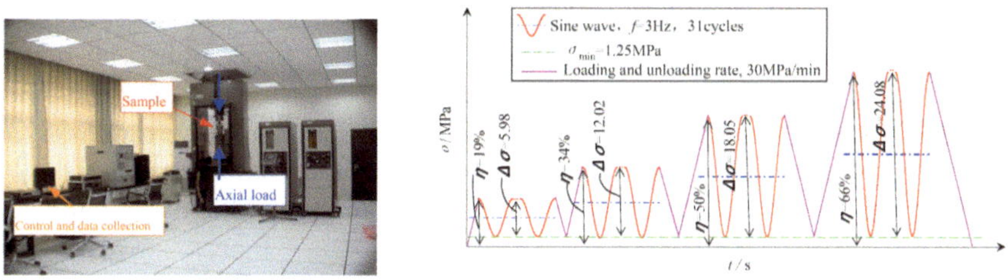

(a) Schematic diagram of tests

(b) Testing process

Fig. 2. Test method

stresses to each grade, respectively, $\Delta\sigma$ is the amplitude of the dynamic stress ($\Delta\sigma = \sigma_{\max} - \sigma_{\min}$), and R is the average uniaxial compressive strength of the testing rock samples. The coefficient $\eta = \sigma_{\max}/R$ is defined as the ratio of the upper stress to the uniaxial compressive strength. The dynamic stress was applied progressively grade by grade. Taking the specification of soil test (The Professional Standards Compilation Group of People's Republic of China SL237– 1999) as a reference, the dynamic stress was applied in 31 cycles during each grade loading, and both the axial and lateral deformations of the samples were measured. The uniaxial compressive strength and the average static elastic modulus of the testing rock were 38.5 MPa and 5.52 GPa, respectively. Hysteresis loops of 4, 8, 12, 16, 20, 24 and 28 cycles for each grade were chosen to analyze the relationships between the rock dynamic parameters, dynamic stress cycle and amplitude.

4 Test Results and Analysis

4.1 Dynamic Elastic Modulus

Under cyclic loading, the dynamic elastic modulus could be denoted as a linear function of the cycle number and a parabolic function of the amplitude stress (Fig. 3). The average slope of the linear function between the dynamic elastic modulus and the cycle number for three samples was 0.0024 ~ 0.0068, and it decreased with the amplitude stress. The comparison showed that within the tested cycle numbers, the dynamic elastic modulus of the sandstone was not affected by the cycle number. The

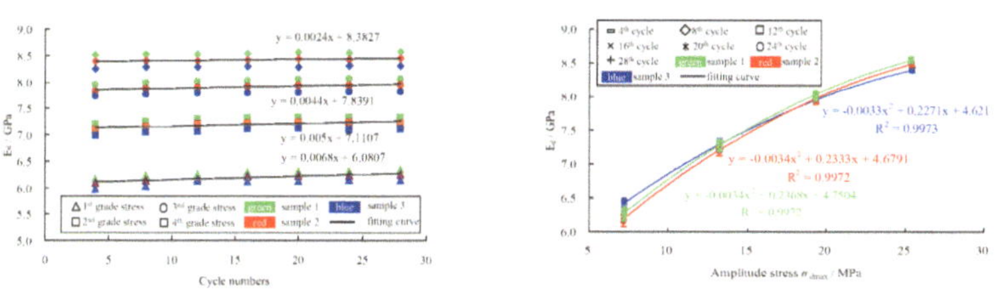

(a) Relationship between E_d and cycle numbers

(b) Relationship between E_d and amplitude stress

Fig. 3. Relationships of cycle number and amplitude stress with the dynamic elastic modulus dynamic elastic modulus of the rocks had a maximum value when subjected to cyclic loading, which increased with the amplitude stress along a parabola with downward opening (Fig. 3(b)). The amplitude stress increased from 7.23 MPa at the 1st grade to 25.33 MPa at the 4th, while the dynamic elastic modulus increased from 5.98 GPa to 8.58 GPa. The results indicate that the influence of the amplitude stress on the dynamic elastic modulus of the fine sandstone was greater than that of the cycle number.

4.2 Dynamic Poisson's Ratio

The larger the amplitude stress was, the larger the relative axial and lateral deformations of the rocks subjected to dynamic loading were (Fig. 4). Under cyclic loading, the average slope of the linear function between the dynamic Poisson's ratio and cycle number for three samples was –0.0002 ~ 0.0002 (Fig. 4(a)). Within the tested cycle numbers, the influence of the cycle number on the dynamic Poisson's ratio could be neglected. The dynamic Poisson's ratio of fine sandstone increased linearly with the amplitude stress (Fig. 4(b)). With the amplitude stress increasing from 7.23 MPa at the 1st grade to 25.33 MPa at the 4th, the dynamic Poisson's ratio increased from 0.10 to 0.29. At a constant loading amplitude stress, the amplitude stress had a large influence on the dynamic Poisson's ratio, while it was not obviously affected by the increase in the cycle number.

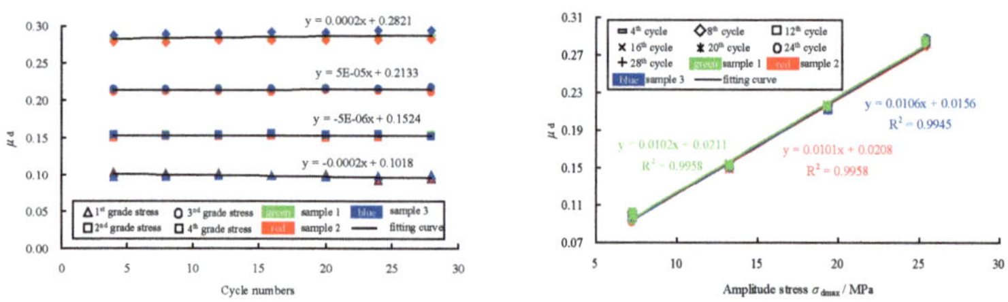

(a) Relationship between μ_d and cycle numbers (b) Relationship between μ_d and amplitude stress

Fig. 4. Relationships of cycle number and amplitude stress with the dynamic Poisson's ratio

4.3 Damping Ratio

The average of damping ratio for three samples could be represented by a linear function in which the slope is 0.0029 ~ 0.0287, and it increased with the cycle number (Fig. 5(a)). The smaller the amplitude stress was, the larger the slope was, which meant that there was a larger influence of the cycle number on the damping ratio. However, when the amplitude stress exceeded a certain value, the influence of the cycle number could be neglected within the tested cycle numbers range. The larger the dynamic stress was, the less the ability of the rocks to absorb and dissipate the external vibration energy was. Unlike the dynamic elastic modulus and Poisson's ratio, the damping ratio could be expressed by a power function of the amplitude stress (Fig. 5(b)) that decreased with the increase in the loading amplitude stress. The damping ratio

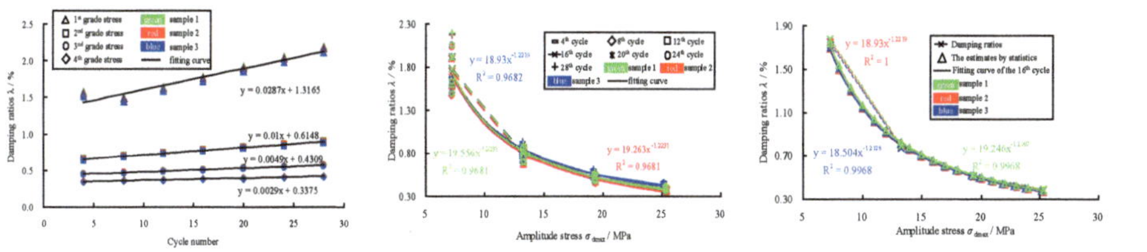

(a) λ and cycle number (b) λ and amplitude stress (c) λ and amplitude stress of the 16th cycle

Fig. 5. Relationships of cycle numbers and amplitude stress with damping ratio

estimated by the statistical expression agreed well with the fitting curve by the power function based on the measured data of the 16th cycle (Fig. 5(c)). Thus, the damping ratio could be obtained based on the hysteresis loop of the intermediate cycle from the examined dynamic stress-strain curves using a 31-cycle low cyclic loading test.

4.4 Damping Coefficient

The average of damping coefficient for three samples could be expressed by a power function of the amplitude stress or a linear function of the cycle number (Fig. 6). The average of damping coefficient increased linearly with the increase of the cycle number (Fig. 6(a)). The smaller the amplitude stress was, the larger the slope was, which means a larger influence of the cyclic number on the damping coefficient. However, when the amplitude stress exceeded a certain value, the influence of the cycle number could be neglected within the tested cycle number range. The damping coefficient decreased by a power function with an increase in the loading amplitude stress (Fig. 6(b)), which was also similar to the behavior of the damping ratio. The damping coefficient estimated by the statistical expression was in good agreement with the fitting curve by the power function based on the data of 16th cycle (Fig. 6(c)). Consequently, the damping coefficient could also be obtained based on the hysteresis loop of the intermediate cycle from the examined dynamic stress-strain curves using a 31-cycle low cyclic loading test. Therefore, the damping coefficient and damping ratio measured using the approach proposed in this paper could be used to describe the damping characteristics of rocks.

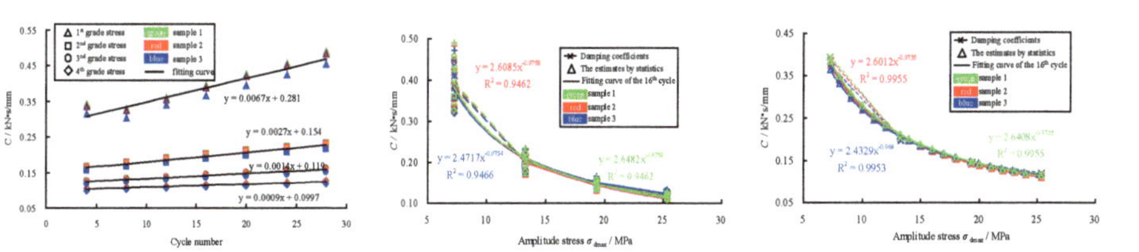

(a) *C* and cycle number (b) *C* and amplitude stress (c) *C* and amplitude stress of the 16th cycle

Fig. 6. Relationships of cycle number and amplitude stress with damping coefficient

5 Conclusions

(1) A solution was derived to evaluate the damping coefficient of rocks under cyclic loading. Then, the deformation and damping characteristics of sandstone were obtained under low cyclic loading. The results suggest that the damping parameters obtained from the dynamic hysteresis loop of the intermediate cycle during a 31-cycle cyclic loading test could provide the damping characteristics of rocks.
(2) The statistical characteristics of both the dynamic elastic modulus and Poisson's ratio increased with the cycle number. However, within the tested cycle number range, the influence of the cycle number could be neglected. The dynamic elastic modulus increased with the amplitude stress along a parabola with a downward opening, while the dynamic Poisson's ratio increased linearly with the amplitude stress.
(3) The damping ratio and coefficient increased with the cycle number linearly, and their increments at low amplitude stresses were larger than those at high stresses, while they decreased in accordance with a power function with the increase of the amplitude stress. The damping ratio and coefficient estimated by the statistical expression were in good agreement with the fitting curve by the power function based on the intermediate cycle data.

Acknowledgments. The authors are grateful for the financial support from the National Natural Science Foundation of China (Grant No. 51374148, 51641405), the Science and Technology Project of Sichuan Province (Grant No. 2017JQ0003). The authors wish to offer their gratitude and regard to the colleagues who contributed to this work.

References

Bagde, M.N., Petroš, V.: Fatigue properties of intact sandstone samples subjected to dynamic uniaxial cyclical loading. Int. J. Rock Mech. Min. Sci. **42**, 237–250 (2005)

Chen, Y.P., Xi, D.Y., Xue, Y.W.: Hysteresis and attenuation of saturated rocks under cyclic loading. Chin. J. Geoph. **47**, 766–774 (2004)

Fathi, A., Moradian, Z., Rivard, P., Ballivy, G.: Shear mechanism of rock joints under pre-peak cyclic loading condition. Int. J. Rock Mech. Min. Sci. **83**, 197–210 (2016)

Ge, X.: Study on deformation and strength behaviour of the large-sized triaxial rock samples under cyclic loading. Rock Soil Mech. **8**, 11–19 (1987)

Ge, X., Jiang, Y., Lu, Y., Ren, J.: Testing study on fatigue deformation law of rock under cyclic loading. Chin. J. Rock Mech. Eng. **22**, 1581–1585 (2003)

Ge, X., Lu, Y.: Discussion on fatigue failure and irreversible deformation of rock under cyclic loading. Chin. J. Rock Mech. Eng. **14**, 56–60 (1992)

He, C.: Modulus and damping under cyclic loading. J. Basic Sci. Eng. **3**, 113–119 (1995)

Jafari, M.K., Pellet, F., Boulon, M., Hosseini, K.A.: Experimental study of mechanical behaviour of rock joints under cyclic loading. Rock Mech. Rock Eng. **37**, 3–23 (2004)

Jiang, X.X., Zhu, Z.D., Shi, C., Hao, Z.Q.: Experimental study on sandstone dynamics of Pengze Nuclear Power Plant. J. Hohai Univ. Nat. Sci. **39**, 78–83 (2011)

Liu, J., Xie, H., Hou, Z.: Dynamic evolution of rock salt under cyclic loading in uniaxial tests. Acta Geotech. **9**, 153–160 (2014)

Nie, M., Xu, J., Ren, H., Liu, J.: Dynamic triaxial experimental research on damping parameters and dynamic elastic parameters for marble. Chin. J. Rock Mech. Eng. **30**, 3989–3994 (2011)

Mo, H.: Study of rock's cyclic experiment and constitutive relation. Chin. J. Rock Mech. Eng. **7**, 215–224 (1988)

Nishi, K., Kokusho, T., Esashi, Y.: Dynamic shear modulus and damping ratio of rocks for a wide confining pressure range. In: Proceedings - 5th Congress of the International Society for Rock Mechanics, pp. 223–226. A.A. Balkema, Rotterdam (1983)

Ni, X.: Quantitative characteristic of meso-damage granite under cyclic loading with different frequencies. Appl. Mech. Mater. **638**, 690–693 (2014)

The Professional Standards Compilation Group of People's Republic of China SL237–1999 Specification of soil test. China Water Power Press, Beijing (1999)

The Professional Standards Compilation Group of People's Republic of China GB/T 50 266-2013 Standard for tests method of engineering rock massas. China Planning Press, Beijing (2013)

Worthington, M.H., King, M.S., Marsden, J.R.: Determining the damping factor of sedimentary rocks required for seismically designed structures. Int. J. Rock Mech. Min. Sci. **38**, 801–806 (2001)

Xi, D., Wang, S., Liu, X.: Nonlinear elastoplastic response of rocks. Chin. J. Rock Mech. Eng. **21**, 772–777 (2002)

Xiao, J., Feng, X., Ding, D., Jiang, F.: Study of hysteresis and damping effects of rock subjected to constant amplitude cyclic loading. Chin. J. Rock Mech. Eng. **29**, 1677–1683 (2010)

Yang, Y., Song, Y., Chu, J.: Experimental study on characteristics of strength and deformation of coal under cyclic loading. Chin. J. Rock Mech. Eng. **26**, 201–205 (2007)

Zhu, M., Zhu, Z., Li, G., Qiu, Y., Chen, Z., Li, J.: Experimental study of dynamic characteristics of granite under cyclic loading. Chin. J. Rock. Mech. Eng. **28**, 2520–2526 (2009)

Zhu, Z.D., Sun, L.Z., Wang, M.Y.: Damping ratio experiment and mesomechanical analysis of deformation failure mechanism on rock under different frequency cyclic loadings. Rock Soil Mech. **31**, 8–12 (2010)

Experimental Study on Elastovisco-Plastic Rheological Properties of Saturated Limestone Under Multi-level Load and Unload Cycles

Yanlin Zhao[1,2(✉)], Liming Tang[1], GuoMing Cheng[3], Qiang Liu[1], and Senlin Xie[1]

[1] Hunan Provincial Key Laboratory of Safe Mining Techniques of Coal Mines, Work Safety Key Lab on Prevention and Control of Gas and Roof Disasters for Southern Coalmines, Hunan University of Science and Technology, Xiangtan 411201, Hunan, China
yanlin_8@163.com
[2] State Key Laboratory of Coal Resources and Safety Mining, China University of Mining and Technology, Xuzhou 221008, China
[3] China Geological Survey, Beijing 100081, China

Abstract. A set of triaxial rheological experiments were performed to obtain elastovisco-plastic properties of saturated Maokou limestone specimens by multi-level load and unload cycles method to explore time-dependent geological disaster in mining engineering. Take saturated Maokou limestone of Meitanba Coal Mining Area in China for example; the elastovisco-plastic deformation properties of this kind of rock were studied. The experimental result are given as follows: (1) instantaneous elastic response is nearly linear, Instantaneous plastic modulus increases with increasing stress; (2) The relation between visco-elastic strain and deviatoric stress is nonlinear, and nonlinear relationship between visco-plastic strain and deviatoric stress is also shown. With the increase of the deviatoric stress level, visco-plastic strain grows rapidly; (3) the steady creep rate increases at an exponential function with deviatoric stress increase. The visco-plastic strain properties of radial creep are similar to that of axial creep. But radial creep strain only is 25%–30% of axial creep strain, radial creep rate is about 35% of axial creep rate. The acquisition of elastovisco-plastic parameters of saturated Maokou limestones provides useful information for study on the time-dependent stability of Maokou limestone tunnel of Meitanba Coal Mining Area in China.

Keywords: Elastovisco-plastic · Rheological experiments
Multi-level load and unload cycles · Time-dependent geological disaster

1 Introduction

Rheological properties of geotechnical engineering are essential for the evaluation of potential geological disaster [1, 2]. The rheological properties of soft rock masses where joints, faults and fractured zones are often encountered have larger influence than those of hard rocks [3]. Experiment is a basic measure to investigate the

L. Zhang et al. (Eds.): GSIC 2018, *Proceedings of GeoShanghai 2018 International Conference: Rock Mechanics and Rock Engineering*, pp. 176–186, 2018.
https://doi.org/10.1007/978-981-13-0113-1_20

rheological mechanical properties of rock. Many experimental achievements of rheological behavior of rocks have been achieved at present. Three types of tests are common in experimental studies on the rheological behavior of rocks, i.e., creep, relaxation, as well as loading tests at different stresses or strain rates [4, 5]. Tsai et al. [6] Presented systematic experimental study on a typical weak sandstone, researched elastovisco-plastic behaviors of the weak sandstone. Fabre et al. [7] Showed the time dependent behavior of three rocks characterized by high proportion of clay particles, and the viscosity of these sedimentary rocks was studied under different loadings. Rheological experiments can be performed either by a multi-level loading and unloading cyclic mode on a single specimen, or by a single level loading and unloading cyclic mode on several specimens. It is a tendency to adopt a multi-level loading and unloading cyclic mode on a single specimen, because a single level loading and an unloading cyclic mode on several specimens usually requires more rock specimens and the in homogeneity of the rock may cause scatter of the experimental data. Laboratory rheological experiments under multi-level loading and unloading cyclic mode can obtain rheological properties of rock specimens at different stress levels, and can help to set up appropriate rheological model and supply related rheological parameters. At present, uniaxial rheological experiments on rock under multi-level load and unload cycles have been performed by some researchers; the elastovisco-plastic properties of some rocks were discovered under uniaxial test [8–10], but the studies on triaxial elastovisco-plastic properties of rock are very lacking. For many rocks engineering, the surrounding rocks are in the three-dimensional stress state. The surrounding rock stress usually alternates between increase and decrease with mining progress. Mechanical behaviors of the surrounding rocks stress can be considered as loading and unloading alternatively due to excavation activity perturb. Rheologies properties of rocks under multi-level load and unload cycles are useful to study time-dependent deformation of geotechnical and mining engineering [11–13]. In this paper, Maokou limestone from the roadway in Meitanba Coal Mining Area in China was chosen as study objects. By means of rheological experiments under multi-level load and unload cycles, the non-linear elastovisco-plastic deformation properties of saturated Maokou limestones in triaxial stress state were studied. The acquisition of triaxial elastovisco-plastic parameters of saturated Maokou limestones provides useful information for study on the time-dependent stability of Maokou limestone tunnel of Meitanba Coal Mining Area in China.

Fig. 1. Specimens before experimenting

2 Triaxial Rheological Experiments Under Multi-level Load and Unload Cycles

Rock specimens taken from the roadway in Meitanba Coal Mining Area in China were Maokou limestone, which in roadway are moist and intercalated with large amount of sediment and fractures are developed. Maokou limestones were cut into 50 mm × 100 mm cylinder standard specimens. The specimens were cut and prepared in the laboratory, in accordance with the experiment specification recommended by International Society for Rock Mechanics (ISRM) and Chinese Rock Experiment Standards (SL 264–2001 and GB/T50266–99). The specimens were placed in water for 30 d to be saturating with water. The photos of specimens before experimenting are shown in Fig. 1. Both conventional triaxial compression experiments and triaxial compression rheological experiments of specimens were carried out using MTS 815 rock test machine in Hunan University of Science and Technology in China as shown in Fig. 2.

Fig. 2. MTS 815 rock test machine and deformation sensor measurement system

Through the study of mechanical experiments, the triaxial compression strength indices and shear strength parameters with confining pressures of 0, 4.0, 8.0, 12.0 MPa, were put forward, which indicate that Maokou limestone presents a confining pressure sensitive behavior. The triaxial compression strength and elastic modulus increase with confining pressure increasing. Strength, elastic modulus, poisson ratio, friction and cohesion of specimens under different confining pressures are listed in Table 1.

Table 1. Triaxial compression strength and shear strength parameter sunder different confining pressures

Confining pressures/MPa	Strength/MPa	Elastic modulus /GPa	Friction coefficient	Cohesion /MPa
0	24.1	17.79	46.26	5.14
4.0	54.7	19.07		
8.0	71.4	21.65		
12.0	101.3	28.34		

Assuming that long-term strength of rock is 70%–80% of triaxial compression strength, in the rheological experiments, to take confining pressures of 4.0 MPa for example, long-term strength of Maokou limestone specimens is assumed to 43.9 MPa [14]. During multi-level load and unload cycles, planning-putting maximal deviatoric stress is 39.9 MPa with confining pressures of 4.0 MPa, which is divided into 6 steps from lower load to higher load in order to study elastovisco-plastic deformation properties of Maokou limestone in triaxial stress state. During the unloading, the confining pressure remains unchanged, however, the axial stress is unloaded to appropriate value according to planning-putting step deviatoric stresses. Figure 3 shows the experimental loading and unloading process.

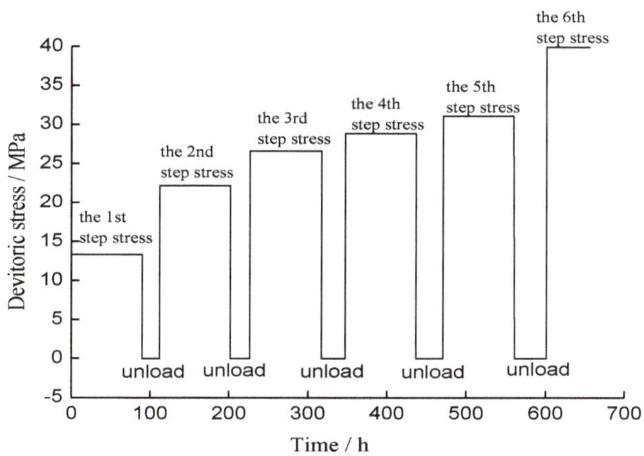

Fig. 3. Multi-level load and unload cycles

Triaxial rheological experiments under multi-level load and unload cycles may observe delayed elastic recovery. And irreversible deformation can be measured. The experimental method can provide a complete set of experimental data to study elastovisco-plastic deformation properties of rock. Figure 4 represents axial rheological curves of specimen M-09 with confining pressures of 4.0 MPa under multi-level load and unloads cycles.

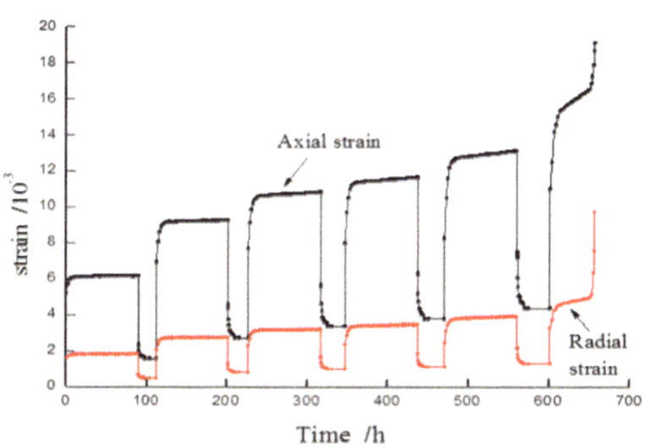

Fig. 4. Rheological curves of specimen M-09

3 Rheological Data Processing

The separation process of instantaneous elastic strain, instantaneous plastic strain, visco-elastic strain and visco-plastic strain from the total deformation is shown in Fig. 5 [15].

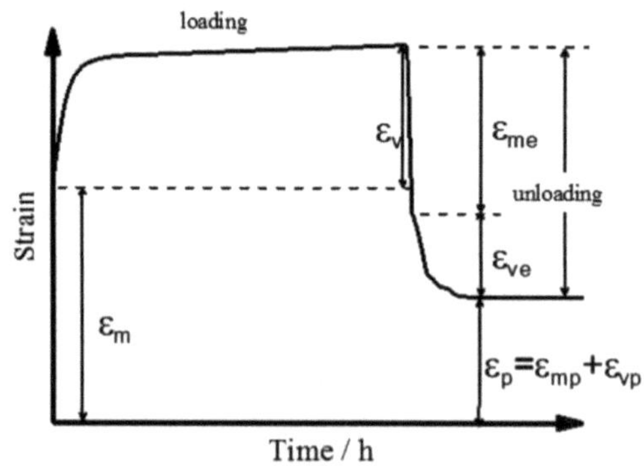

Fig. 5. Diagrammatic sketch of the total deformation separation process

Total strain $\varepsilon^{(i)}$ is made up of instantaneous elastic strain $\varepsilon_{me}^{(i)}$, instantaneous plastic strain $\varepsilon_{mp}^{(i)}$, visco-elastic strain $\varepsilon_{ve}^{(i)}$ and visco-plastic strain $\varepsilon_{vp}^{(i)}$. The total strain of rock can be expressed as follow:

$$\varepsilon^{(i)} = \varepsilon_m^{(i)} + \varepsilon_v^{(i)} = \varepsilon_{me}^{(i)} + \varepsilon_{mp}^{(i)} + \varepsilon_{ve}^{(i)} + \varepsilon_{vp}^{(i)} \tag{1}$$

Instantaneous strain $\varepsilon_m^{(i)}$ at the *ith* step deviatoric stress level is composed of two parts: instantaneous elastic strain $\varepsilon_{me}^{(i)}$ and instantaneous plastic strain cumulation $\varepsilon_{mp}^{(i)}$. Instantaneous strain $\varepsilon_m^{(i)}$ is obtained from Eq. (2) considering the history of loading.

$$\varepsilon_m^{(i)} = \varepsilon_{me}^{(i)} + \varepsilon_{mp}^{(i)} \tag{2}$$

Creep strain $\varepsilon_v^{(i)}$ at the *ith* step deviatoric stress level is composed of two parts: visco-elastic strain $\varepsilon_{ve}^{(i)}$ and visco-plastic strain $\varepsilon_{vp}^{(i)}$. Creep strain $\varepsilon_v^{(i)}$ is obtained from Eq. (3) considering the history of loading.

$$\varepsilon_v^{(i)} = \varepsilon_{ve}^{(i)} + \varepsilon_{vp}^{(i)} \tag{3}$$

Visco-elastic strain $\varepsilon_{ve}^{(i)}$ at the *ith* deviatoric stress level is equal to the delayed elastic recovery strain at the unloading moment on the assumption that visco-elastic strain curve has the same route with the unloaded delayed elastic recovery curve which can be obtained by unloading experiment. Visco-plastic strain $\varepsilon_{vp}(t)$ is irreversible viscous strain, which is expressed as follows.

$$\varepsilon_{vp}(t) = \varepsilon_v(t) - \varepsilon_{ve}(t) \quad (4)$$

Where $\varepsilon_v(t)$, $\varepsilon_{ve}(t)$ and $\varepsilon_{vp}(t)$ represent creep strain, visco-elastic strain and visco-plastic strain at different time, respectively.

Visco-plastic strain $\varepsilon_{vp}(t)$ characterizes primary creep and steady creep. But visco-elastic strain $\varepsilon_{ve}(t)$ only characterizes primary creep. $\varepsilon_v(t)$, $\varepsilon_{ve}(t)$ and $\varepsilon_{vp}(t)$ at the 5th step deviatoric stress level are shown in Fig. 6.

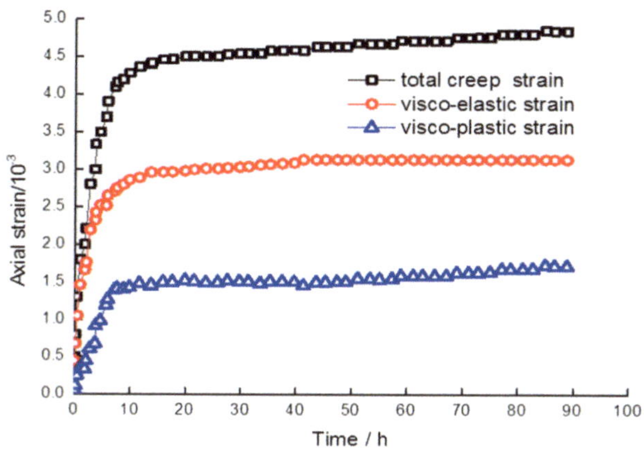

Fig. 6. Separation of creep strain at the 5th step deviatoric stress level

Because the accelerated creep results in the delayed failure of the specimen, visco-elastic strain and visco-plastic strain at the 6th step deviatoric stress level cannot be separated when the accelerated creep develops. In this paper, we consider that after failure of the specimen, its deformation is irreversible, so it may be taken for granted that for the specimen failure, creep deformation is attributed to visco-plastic deformation. Experimental data of specimen M-09 under multi-level load and unload cycles are list in Table 2.

Table 2. Experimental data of axial elastovisco-plastic strains under multi-level load and unload cycles

Steps	$\sigma_1 - \sigma_3$ (MPa)	ε_m ($\times 10^{-3}$)	ε_v ($\times 10^{-3}$)	ε_{me} ($\times 10^{-3}$)	ε_{mp} ($\times 10^{-3}$)	ε_{ve} ($\times 10^{-3}$)	ε_{vp} ($\times 10^{-3}$)
1	13.305	5.221	0.953	3.796	1.425	0.784	0.169
2	22.175	6.929	2.336	4.695	2.234	1.825	0.511
3	26.610	7.684	3.126	5.120	2.564	2.311	0.815
4	28.828	7.990	3.675	5.311	2.679	2.574	1.101
5	31.045	8.274	4.847	5.485	2.789	3.135	1.712
6	39.916	11.00	8.125	–	–	–	5.482

4 Elastovisco-Plastic Strain Properties Analysis

4.1 Instantaneous Elastic and Instantaneous Plastic Strains

Instantaneous strain consists of instantaneous elastic strain and instantaneous plastic strain considering the loading history. Instantaneous elastic strain and instantaneous plastic strain increase with the increase of deviatoric stress level. Figure 7 shows instantaneous strain characteristic curves of specimen at different deviatoric stress levels. The relation between deviatoric stress and instantaneous strain approximately abides by linearity relations.

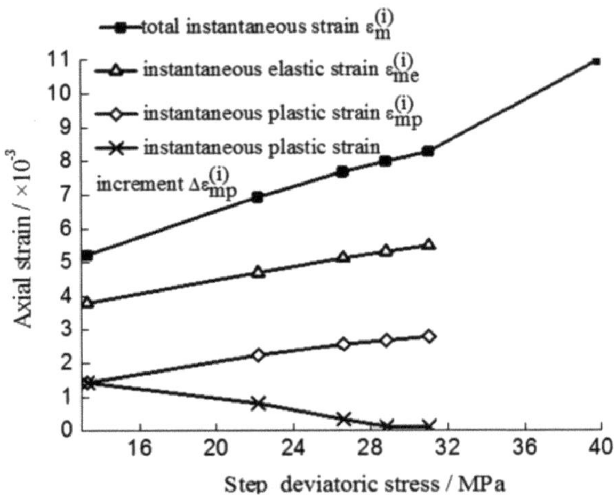

Fig. 7. Curves of step deviatoric stress- axial instantaneous strain relation

The ratio of deviatoric stress and instantaneous elastic strain is defined as intransient elastic modulus $E_{me}^{(i)}$

$$E_{me} = \frac{\Delta\sigma}{\Delta\varepsilon_{me}} \quad (5)$$

The relation between deviatoric stress and instantaneous elastic strain approximately abides by linearity relations. From Eq. (5) It can be obtained that the instantaneous plastic modulus $E_{me} = 10.39\,\text{GPa}$.

The ratio of deviatoric stress increment $\Delta\sigma^{(i)}$ and instantaneous plastic strain increment $\Delta\varepsilon_{mp}^{(i)}$ is defined as instantaneous plastic modulus $E_{mp}^{(i)}$

$$E_{mp}^{(i)} = \frac{\Delta\sigma^{(i)}}{\Delta\varepsilon_{mp}^{(i)}} \quad (6)$$

From Eq. (6) it can be obtained that the instantaneous plastic modulus increases with the increase of deviatoric stress level. Instantaneous plastic modulus of the 1st deviatoric stress level $E_{mp}^{(1)} = 10.96\,\text{GPa}$. While instantaneous plastic modulus of the 4th deviatoric stress level $E_{mp}^{(4)} = 20.16\,\text{GPa}$. This suggests the ability to resist instantaneous plastic deformation enhances with the increase of deviatoric stress level. The fact that instantaneous plastic modulus at the 1st deviatoric stress level is small than that at the 4th deviatoric stress level is attributed to the rock crack closure effect, because of rock specimen with large amount of sediment and fractures are developed.

4.2 Visco-Elastic and Visco-Plastic Strains

Creep strain ε_v is composed of visco-elastic strain ε_{ve} and visco-plastic strain ε_{vp}. There is a nonlinear relationship between visco-elastic strain, visco-plastic strain and deviatoric stress, as shown in Fig. 8, under multi-level load and unload cycles. Creep stain increment tends to increase under unit deviatoric stress increment, which suggests creep deformation is more and more obvious with deviatoric stress increasing. Figure 8 shows that the ratio of visco-plastic strain $\varepsilon_{vp}^{(i)}$ to creep strain $\varepsilon_v^{(i)}$ at various stress levels varies from 0.18 to 0.67, and has increasing tendency with deviatoric stress increasing. This means that the possessive proportion of visco-plastic strain in creep strain will increase with deviatoric stress increasing.

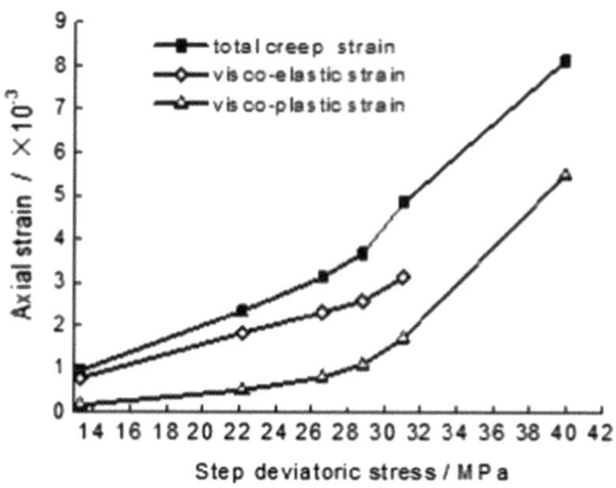

Fig. 8. Curves of step deviatoric stresses- axial creep strain relation

At the 6th step deviatoric stress level, specimen M-09 entered accelerated creep phase after the duration of 55.4 h. Accelerated creep last nearly 2.1 h, which is a relatively short period, a momentary specimen failure appeared when the axial strain reached 19.1274×10^{-3}. Accelerated creep characterized irreversible visco-plastic deformation.

4.3 Steady Creep Strain Rate

Rock rheological experiment curves show three stages of creep: primary creep, steady creep and accelerated creep. In steady creep stage, creep rate remains nearly unchanged, keeping fluctuates limitedly. Figure 9 shows the variation of steady creep rate with respect to deviatoric stress for specimen M-09 with confining pressures of 4.0 MPa. The larger deviatoric stress, the greater the steady creep rate is. The relation between the steady creep rate and deviatoric stress is nonlinear. Because of the nonlinear relation between the steady creep rate and deviatoric stress, it is necessary to introduce the non-stationary viscous coefficient in creep model of rock. From Fig. 9 it can be represented that the steady creep rate increases at an exponential function with deviatoric stress increasing, with a correlation equation as follow.

$$\dot{\varepsilon} = 1.72 \times 10^{-7} e^{0.165(\sigma_1 - \sigma_3 - 13.31)} \tag{7}$$

For radial creep, fluctuation effect of radial creep strain curves due to rock load fragmentation is less obvious than that of axial creep strain curve. The elastovisco-plastic strain properties of radial creep are similar to that of axial creep. But radial creep strain only is 25%–30% of axial creep strain. Radial creep rate is about 35% of axial creep rate. The elastovisco-plastic strain properties of radial creep will not be is introduced particularly here.

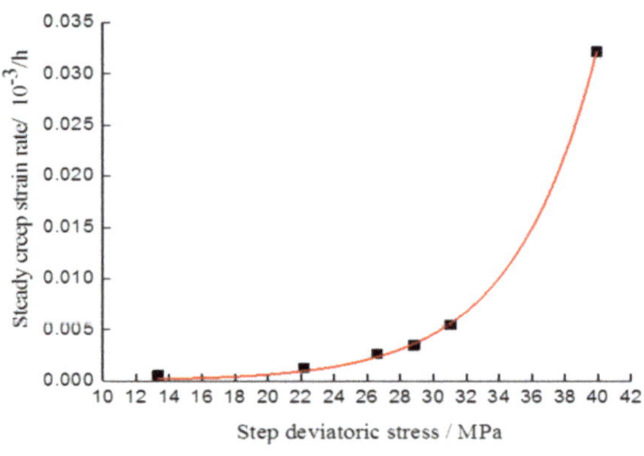

Fig. 9. Axial steady creep rate under different deviatoric stress levels

5 Conclusion

(1) A new experimental data processing method of separation of elastovisco-plastic strains under multi-level load and unload cycles is suggested.

(2) Instantaneous elastic response is nearly linear, Instantaneous plastic modulus is increases with increasing stress; The relation between visco-elastic strain and deviatoric stress is nonlinear, and non-linear relationship between visco-plastic strain and deviatoric stress is also shown; With the increase of the deviatoric stress level, visco-plastic strain grows rapidly; the steady creep rate increases at an exponential function with deviatoric stress increase.

(3) The elastovisco-plastic strain properties of radial creep are similar to that of axial creep. But radial creep strain only is 25%–30% of axial creep strain, radial creep rate is about 35% of axial creep rate.

Acknowledgments. This research is supported by the CRSRI Open Research Program (CKWV2017508/KY), the National Natural Science Foundation of China (No. 51774131, 41172266), the Natural Science Foundation of Hunan province (No. 2015JJ2067), the Open Projects of State Key Laboratory of Coal Resources and Safe Mining, CUMT (SKLCRSM 16KF12).

References

1. Horsrund, P., Holt, R.M., Sonste Bo, E.F.: Time dependent borehole stability: laboratory studies and numerical simulation of different mechanism in shale. In: Proceedings of Eurock SPE/ISRM 1994, pp. 259–266. Balke ma, Rotterdam (1994)
2. Zhao, Y.L., Zhang, L.Y., Wang, W.J., Wan, W., Li, S.Q.: Creep behavior of intact and cracked limestone under multi-level loading and unloading cycles. Rock Mech. Rock Eng. **2017**, 1–16 (2017)
3. Zhang, Z.L., Xu, W.Y., Wang, W.: Triaxial creep tests of rock from the compressive zone of dam foundation in Xiangjiaba Hydropower Station. Int. J. Rock Mech. Min. Sci. **50**(1), 133–139 (2012)
4. Zhao, Y.L., Wang, Y.X., Wang, W.J., Wan, W., Tang, J.: Modeling of non-linear rheological behavior of hard rock using triaxial rheological experiment. Int. J. Rock Mech. Min. Sci. **93**, 66–75 (2017)
5. Li, Y.S., Xia, C.C.: Time dependent tests on intact rocks in uniaxial compression. Int. J. Rock Mech. Min. Sci. **37**(1), 467–475 (2000)
6. Tsai, L.S., Hsieh, Y.M., Weng, M.C.: Time-dependent deformation behaviors of weak sandstones. Int. J. Rock Mech. Min. Sci. **45**(2), 144–154 (2008)
7. Fabre, G., Pellet, F.: Creep and time dependent damage in argillaceous rocks. Int. J. Rock Mech. Min. Sci. **43**(1), 950–960 (2006)
8. Li, Y.S., Xia, C.C.: Time-dependent tests on intact rocks in uniaxial compression. Int. J. Rock Mech. Min. Sci. **37**(3), 467–475 (2000)
9. Gatelier, N., Pellet, F., Loret, B.: Mechanical damage of an anisotropic porous rock in cyclic triaxial experiments. Int. J. Rock Mech. Min. Sci. **39**(3), 335–354 (2002)
10. Li, K., Zheng, D., Huang, W.: Experiment research on shear characteristics of sandstone considering cyclic drying-wetting effect. Disaster Adv. **6**(S1), 83–87 (2013)
11. Zhao, Y.L., Zhang, L.Y., Wang, W.J., Wan, W., Ma, W.H.: Separation of elastovisco-plastic strains of rock and a nonlinear creep model. Int. J. Geomech. (2018). https://doi.org/10.1061/(ASCE)GM.1943-5622.0001033
12. Zhao, H.B., Du, Q.H., Jiang, J.J., Li, Q.C., Zhang, Z.X.: Experimental study on evolution of mechanical parameters of coal containing gas under unloading confining pressure. Disaster Adv. **5**, 37–41 (2012)
13. Xu, T., Tang, C.A., Zhao, J.: Modeling of rheological deformation of inhomogeneous rock and associated time-dependent response of tunnels. Int. J. Geomech. **12**(2), 147–159 (2011)

14. Zhang, Y., Xu, W.Y., Gu, J.J.: Triaxial creep tests of weak sandstone from fracture zone of high dam foundation. J. Cent. South Univ. **20**(9), 2528–2536 (2013)
15. Xia, C.C., Zhong, S.Y.: Experimental data processing method in consideration of influence of loading history on rock specimen deformation. J. Cent. South Min. Metall. **20**(2), 18–24 (1989)

Experimental Study on Physical and Mechanical Properties of Iron Tailing Materials

Jianting Du[1,2], Liming Hu[1(✉)], Dantong Lin[1], and Lin Zhang[1]

[1] Tsinghua University, Beijing 100084, China
gehu@tsinghua.edu.cn
[2] PowerChina Kunming Engineering Corporation Limited,
Kunming 650051, China

Abstract. The physical and mechanical properties of tailing materials are important for the design and construction of tailings dams. In the present study, laboratory experiments were performed to investigate the geotechnical properties of two different iron tailing materials (coarse and fine). Compared to the coarse tailings, the fine tailings showed greater coefficients of compressibility and smaller permeability, shear strength. With the decrease in void ratio, the oedometric modulus increased and the permeability decreased. The relationships between void ratio and coefficient of compressibility, as well as hydraulic conductivity, can be fitted with logarithmic functions. Moreover, we found the coefficient of consolidation C_v gradually increased for both the coarse and fine tailings during the compression process. The shear strength of the two tailing materials were measured though direct shear tests and triaxial tests. The results of friction angle and cohesion obtained from the two test methods were in accordance with each other. This study provided useful information to understand the geotechnical behaviors of mine tailings and fundamental data for the scientific design and safety evaluation of tailings dams.

Keywords: Iron tailing materials · Physical and mechanical properties
Consolidation characteristics · Triaxial tests

1 Introduction

The utilization and exploitation of mineral resources produce large amount of mine tailings which are commonly stored in a slurry form and pumped into a sedimentation pond. With the development of mining technology and the increase in the demand for minerals and metals, the volume of tailings being generated increased dramatically over the last few decades. The tailings generally cause lots of environmental and ecological issues, because they always contain plenty of toxic substances, heavy metals, and chemicals added during mineral processing. In addition, due to the high water content and small grain size, the mechanical stability of the tailings mass is poor. Therefore, the disposal and storage of tailings has become increasingly important in mining industry.

In order to maintain the normal mining operations, several tailing disposal methods are propose, including disposal of dry or thickened tailings in free-standing piles or

L. Zhang et al. (Eds.): GSIC 2018, *Proceedings of GeoShanghai 2018 International Conference: Rock Mechanics and Rock Engineering*, pp. 187–194, 2018.
https://doi.org/10.1007/978-981-13-0113-1_21

impoundments, backfilling underground mine workings, subaqueous disposal. The most common method is the disposal of tailings slurry in impoundments that involves the transport of tailing mass in slurry pipes to an impoundment and thereafter consolidation under their own weight and compaction to form a tailing dam. For addressing the high global demand for metals and minerals, the size and number of tailing dams increased rapidly with the increase in mining activity. There are over 18,400 tailing dams worldwide [1]. Meanwhile the height and storage capacity of tailing dams have increased continuously during the past decades, which inevitably leads to the increase in the risk of tailing dam failure [1–4]. The database of International Commission of Large Dams and United Nations Environment Programme showed that several hundreds of tailings dam have failed since 1910. There were a total of 198 tailing dam failure events before the year 2000 and 20 failure events from 2000 to 2010 [1]. Eleven other failure events were reported from 2010 to 2015 [5].

The characteristics of the tailings used for the construction of tailing dams are very important for the design and operation of tailing dams, which will greatly enhance the safety of tailing dams. Many investigations have been conducted to study the mechanical properties of mine tailings. The physical-mechanical properties of tailing soils from Yang-la Copper Mine were investigated for providing detailed information for engineering design [6]. The strength characteristics of tailing sand were studied by conducting consolidation drained triaxial experiments and compression tests [7]. Physicochemical properties [8], permeability characteristics [9] and mechanical properties [10] of iron tailings were studied by laboratory investigations. A comprehensive understanding of physical and mechanical properties of mine tailings is critical for the design of tailing dams.

In this paper, a series of laboratory experiments were conducted to investigate the physical and mechanical properties of iron tailings obtained from Daxigou iron mine in Shanxi province, China. The characteristics of the tailing materials, including grain size distribution, hydraulic permeability, compression and strength, were measured. The laboratory tests provided essential information for the numerical analysis and scientific design of the tailing dam.

2 Materials and Methods

2.1 Tailings

The test tailing materials were from Daxigou iron mine in China. As the produced tailings were transported into a sedimentation pond, two kinds of tailings were generated, i.e., coarse tailings and fine tailings. A series of laboratory experiments were performed on these two different iron tailing materials (coarse and fine).

2.2 Test Methods

Several conventional laboratory tests were conducted to investigate the mechanical properties of the tailings, including specific gravity, grain size distribution, Atterberg limits, and hydraulic conductivity. Consolidation tests, triaxial tests, and cyclic triaxial tests were carried out to investigate the compression characteristics and strength of the tailings. The pycnometer method was chosen to measure the specific gravity. The sieving method, hydrometer method and laser particle size analysis were combined for measuring grain size distribution. The Atterberg limits were measured with the fall cone test. The permeability of the two tailings was studied with a flexible wall permeameter under different consolidation stresses (50, 100, 200, 400, and 600 kPa). The consolidation test was performed in a one-dimensional oedometer with the diameter of 8 cm according to American Society for Testing and Materials (ASTM) standard D2435M-11 [11]. Direct shear tests were conducted on tailings samples to measure the shear strength of the tailing materials. Normal stresses of 100, 200, 300 and 400 kPa were applied to the samples prior to shearing [12]. The unsaturated triaxial testing system (VJTech, Reading, Berkshire, U.K.) was used to conduct the consolidated undrained (CU) triaxial compression tests on the tailings according to ASTM standard D4767-11 [13].

3 Mechanical Properties of the Tailings

3.1 Characterization of the Tailings

The basic geotechnical properties of the two tailings are summarized in Table 1 and Fig. 1 displays the grain size distributions. The specific gravity of five samples for each tailing were measured and the average specific gravity of coarse and fine iron tailings are 3.127 and 3.183 respectively, which are much greater than those of natural soils. We measured the grain size distribution of coarse tailing particles whose diameters were more than 0.075 mm by the sieving method, and for the measurement of fine tailing materials and coarse tailing particles whose diameters were less than 0.075 mm, both the hydrometer method and laser particle size analysis were used for mutually verifying and validating the results from the two methods. The results of the hydrometer method and

Table 1. Geotechnical properties of coarse and fine iron tailings

Properties	Coarse tailings	Fine tailings
Specific gravity, G_s	3.127	3.183
Maximum dry density, ρ_{dmax} (g/cm^3)	2.32	2.19
Minimum dry density, ρ_{dmin} (g/cm^3)	1.40	1.26
Optimum water content, w_{op} (%)	——	15.66
Liquid limit, w_L (%)	20.99	25.06
Plastic limit, w_p (%)	10.63	13.17
Plasticity index, I_p (%)	10.36	11.89
D_{10} (mm)	0.013	0.012
D_{30} (mm)	0.040	0.020
D_{50} (mm)	0.088	0.038
D_{60} (mm)	0.107	0.049
Coefficient of uniformity, C_u	8.23	4.08
Coefficient of curvature, C_c	1.15	0.68

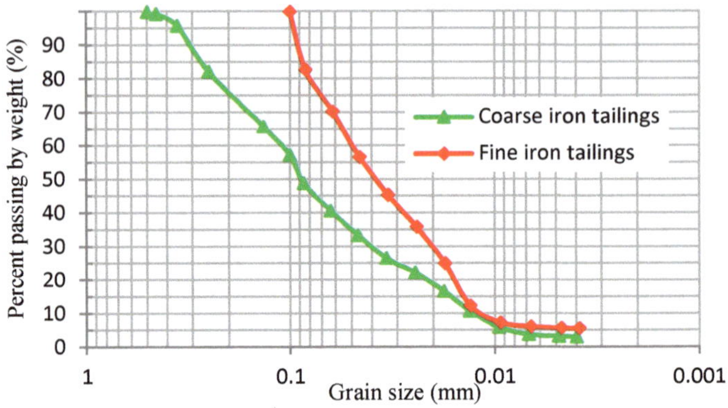

Fig. 1. Grain size distribution of the coarse and fine tailings

the laser particle size analysis were in good accordance with each other. Figure 1 indicates that the mass percentage of grain size less than 0.075 mm is 45.7% for the coarse iron tailings, 77.4% for the fine iron tailings. According to the coefficient of uniformity and coefficient of curvature results, the coarse iron tailings are well-graded soils, whereas the fine iron tailings present a poorly-graded grain size distribution. Based on the testing results of Atterberg limits and grain size distribution, the coarse iron tailings are classified as silty sand (SM), and the fine iron tailings are classified as lean clay (CL) according to the Standard for Engineering Classification of Soil [14].

3.2 Consolidation Test

One-dimensional consolidation tests were conducted on coarse and fine tailings samples under saturated conditions in order to determine compressibility characteristics. Three samples were measured for each tailing material and the results of different samples of each tailing agreed with each other well. With the increase in vertical stress, the void ratio (e) decreases. A logarithmic function can be used to fit the compression curves of the coarse and fine tailings [15].

$$e = e_0 - C_c \times \log(p/p_0) \quad (1)$$

in which $p_0 = 1$ kPa; e_0 = initial void ratio corresponding to p_0; and C_c = compression index. The compression index (C_c) is 0.111 and 0.127 for the coarse and fine iron tailings. According to the values of C_c, both the coarse and fine iron tailings are characterized as a moderate compressible soil. And the fine tailings present a higher compressibility than the coarse tailings.

According to the results of the consolidation tests, the relationships between void ratio (e) and oedometric modulus (E_s) were fitted with a logarithmic function as shown in Fig. 2. Compared to the fine tailings, the coarse tailings present a greater E_s. With the increase in the vertical stress, the oedometric modulus increases, and for fine iron tailings, the increase of oedometric modulus is greater than the coarse iron tailings'. This indicates that the mechanical properties will be improved as the tailing materials

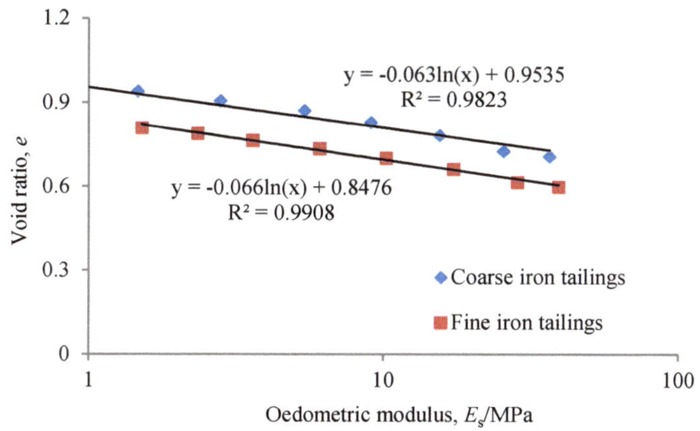

Fig. 2. Relationship between the oedometric modulus and void ratio of the tailings

have consolidated. According to the fitting results in Fig. 2, the relationship between E_s and e can be described with the following equation:

$$e = e_0 - N \times \log(E_s/E_{s0}) \tag{2}$$

in which E_{s0} (MPa^{-1}) = oedometric modulus corresponding to p_0 and initial void ratio e_0; N = the factor describing the change rate of E_s with the change in void ratio. The fitting results of the two fine tailings can be expressed as

$$\begin{aligned} &\text{Coarse iron tailings: } e = 0.88 - 0.145 \times \log\left(E_s/0.612\ \text{MPa}\right) \\ &\text{Fine iron tailings: } e = 1.11 - 0.152 \times \log\left(E_s/\,0.0833\ \text{MPa}\right) \end{aligned} \tag{3}$$

3.3 Hydraulic Conductivity

Figure 3 further shows the relationship between void ratio and hydraulic conductivity of the two tailings. The hydraulic conductivities of the two tailings increase with the increase in void ratio, and the coarse iron tailings present a higher hydraulic conductivity than the fine iron tailings. The change in hydraulic conductivity with void ratio can be further fitted with a logarithmic function as

$$e = e_0 + M \times \log(k/k_0) \tag{4}$$

where k_0 = initial hydraulic conductivity under initial void ratio (cm/s). The fitting results of the two fine tailings can be expressed as

$$\begin{aligned} &\text{Coarse iron tailings: } e = 0.88 + 0.61 \times \log(k/5.39\,\text{cm/s}) \\ &\text{Fine iron tailings: } e = 1.11 + 0.49 \times \log(k/8.28\,\text{cm/s}) \end{aligned} \tag{5}$$

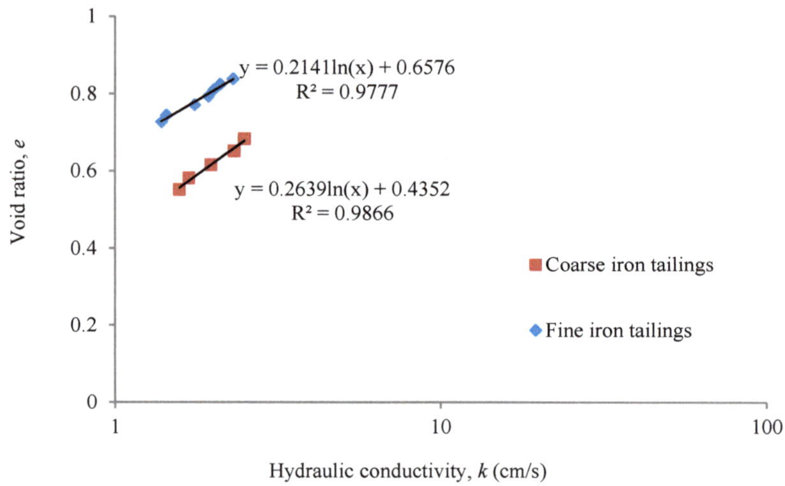

Fig. 3. Relationship between void ratio and hydraulic conductivity of the two tailings

The coefficient of consolidation can be derived from Eqs. (2) and (4) as

$$C_v = \frac{k \times E_S}{\gamma_w} = C_{v0} \times 10^{(1/N - 1/M)(e_0 - e)} \tag{6}$$

In which γ_w = unit weight of water; and C_{v0} = initial coefficient of consolidation, which can be expressed as $k_0 \times E_{s0}/\gamma_w$. The fitting results of Eqs. (2) and (4) are summarized in Table 2. For both the coarse and fine iron tailings, the value of M is greater than that of N, which means C_v increases during the test. This is consistent with the results calculated from the consolidation tests. The classical consolidation theory proposed in 1943 assumed a constant coefficient of consolidation [16]. In order to consider more realistic assumptions about soil behavior including the nonlinear variation of compressibility and hydraulic conductivity as described in Eqs. (2) and (4), a solution for a one-dimensional consolidation problem was derived [17]. Based on the above analysis of E_s and k for the two tailings, this derived theory is recommended.

Table 2. Fitting results of Eqs. (2) and (4)

Tailing type	e_0	E_{s0} (MPa)	k_0 (cm/s)	M	N
Coarse tailings	0.88	0.612	5.39	0.61	0.145
Fine tailings	1.11	0.0833	8.28	0.49	0.152

3.4 Shear Strength

The shear strength is a very important index for evaluating the stability of tailing dams. The friction angle and cohesion were calculated from the results of the CU and direct shear tests as listed in Table 3, the strength indices of the coarse iron tailings are greater

Table 3. Friction angle and cohesion calculated from CU and direct shear tests

Tailing type	CU tests				Direct shear tests			
	ϕ_{cu} (kPa)	c_{cu} (°)	ϕ' (kPa)	c' (°)	ϕ_s (kPa)	c_s (°)	ϕ_{cq} (kPa)	c_{cq} (°)
Coarse tailings	37	0	39	0	37.5	0.0	36.7	0
Fine tailings	29	3	34	0	29.9	17.0	27.4	15.5

than those of the fine iron tailings. And the results of coarse tailings obtained from the two test methods were in good accordance with each other, whereas the effective cohesion from CU tests is smaller than that from the direct shear tests.

4 Conclusions

A series of laboratory experiments were conducted to examine the physical and mechanical characteristics of coarse and fine iron tailings from the Daxigou iron mine in China. The basic geotechnical properties including specific gravity, grain size distribution, plasticity, permeability, and compressibility were investigated, and triaxial compression and direct shear tests were performed to analyze the strength of the two tailings. The following conclusions were obtained:

1. The coarse iron tailings were classified as silty sand (SM), and the fine iron tailings were classified as lean clay (CL). The fine tailings showed greater coefficients of compressibility and smaller permeability, strength, and cyclic resistance compared to the coarse tailings, respectively.
2. With the decrease in void ratio, the oedometric modulus increased and the permeability decreased. The relationships between void ratio and coefficient of compressibility, as well as hydraulic conductivity can be fitted with logarithmic functions.
3. During the compression process, the coefficient of consolidation C_v gradually increased for both the coarse and fine tailings. Therefore, the acceleration of the consolidation process caused by the increase in C_v should be considered.
4. The results of CU and direct shear tests showed that the strength indices of the coarse iron tailings are greater than those of the fine iron tailings, and the results of coarse tailings obtained from the two test methods were in good accordance with each other, whereas the effective cohesion from CU tests is smaller than that from the direct shear tests.

References

1. Azam, S., Li, Q.: Tailings dam failures: a review of the last one hundred years. Geotech. News **28**(4), 50–54 (2010)
2. Davis, M.P.: Tailings impoundment failures: are geotechnical engineers listening. Geotech. News **20**(3), 31–36 (2002)

3. Psarropoulos, P.N., Tsompanakis, Y.: Stability of tailings dams under static and seismic loading. Can. Geotech. J. **45**(5), 663–675 (2008)
4. Ferdosi, B., James, M., Aubertin, M.: Effect of waste rock inclusions on the seismic stability of an upstream raised tailings impoundment: a numerical investigation. Can. Geotech. J. **52**(12), 1930–1944 (2015)
5. WISE 2015 (World Information Service on Energy 2015). http://www.wise-uranium.org/mdaf.html. Accessed 22 Dec 2017
6. Yin, G.Z., Yang, Z.Y., Wei, Z.A., Tan, Q.W.: Physical and mechanical properties of YangLa-copper's tailings. J. Chongqing Univ. **30**(9), 117–122 (2007). (in Chinese)
7. Yang, K., Lü, S.R., Zhang, Y.Y.: Experimental study of strength characteristics of tailing sand in tailings dam. Metal Mine **2**, 166–170 (2014). (in Chinese)
8. Wolff, A.P., da Costa, G.M., de Castro Dutra, F.: A comparative study of ultra-fine iron ore tailings from Brazil. Miner. Process. Extr. Metall. Rev. **32**(1), 47–59 (2010)
9. Hai, L., Liang, B., Dai, H.F., Li, H.L., Zhou, X.: Testing study on the permeability features of iron tailings. J. Safety Environ. **13**(4), 207–210 (2013). (in Chinese)
10. Li, R.H., Wang, J., Zhou, Z.H., Liu, T.J.: Research of iron tailings application in the highway engineering. Mining Eng. **5**(5), 52–54 (2007). (in Chinese)
11. American Society for Testing and Materials: Standard test methods for one-dimensional consolidation properties of soils using incremental loading (D2435 M-11), West Conshohocken, PA (2011)
12. Ministry of Water Resource of P. R. China: Standard for Soil Test Method (GB/T 50123-1999). 1st edn. China Planning Press, Beijing (1999). (in Chinese)
13. American Society for Testing and Materials: Standard test methods for consolidated undrained triaxial compression test for cohesive soils (D4767-11). West Conshohocken, PA (2011)
14. Ministry of Water Resource of P.R. China: Standard for Engineering Classification of Soil (GB/T50145-2007). 1st edn. China Planning Press, Beijing (2008). (in Chinese)
15. Wong, R.C., Mills, B.N., Liu, Y.B.: Mechanistic model for one-dimensional consolidation behavior of nonsegregating oil sands tailings. J. Geotech. Geoenviron. Eng. **134**(2), 195–202 (2008)
16. Terzaghi, K.: Theoretical Soil Mechanics. Wiley, New York (1943)
17. Lekha, K.R., Krishnaswamy, N.R., Basak, P.: Consolidation of clays for variable permeability and compressibility. J. Geotech. Geoenviron. Eng. **129**(11), 1001–1009 (2003)

Experimental Study on the Rheological Property of Compacted Clay and Its Influence on the Stress and Deformation of the Core-Wall Dam

Zhi-zhou Geng[1,2], Kai Xu[1,2(✉)], Zhiqiang Wu[1,2], and En-yue Ji[1,3]

[1] Geotechnical Engineering Department, Nanjing Hydraulic Research Institute, Nanjing 210024, China
kxu@nhri.cn

[2] State Key Lab of Hydrology-Water Resources and Hydraulic Engineering, Nanjing 210024, China

[3] Key Laboratory of Ministry of Education for Geomechanics and Embankment Engineering, Hohai University, 01 Xikang Road, Nanjing 210098, People's Republic of China

Abstract. The secondary consolidation rate of compacted clay and rockfills has been proven to be relatively close, however, only the rheological property of the rockfill is considered in commonly used finite element analysis of core-wall dams. Therefore, the rheological property of the compacted clay and its influence on the stress and deformation of the Core-wall dam is discussed in this paper. The coefficient of secondary consolidation of unsaturated clay is significantly larger than that of saturated clay, and when the clay is saturated, the coefficient of secondary consolidation increases with the increasing loading. Both the unidirectional compression and triaxial compression rheological tests can be used to determine the rheological parameters and the results are relatively close. The finite element analysis indicates that the rheological deformation of clay core-wall has a considerable influence on the stress and deformation of the dam, e.g., causing higher stress levels and larger settlement especially around the core-wall. Hence, the rheological property of clay core-wall cannot be ignored in finite element analysis.

Keywords: Core-wall dam · Compacted clay · Unidirectional compression test Triaxial compression test · Secondary consolidation coefficient

1 Introduction

The earth and rockfill dam is widely distributed around the world for its locally-produced raw materials, simple structure, quick construction and the advantages of lower requirement to terrain conditions [1–3]. The safety of the dam is significantly affected by the resist hydro fracture ability and arching effect of core-wall [4], which relates closely to the mechanical properties of the used soils. As the main materials of the earth and rockfill dam, researchers have put too much attention on the stress and deformation properties of compacted clay and rockfills [5–8]. For the rheological

L. Zhang et al. (Eds.): GSIC 2018, *Proceedings of GeoShanghai 2018 International Conference: Rock Mechanics and Rock Engineering*, pp. 195–203, 2018.
https://doi.org/10.1007/978-981-13-0113-1_22

property of rock and soil, some researchers did many meaningful works [9–12]. However, the rheological property of the clay core-wall and its influence on the safety of the dam was ignored. As a result, only the rheological property of rockfills was considered in current commonly used finite element analysis.

It has been pointed out that the secondary consolidation rate of compacted clay and rockfills are relatively close [2]. In order to investigate the secondary consolidation coefficient and its influence on the stress and deformation of the dam, a series of unidirectional compression and triaxial compression rheological tests are carried out. Then the rheological parameters are obtained from the Shen Zhujiang Model. An earth and rockfill dam is taken as the example for the finite element analysis, and the influence of rheological property of the clay core-wall on the dam is discussed.

2 Property of Materials and Test Program

2.1 Property of Materials

The clay material is from the core-wall of Xiaolangdi earth and rockfill dam, and the main mineral is illite. In addition, the non-clay minerals are feldspar, quartz and a small content of oxide. The basic parameters of the core are shown in Table 1.

Table 1. Basic parameters of the core material

G_s	w_L (%)	w_P (%)	ρ_{dmax} (g/cm^3)	w_{opt} (%)	Contents (%)		
					Sand	Silt	Clay
2.72	35	20	1.72	13.1	29	22	49

where G_s is the specific gravity, L_L is the liquid limit, L_P is the plastic limit, ρ_{dmax} is the maximum dry density, w_{opt} is the optimum water content.

2.2 Test Program

Unidirectional compression and triaxial compression rheological tests are carried out in this paper. The diameter and height of the sample in unidirectional and triaxial compression tests are 6.2 cm, 2 cm and 3.9 cm, 8 cm, respectively. The axial loading steps of unidirectional compression are 50, 100, 200, 400, 800, 1600 and 3200 kPa, and each stress step will take turns to be loaded and continued for 7 days or more. It is noted that the original specimen is unsaturated. Among them, 50, 400 and 800 kPa, are designed to be the boundary of unsaturation and saturation clay (called p_c), as shown in Table 2. For example, if p_c is 400 kPa, which means the clay sample will be saturated by adding water. Consequently, when the loading stress is less than 400 kPa, the sample is unsaturated, otherwise, the sample is saturated.

Two loading stress levels were designed in triaxial compression tests, which are $S = 0.7$ and 0.9 with the same confining pressure of 400 kPa, as shown in Table 2. The stress level, S, is defined as:

$$S = \frac{(\sigma_1 - \sigma_3)}{(\sigma_1 - \sigma_3)_f} \tag{1}$$

where $(\sigma_1 - \sigma_3)_f$ is the deviatoric stress on failure, and can be calculated by:

$$(\sigma_1 - \sigma_3)_f = \frac{2c \cdot \cos\varphi + 2\sin\varphi \cdot \sigma_3}{1 - \sin\varphi} \tag{2}$$

where c is the cohesion, and φ is the internal frictional angle.

Table 2. Schemes of rheological tests

Unidirectional compression		Triaxial compression		
No.	p_c (kPa)	No.	σ_3 (kPa)	S
1	50	4	400	0.7
2	400	5	400	0.9
3	800			

3 Testing Results

3.1 Unidirectional Compression Tests

The solidification process of the unidirectional compression can be divide into two parts, which are the main consolidation and secondary consolidation, and the classical e-lg$t$ curve is illustrated in Fig. 1.

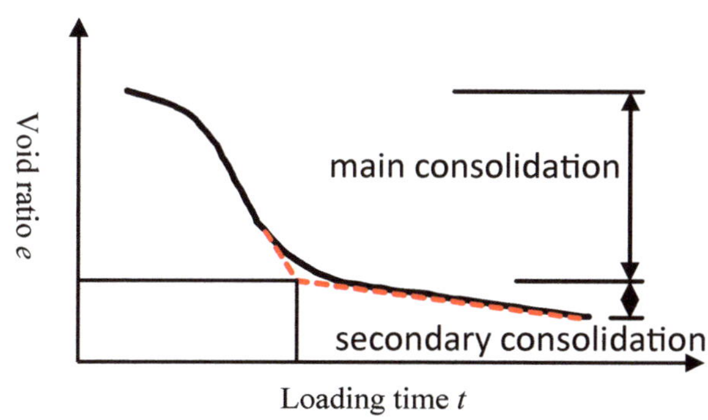

Fig. 1. Classical compression curve of the soil.

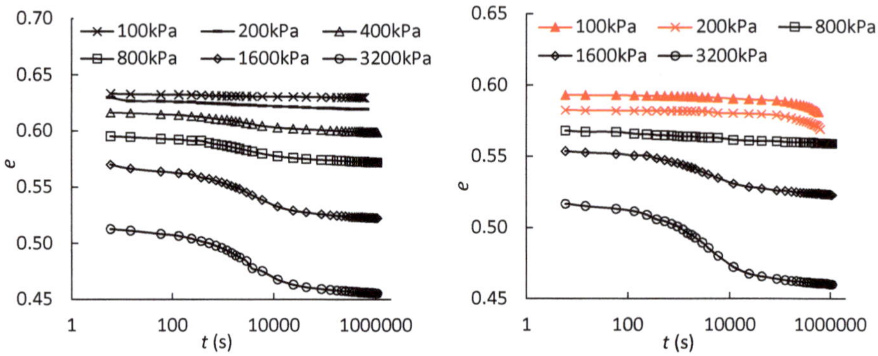

(a) Smple No.1 (p_c=50kPa) (b) Smple No.2 (p_c=400kPa)

Fig. 2. Relation curves of e and lg$t$.

The curve of secondary consolidation in Fig. 1 is thought to be a line in $e \sim \lg t$ coordinates, and the coefficient of secondary consolidation, C_α, is the slope of the line.

Taking sample No. 1 and No. 2 as the examples, and the testing curves are shown in Fig. 2. It can be seen from Fig. 2 that: (1) when the sample is saturated, the $e \sim \lg t$ curve is similar with the classical curve as shown in Fig. 1, which means that the consolidation deformation tends to be stable. (2) when the sample is unsaturated, the consolidation deformation shows no sign of stabilization, indicating that the samples need a longer time to reach the consolidation stability when the clay is unsaturated.

In order to get the coefficients of secondary consolidation under variable axial stresses, the secondary part of the curves are fitted by a line, with the fitting parameters shown in Table 3.

Table 3. Secondary compression coefficients of specimens.

Sample	p_c (kPa)	Loading step (kPa)					
		100	200	400	800	1600	3200
No. 1	50	0.180	0.707	0.896	1.00	1.28	1.50
No. 2	400	4.11	4.32	—	0.623	1.16	1.40
No. 3	800	4.14	4.84	2.74	—	0.807	1.36

It can be seen from Table 3 that the coefficient of secondary consolidation of unsaturated clay is significantly larger than that of saturated clay. On the other hand, when the clay is saturated, the coefficient of secondary consolidation increase with increasing loading.

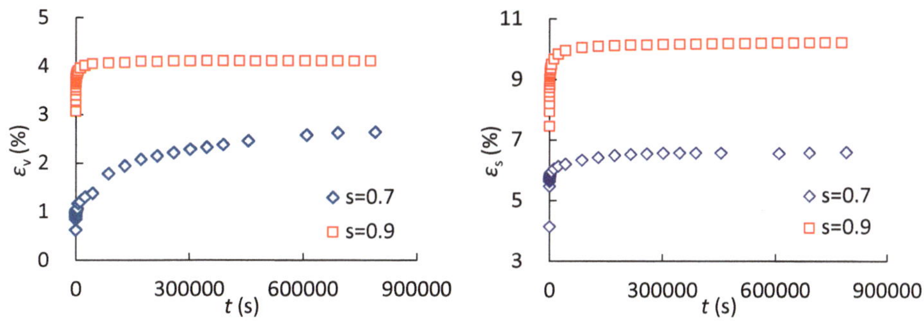

(a) ε_{v} versus t (b) ε_{s} versus t

Fig. 3. Relation curves of rheological strains versus loading time.

3.2 Triaxial Compression Tests

The testing data of triaxial compression tests were plotted in Fig. 3, the strians, e.g., volumetric strain (ε_{v}) and generalized shear strain (ε_{s}), increase with the increasing time at a decrease rate, and tend to be stable. Furthermore, when the stress level is high, the tendency is more significant. In other words, although the stress level is high, the time spending on the stable secondary consolidation is relatively small.

4 Finite Element Analysis

4.1 Parameters of the Rheological Model

The three-parameter rheological model, proposed by Shen [13], has been widely used in the finite element analysis of soil and rockfill dams. It has been proven to have a better adaptability and parameter stability, and the rheological strain was suggested to be expressed as:

$$\varepsilon(t) = \varepsilon_f(1 - e^{-ht}) \tag{3}$$

Where ε(t) is the current accumulated rheological strain, ε_{f} is the final accumulated rheological strain, h is a model parameter.

Since the final accumulated rheological strains are different for ε_{v} and ε_{s}, they were suggested to be expressed as:

$$\varepsilon_{vf} = b(\sigma_3/p_a) \tag{4}$$

$$\varepsilon_{sf} = d\frac{S}{1 - S} \tag{5}$$

where b and d are another two model parameters. Thus, h, b and d are the so-called three parameters.

The confining pressure of unidirectional compression tests can be calculated by the lateral pressure coefficient of the soil K_0, and the confining pressure can be expressed as $\sigma_3 = K_0\sigma_1$. The basic parameters of this clay were derived by conventional triaxial test, which are $c = 31.9$ kPa, $\varphi = 26.7°$ and $K_0 = 0.55$. Then the confining pressures of different axial loadings are shown in Tabel 4, as well as $(\sigma_1 - \sigma_3)_f$ and S according to Eqs. (2) and (1), respectively. As a result, the model parameters calculated by unidirectional compression tests and triaxial tests are listed in Table 4.

Table 4. Parameters b, d and h from the tests.

	Unidirectional compression				Triaxial
σ_1 (kPa)	400	800	1600	3200	$S = 0.7$
σ_3 (kPa)	220	441	881	1762	400
$(\sigma_1 - \sigma_3)_f$ (kPa)	400	764	1467	2765	—
S	0.45	0.47	0.49	0.52	0.7
b	0.000825	0.0007	0.000438	0.000397	0.000318
d	0.0027	0.003	0.0048	0.0048	0.0044
h	0.00115	0.002003	0.002522	0.004636	0.00503

It should be noted that, in the unidirectional compression rheological tests, the parameter b decreases with the increasing loading stress and the parameters d and h show an increase tendency with the increasing loading. Comparing with the two methods, the three parameters b, d and h are in the same order of magnitude, and the results are close, which shows that both methods can be used to determine the parameters.

At present, in the design of high earth dam, only the rheological deformation of the rockfills are considered. As discussed above, the coefficient of secondary consolidation of the clay from the core-wall of earth dam is in the range of 0.00018–0.0015, while that of the coarse-grained soil under confining pressure of 800 kPa is 0.001 [2]. It indicates that the rheological coefficient of compaction clay and rockfills are close, thus it is necessary to consider the rheological deformation of the core-wall in the calculation of stress and deformation of high earth and rockfills dam.

4.2 FEM Results

An earth and rockfill dam is taken as the example, and two schemes of the finite element analysis are set as: (a) Scheme-1, rheological deformation of both rockfills and clay core-wall are considered; (b) Scheme-2, only rheological deformation of rockfills is considered. The mesh of the dam is illustrated as Fig. 4, and the rheological parameters are listed in Table 5.

The displacement increment caused by the rheological deformation of clay core-wall is defined as the difference values calculated by Scheme-1 and Scheme-2. Additionally, the contour diagrams of the displacement increment are illustrated in Figs. 5 and 6. As can be seen from the horizontal displacement increment contour

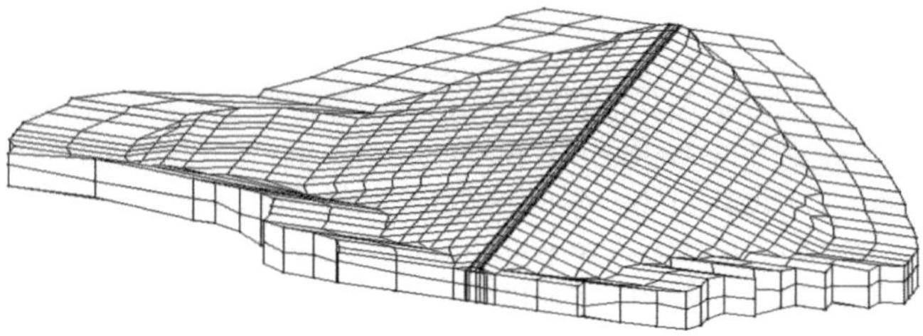

Fig. 4. Three dimensional mesh of the dam.

Table 5. Rheological parameters of Shen Zhujiang model.

Material	*b*	*h*	*d*
Rockfills (upstream)	0.0003	0.007	0.006
Filter	0.0003	0.007	0.006
Rockfills (downstream)	0.0005	0.007	0.006
Clay core-wall	0.0012	0.003	0.004

diagram in Fig. 5 that the values are negative at downstream and positive at upstream, with the maximum positive value of 9.46 cm. It indicates that the rheological deformation of clay core-wall has a significantly effect on the horizontal deformation of the dam.

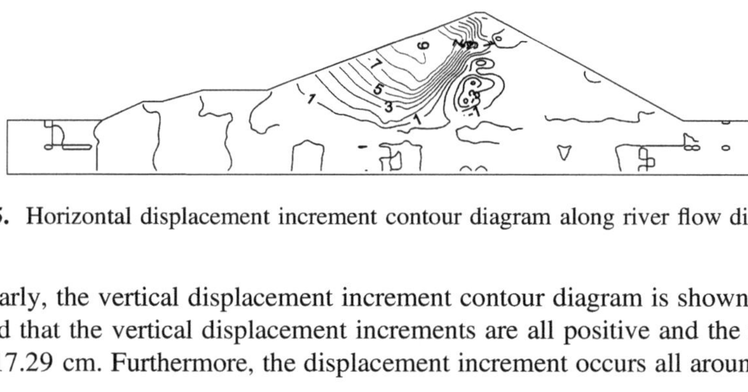

Fig. 5. Horizontal displacement increment contour diagram along river flow direction

Similarly, the vertical displacement increment contour diagram is shown in Fig. 6. It is noted that the vertical displacement increments are all positive and the maximum value is 17.29 cm. Furthermore, the displacement increment occurs all around the clay core-wall.

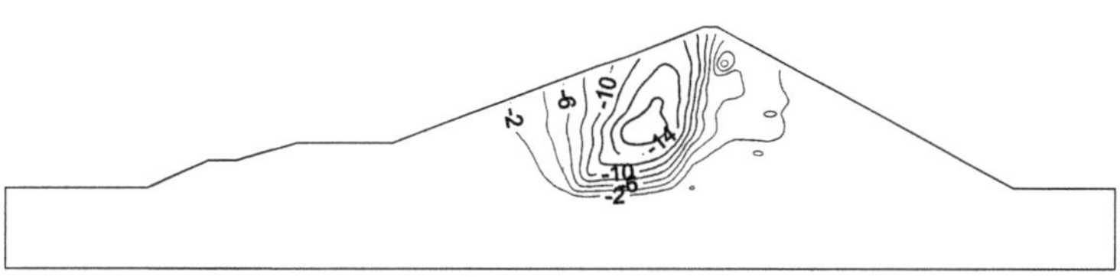

Fig. 6. Vertical displacement increment contour diagram

It is also observed that the maximum stress levels occur at the bottom of the clay core-wall, and the stress levels calculated by Scheme-1 are larger than that of Scheme-2, that is, when the rheological deformation of clay core-wall is considered, the stress levels are larger than before. In other words, the current finite element analysis method (Scheme-2) tends to be more safety than Scheme-1. As a conclusion, the rheological deformation of clay core-wall has a considerable influence on the stress and deformation on the dam, and thus cannot be ignored in actual dam design.

5 Conclusion

It is shown from the rheological tests that the coefficient of secondary consolidation of unsaturated clay is significantly larger than that of saturated clay. When the clay is saturated, the coefficient of secondary consolidation increases with increasing loading stress. Both the unidirectional compression and triaxial compression rheological tests can be used to determine the rheological parameters and the results are relatively close. The finite element analysis indicates that the rheological deformation of clay core-wall has a considerable influence on the stress and deformation on the dam, e.g., causing higher stress levels and settlement especially around the core-wall. As a conclusion, the rheological property of clay core-wall cannot be ignored in actual dam design.

References

1. Guo, W.L., Zhu, J.G., Peng, W.M.: Study on dilatancy equation and generalized plastic constitutive model for coarse-grained soil. Chin. J. Geotech. Eng. (2017). http://kns.cnki.net/kcms/detail/32.1124.TU.20170630.1121.008.html. (in Chinese)
2. Feng, B.L., Niu, J.D., Yang, Q.F.: The coefficient of secondary consolidation of soft clay determined by full automatic consolidation test system. Geotech. Invest. Surv. **39**(3), 11–14 (2011). (in Chinese)
3. Wang, J.J., Zhang, H.P., Tang, S.C., Liang, Y.: Effects of particle size distribution on shear strength of accumulation soil. J. Geotech. Geoenviron. Eng. **139**(11), 1994–1997 (2013)
4. Chen, L.H., Chen, Z.Y., Zhang, J.P., et al.: Study on high pore pressure in clay core-wall of earth dam in Xiaolangdi Project. J. Hydraul. Eng. **36**(2), 219–224 (2005). (in Chinese)
5. Murdoch, L.C.: Mechanical analysis of idealized shallow hydraulic fracture. J. Geotech. Geoenviron. Eng. **128**(6), 488–495 (2002)
6. MacDonald, T.C., Langridge-Monopolis, J.: Breaching charateristics of dam failures. J. Hydraul. Eng. **110**(5), 567–586 (1984)
7. Feng, X.Y., Xu, Z.P.: Centrifugal model study on mechanism of hydraulic fracturing of clay core-wall in rockfill dams. J. Hydraul. Eng. **40**(10), 1259–1263 (2009)
8. Guo, W.L., Zhu, J.G., Yin, J.H., et al.: Investigation into the effects of the thickness of a hollow-cylinder soil specimen on the stress distributions in triaxial torsional shear testing. Geotech. Test. J. **39**(5), 786–794 (2016)
9. Li, Y., Xia, C.C.: Time-dependent tests on intact rocks in uniaxial compression. Int. J. Rock Mech. Min. Sci. **37**(3), 467–475 (2000)
10. Xia, C.C., Wang, X.D., Xu, C.B., et al.: Method to identify rheological models by unified rheological model theory and case study. Chin. J. Rock Mech. Eng. **27**(8), 1594–1600 (2008). (in Chinese)

11. Zhao, Y.L., Cao, P., Wang, W., et al.: Viscoelasto-plastic rheological experiment under circular increment step load and unload and nonlinear creep model of soft rocks. J. Cent. South Univ. Technol. **16**(3), 488–494 (2009)
12. Zhao, Y.L., Cao, P., Wang, W.: Rock cracks subcritical propagation test and compression-shear rheological fracture model. J. Cent. South Univ. (Sci. Technol.) **45**(1), 276–286 (2014)
13. Shen, Z.J.: Back analysis of deformation of Lubuge earth core rockfill dam. Chin. J. Rock Mech. Eng. **16**(3), 1–13 (1994). (in Chinese)

Influence of Different Loading Angles with Respect to the Stratification on Tensile Strength of Black Shale

Xiao-shan Shi[1,2], Jing-rui Niu[1,2(✉)], Da-an Liu[1,2], Tie-wu Tang[1,2], Bonan Wang[3], and Weige Han[1,2]

[1] Key Laboratory of Shale Gas and Geoengineering, Institute of Geology and Geophysics, Chinese Academy of Sciences, Beijing 100029, China
{xiaoshanshi,nuijingrui}@mail.iggcas.ac.cn
[2] University of Chinese Academy of Sciences, Beijing 100049, China
[3] School of Mechanics and Civil Engineering, China University of Mining and Technology, Beijing 100083, China

Abstract. The investigation on anisotropy of shale in mechanical and psychical characteristics is of great significance to successful implementation of hydraulic fracturing in gas shale exploitation. In this paper, the effect of shale anisotropy on tensile strength was studied via Brazilian splitting tests (BD) at different loading angles (the angle between loading axis and the stratification). The specimens were prepared along different angles (0°, 30°, 45°, 60°, 90°) with respect to the stratification of black shale (BS). The results indicate that the tensile strength for samples with different loading angles exhibits obvious anisotropy. It was noted that complex tensile fractures can be formed under larger loading angels. Studying the tensile strength anisotropy of shale can provide some theoretical guidance for optimization of fracturing scheme and fracture evaluation.

Keywords: Black shale · Failure pattern · Tensile strength anisotropy
Brazilian test · Loading angle

1 Introduction

Shale often behaves as obvious anisotropy due to depositional environment and the pre-existing micro-cracks induced by long-term geological processes. Characterizing the effect of anisotropy of shale on the tensile strength is of great significance in hydraulic fracturing engineering. Stratification is the most significant features in shale, which induces anisotropic characteristics for shale [1–5]. As reported by previous research [6–9], shale shows high degree of anisotropy with parallel or perpendicular to the bedding, not only in velocity, elastic modulus, Poisson's ratio, diffusivity and porosity, but also in the strength and fracture mechanism. The mechanical anisotropy of shale was induced by the organized distribution of platy clay minerals and compliant organic materials [10–12]. These anisotropic characteristics have significant influence on the hydraulic fracturing, borehole stability, perforation direction chosen and volume

L. Zhang et al. (Eds.): GSIC 2018, *Proceedings of GeoShanghai 2018 International Conference: Rock Mechanics and Rock Engineering*, pp. 204–211, 2018.
https://doi.org/10.1007/978-981-13-0113-1_23

fracturing and so on. Therefore, it is highly necessary to investigate the mechanical anisotropy of shale under the influence of stratification, such as, the tensile strength, compressive strength and fracture toughness. Some studies showed that the stratification also had a great influence on the tensile strength. Jung [13] conducted experiments of deformation and strength anisotropy of Asan gneiss, Boryeong shale, and Yeoncheon schist under different loading angles. Claessona [14] researched the stress field and tensile strength of anisotropic rocks by using an analytical solution method. It can be seen from the researches mentioned above that limited investigations exist on the tensile strength anisotropy of shale induced by the different loading angles with respect to the stratification.

Commonly, there are two ways for measuring rock tensile strength, direct tensile test and indirect tensile test, respectively. However, sample preparation is much more difficult than for indirect tensile test. In addition, very often phenomenon of stress concentration results in damage near sample ends and the incorrect interpretation of outcomes. Indirect methods, such as, Brazilian disk test [15–17], bending test [17] and ring test [16, 18] methods, have been developed to determine the tensile strength. These methods simplify loading mode and easily cause tensile stress in the sample by far-field compression. In the present study, BD test was used to investigate the influence of stratification on the tensile strength. The Cambrian Silurian black shale was drilled from Shizhu County, Chongqing in China, and the specimens were prepared along different angles (0°, 30°, 45°, 60°, 90°) with respect to the stratification. The paper consists of four sections in total. Section 2 presents the experiments methodology including the experiment instrument system, specimen preparation, and tensile stress calculation, followed by result and discussion in Sect. 3. Section 4 summarizes the main conclusions.

2 Experiment Methodology

2.1 Experiment Preparation

The BD test was conducted by a conventional servo-control material test system with a loading rate of 200 N/s. The Cambrian Silurian Black shale, without weathering and disturbance, was collected from Shizhu County, Chongqing city of China (shown in Fig. 1). The mineralogical composition of BS obtained by using X-ray diffraction technique was shown in Table 1.

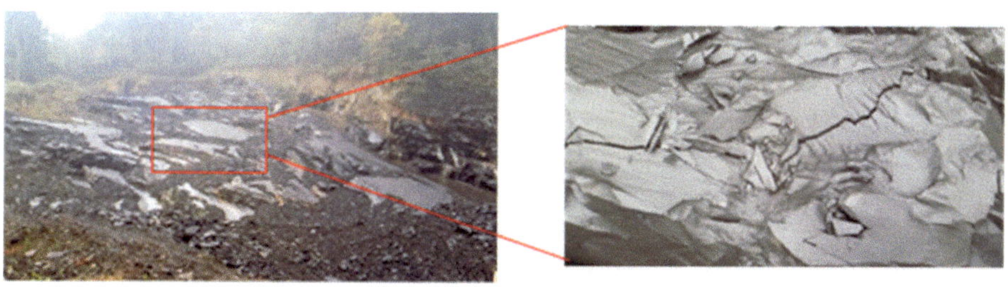

Fig. 1. Sample on site.

Table 1. Mineral composition of BS.

Composition	Quartz	Clay	Feldspar	Pyrite	Calcite
Content (weight %)	49.3	33.6	10.9	3.6	2.6

For BD test, 50 mm - diameters cylinders are cored from a rock block of BS with parallel (along the sampling 1 direction in Fig. 2) and are then sliced to discs with nominal thickness of 25 mm. All samples are polished to a surface roughness better than 0.5% of the sample thickness. Figure 3 shows the loading schematics of the BD test. To investigate the effect of the stratification on the tensile strength, the loading angle β between the loading direction and the stratification varies. For specimens along the sampling 1 direction, the loading angle β equals to 0°, 30°, 45°, 60° and 90° (shown in Fig. 3b). Via the basic mechanical testing, the elasticity modulus and Poisson's ratio of the BS at different angles with respect to the stratification are listed in Table 2.

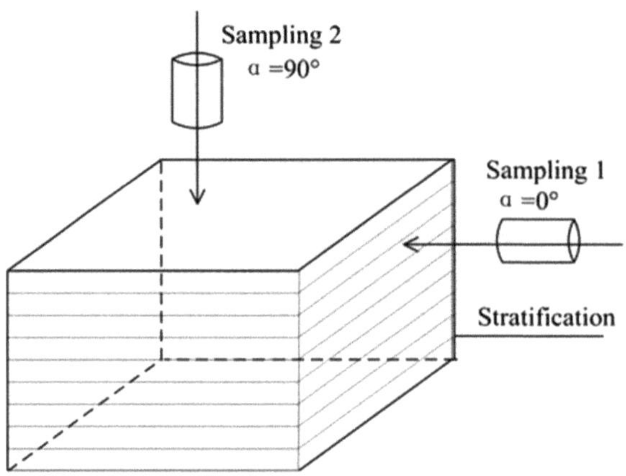

Fig. 2. Sampling direction

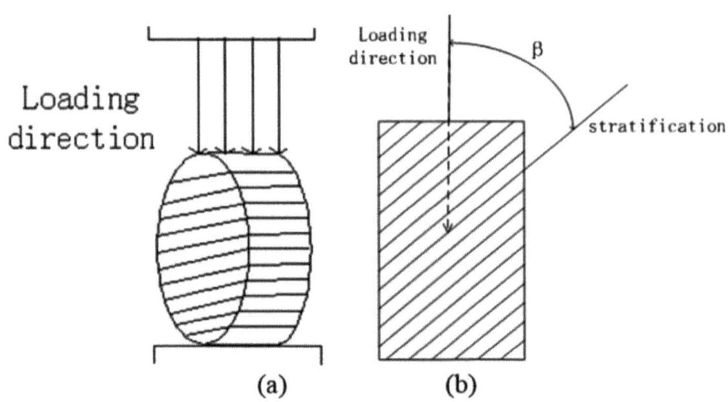

Fig. 3. Schematics of the BD test

Table 2. Elastic modulus and Poisson's ratio of BS.

β (°)	Young's modulus (GPa)	Poisson's ratio
0	24.90	0.26
30	21.72	0.22
45	20.01	0.24
60	16.68	0.27
90	14.09	0.32

2.2 Tensile Stress Calculation

The mechanical property of BS shows high degree of anisotropy with perpendicular to the stratification, while shows isotropic feature with parallel to the bedding, so it can be considered a transversely isotropic material [15]. However, many researchers neglected the effect of stratification on the properties of rocks and used the tensile strength calculation for isotropic rocks. McLamore et al. [19] research mechanical behavior of anisotropy sedimentary rocks. Istvan et al. [20] conducted a series studies about rock mechanics experiments of layered rocks. Claesson and Bohloli [14] used an analytical solution to get the stress field and tensile strength by Brazilian test under different loading angles, then verified this method through experiments. In the present study, the formula proposed by Claesson and Bohloli [14] was used to calculate the tensile strength for BS.

$$\sigma_{pt}(0,0) = \frac{2 * P}{\pi DH}[\left(\sqrt[4]{E/E'}\right)^{cos(\beta)} - \frac{cos(4\beta)}{4}(b-1) \tag{1}$$

$$b = \frac{\sqrt{EE'}}{2}\left(\frac{1}{G'} - \frac{2v'}{E'}\right) \tag{2}$$

Where $\sigma_{pt}(0,0)$ denotes the tensile stress on the point of central fracture, P is the peak force; β is the loading angle between the stratification and loading direction; D, H and E are the diameter, thickness and Elastic Modulus of the specimen along the stratification, respectively; G', E' and v' denote shear modulus, elastic modulus and Poisson ratio perpendicular with the stratification; and b is a function of E, E', v' and G'.

So, we use the equation of (2) and (3) to get the tensile strength under different loading angles.

For isotropic materials in the Brazilian tests, the formula is used to get the tensile strength as follows:

$$\sigma_t = \frac{2 * P}{\pi DH} \tag{3}$$

Therefore, $K(\beta)$ was named as a coefficient which is a function of β, E and E':

$$K(\beta) = \left(\sqrt[4]{E/E'}\right)^{cos(\beta)} - \frac{cos(4\beta)}{4}(b-1) \quad (4)$$

$$\sigma_{pt}(0,0) = \sigma_t * K(\beta) \quad (5)$$

3 Result and Discussion

3.1 Tensile Stress Results

The tensile strength can be calculated via the Eq. (4) and all values are tabulated in the Table 3. Figure 4 exhibits the variation of the tensile strength under different loading angles. For each loading angle, the averaged value of the tests is used. From the Fig. 4, it can be seen that the tensile strength increases with the increases of the loading angle. The specimen with $\beta = 0°$ has the lowest tensile strength of 8.01 MPa, i.e. the BS specimen with the loading axis (the same as disc fracture plane) along the stratification yields the lowest tensile strength. The BS specimens with $\beta = 30°$, 45° or 60° have the middle tensile strength of 8.24 MPa, 9.17 MPa and 9.64 MPa, respectively. The specimen with $\beta = 90°$ has the highest tensile strength of 12.84 MPa. The highest value (12.84 MPa from the specimen with $\beta = 90°$) is 1.6 times higher than the lowest one (8.01 MPa from the specimen with $\beta = 0°$). This phenomenon indicates that the tensile strength anisotropy induced by different loading angles is obvious for BS.

Table 3. Tensile strengths under different loading direction.

No.	H (mm)	D (mm)	β (°)	P (KN)	K (β)	σ_{pt} (KPa)
0-1	25.30	49.80	0	14.35	1.071	7.76
0-2	25.33	49.60		15.11		8.20
0-3	25.17	49.80		15.07		8.19
0-4	25.29	49.90		14.61		7.89
30-1	25.30	49.83	30	16.99	0.955	8.20
30-2	25.23	49.70		16.75		8.12
30-3	25.20	49.63		17.07		8.30
30-4	25.18	49.69		17.18		8.35
45-1	25.37	49.73	45	16.5	1.127	9.38
45-2	25.47	49.80		15.86		8.97
45-3	25.57	49.73		16.62		9.38
45-4	25.36	49.87		15.77		8.95
60-1	25.23	49.50	60	23.25	0.846	10.03
60-2	25.40	49.80		23.02		9.81
60-3	25.60	49.80		22.91		9.68
60-4	25.49	49.61		21.18		9.03

(continued)

Table 3. *(continued)*

No.	H (mm)	D (mm)	β (°)	P (KN)	K (β)	σ_{pt} (KPa)
90-1	25.33	49.60	90	25.9	0.962	12.62
90-2	25.17	49.67		26.5		12.98
90-3	25.27	49.63		25.97		12.68
90-4	25.47	49.61		26.99		13.08

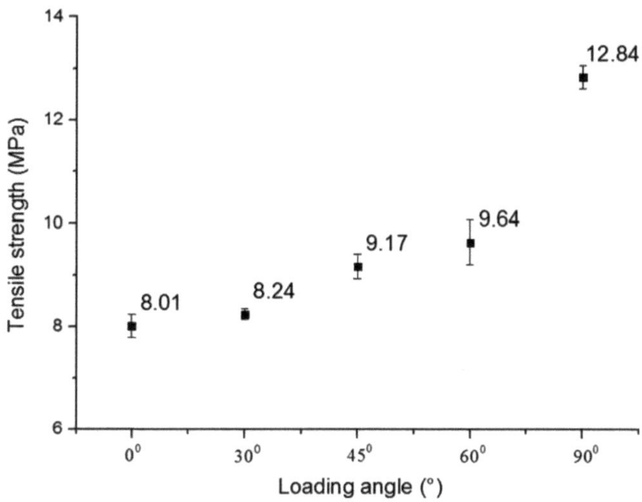

Fig. 4. The variation of the tensile strength under different loading angles.

3.2 Failure Pattern Analysis

It's well known that isotropic materials are almost split cleanly into two halves along the loading axis in the BD test. Figure 5 shows typical recovered samples after BD test. For specimen with β = 0°, the tensile failure occurs and a main failure crack propagates along the loading direction until the final failure. While for specimen with β = 30°, not only the tensile failure happens along the loading direction, but also the slippage crack occurs along the stratification. The crack initiates and propagates from the central point of specimen and then deviates towards the stratification plane. When β = 45°, 60° and 90°, the main crack along the loading direction and tensile failure occurs. At meantime, some second cracks accompanied. It can be seen that more complicated cracks occur when the loading angle increases.

The cementation strength of the stratification is weak and many micro-cracks exist parallel to the stratification plane, resulting in weak strength along this direction. In other words, it is easy for the crack to propagate along the stratification. It is hard to propagate vertically to the bedding plane because the crack overcomes double cementation strength (i.e., the cementation strength on the stratification and between different layers). Therefore, the failure plane is parallel to the stratification when β = 0° while it is perpendicular to the stratification when β = 90°. Therefore, the specimen

Fig. 5. Failure patterns of recovered specimen in the BD tests

with β = 0° has the lowest tensile strength while the specimens with β = 90° yields the highest tensile strength. And the specimen with β equal to 45° and 60° own the middle tensile strength.

4 Conclusions

In this work, we investigated the tensile strength anisotropy of BS with varied loading angles. The BD test was conducted with a conventional servo-control material test system (MTS-CMT5105). The results demonstrate that the tensile strength of BS shows obvious anisotropy under different loading angles. The highest tensile strength value is 1.6 times higher than the lowest one, which reveals that the tensile strength anisotropy of BS induced by different loading angles with respect to the stratification should be considered in mechanical analysis for BS. In addition, the failure pattern of BS specimens under different loading angles has anisotropic trait. When the loading angles equal to 0° and 45°, 60° and 90°, the specimen is subject to tensile failure. While for specimens with loading angles of 30°, the tensile – shear failure happens. And the larger loading angles are, the more complicated cracks occur. The results can provide some references for the hydraulic fracturing in gas shale engineering.

Acknowledgements. The research reported in this paper was supported by the National Natural Science Foundation of China, study of Rock Fracture under Hydraulic External Force Disturbance (No. 41172270), and supported by the Strategic Priority Research Program of Chinese Academy of Sciences, Research and Development of Test Equipment about Fracture, Seepage and Microseisms (No. XDB10050200) and Science Foundation for Youths (No. 41602330).

References

1. Jaeger, J.C.: Shear failure of anisotropic rock. Geol. Mag. **97**(1), 65–72 (1960)
2. Taliercio, A., Landriani, G.S.: A failure condition for layered rock. Int. J. Rock Mech. Mining Sci. **25**(5), 299–305 (1988)

3. Ramamurthy, T.: Strength, modulus responses of anisotropic rocks. Compressive Rock Eng. **1**(1), 313–329 (1993)
4. Tien, Y.M., Kuo, M.C.: A failure criterion for transversely isotropic rocks. Int. J. Rock Mech. Min. Sci. **38**(3), 399–412 (2001)
5. Nasseri, M.H.B., Rao, K.S., Ramamurthy, T.: Anisotropic strength and deformational behavior of Himalayan schists. Int. J. Rock Mech. Min. Sci. **40**(1), 3–23 (2003)
6. Lee, Y.K., Pietruszczak, S.: Application of critical plane approach to the prediction of strength anisotropy in transversely isotropic rock masses. Int. J. Rock Mech. Min. Sci. **45**(4), 513–523 (2008)
7. Wan, Y., et al.: An experimental investigation of diffusivity and porosity anisotropy of a Chinese gas shale. J. Nat. Gas Sci. Eng. **23**, 70–79 (2015)
8. Kuilaa, U., Ewhurstb, D.N., Igginsb, A.F.: Stress anisotropy and velocity anisotropy in low porosity shale. Tectonophysics **503**(1–2), 34–44 (2011)
9. Niandou, H., Shao, J.F., Henry, J.P.: Laboratory investigation of the mechanical behaviour of Tournemire shale. Int. J. Rock Mech. Min. Sci. **34**(1), 3–16 (1997)
10. Sondergeld, C.H., et al.: Microstructural studies of gas shales. In: SPE Unconventional Gas Conference. Society of Petroleum Engineers, Pittsburgh (2010)
11. Sondergeld, C.H., Rai, C.S.: Elastic anisotropy of shales. Lead. Edge **30**, 324–331 (2011)
12. Sondergeld, C.H., et al.: Ultrasonic measurement of anisotropy on the Kimmeridge shale. In: 70th Annual International Meeting, SEG, Expanded Abstracts, pp. 1858–1861 (2000)
13. Jung, W.C., Hanna, K., Seokwon, J.: Deformation and strength anisotropy of Asan gneiss, Boryeong shale, and Yeoncheon schist. Int. J. Rock Mech. Min. Sci. **50**, 158–169 (2012)
14. Claessona, J., Bohloli, B.: Brazilian test: stress field and tensile strength of anisotropic rocks using an analytical solution. Int. J. Rock Mech. Min. Sci. **39**, 991–1004 (2002)
15. McLamore, R., Gray, K.E.: The mechanical behavior of the anisotropic sedimentary rocks. J. Eng. Ind. **89**, 62–76 (1967)
16. Istvan, J.A., et al.: Rock mechanics forgas storage in bedded salt caverns. Int. J. Rock Mech. Min. Sci. **34**, 3–4 (1997)

Limit Strain Method for Determining Long-Term Strength Based on Isostrain-Rate Creep Curve

Zhen Wang[1], Mingrong Shen[1,2], Qingzhao Zhang[1,2](✉), Linlin Gu[3], and Zejun Luo[1]

[1] Department of Geotechnical Engineering, Tongji University, Shanghai 200092, China
zqz0726@163.com
[2] Key Laboratory of Geotechnical and Underground Engineering of Ministry of Education, Tongji University, Shanghai 200092, China
[3] Department of Civil Engineering, Qingdao University of Technology, Qingdao 266520, China

Abstract. Long-term strength is one of important behaviors in predicting the long-term stability of rock mass. In this paper, conventional compression tests and muti-stage creep tests were conducted and limit strain method for determining the long-term strength were proposed based on isostrain-rate curve. The results demonstrated that the long-term strength determined by the limit strain method could meet the range of long-term strength could solved by transition creep method, and the results was reasonable.

Keywords: Creep · Long-term strength · Limit strain method Transition creep method

1 Introduction

All rocks may show a mechanical response that is time-dependent. The failure strength of the rocks will tend to decrease over time and the deformation also increases even under constant long-term loading [1]. In many applications, the failure of the rock mass under loading is not a sudden brittle failure but instead damaged after years or decades [2].

Long-term strength is one of rock's important time-dependent properties, and it can be defined as the maximum stress that would not cause failure of rock under a long-term loading [3, 4]. According to the definition of long-term strength, the most direct method to determine the long-term strength is to use a single-stage constant load test under a series of different stress levels, which can directly obtain the long-term strength value. Although this method is reasonable in theory, it is difficult to be used in actual operation as the failure stress is unpredictable. Therefore, the indirect method were proposed and a number of methods to determine the long-term strength have been proposed [5–7]. The tautochrone method and transition creep method were most widely used methods in present researches. However, related research demonstrated that tautochrone inflection point has numerous variation laws and can't always determine the long-term strength accurately [2]. Transition creep method is also proposed by

L. Zhang et al. (Eds.): GSIC 2018, *Proceedings of GeoShanghai 2018 International Conference: Rock Mechanics and Rock Engineering*, pp. 212–219, 2018.
https://doi.org/10.1007/978-981-13-0113-1_24

many researchers [7, 8], and the results of this method are a range of values depending on the stress step of creep test, which makes it rather difficult using such method to obtain accurate long-term strength. So it's necessary to explore some reasonable methods to determine the long-term strength.

In this paper, the conventional tests and muti-stage creep tests were conducted on green schist. The isostrain-rate curve was plotted and the long-term strength were solved by transition creep method and limit strain method.

2 Methodology

2.1 Test Equipment

A triaxial rehological test machine TLW-2000 were used for triaxial compression test and traxial muti-stage creep tests. The maximum vertical axial compression load is 2000 kN, and the maximum confining pressure can be applied to 50 MPa. The strain in each directions were measured by two displacement transducers and the measuring accuracy of deformation is ±1% [2] (Fig. 1).

Fig. 1. Test equipments

2.2 Samples

The study is conducted using green schist from Jinping II hydropower station in Sichuan Province, China [9]. The samples were obtained from the diversion tunnel which is 1600 m below ground. The samples prepared for tests are 50 mm in diameter, 100 mm in height according to ASTM [10].

To get further information of the sample, the uniaxial compression test and triaxial compression tests were conducted. The uniaxial compressive strength, elastic modulus

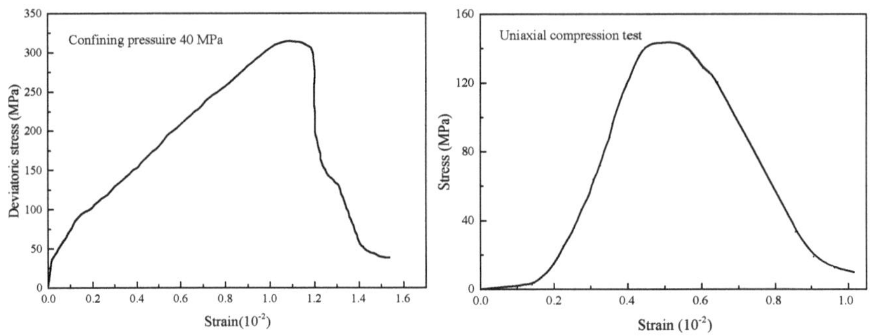

Fig. 2. Complete stress-strain curves of triaxial compression test

and Poisson's ratio were 134.3 MPa, 9.39 GPa and 0.25, respectively. The cohesion of the greenschist samples was 31.2 MPa and the friction angle was 42.3°. The complete strain-stress curves were plotted in Fig. 2.

2.3 Tests Procedure

The triaxial and uniaxial muti-stage creep tests were initially loaded at a constant strain rate 0.15 MPa/s, until loading was stopped at a predetermined axial stress, and the axial load was held constant for 72 h and then loaded to the next predetermined axial stress. During the tests, the temperature was controlled at 20 ± 1 °C and the humidity was maintained at 50 ± 2% using a constant temperature humidistat. The stress and strain were recorded, automatically [2].

3 Results and Discussion

3.1 Solution of Long-Term Strength

(1) Transition creep method

During the creep process, the samples don't fail and only exhibit transient creep stage if the applied stress is below a stress threshold value. Once the applied stress surpass the threshold stress, clear steady creep stage can be observed, and finally termed to the accelerating creep stage and caused the failure. The transition creep method defined the threshold stress as the long-term strength. Although it is impossible to record the exact value of this stress threshold at which transient creep occurs, an interval can be obtained from the muti-stage creep tests.

Take the muti-stage creep test under confining pressure 40 MPa as an example. As shown in Fig. 3, the curve termed to the steady creep stage after transient creep stage. it shows the obvious steady creep rate characteristics, and the creep rate kept almost constant in the last 24 h in Fig. 3. Therefore, average creep rate of the last 24 h can be defined as steady creep rate which can be used for determining the long-term strength, approximately. When the axial stress was 290 MPa or lower than it, the steady creep

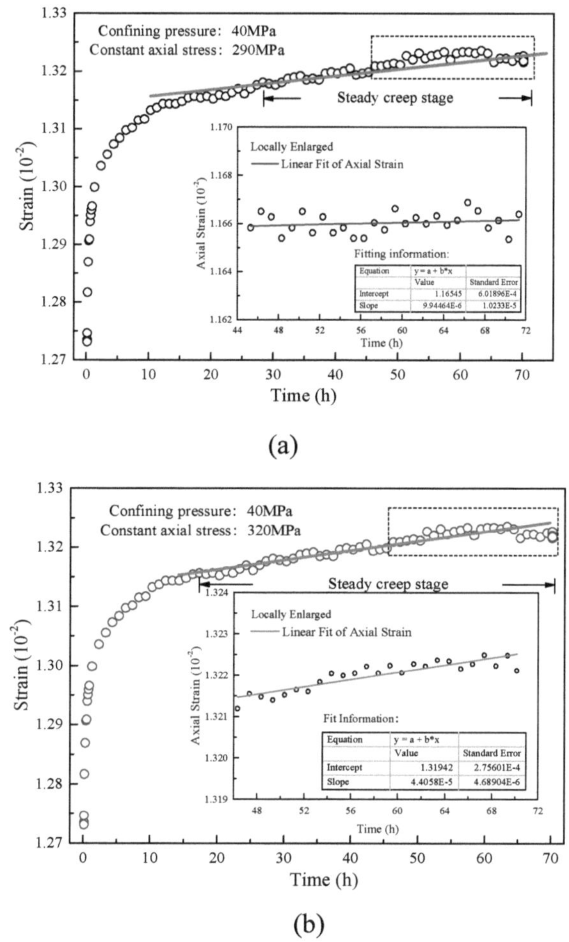

Fig. 3. Some creep curves under confining pressure 40 MPa in muti-stage creep test

rate of last 24 h was close to 0, and the fitting curve in Fig. 3(a) was almost horizontal. Once the axial stress surpass 290 MPa, the steady creep rate increased greatly. As shown in Fig. 3(b), the slope of the fitting curve increased obviously when the axial stress was 320 MPa.

The steady creep rate of each stress level at last 24 h in muti-stage creep tests were plotted in Fig. 4. We can use it to estimate the range of long-term strength according to the transition creep method, as shown in Fig. 4. In Fig. 4, the steady creep rate increased greatly when the axial stress surpass 290 MPa, and we can defined that the range of long-term strength was 290 MPa to 320 MPa. We can also determine the range of long-term strength under uniaxial creep test is 94–105 MPa.

The long-term strength can be also determined by creep properties [2], and the results were also [290, 320], [94, 105].

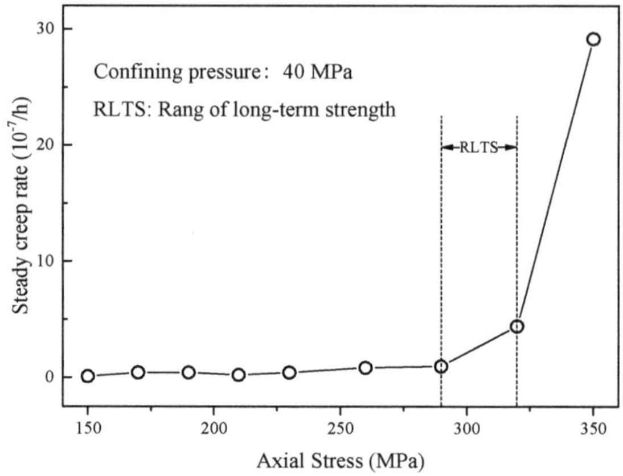

Fig. 4. Relationship between steady creep rate and axial stress

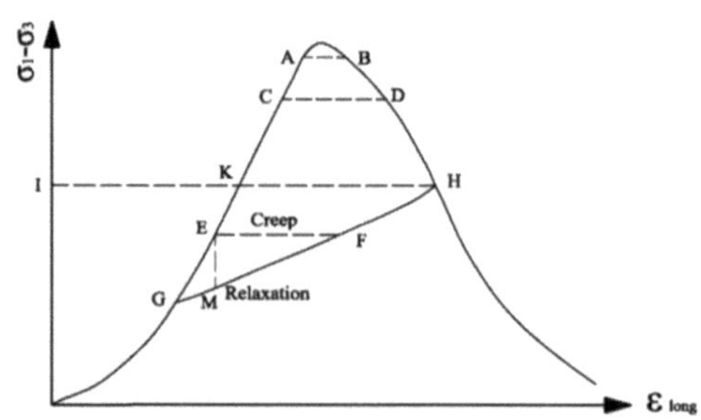

Fig. 5. Limit strain method for determining the long-term strength

(2) Limit strain method

The concept of limit strain curve was also proposed by Bérest et al. [11]. It was defined as the stress-stain curve when the loading strain rate is slow enough to allow the development of strain in real time. The stress value of point H can be called long-term strength, because it is the minimum stress (threshold) that can cause the failure of rock in long period of time. Therefore, if the locus of limit strain could be calculated, then point H can be solved combining with the complete stress-stain curve, and the long-term strength could be determined (Fig. 5).

As an example, the long-term strength of green schist is determined by the limit strain method as follows:

Isostrain-Rate Creep Curves

The stress (σ), strain(ε) which equal in strain rate(v) were taken to plot an isostrain-rate creep curve in the stress-strain coordinate system. Figure 6 shows the process of plotting isostrain-rate creep curve, and curves under different strain rate can be called the isostrain-rate creep curves cluster [2].

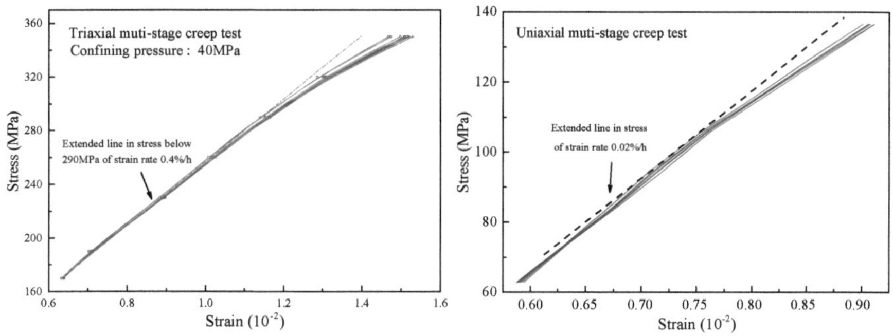

Fig. 6. Iso-strain rate curves [2]

The isostrain-rate creep curves were plotted using the method mentioned above which contains ten different strain rates as follows: 0.4%/h, 0.2%/h, 0.1%/h, 0.05%/h, 0.005%/h, 0.002%/h, 0.001%/h, 0.0005%/h, 0.0004%/h, 0.0002%/h. As shown in Fig. 6, all the curves approach to linear relationship when stressed below 290 MPa and 94 MPa. The morphology of curves turned to nonlinear when the specimen was stressed above a value larger than 290 MPa and 94 MPa, and the characteristics of nonlinear turned to more and more obvious with the increasing of stress.

Solution of Limit Strain Curve

The isostrain-rate curve when the stress was lower than the threshold, the curve were all approach to a straight line, as shown in Fig. 6. The linear section of isostrain-rate curve can be described with the formula as follows:

$$\sigma = A\varepsilon + B \quad (1)$$

where:

σ = stress, MPa, and ε = strain, %. The correlation coefficient of fitting using the formula (1) are all higher than 0.9997. The parameter A and B in different strain rate were plotted in Fig. 7. The curves under different strain rates are similar in shape, and limit strain curves (strain rate is 0) can be calculated by the regularity between curves as shown in Fig. 7.

Therefore, when the confining pressure was 40 MPa, the equation of limit strain curve was $\sigma = 22.3 \times 10^3\varepsilon + 28.32$, and the limit strain curve under uniaxial condition can be described as $\sigma = 26.3 \times 10^3\varepsilon - 92.62$.

Long-Term Strength of Green Schist

In order to reduce the error caused by the difference between rock samples, the vertical axis in Fig. 8 is the ratio of stress and peak strength. As shown in Fig. 8, the long-term strength of green schist is the stress at point *H* which is the intersection point of complete stress-strain curve and limit strain curve, and the results was 78.3% of peak strength under confining pressure 40 MPa, and 64% in uniaxial tests.

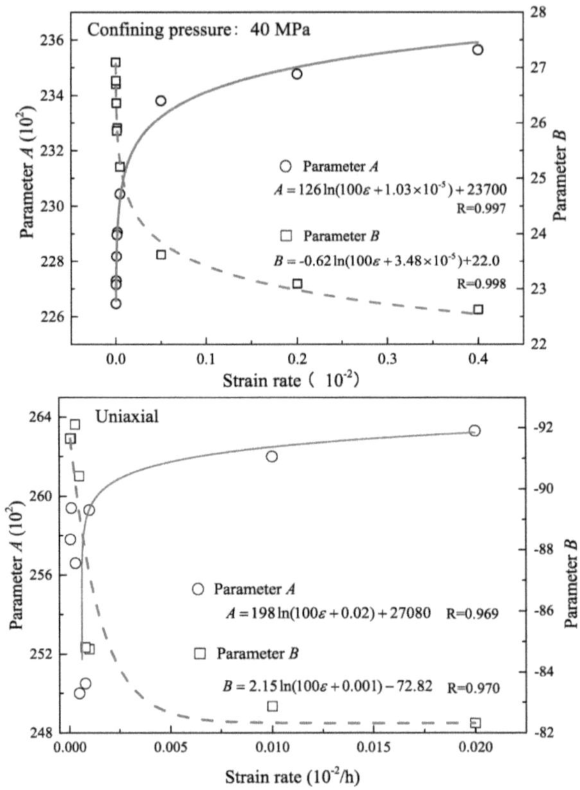

Fig. 7. Relationships between parameter and stain rate in formula (1)

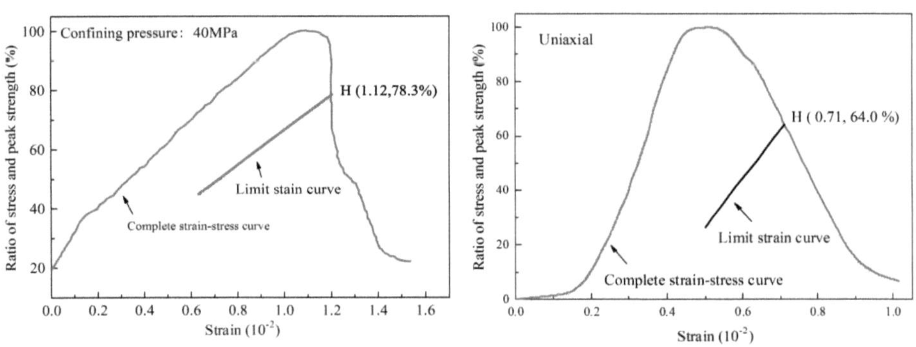

Fig. 8. Long-term strength determined by limit strain method

4 Conclusion

In this paper, we proposed a method for determining the long-term strength based on the isostrain-rate creep curve. To verify this method, conventional compression tests and muti-stage creep tests were performed. Based on the isostrain-rate curve. The following conclusions can be drawn: The long-term strength of green schist under uniaxial and triaxial conditions determined by limit strain method were 94.1 MPa, and 300 MPa. The results agrees with the range value estimated by the transition creep method. The results are close to each other, and also meet the empirical value of long-term strength.

Acknowledgments. This work is supported Doctoral Fund of Shandong Province (Grant No. ZR2017BEE071), China Postdoctoral Science Foundation (Grant No. 2017M612227), National Natural Science Foundation of China(No. 41602287)

References

1. Cristescu, N.D., Hunsche, U.: Time Effect in Rock Mechanics. Wiley, New York (1997)
2. Wang, Z., Shen, M., Gu, L., Zhang, F.: Creep behavior and long-term strength characteristics of greenschist under different confining pressures. Geotech. Test. J. https://doi.org/10.1520/GTJ20170143. ISSN 0149-6115
3. Zhao, Y., Ping, C., Wen, Y., Wang, Y., Chai, H.: Elastovisco-plastic rheological experiment and nonlinear rheological model of rocks. Chin. J. Rock Mech. Eng. **27**(3), 477–486 (2008)
4. Liu, X.: Rheology of Rocks, pp. 194–207. Geological Publishing House, Beijing (1994)
5. Szczepanik, Z., Milne, D., Kostakis, K., Eberhardt, E.: Long term laboratory strength tests in hard rock. In: 10th ISRM Congress. International Society for Rock Mechanics (2003)
6. Cui, X.H., Fu, Z.L.: Experimental study on rheology properties and long-term strength of rocks. Chin. J. Rock Mech. Eng. **25**(5), 1021–1024 (2006)
7. Liu, A., Shen, M.R., Jiang, J.C.: Investigation of the shear stress relaxation characteristics of a structural plane using the isostress cyclic loading method. Geotech. Test. J. **38**(2), 219–228 (2015)
8. Shen, M.R., Chen, H.J.: Testing study of long-term strength characteristics of red sandstone. Rock Soil Mech. **32**(11), 3301–3305 (2011)
9. Shen, M., Zhang, Q.: Study of the shear creep characteristic of green schist discontinuity. Chin. J. Rock Mech. Eng. **29**(6), 1149–1155 (2010)
10. ASTM D7070-08: Standard Test Method for Creep of Rock Core Under Constant Stress and Temperature, USA. ASTM (2008)
11. Bérest, P., Bergues, J., Duc, N.M.: Comportement des roches au cours de la rupture: application á l'interprétation d'essais sur des tubes épais. Rev franc-aise Géotech **9**, 5–12 (1979)

Effects of Normal Stress and Clay Content on the Frictional Properties of Reservoir Rocks Under Fully Saturated Conditions

Mengke An[1,2], Fengshou Zhang[1,2](✉), Lianyang Zhang[3](✉), and Yi Fang[4]

[1] Department of Geotechnical Engineering, Tongji University, Shanghai 200092, China
fengshou.zhang@tongji.edu.cn
[2] Key Laboratory of Geotechnical and Underground Engineering of Ministry of Education, Tongji University, Shanghai 200092, China
[3] Department of Civil Engineering and Engineering Mechanics, University of Arizona, Tucson, AZ 85721, USA
lyzhang@email.arizona.edu
[4] Department of Energy and Mineral Engineering, G3 Center, EMS Energy Institute, The Pennsylvania State University, University Park, PA 16802, USA

Abstract. The large magnitude micro-seismicity observed during the process of hydraulic fracturing is likely to be associated with the activation of pre-existing faults. To better understand the mechanism of fault stability, shear tests were performed on powdered gouge of shale and sandstone reservoir rocks from four different oilfields in China using the double direct shearing geometry, at three successive normal stresses, 10, 20 and 40 MPa, and under fully saturated conditions, aiming to investigate the effects of normal stress and clay contents on frictional strength and stability of faults. The velocity step tests were performed from 0.1 to 100 μm/s to assess the frictional stability by calculating the value of $a - b$. It is observed that the frictional strength decreases with higher clay content and a transition from velocity weakening behavior ($a - b < 0$) to velocity strengthening behavior ($a - b > 0$) at a clay content of 15–20%. The frictional strength generally decreases slightly with higher normal stress. The critical slip displacement D_c shows an increasing trend with the increase of clay content at both low and high normal stress, while an opposite trend appears under medium normal stress.

Keywords: Friction · Stability of faulting · Minerology · Stress conditions Reservoir rocks

1 Introduction

Hydraulic fracturing technique has been widely applied to the exploitation of unconventional oil and gas resources due to its effectiveness in enhancing oil and gas recovery, especially for low permeability reservoirs [1]. However, a series of felt earthquakes in North America and Europe are believed to be highly correlated with the hydraulic fracturing operations, such as injection of high pressure fluids and deep disposal of

L. Zhang et al. (Eds.): GSIC 2018, *Proceedings of GeoShanghai 2018 International Conference: Rock Mechanics and Rock Engineering*, pp. 220–231, 2018.
https://doi.org/10.1007/978-981-13-0113-1_25

wastewater [2, 3]. One possible mechanism of hydraulic fracturing induced seismicity is that the stress changes can activate the underlying pre-existing faults [4, 5]. Therefore, a careful understanding of the potential factors for fault stability and activation in reservoir systems is significant for assessing seismic hazard during the lifetime of hydraulic fracturing.

Clay minerals, common components in mature natural faults, are considered to greatly affect the macroscopic fault strength and stability [6, 7]. The research work of Shimamoto and Logan [8] showed that the fault strength can be significantly altered with clay contents as low as 15–20%. Takahashi et al. [9] reported that a sharp drop of friction coefficient could be observed at 50% of clay content. Crawford et al. [10] conducted triaxial shearing tests on synthetic quartz and kaolinite under hydrostatic conditions and showed that the frictional strength of fault gouge decreases continuously with increasing clay contents. Tembe et al. [11] explored the effects of binary and ternary mixtures of quartz, illite, and montmorillonite on the frictional sliding of simulated faults. The experimental results demonstrated that illite/quartz mixture shows different properties compared with montmorillonite mixtures. Later, the experiments of Ikari et al. [12] and Tesei et al. [13] also indicated that the phyllosilicate-rich fault gouge is less likely to promote unstable slip and accelerate the earthquake nucleation. Although the effect of clay minerals on the fault friction behavior has been investigated under a broad range of laboratory conditions, little research work has been carried out on the natural reservoir rock samples under fully saturated conditions.

Normal stress is another key parameter for controlling the behavior of frictional stability and strength [14]. For example, Saffer et al. [15] reported that the friction of semectite clay is closely related with the normal stress conditions. Under low normal stress (<30 MPa), the semectite clay exhibits velocity weakening behavior, while under higher normal stress (>50 MPa) only velocity strengthening behavior can be observed. Saffer and Marone [16] pointed out the effects of normal stress on frictional properties of smectite- and illite-rich gouge are quite different. The results of Ikari et al. [17] revealed that high effective stress is crucial for fault instability and seismic nucleation. Hartog et al. [18] performed biaxial tests on illite shale fault gouge under different normal stresses and found that the increased velocity strengthening behavior can be observed at higher normal stress. However, the effect of normal stress on the frictional properties of natural reservoir rocks especially under fully saturated conditions has not been fully understood yet.

In this study the double direct shearing tests are carried out on natural rock samples from different oilfields in China in order to investigate the effects of normal stress and clay content on the frictional strength and stability of faults. Different from most of the documented experiments in the literature which are conducted with dried rock samples, all our experiments are performed under fully saturated conditions. The rate- and state-friction equations and Dieterich evolution law are adopted to evaluate experimental results and calculate friction parameters of $a - b$ and other constitutive parameters.

The arrangement of this paper is as follows. Section 2 describes the experimental methods and inverse modelling technique for constitutive friction parameters and Sect. 3 presents the experimental results under fully saturated conditions. Next, the effects of normal stress on friction parameters and implications for natural faults' behavior are discussed in Sect. 4. Finally, the main findings are summarized in Sect. 5.

2 Methods

2.1 Rock Samples

A total of eight reservoir rocks samples from four different oilfields in China were collected for shearing tests, including four types of shale and four types of sandstone, as shown in Fig. 1. Among them, the samples Longmaxi Shale-1, Longmaxi Shale-2, Longmaxi Shale-3 and Ordos Sandstone were from rock outcrop, while others were taken from the core samples at a depth between ~3300 m and ~5200 m. The mineral compositions of these reservoir samples were determined by XRD (X-ray diffraction) and is presented in Table 1. Typical minerals in these reservoir rock samples incorporate clay minerals, quartz, albite, orthoclase and calcite, with a bit of dolomite and pyrite content in shale. It is worth noting that the clay contents range from 8% to 26% among eight rock samples and the clay content in shale are generally higher than that in sandstone. All rock samples were crushed and sieved to obtain a particle size less than 150 μm before the shearing tests.

Fig. 1. Examples of rock sample materials for shearing tests. From left to right, from top to bottom: Longmaxi Shale-1, Longmaxi Shale-2, Longmaxi Shale-3, Xuwu Shale, Ordos Sandstone, Junggar Sandstone-1, Junggar Sandstone-2 and Shengli Sandstone.

Table 1. Mineral compositions of tested reservoir rock samples.

Name	Clay minerals	Quartz	Albite	Orthoclase	Calcite	Dolomite	Pyrite
Longmaxi Shale-1	10%	44%	1%	1%	19%	23%	2%
Longmaxi Shale-2	23%	45%	10%	5%	9%	4%	4%
Longmaxi Shale-3	18%	39%	7%	3%	23%	6%	4%
Xuwu Shale	26%	50%	12%	2%	6%	4%	–
Ordos Sandstone	14%	42%	20%	9%	15%	–	–
Junggar Sandstone-1	12%	53%	22%	7%	6%	–	–
Junggar Sandstone-2	17%	47%	26%	10%	–	–	–
Shengli Sandstone	8%	33%	13%	5%	8%	33%	–

2.2 Experimental Setup and Procedure

Double direct shearing experiments were performed with the static biaxial loading system at room temperature and fully saturated conditions, as shown in Fig. 2. Two 5-mm-thick

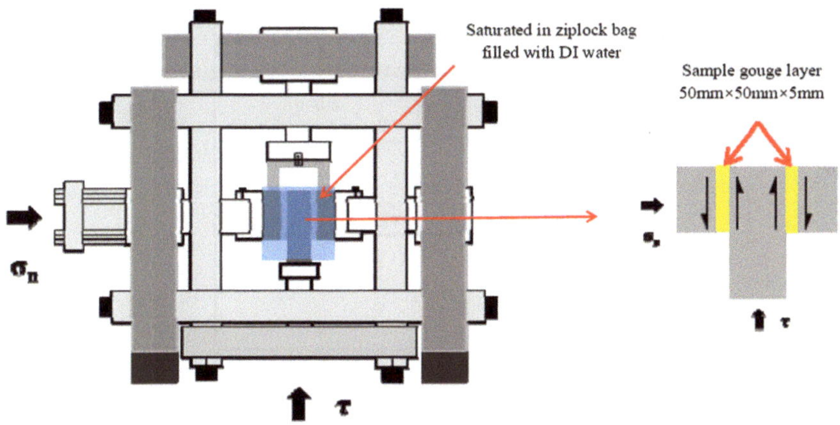

Fig. 2. Setup of the double direct shearing test.

sample gouge layers were sandwiched between three roughened steel forcing blocks. The two outer steel blocks have the dimensions of $5 \times 5 \times 2$ cm^3 and the size of central steel block is $10 \times 5 \times 3$ cm^3 so that a constant area of 50×50 mm^2 was maintained thought the shearing tests [19, 20]. The normal stress was applied by the horizontal piston with 1 N resolution, while the vertical piston could be precisely controlled to obtain a constant shearing velocity. In each experiment, the fault sample was put in a ziplock bag filled with distilled water and placed on the loading platform. After that, we applied a small normal force on the whole sample and kept it for about 12 h to ensure that the sample gouge layers were fully saturated.

The loading process can be distinguished into three different stages according to the normal stress. At the first stage, we applied a constant normal stress of 10 MPa to the whole sample and the shearing load was obtained by forcing the vertical piston with the constant shearing rate of 10 μm/s. When the steady state of friction was achieved (about 4 mm shearing displacement for all reservoir samples), the velocity steps were performed with the shearing velocity changing from 0.1 μm/s to 100 μm/s until the shearing displacement reached 7.5 mm. The shearing displacement was 0.5 mm for each velocity step throughout the whole tests. At the second stage, the normal stress was raised up to 20 MPa and the shearing rate was set at 10 μm/s until the shearing displacement reached 8.5 mm. The velocity steps were also conducted the same as the first stage until the shearing displacement was 12 mm. The final stage was similar with the second stage except that the normal stress was changed to 40 MPa. The total shearing displacement was 16.5 mm for the whole test procedure.

2.3 Data Analysis

The rate and state constitutive friction (RSF) law [21, 22] were employed to do inverse modeling of the experimental data. Some of the experimental data were fit well with

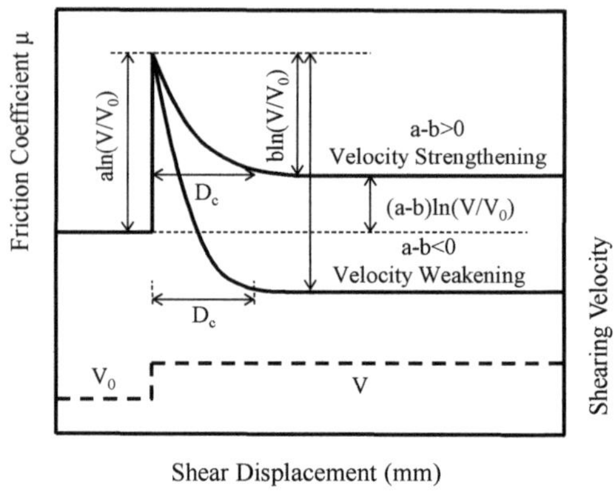

Fig. 3. The schematic diagram for idealized rate and state friction law.

one state variable RSF law, while others needed to be modeled by two state variable RSF law. In the theory of two state variables RSF law, the friction μ has the form of:

$$\mu = \mu_0 + aln(V/V_0) + b_1 ln(V_0\theta_1/D_{c1}) + b_2 ln(V_0\theta_2/D_{c2}) \quad (1)$$

where μ is steady state friction coefficient at velocity V, μ_0 is the reference steady state friction coefficient at velocity V_0 ($V > V_0$), a, b_1 and b_2 are friction constitutive parameters reflecting the effects of instantaneous and displacement changes in friction, θ_1 and θ_2 are state variables, D_{c1} and D_{c2} are the critical slip distance needed for a new steady state happens. The elastic coupling equation of μ is given by:

$$d\mu/dt = K(V_{lp} - V) \quad (2)$$

where K is the stiffness of the loading system and the method for the determination of K is described in Leeman et al. [23], V_{lp} represents the load point velocity.

Here we use the Dieterich law [21] to describe the evolution of state variable θ which expressed as,

$$d\theta_i/dt = 1 - (V\theta_i - D_{ci}),\ (i = 1, 2) \quad (3)$$

The value of a − b is mainly used for indicating the friction stability for a finite velocity step, expressed by,

$$a - b = \Delta\mu_{ss}/\Delta lnV \quad (4)$$

For the case of two state variables, we define $b = b_1 + b_2$ and $D_c = D_{c1} + D_{c2}$ for data analysis [12, 24]. The schematic diagram of idealized friction law is shown in Fig. 3 where the positive value of a − b indicates velocity-strengthening friction behavior and the negative value of a − b represents velocity-weakening friction behavior [25].

3 Results

The friction-displacement curves for all reservoir rock samples under three consecutive normal stresses are shown in Fig. 4. For better illustration, we plot the friction-displacement curves of shale and sandstone rock samples in Fig. 4(a) and (b),

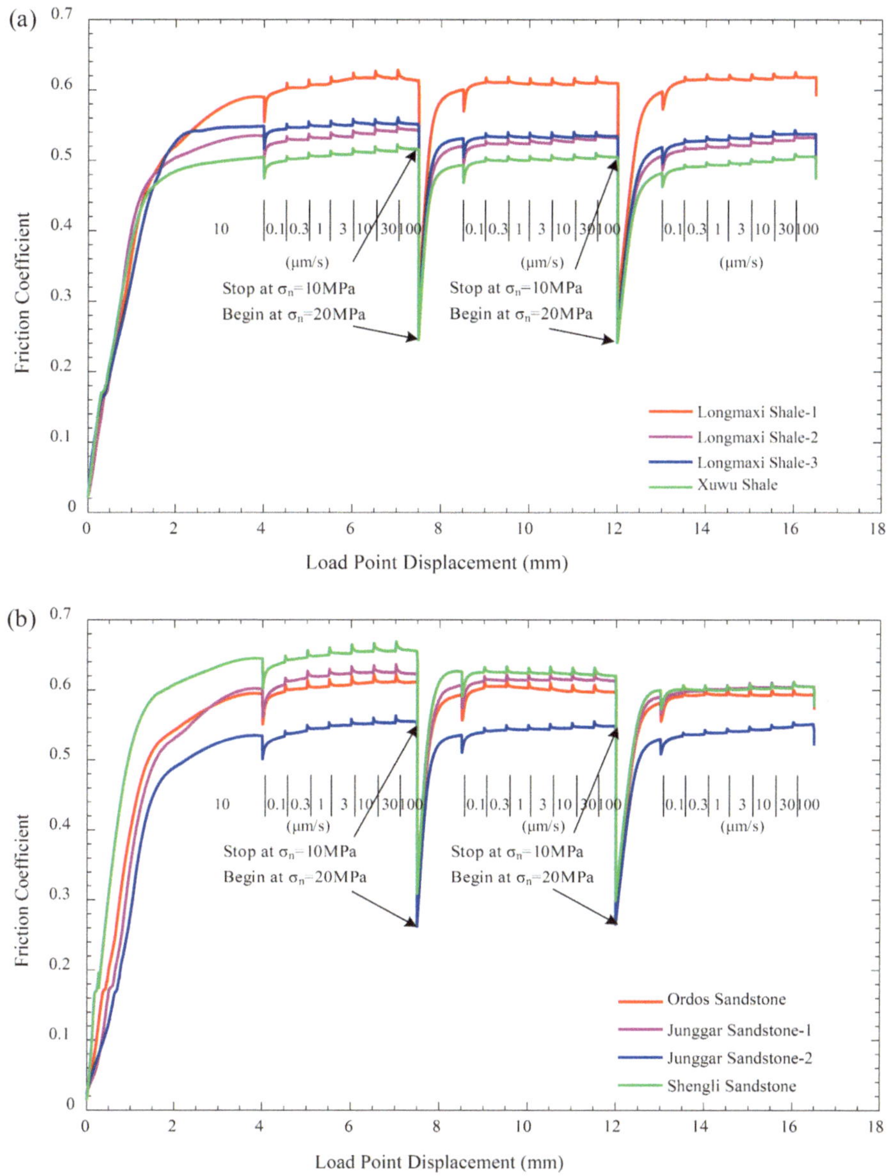

Fig. 4. Friction-displacement curves for all reservoir samples under normal stresses of 10 MPa, 20 MPa and 40 MPa, respectively. (a) Four kinds of shale. (b) Four kinds of sandstone.

respectively. A clear disparity of friction coefficient of eight reservoir rock samples can be observed in these two pictures. The friction coefficient of four kinds of shale ranges from 0.48–0.64 under normal stresses of 10 MPa, 20 MPa and 40 MPa, and the rock

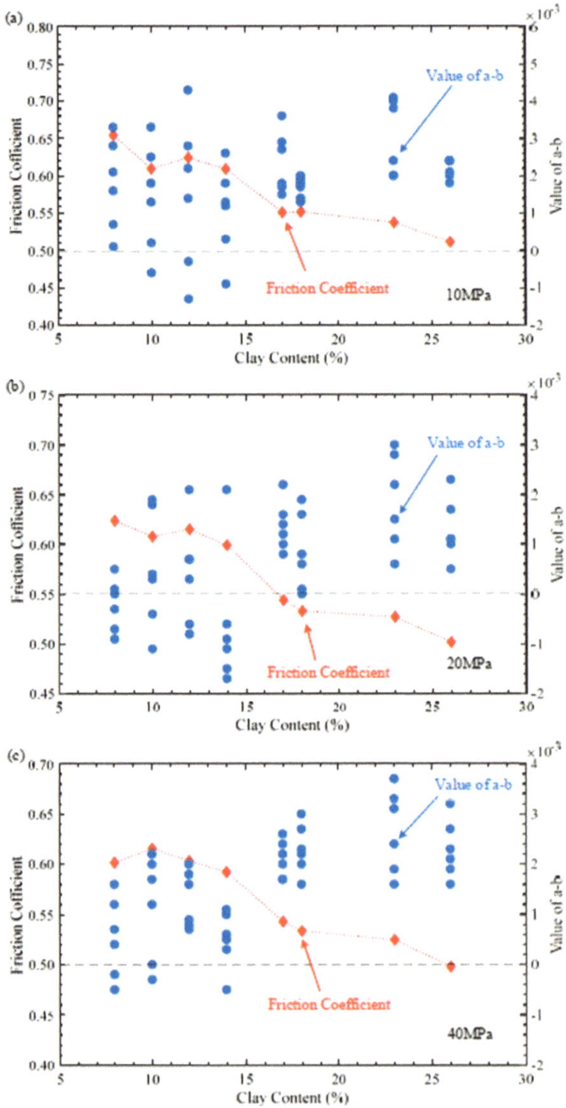

Fig. 5. Friction coefficient and distribution of value of a − b as functions of the clay content under different normal stresses. The red rhombuses represent the friction coefficient and the blue circles show the distribution of value of a − b. (a) Under normal stress of 10 MPa. (b) Under normal stress of 20 MPa. (c) Under normal stress of 40 MPa.

sample Longmaxi Shale-1 shows the highest friction coefficient because of its relatively lower content of clay compared with other shale. On the contrary, the rock sample Xuwu shale exhibits the lowest friction coefficient of about 0.5 on account of its highest clay content among all reservoir samples. The friction coefficient of four types of sandstone ranges from 0.52–0.68 under three different normal conditions, and due to the lowest clay content, the friction coefficient of Shengli Sandstone is the highest compared with other sandstone. Interestingly, the friction coefficient of Shengli Sandstone shows apparently monotonously decreasing trend with the increasing normal stress, which is possibly affected by its particle size distribution and this will be discussed later.

We adopt the friction coefficient under shearing displacement of 6.25 mm, 10.75 mm and 15.25 mm as friction coefficient under 10 MPa, 20 MPa and 40 MPa, respectively. The friction coefficient and value of $a - b$ as functions of clay content under three different normal stresses are shown in Fig. 5. We observe a significant decreasing trend of friction coefficient and an increasing trend of $a - b$ with the increase of clay content of reservoir rocks from different oilfields under normal stresses of 10 MPa, 20 MPa and 40 MPa, similar with the results of Kohli and Zoback [26].

Under normal stress of 10 MPa, the value of $a - b$ ranges from −0.0014 to 0.0043 and velocity weakening ($a - b < 0$) behavior can be observed with clay content less than 15%. This is a bit different from the results of Kohli and Zoback [26] on dried natural shale reservoir rock samples. The value of $a - b$ is between −0.0017 and 0.0031 under normal stress of 20 MPa. It is worth noting that the transition from velocity weakening ($a - b < 0$) behavior to velocity strengthening ($a - b > 0$) behavior occurs with 20% content of clay minerals under 20 MPa normal stress, a little higher than that at 10 MPa normal stress. When the normal stress is raised to 40 MPa, the value of $a - b$ is mainly in the range from −0.0006 to 0.0037, and the transition from velocity weakening to velocity strengthening behavior is nearly identical with the condition of 10 MPa with 15% clay content.

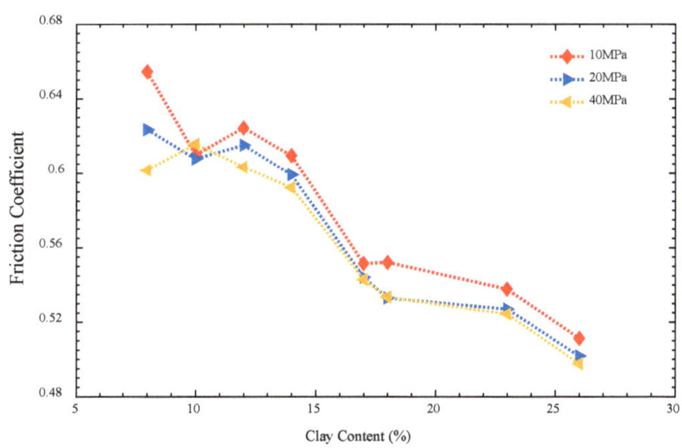

Fig. 6. The friction coefficient as a function of clay content under three normal stresses.

4 Discussion

The normal stress is thought to be a significant factor for the frictional strength and stability of faults [24]. The friction coefficient as a function of clay content under three successive normal stresses is presented in Fig. 6. The friction coefficient under normal stress of 20 MPa and 40 MPa is nearly identical, but they are a bit lower than that under 10 MPa normal stress. This is because the rock powders need to be crushed to reach a steady state and form the sliding plane under the normal stress of 10 MPa.

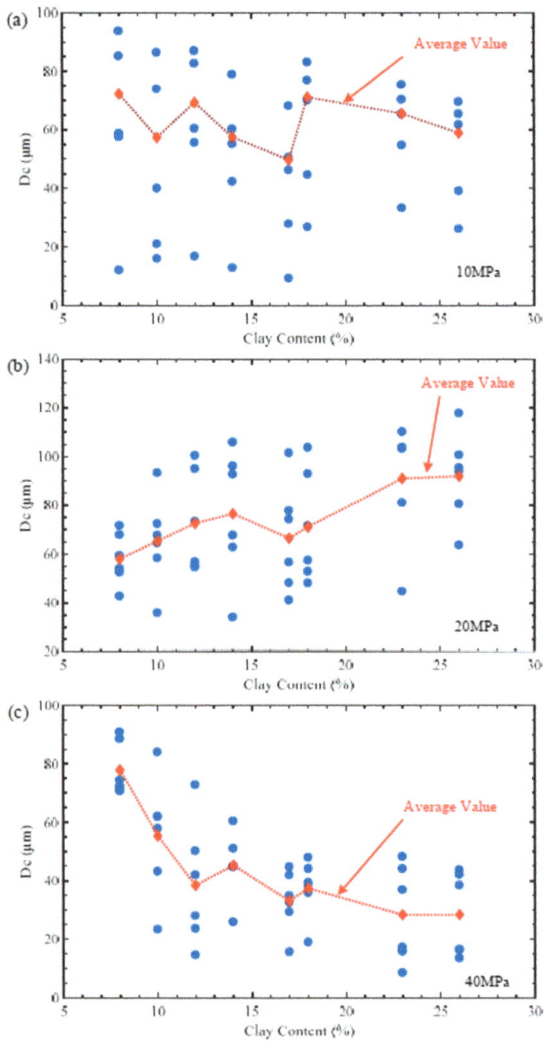

Fig. 7. The critical slip displacement as a function of clay content under different normal stresses. The red rhombuses represent the average value of D_c and the blue circles illustrate the distribution of D_c with a certain clay content. (a) 10 MPa. (b) 20 MPa. (c) 40 MPa.

Therefore, the friction tests under 20 MPa are conducted on the basis of the powders of 10 MPa, which leads to the friction strength under 20 MPa typically lower than that under 10 MPa. In addition, the rock powders has been crushed to a critical value and the sliding plane has formed under the normal stress of 20 MPa. This can partially account for the reason that the rock samples show similar friction coefficient under normal stress of 20 MPa and 40 MPa.

The critical slip displacement Dc represents the distance required to reach a new steady state in velocity stepping tests and have important implications for determining the dimension of rupture nucleation and the amount of pre- and post-seismic slip [27]. The critical slip displacement as a function of clay content is depicted in Fig. 7. The values of D_c range from 10 to 95, from 35 to 120 and from 10 to 90 under 10 MPa, 20 MPa, and 40 MPa normal stress. Most importantly, the values of D_c show decreasing trend with higher clay content under normal stress of 10 MPa and 40 MPa, while the values of D_c show increasing trend under normal stress of 20 MPa. This phenomenon reflects that the normal stress has complex effects on the critical displacement.

The observed transition from velocity weakening behavior ($a - b < 0$) to velocity strengthening behavior ($a - b > 0$) for shale and sandstone reservoir rocks at the clay content of 15–20% has important implications for the stability of natural pre-existing faults in reservoirs. The recently observed large magnitude micro-seismicity is shown to be closely correlated with the activation of the pre-existing faults. The observed results can be helpful for assessing the stability of faults by analyzing the mineral compositions and stress conditions of reservoir rocks at different depth.

5 Conclusions

Different from conventional experiments on dried synthetic fault gouge, we conducted double direct shearing tests on natural shale and sandstone reservoir rock powder samples to primarily explore the key parameters that control the frictional strength and stability of faults, under fully saturated conditions and at room temperature. The following conclusions can be drawn from the study:

(1) The frictional strength of shale and sandstone fault gouge shows a decreasing trend with higher clay content.
(2) A transition from velocity weakening to velocity strengthening behavior can be observed at a clay content of 15–20% under water-saturated conditions.
(3) The frictional strength of the tested samples generally decreases slightly with higher normal stress.
(4) The normal stress has complex effects on the critical slip displacement D_c. Under normal stresses of 10 MPa and 40 MPa, D_c decreases with higher clay content, while an opposite trend is observed under 20 MPa normal stress.

Acknowledgement. This research is supported by the National Natural Science Foundation of China (No. 41672268).

References

1. Rubinstein, J.L., Mahani, A.B.: Myths and facts on wastewater injection, hydraulic fracturing, enhanced oil recovery, and induced seismicity. Seismol. Res. Lett. **86**(4), 1060–1067 (2015)
2. Ellsworth, W.L.: Injection-induced earthquakes. Science **341**(6142), 1225942/1–1225942/7 (2013)
3. Elst, N.J.V.D., Savage, H.M., Keranen, K.M., Abers, G.A.: Enhanced remote earthquake triggering at fluid-injection sites in the Midwestern United States. Science **341**(6142), 164–167 (2013)
4. Mcgarr, A., Bekins, B., Burkardt, N., Dewey, J., Earle, P., Ellsworth, W., Ge, S., Hickman, S., Holland, A., Majer, E., Rubinstein, J., Sheehan, A.: Coping with earthquakes induced by fluid injection. Science **347**(6224), 830–831 (2015)
5. Bao, X., Eaton, D.W.: Fault activation by hydraulic fracturing in western Canada. Science **354**(6318), 1406–1409 (2016)
6. Vrolijk, P., van der Pluijm, B.A.: Clay gouge. J. Struct. Geol. **21**(8), 1039–1048 (1999)
7. Niemeijer, A.R., Spiers, C.J.: Velocity dependence of strength and healing behaviour in simulated phyllosilicate-bearing fault gouge. Tectonophysics **427**(1–4), 231–253 (2006)
8. Shimamoto, T., Logan, J.M.: Effects of simulated clay gouges on the sliding behavior of Tennessee sandstone. Tectonophysics **75**(3–4), 243–255 (1981)
9. Takahashi, M., Mizoguchi, K., Kitamura, K., Masuda, K.: Effects of clay content on the frictional strength and fluid transport property of faults. J. Geophys. Res. Solid Earth **112**(B8), 59–69 (2007)
10. Crawford, B.R., Faulkner, D.R., Rutter, E.H.: Strength, porosity, and permeability development during hydrostatic and shear loading of synthetic quartz-clay fault gouge. J. Geophys. Res. Solid Earth **113**(B3), 133–144 (2008)
11. Tembe, S., Lockner, D.A., Wong, T.: Effect of clay content and mineralogy on frictional sliding behavior of simulated gouges: binary and ternary mixtures of quartz, illite, and montmorillonite. J. Geophys. Res. Solid Earth **115**(B3), 153–164 (2010)
12. Ikari, M.J., Niemeijer, A.R., Marone, C.: The role of fault zone fabric and lithification state on frictional strength, constitutive behavior, and deformation microstructure. J. Geophys. Res. Solid Earth **116**(B8), 125–157 (2011)
13. Tesei, T., Collettini, C., Carpenter, B.M., Viti, C., Marone, C.: Frictional strength and healing behavior of phyllosilicate-rich faults. J. Geophys. Res. Solid Earth **117**(B9), B09402/1–B09402/13 (2012)
14. Mair, K., Marone, C.: Friction of simulated fault gouge for a wide range of velocities and normal stresses. J. Geophys. Res. Solid Earth **104**(B12), 28899–28914 (1999)
15. Saffer, D.M., Frye, K.M., Marone, C., Mair, K.: Laboratory results indicating complex and potentially unstable frictional behavior of smectite clay. Geophys. Res. Lett. **28**(12), 2297–2300 (2001)
16. Saffer, D.M., Marone, C.: Comparison of smectite- and illite-rich gouge frictional properties: application to the updip limit of the seismogenic zone along subduction megathrusts. Earth Planet. Sci. Lett. **215**(1–2), 219–235 (2003)
17. Ikari, M.J., Saffer, D.M., Marone, C.: Frictional and hydrologic properties of clay-rich fault gouge. J. Geophys. Res. Solid Earth **114**(B5), B05409/1–B05409/18 (2009)
18. Hartog, S.A.M.D., Peach, C.J., Winter, D.A.M.D., Spiers, C.J., Shimamoto, T.: Frictional properties of megathrust fault gouges at low sliding velocities: New data on effects of normal stress and temperature. J. Struct. Geol. **38**(5), 156–171 (2012)

19. Frye, K.M., Marone, C.: Effect of humidity on granular friction at room temperature. J. Geophys. Res. Solid Earth **107**(B11), ETG 11-1–ETG 11-13 (2002)
20. Fang, Y., Hartog, S.A.M.D., Elsworth, D., Marone, C., Cladouhos, T.: Anomalous distribution of microearthquakes in the Newberry Geothermal Reservoir: mechanisms and implications. Geothermics **63**, 62–73 (2016)
21. Dieterich, J.H.: Modeling of rock friction: 1. Experimental results and constitutive equations. J. Geophys. Res. Solid Earth **84**(B5), 2161–2168 (1979)
22. Ruina, A.: Slip instability and state variable friction laws. J. Geophys. Res. Solid Earth **881** (B12), 10359–10370 (1983)
23. Leeman, J.R., Saffer, D.M., Scuderi, M.M., Marone, C.: Laboratory observations of slow earthquakes and the spectrum of tectonic fault slip modes. Nat. Commun. **7**, 11104/1–11104/6 (2016)
24. Carpenter, B.M., Collettini, C., Viti, C., Cavallo, A.: The influence of normal stress and sliding velocity on the frictional behaviour of calcite at room temperature: insights from laboratory experiments and microstructural observations. Geophys. J. Int. **205**(1), 548–561 (2016)
25. Gu, J.C., Rice, J.R., Ruina, A.L., Tse, S.T.: Slip motion and stability of a single degree of freedom elastic system with rate and state dependent friction. J. Mech. Phys. Solids **32**(3), 167–196 (1984)
26. Kohli, A.H., Zoback, M.D.: Frictional properties of shale reservoir rocks. J. Geophys. Res. Solid Earth **118**(9), 5109–5125 (2013)
27. Marone, C., Kilgore, B.: Scaling of the critical slip distance for seismic faulting with shear strain in fault zones. Nature **362**(6421), 618–621 (1993)

Microscopic Experimental Study on Failure Mechanism of Rock-Like Material with Flaw Pairs Under Uniaxial Compression

Cheng Zhao[1,2(✉)], Jialun Niu[2], Chunfeng Zhao[2,3], and Rui Zhang[2]

[1] Tibet University Institute of Technology, Lhasa, China
[2] Department of Geotechnical Engineering, Tongji University, Shanghai, China
zhaocheng@tongji.edu.cn
[3] Department of Geotechnical Engineering and Key Laboratory of Geotechnical and Underground Engineering of Ministry of Education, Tongji University, Shanghai, China

Abstract. The micro-scale damage evolution of the rock-like specimens with prefabricated flaw pairs under uniaxial compression were studied by scanning electron microscopy. The morphologies and mechanical properties on the surfaces of failure specimens were compared with the specimens failed under single tension and single shear respectively. By analyzing the numerical simulation results, the characteristics of cracks propagation and stress distribution on the fractured surfaces were studied. The results show that the tensile stress plays a leading role in the propagation of wing cracks and anti-wing cracks during the failure of the double flawed specimens with the inclination angle of 45°. The rock bridges have little effect on the propagation of wing cracks outside the outer tips of flaw pairs, but have influence on the initiation and propagation of anti-wing cracks. In the double flawed specimens with the inclination angle of 60°, the coalescence types are related to the angles of the rock bridges. The failure characteristics near the crack tips are different when the rock bridge angles change. With the cracks overlap and coalesce, the proportion of tensile failure increases gradually.

Keywords: Uniaxial compression · Scanning electron microscopy
Flaw pairs · Failure mechanism

1 Introduction

The microscopic morphological features of the fracture surfaces formed by the failure of rock mass, to a certain extent, reflect the stress distribution and the failure behaviors of rock mass during the whole growth process of the cracks. Therefore, the mechanical properties of failure in rock mass can be studied by analyzing the geometrical images of fracture surfaces obtained by scanning electron microscopy (SEM) tests. This method provides an effective way to establish the relationship between macroscopic failure process and microscopic failure mechanism.

The SEM imaging technique has been successfully used to study the characteristics of crack evolution and rock damage [1–5]. In addition, Wu et al. [6] used optical

L. Zhang et al. (Eds.): GSIC 2018, *Proceedings of GeoShanghai 2018 International Conference: Rock Mechanics and Rock Engineering*, pp. 232–241, 2018.
https://doi.org/10.1007/978-981-13-0113-1_26

microscopy and SEM to observe the failure modes of sandstone under different confining pressures. Wong et al. [7] used SEM to study the white patches and their evolution to macroscopic tensile cracks and shear cracks in marble. They also studied the microscopic initiation of hair-line tensile cracks in gypsum specimens. In the experimental study of rock failure, the SEM method was often combined with other methods, especially the digital image correlation (DIC) method. Wang et al. [8] carried out the investigation at the micrometric scale of the composite microstructure of the rock by using environment scanning electron microscopy (ESEM) and DIC techniques. Sone et al. [9] used SEM and DIC method to observe the micromechanical deformation occurring during fracturing and the creep deformation in organic-samples from the Alum shale. Guillaume et al. [10] used SEM and broad ion beam (BIB) polishing to study the evolution of microstructure in samples of clay. However, most of them have studied quantitatively for the evolution of cracks on surface. There are few studies concentrated on the variation of mechanical characteristics on the basis of qualitative analysis of microscopic characteristics.

In this paper, the fracture surfaces of the rock-like specimens with flaw pairs were scanned by electron microscopy. The morphologies and mechanical properties of the fracture surfaces were compared with the fracture surfaces under single tension and single shear. The characteristics of the fracture surfaces and stress distribution were observed and analyzed. Based on the numerical simulation results, the mechanical mechanisms of cracks evolution in different specimens with flaw pairs were studied.

2 Experiments

2.1 Specimens Preparation

The size of specimens are 100 mm × 40 mm × 10 mm (length × width × thickness), the specimen is shown in Fig. 1, where α is the inclination angle and β is the rock bridge angle. The length of the flaws are 14 mm and the length of rock bridges are 21 mm. Because when the length of the flaw projected on the horizontal plane is less than 1/3 of the width of the specimen, the cracks propagation path would not be affected by the left and right boundaries of specimens.

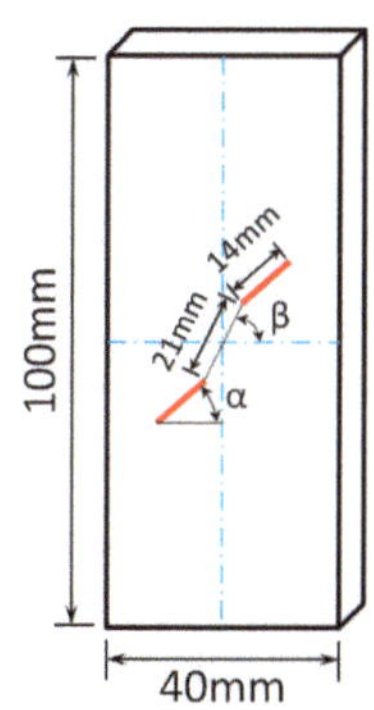

Fig. 1. Rock-like specimen used in uniaxial compression

In order to study the geometrical morphologies of the cross-sections of the rock-like material under single tension and single shear, the Brazilian discs splitting test and shearing test were conducted. The samples were cut to blocks and the fracture surfaces were sprayed with conductive films. Different samples are shown in Fig. 2.

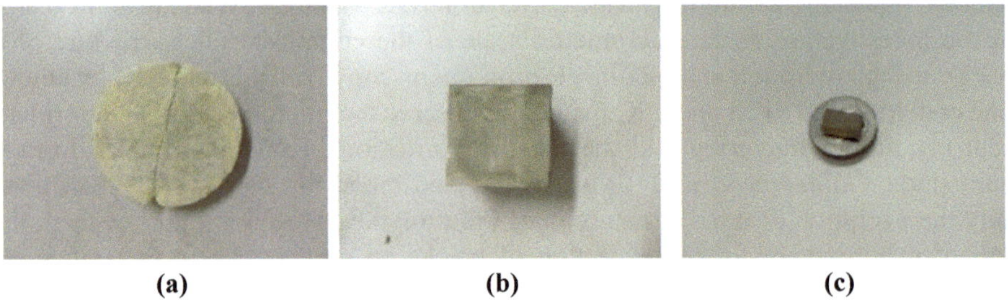

Fig. 2. Samples prepared for scanning. **a** Sample in Brazilian discs splitting test. **b** Sample in shearing test. **c** Sample used in SEM method

2.2 SEM Method

The typical tensile fracture surfaces by SEM in splitting test are shown in Fig. 3. The main features include: flocculent particles distributed at the fracture surfaces, the surfaces were roughness in magnification of 500 times, the holes were complete and the boundary lines were obvious.

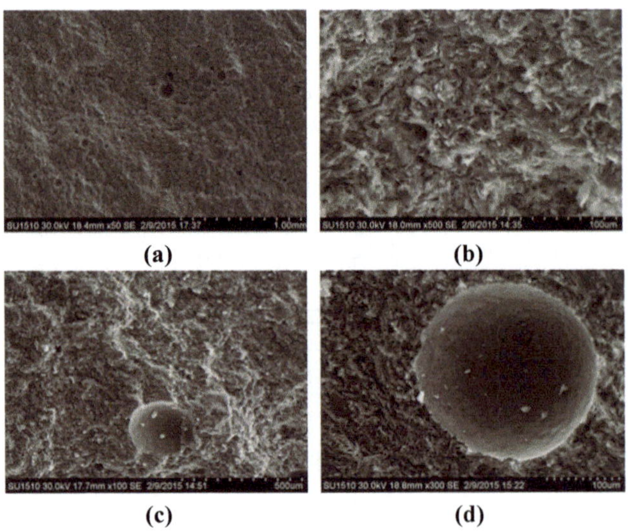

Fig. 3. The typical fracture surfaces under single tension. **a** Full view of surface: particles were "flocculent", holes were complete. **b** Flocculent structure: the most prominent feature of tensile failure section. **c** Terrace structure: uneven distribution of terraced fields, there were small amount of crystal debris distributed around the interlayer lines. **d** Hole: the holes were complete and the surrounding granules were flocculent

The typical shear fracture surfaces are shown in Fig. 4. The main features include: the cross sections contain complex and variable profiles, the shear zones were smooth, the holes were not complete, the boundary lines were not obvious and the surrounding surfaces were distributed smoothly.

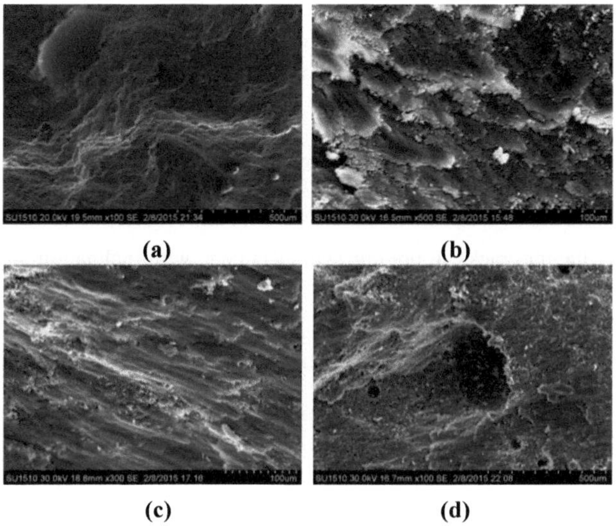

Fig. 4. The typical fracture surfaces under single shear. **a** Terraced structure: the shear effect between different platforms was significant. **b** Shearing layered structure: the upper and lower surfaces were failed by shearing. **c** Smooth shearing zone: single shear area, has a clear texture like river. **d** Hole: failed under both tensile and shear stress, there were "cut hats" around the holes

The scanned failure surfaces of the rock-like specimens can reflect the failure characteristics and the stress distribution in the process of damage growth. The failure surfaces in specimens with different inclination angles and rock bridge angles were scanned in the whole process of cracks propagation. The results were compared with the specimens under single tensile and single shear stress. The characteristics of stress distribution and damage growth during the crack evolution were studied according to the features of the fracture surfaces.

3 Results and Discussion

3.1 Specimens with Inclination Angle of 45°

Pre-cracked specimens with flaw pairs were loaded to failure in uniaxial compression tests. Some typical fracture surfaces were observed by SEM method. Figure 5 shows different cracks in specimens with same inclination angle of 45° and different rock bridge angles. The scanned wing cracks are circled with solid-line ellipses, and anti-wing cracks are circled with dotted ellipses.

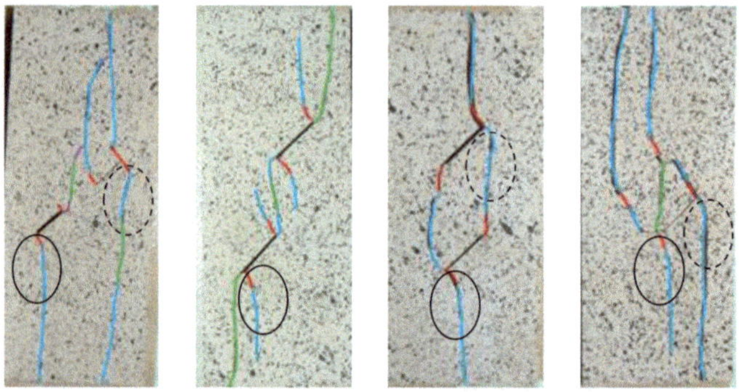

(a) β = 45 ° (b) β = 90 ° (c) β = 120 ° (d) β =180 °

Fig. 5. Different cracks in pre-cracked specimens with inclination angle of 45°

Wing Cracks. Figure 6 shows the fracture surfaces in specimens with inclination angle of 45° and the rock bridge angle of 45°, 90°, 120° and 180°, respectively. The morphologies and damage characteristics during the whole process of wing cracks propagation are summarized as follows:

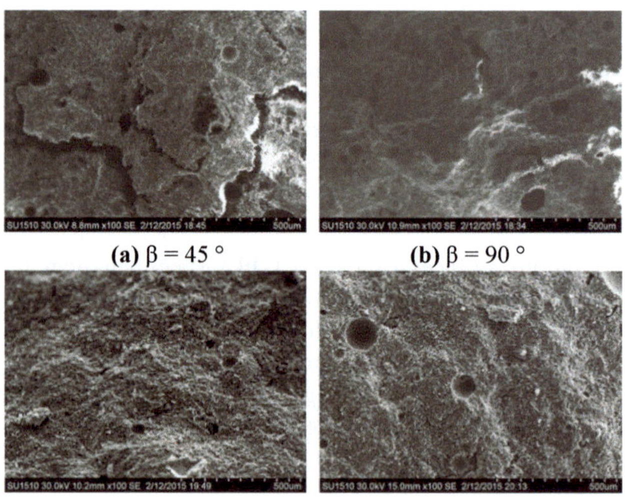

(a) β = 45 ° (b) β = 90 °

(c) β = 120 ° (d) β =180 °

Fig. 6. Fracture surfaces of wing cracks in specimens with inclination angle of 45°

β = 45°: In the early stage of cracks propagation, most of the cross surfaces were flocculent. There were "layered, hill-like" structures and a small part of shear traces in the middle and late stages. β = 90°: The "stepped" structures were obvious at initial surfaces, while surfaces at the middle and late stages were mostly "layered" structures. β = 120°: Flocculent structures were significant and most of holes were complete at the whole process of crack propagation. β = 180°: A part of the areas were "hill-like" structures, and flocculent structures of particles were significant.

In addition, the fracture surfaces of wing cracks were rough and changed little at different stages. The tensile stress played a leading role in the evolution of wing cracks. The rock bridges had little effect on the propagation and damage growth of wing cracks between flaw tips and boundaries of specimens.

Anti-wing Cracks. The different fracture surfaces of anti-wing cracks in specimens with inclination angle 45° are shown in Fig. 7. The morphologies characteristics are summarized as follows:

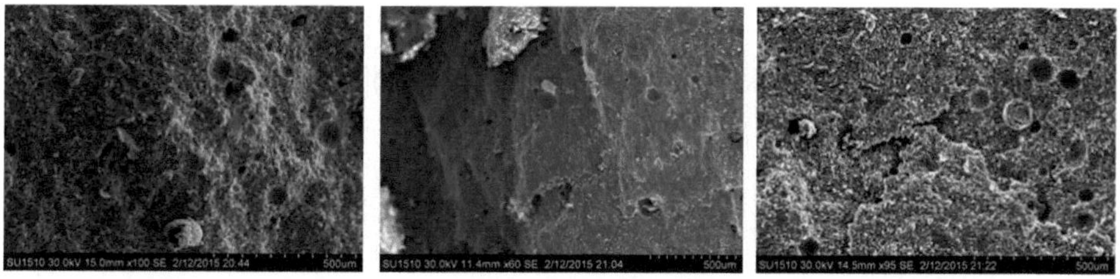

(a) β = 45 ° **(b)** β = 120 ° **(c)** β = 180 °

Fig. 7. Fracture surfaces of anti-wing cracks in specimens with inclination angle of 45°

β = 45°: Mountain-like structures widely distributed and there were large amount of flocculent particles, the holes were complete. β = 120°: The fracture surfaces were mostly step-like. Most of the particles were flocculently distributed in the late stage of propagation. β = 180°: The sheet-like and mountain-like structures were significant and most of the holes were complete.

The characteristics of fracture surfaces within anti-wing cracks were similar to the wing cracks, which are basically tensile failure. Rock bridge areas had a certain impact on the evolution of anti-wing cracks at the flaw tips. For example, it was difficult to generate anti-wing cracks when the rock bridge angle was 90°.

3.2 Specimens with Inclination Angle of 60°

Figure 8 shows different cracks in double flawed specimens with same inclination angle of 60° and different rock bridge angles. The scanned cracks in the vicinity of rock bridge areas are circled with solid-line ellipses. The specimens with rock bridge angle of 60°, 90° and 120° are selected to analyze. For the sake of simplicity, different specimens are represented by "α - β".

60°-60° Specimen. As shown in Fig. 9, the fracture surfaces near the tips of flaw pairs are smooth and the cross sections are mostly river-like and terraced when the rock bridge angle is 60°. The particles increased and the boundaries of the holes turned to be obvious with the growth of cracks. There were many shear cracks in the vicinity of the initiation position, and the coalescence type of rock bridge area was shear coalescence.

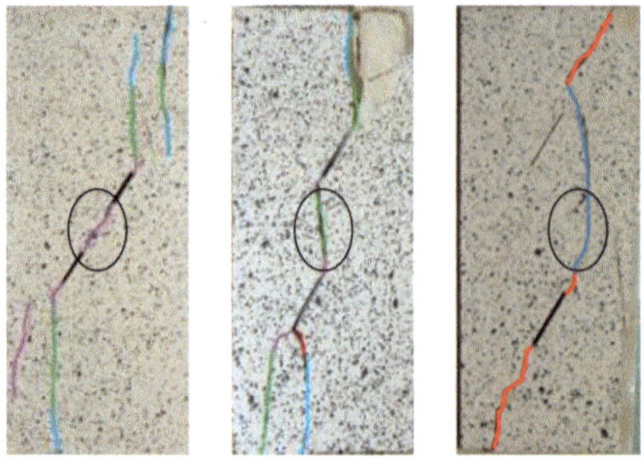

(a) β = 60 ° (b) β = 90 ° (c) β = 120 °

Fig. 8. Different cracks in pre-cracked specimens with inclination angle of 60°

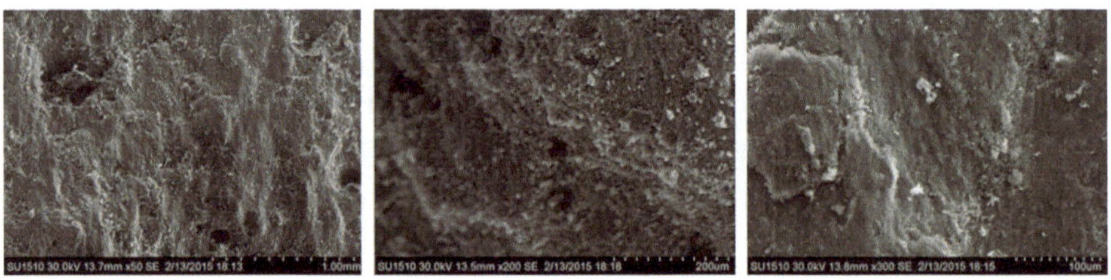

Fig. 9. Fracture surfaces in specimen with α = 60° and β = 60°

With the further development of the damage, the proportion of tensile failure was gradually increased.

Numerical experiments based on extending discrete element method were conducted to study the mechanical characteristics quantitatively. The strain criterion was embedded in UDEC to simulate the failure progress of specimens with flaw pairs. This numerical method has been verified to be appropriate for simulating the crack initiation

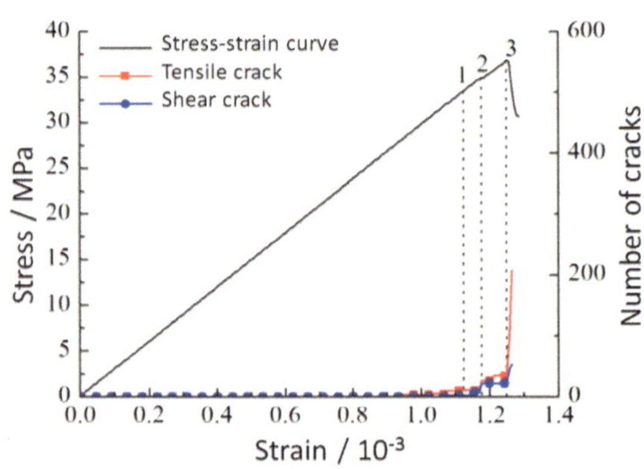

Fig. 10. Stress-strain curve, number of tensile and shear cracks in 60°-60° specimen

stress, peak strength and crack growth morphology of the specimens in lab test. The parameters used in simulation were in the same with the paper [11] by the authors of the present paper.

Figure 10 shows the stress-strain curve, the number of tensile cracks and shear cracks in 60°-60° specimen. After the initiation, the tensile and shear cracks almost propagated simultaneously (point 1). The shear cracks between inner tips connected to each other soon after (point 2). The cracks at outer tips began to propagate after the rock bridge was coalescence. The number of tensile and shear cracks were approximate. After the stress gradually reached to the peak (point 3), the number of tensile cracks increased rapidly until the specimen was failure.

60°-90° Specimen. During the initiation and coalescence, the fracture surfaces in 60°-90° specimen were in same with that of 60°-60° specimen, as shown in Fig. 11. But most of the fracture surfaces in the rock bridge area were rough, the holes were complete and the boundary lines were obvious, which indicate the characteristic of tensile failure.

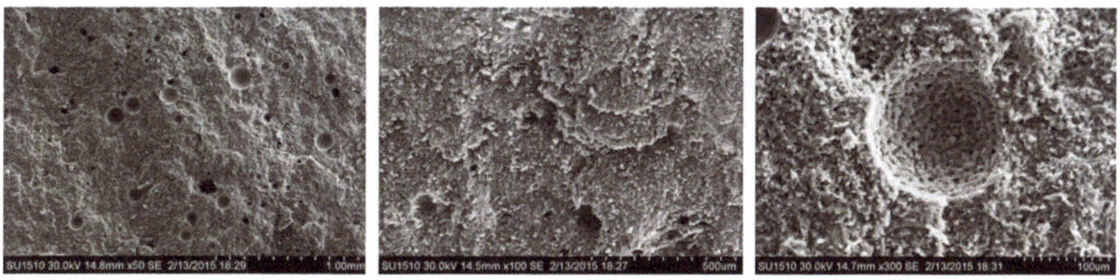

Fig. 11. Fracture surfaces in specimen with $\alpha = 60°$ and $\beta = 90°$

Figure 12 shows the numerical simulation results of 60°-90° specimen. The cracks initiated at the inner and outer tips and the propagation rate at the inner tips was slower than that of the outer tips (point 1). The lengths of tensile cracks at the rock bridge area were relatively short compared to the cracks at outer tips of flaw pairs before

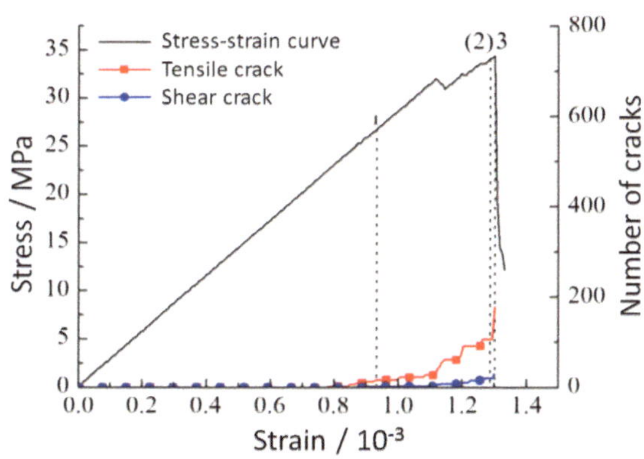

Fig. 12. Stress-strain curve, number of tensile and shear cracks in 60°-90° specimen

approaching coalescence (point 2). The number of tensile cracks in the rock bridge area was much larger than the shear cracks. The tensile cracks coalesced at the rock bridge area (point 3) and then suddenly penetrated the entire specimen.

60°-120° Specimen. The fracture surfaces in specimen with rock bridge angle of 120° are shown in Fig. 13. There was no coalescence at the rock bridge area. Most of the fracture surfaces between the outer tips and two ends of the specimen were failed under tensile stress. There were a few shear cracks near the flaw tips. The characteristics of damage growth were similar to the specimens with rock bridge angles of 60° and 90°.

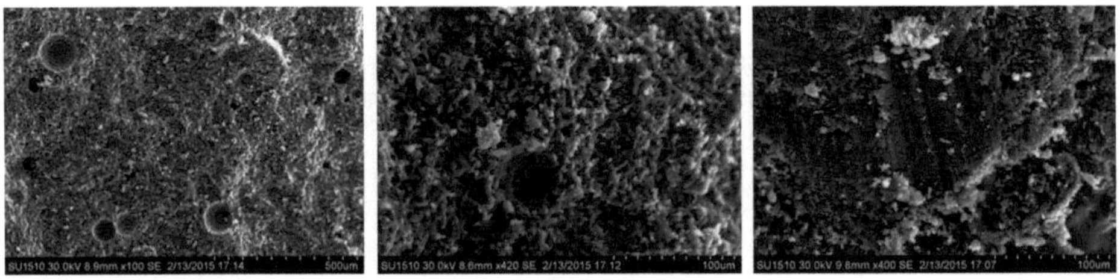

Fig. 13. Fracture surfaces in specimen with $\alpha = 60°$ and $\beta = 120°$

4 Conclusions

The failure surfaces of the specimens with flaw pairs were scanned by electron microscopy, and the extending discrete element numerical method was also used to study the mechanism of cracks propagation and damage growth. The main conclusions are as follows:

(1) The fracture surfaces of the specimens with flaw pairs in the inclination angle of 45° are rough. The differences of morphological features are not significant during the different stages of the cracks propagation, and the tensile stress plays a leading role in the evolution of wing cracks and anti-wing cracks. The angle of the rock bridge has little effect on the propagation of wing cracks between the crack tips and the boundaries of the specimen, which has a certain influence on the evolution of the anti-wing cracks at the tips of the flaw pairs. It is difficult to generate anti-wing cracks when the rock bridge angle is 90°.

(2) In the double flawed specimens with the inclination angle of 60°, the coalescence types of the rock bridge area are affected by the angle of the rock bridge. When the rock bridge angles are 60°, 90° and 120°, the coalescence types are shear coalescence, tensile coalescence and no coalescence, respectively. The failure characteristics of the flaw tips are different when the angle of the rock bridge is different. The shear cracks are mainly concentrated near the tips of the flaw pairs. The proportion of tensile cracks gradually increased along with the coalescence of cracks.

Acknowledgements. The research presented in this paper has been supported by the National Natural Science Foundation of China (No. 41202193 and No. 41572262), Innovation Program of Shanghai Municipal Education Commission (No. 15ZZ016), and Shanghai Rising-Star Program (No. 17QC1400600).

References

1. Sprunt, E.S., Brace, W.F.: Direct observation of microcavities in crystalline rocks. Int. J. Rock Mech. Min. Sci. Geomech. Abstr. **11**, 139–150 (1974)
2. Tapponnier, P., Brace, W.F.: Development of stress-induced microcracks in Westerly granite. Int. J. Rock Mech. Min. Sci. Geomech. Abstr. **13**, 103–112 (1976)
3. Kranz, R.L.: Crack growth and development during creep of Barre Granite. Int. J. Rock Mech. Min. Sci. Geomech. Abstr. **16**, 23–35 (1979)
4. Lu, Y.F., Zhang, M.Y., Ge, X.R.: Electron-microscopic analysis on marble specimens of static and cyclic loading tests. Rock Soil Mech. **11**(4), 75–80 (1990)
5. Sagong, M., Bobet, A.: Micro-fractographic characterization of tensile and shear cracks. In: Proceedings of the Soil and Rock America Symposium, Cambridge, Massachusetts, pp. 937–944 (2003)
6. Wu, X.Y., Baud, P., Wong, T.F.: Micromechanics of compressive failure and spatial evolution of anisotropic damage in Darley dale sandstone. Int. J. Rock Mech. Min. Sci. **37**(1), 143–160 (2000)
7. Wong, L.N.Y., Einstein, H.H.: Crack coalescence in molded gypsum and carrara marble: part 2-Microscopic observations and interpretation. Rock Mech. Rock Eng. **42**, 513–545 (2009)
8. Wang, L.L., Bornert, M., Chanchole, S., et al.: Micro-scale experimental investigation of the swelling anisotropy of the Callovo-Oxfordian argillaceous rock. Clay Miner. **48**, 391–402 (2013)
9. Sone, H., Morales, L.F., Dresen, G.: Microscopic observations of shale deformation from in-situ deformation experiments conducted under a scanning electron microscope, pp. 15–27. American Rock Mechanics Association (2015)
10. Guillaume, D., Nadine, H., Janos, L., et al.: Deformation in cemented mudrock (Callo-vo-Oxfordian Clay) by microcracking, granular flow and phyllosilicate plasticity: insights from triaxial deformation, broad ion beam polishing and scanning electron microscopy. Solid Earth **8**, 291–305 (2017)
11. Zhao, C., Ma, C., Zhao, C.F., et al.: Crack propagation simulation of rock-like specimen using strain criterion. Eur. J. Environ. Civ. Eng. 1–18 (2017). https://doi.org/10.1080/19648189.2017.1359677

Modeling Micro-crack Coalescence in Marble Containing a Single Flaw Under Uniaxial Compression

Dan Huang(✉) and Xiao-qing Li

Department of Civil Engineering and Mechanics,
Huazhong University of Science and Technology, Wuhan, China
303374142@qq.com

Abstract. Due to the experimental limitations and the complexity of rock texture, the numerical simulation was employed to research the cracking process of marble sample containing a single flaw under uniaxial vertical compression. Parallel bond model (BPM), a type of bonded-particle model was used to represent the intact rock, while the smooth joint model (SJM) recently proposed has emerged as a promising tool for simulating the preexisting flaws. The two parameters of the flaw was considered in the paper, the flaw's inclinations β varied from 0° to 90°measured from the horizontal, and the flaw's lengths L varied from 10 mm to 80 mm. According to the position of the micro-crack, sequence to observe what is happening to crack initiation, propagation and coalescence. The progress was divided into two stages, the evolution of micro-crack and the stress distribution of the model were considered to adopt to analysis the failure progress of the model. Simulation results indicated that the flaw's inclinations and flaw's lengths showed an obvious effect on the unconfined compressive strength behavior of the sample. The flaw's inclinations were changed the fracture angle of the model and the fracture mode of the model. The flaw's inclinations There was a critical flaw length in the simulation between 30 mm and 40 mm of this sample, which distinguished the existing of flaw whether the main reason for the failure of rock.

Keywords: Discrete element method · Smooth joint model
Synthetic rock mass · Micro-crack coalescence

1 Introduction

Cracking processes are the crack initiation, propagation, interaction as well as eventual coalescence processes in rock. The intact rock compressive experiments include some observation techniques, such as acoustic emission (AE) [1] and computerized tomography (CT) scan [2, 3] cannot fully explain the detailed cracking processes. To better illustrate the fundamental cracking processes involved, the pre-cracked samples under compression have been extensively and experimentally studied on different natural rocks [4, 5] and rock-like materials [6, 7].

Through the physical experimental studies, some researcher summarized the modes of coalescences of micro-cracks. Wong [8] classified the coalescence pattern into nine

L. Zhang et al. (Eds.): GSIC 2018, *Proceedings of GeoShanghai 2018 International Conference: Rock Mechanics and Rock Engineering*, pp. 242–253, 2018.
https://doi.org/10.1007/978-981-13-0113-1_27

categories, which pertained to three main modes of crack coalescence: (1) shear crack coalescence, (2) mixed shear/tensile crack coalescence and (3) wing tensile crack coalescence. Bobet and Einstein [9] proposed five coalescence categories in the uniaxial compressive test and two additional coalescence categories in the biaxial compressive test. Sagong and Bobet [10] classified the crack types into tensile crack, quasi-coplanar shear crack and oblique shear crack. Nine coalescence categories were classified by Wong and Einstein [11] based on the basis of the various combinations of these cracks, and seven crack types based on the crack nature(tensile/shear) and crack trajectories were classified by Wong and Einstein [12].

In addition to the physical experimental studies, the discrete element method (DEM) has come up as a powerful tool for exploring the mechanical behavior of intact and jointed rock masses at both laboratory-scales and field-scales. a new discrete element simulation method, called the synthetic rock mass (SRM) approach, has been developed by Mas Ivars [13] for the more realistic simulation of jointed rock masses. In this approach, intact rock was represented by BPM, in which the major advantage was replacing the complex empirical constitutive behavior by simple particle contact logic. For the superiority, BPM has been extensively applied for rock failure analysis for the past decade [14–16]. The pre-existing cracks were represented by SJM, which simulated the behavior of a smooth interface created by a joint plane, reflecting the predominant influence of joint orientation and inclination on the joint and hence the whole rock mass behavior by neglecting the local particle contact geometry surrounding the joint plane.

The structure of this paper is as follows. In Sect. 2, the methodology of the paper about the conception of the BPM and SJM are listed. In Sect. 3, the background and related calibration work of the BPM by the pure rock of the marble were studied. In Sect. 4, the behavior of rock containing one flaw is studied, containing two kinds of test. The one is the BPM containing different dip angle flaw, the second is the BPM containing different flaw length. In Sect. 5, discusses the key finding of the present numerical study, and make a conclusion to end of the paper.

2 Methodology

2.1 BPM (Parallel Bond Model)

The Particle Flow Code in three dimensions (PFC^{3D}, [17]) is a discrete element code that represents a rock mass as an assemblage of spheres connected by planar walls. The BPM represents the rock as rigid circular (2D) or spherical (3D) particles (grains) bonded together at their contacts by parallel bonds(cement). The calibration process involves the trial-and-error adjustment of the following micro-parameters until the model prediction matches the measured response of marble samples. In the present numerical study, specimens are loaded under uniaxial vertical compression in a displacement controlled manner. In other words, the loading platens are moving toward each other at a constant velocity. A sufficiently low load of 0.08 m/s (time step is about 4.24×10^{-8} s for the model) is applied to ensure that the specimen remains in a quasi-static equilibrium throughout the test. It takes approximately 140,000 steps to load one specimen to complete failure.

2.2 SJM (Smooth Joint Model)

In this paper, flaws are created by SJM, which is a more reasonable expression to represent flaws. In 3D SJM, a flaw plane is assumed to have a circular shape and its geometry is fully defined by three parameters, namely, dip angle, dip direction and flaw lengths. In this paper, a single pre-existing flaw is represented by inclination angle β (same as dip angle), flaw length L, not considered dip direction (equal to 0), the sketch of sample is shown in Fig. 1a. The SRM sample containing one flaw is shown in Fig. 1b, the particle is expressed in yellow ball, the flaw plane is represented in red disk, and the smooth joint contact is represented in black disk.

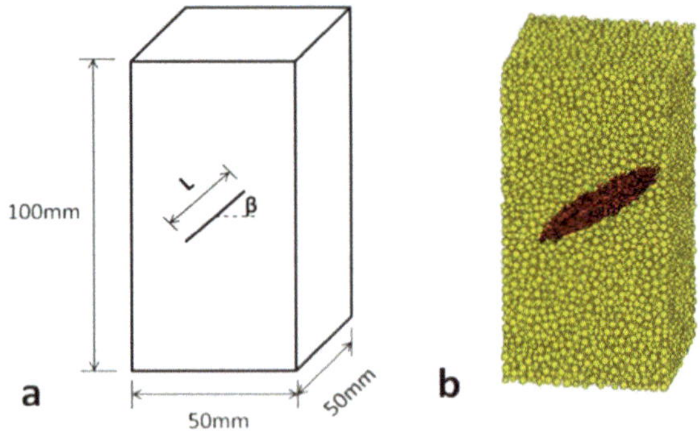

Fig. 1. (a) Sketch of SRM containing one flaw, (b) sample of SRM.

2.3 Micro-crack

In the BPM, a micro-crack formed when the parallel bond between adjacent parent particles was broken. The number and location of potential micro-cracks was limited by the number and location of the bonds in the initial specimen. Each micro-crack was assumed to be a cylinder whose axis lied along the line joining the two parent particles. The tensile crack was shown as a red cylinder and the shear crack was shown as a black cylinder. The geometry and location of the cylinder were fully described by the thickness, radius, unit-normal, and centroid location. The thickness equaled the gap between the two parent particles. The radius was given by the intersection of the cylinder bisection plane, with a membrane stretched tightly between the two parent particles. The unit-normal was directed along the line joining the centers of the two parent particles. And the centroid lied along the line between the centers of the two parent particles and bisected the gap between the two parent particles.

3 Behavior of Pure Rock

In the current study, the BPM was constructed and calibrated against the laboratory test results of marble samples extracted from the site of Jinping II hydropower station located at Yalong River of Sichuan, China [18]. The BPM model had a dimension of

$50 \times 50 \times 100\ mm^3$, which was the same size as that of the real marble sample. I used a cubic model to prepare for future large-scale discrete fracture model. The discrete fracture model contains several groups of fractures, and the rock mass volume was large. The macro-properties of Jinping marble used for model calibration included Young's modulus E, unconfined compressive strength and Poisson's ratio. This parametric study yielded a set of best-fit parameters as listed in Table 1, which resulted in the best predictions of the laboratory measured macro-properties of Jinping marble samples from uniaxial compression tests, as listed in Table 2.

Table 1. The micro-parameters of BPM numerical model

Particle parameters		Parallel bond parameters	
E_c	35 GPa	$\bar{E}_c$	35 GPa
k_s/k_n	0.4	$\bar{k}_n/\bar{k}_s$	0.4
μ	0.5	$\bar{\sigma}_c$	124e6 ± 24.8e6
R_{max}/R_{min}	1.66	$\bar{\tau}_c$	124e6 ± 24.8e6
R_{min}	1 mm	$\bar{\lambda}$	1.0

Table 2. Comparison of material properties of physical experiment and numerical study

	Physical experiment	BPM
Yong's modulus, E(GPa)	31.6	33.1
Poisson's ratio, γ	0.5	0.42
Uniaxial compression strength, UCS(MPa)	140	136.8

The corresponding stress-strain relationship from the simulation was also compared with the laboratory measurement in Fig. 2a, with the shear strain distribution within the sample after its failure shown in Fig. 2b. It is seen in Fig. 2 that the stress-strain curve resulting from the best-fit parameters matched the experimental curve very well, accompanied by a clear shear band cutting through the sample.

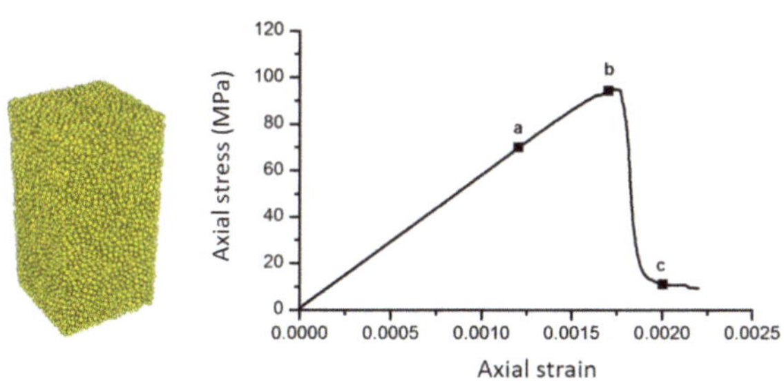

Fig. 2. (a) UC test of best match, and (b) shear strain location of SJ model.

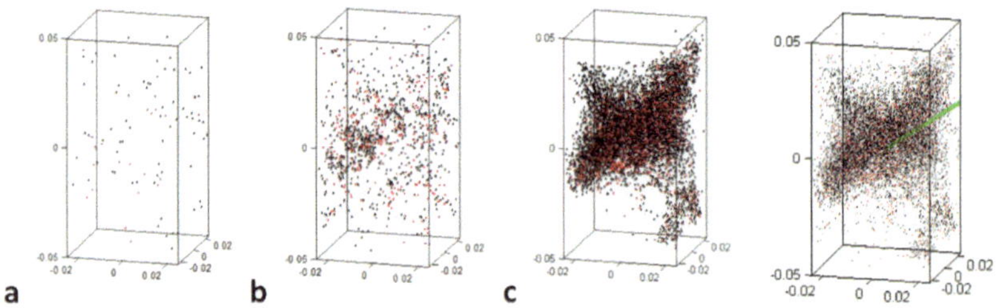

Fig. 3. Micro-crack distribution and the determination of the fracture angle

Extract the position of the micro-crack of the failure stage, and do the principal component analysis, the first principal component of the vector is shown in Fig. 3, then match the fracture angle is 64.22°. And the Elastic mechanical calculation value is $\alpha = 45° + \psi/2 \approx 56.5°$. Finally, the fracture angle is about 60°.

In the next part of this paper, the BPM of Jinping marble with the best-fit parameters established above will be used for research on the effects of pre-existing joints on the rock mass behavior.

4 Behaviours of Rock Containing Single Flaw

4.1 Effects of Dip Angle

In the present numerical research, there are five different SRM specimens containing differently oriented single flaw of β equal to 0°, 30°, 45°, 60° and 90°, with the flaw length fixed in the 40 mm. The stress-strain curve of five samples in the unaxial compression test and the relationship between inclination angle and axial stress are shown in Fig. 4. It can be seen that the inclination angle has an influence on the mechanics behavior of samples. The sample containing the inclination angle β of 90° has the biggest peak strength, while the sample containing the inclination angle β of 45° has the smallest peak strength. Figure 4b showed that, the rock strength is high

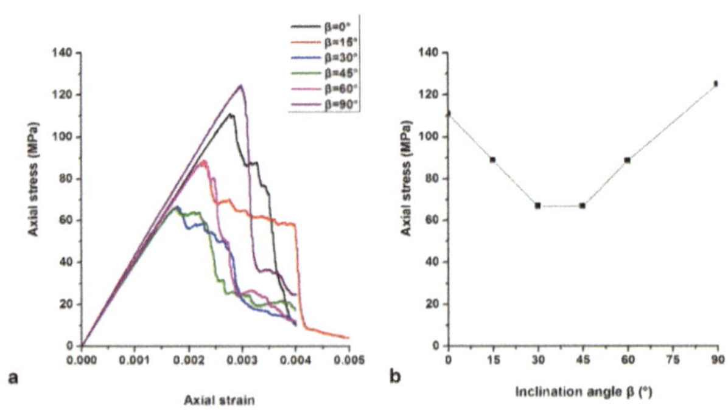

Fig. 4. (a) Stress-strain curve of SRM with different inclination angle under uniaxial compression test; (b) The relationship between inclination angle and the axial stress.

when the dip angle is 90° or 0°, which is similar to the intact rock. The rock strength is low when the dip angle is 30° or 45°.

According to the distribution of micro-cracks, the failure process of rock was the main cracks appearing, the secondary cracks appearing, and the failure. So the development process of micro-cracks was divided into two stages, the first stage was the main cracks appearing, the second stage was the secondary cracks appearing and failure. The model of dip angle increasing from 15° to 60° were chosen. In the first stage, the distribution of micro-crack of different model is shown in the Fig. 5 a, b, c, d represent the dip angle is increases from 15° to 60°. When the dip angle is 15°, 30°, 45°, the micro-cracks are along the joint point initiation, extending in the direction of principal stress. When the dip angle is 60°, the micro-cracks are along the joint point of crack initiation and extension in the direction of joint.

The shear stresses of the different models in the first stage are shown in the Fig. 6. When the dip angle is 15°, 30°, 45°, micro crack is initiate from the tip of the joint and extends to the main stress direction, and the biggest shear value appeared in the side of the micro-crack close to the model boundary. When the dip angle is 45°, micro-crack is initiation from the tip of the joint and extend to the joint direction, and the biggest shear value appeared in the extension of the joint plane.

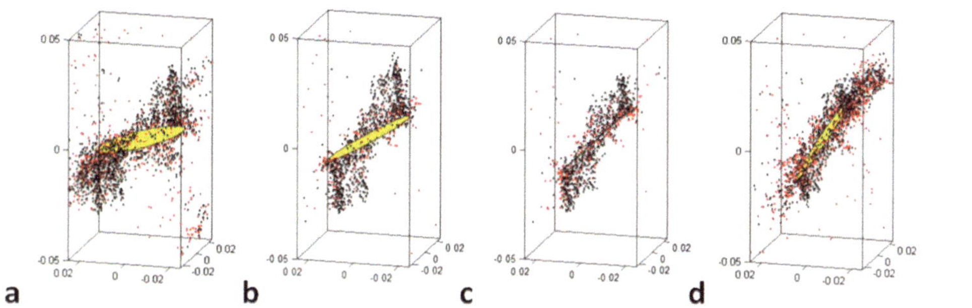

Fig. 5. The first stage of micro crack development of different model

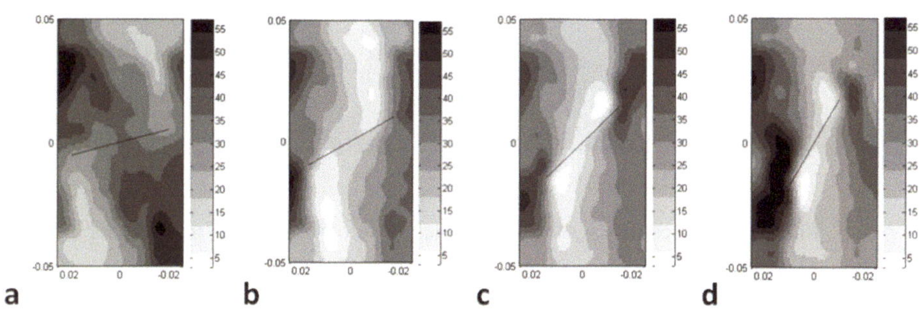

Fig. 6. The shear stress of different model in the first stage

In the second stage, the distribution of micro-crack of different model is shown in the Fig. 7, a, b, c, d represents the dip angle increasing from 15° to 60°. It can be seen, when the dip angle is 15°, the micro-crack is performance for split and break. When the dip angle is 30°, 45°, 60°, the evolution of micro-crack has two trends. The first is micro-crack extending along the joint direction on the both sides, and the second is wing crack obviously, which is perpendicular to the joint direction.

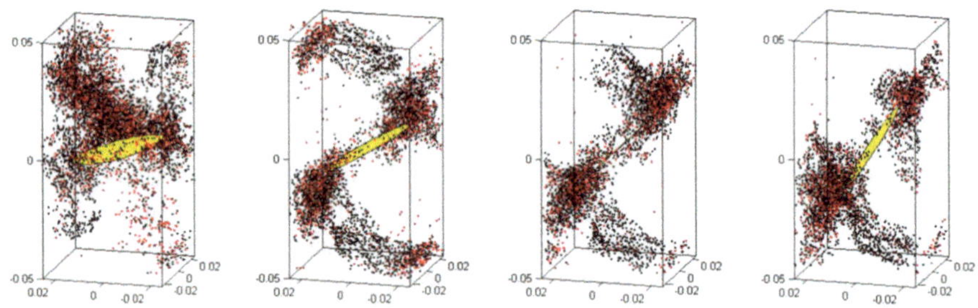

a b c d

Fig. 7. The second stage of micro-crack development of different model

The shear stress of the different model in the second stage are shown in the Fig. 8. In the Fig. 8, when the dip angle is 15°, the failure mode is split and break, and the biggest shear value appeared in another direction of the split failure. When the dip angle is 30°, micro-crack is wing crack, and the biggest shear value is appeared in the undamaged areas. When the dip angle is 45° and 60° respectively, micro-crack extend from the joint plane and wing crack, and the biggest shear value appeared in the extension of the joint plane.

a b c d

Fig. 8. The second stage of micro-crack development of different model

4.2 Effects of Flaw Length

In the present numerical study, there are six different SRM specimens containing differently single flaw L equal to 10 mm, 20 mm, 30 mm, 40 mm, 60 mm and 80 mm.

Geometry of flaws are definite by flaw inclination angle β = 60°, flaw length L and center point. The inclination angle is a fixed value for suitable analysis, but it's just a little bit better, and other angles can draw a similar conclusion. The stress-strain curve of six samples in the unaxial compression test and the relationship between the flaw length and axial stress is shown in Fig. 9. When the flaw length was less than 30 mm, the peak strength were high and near the intact sample; When the flaw length was more than 30 mm, the peak strength were declined obviously; When the flaw length was 80 mm, the flaw were through the whole sample, the peak strength were very low and there was little capability.

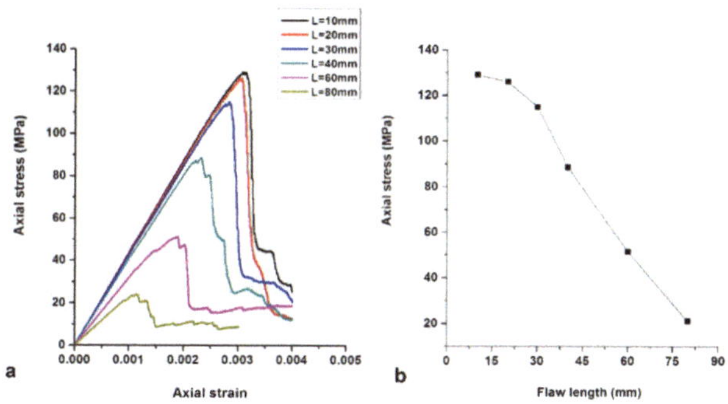

Fig. 9. **(a)** Stress-strain curves at SRM with different flaw length under Uniaxial compression test; (b)The relationship between flaw length and axial stress.

In the first stage, the distribution of micro-crack of different model is shown in the Fig. 10, a, b, c represent the dip lengths of 10 mm, 30 mm, 80 mm. It can be seen, when the dip length was 10 mm, the micro-cracks are dispersion. When the dip length was 30 mm, the micro cracks appeared along the joint plane. When the dip length was 80 mm, the joint was through, the micro cracks appeared in the end of the joint plane.

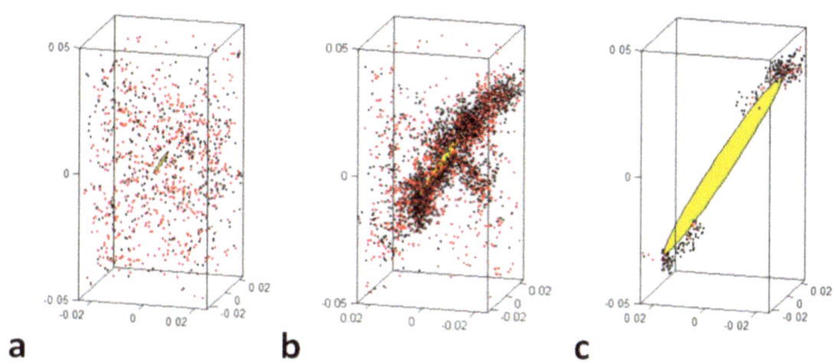

Fig. 10. The first stage of micro-crack development of different model

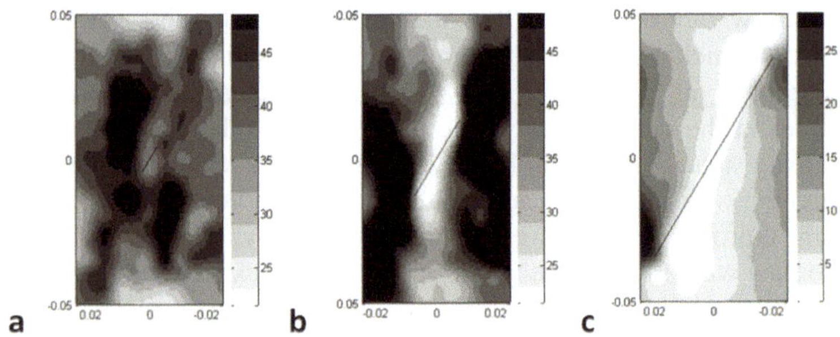

Fig. 11. The shear stress of different model in the first stage

The shear stresses of different models in the first stage are shown in the Fig. 11. It can be seen, when the dip length was 10 mm, the biggest shear value appeared in the joint tip and extended to the main stress direction and the joint extension direction. When the dip length was 30 mm, the lowest shear value appeared in the joint tip were extended to the main stress direction, the biggest shear value appeared in the side of the micro-crack close to the model boundary. When the dip length was 80 mm, the stress value was lower than that of other models, and the biggest shear value appeared in the joint boundary.

In the second stage, the distribution of micro-crack of different model is shown in the Fig. 12, a, b, c represents the dip length of 10 mm, 30 mm,60 mm. It can be seen, when the dip length is 10 mm, the fracture band was not obvious. When the dip length was 30 mm, the micro-cracks extended to the direction of the main stress direction. When the dip length is 80 mm, the micro-cracks appeared in the tip of the joint and followed the joint plane.

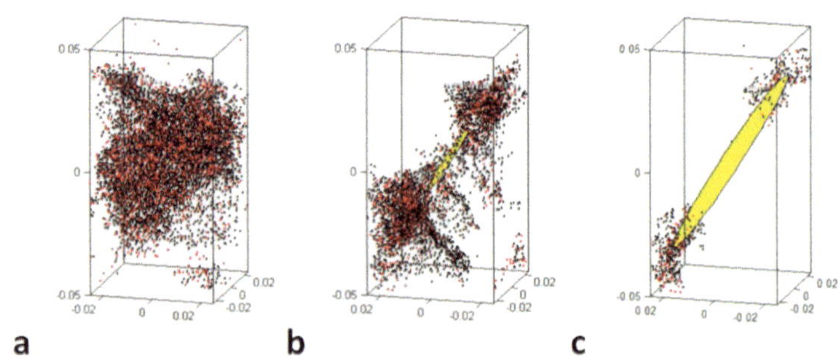

Fig. 12. The second stage of micro-crack development of different model

The shear stresses of the different models in the second stage are shown in the Fig. 13.When the joint length is 10 mm, the second stage of the micro-crack formed in the fracture zone, where the most concentrated area was still the joint plane direction. When the joint length was 30 mm and 80 mm, the biggest shear value is similar to the first stage, but the number was lower than the first stage.

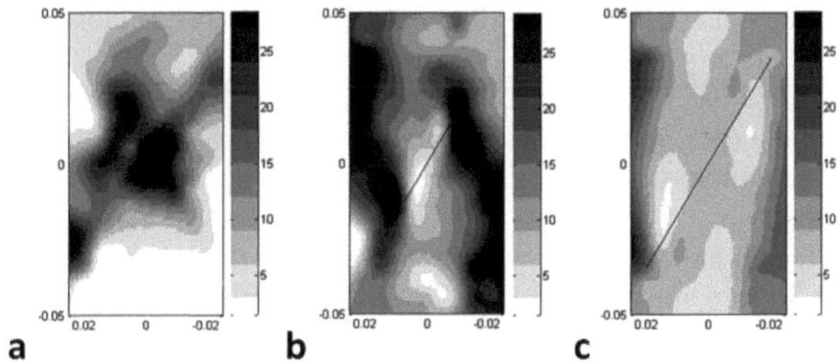

Fig. 13. The second stage of micro-crack development of different model

5 Discussion

One of the main objectives of all these tests was to observe and characterize the cracks emanating from the flaws. The processes of failure of rock was the flaw slippage, wing crack initiation, secondary crack initiation, crack coalescence, and then failure. Combining with micro-crack evolution rule and shear stress distribution, it can be found: the area of micro-cracks appear, shear stress was small. The first stage the micro progress of the test was the stress concentration, and the second stage was the micro-cracks appears, and the third stage was stress reduced to zero at the position of the micro-cracks, and the fourth stage was the stress redistributed. The micro cycle was finished, the next cycle was carried out. So there were couple effect between micro-crack appear and stress concentration.

According to the distribution of the micro-crack and the shear stress, there were four main modes of crack coalescence: (1) wing tensile crack coalescence, such as the model with inclination angle of 30°and 45°, (2) splitting shear crack coalescence, such as the model with inclination angle of 15°, (3) micro-crack scattered through such as the model with flaw length of 10 mm and (4) slip tensile/shear crack coalescence, such as the model with flaw length of 60 mm, which represented the fracture morphology affecting the penetration of rocks. And the quantitative relationship between the fracture morphology and the crack coalescence can be summarized through the future study.

6 Conclusions

This paper presents a PFC numerical simulation study based on the smooth joint model approach to model the cracking processes from a pre-existing flaw under a uniaxial compressive stress field. The cracking process and stress distribution during the uniaxial compressive test of the two kinds of marble was analyze in the paper.

(1) The pure rock has the fracture plane angle, which is between 40° and 48° in the DEM simulation. For the model with one joint, the joint dip angle has changed the fracture plane of the model. When the joint length is more than 30 mm, the

fracture plane angle is close to dip angle, and the other is close to the pure rock fracture plane angle.

(2) The dip angle has a big influence on the failure of the rock, especially the failure pattern of the rock. In the uniaxial compression test of the model with different dip angle, the micro-crack is divided into two stages, which has a similar characteristic.

(3) The dip length has a big influence on the failure of the rock, when the dip length is less than 30 mm, the micro-crack is dispersion similar to the intact rock. When the dip length is more than 30 mm, the longer the joint is, the lower the strength is, and the micro-cracks are lies on the extension of the joint plane. The micro-crack is also divided into two stages, which has a similar characteristic.

Acknowledgments. The authors thank the ITASCA (Wuhan) for the support of the Jinping II hydropower station project.

References

1. Hazzard, J.F., Young, R.P.: Simulating acoustic emissions in bonded-particle models of rock. Int. J. Rock Mech. Min. Sci. **37**(5), 867–872 (2000)
2. Feng, X.T., Chen, S.L., Zhou, H.: Real-time computerized tomography (CT) experiments on sandstone damage evolution during triaxial compression with chemical corrosion. Int. J. Rock Mech. Min. Sci. **41**(2), 181–192 (2004)
3. Kawakata, H., Cho, A., Kiyama, T., Yanagidani, T., Kusunose, K., Shimada, M.: Three-dimensional observations of faulting process in Westerly granite under uniaxial and triaxial conditions by X-ray CT scan. Tectonophysics **313**(3), 293–305 (1999)
4. Wong, L.N.Y., Einstein, H.H.: Fracturing behavior of prismatic specimens containing single flaws. In: Proceedings of the 41st US symposium on rock mechanics, Golden, CO (paperARMA/USRMS06-899) (2006)
5. Li, Y.P., Chen, L.Z., Wang, Y.H.: Experimental research on precracked marble under compression. Int. J. Solids Struct. **42**(9–10), 2505–2516 (2005)
6. Lajtai, E.Z.: Brittle fracture in compression. Int. J. Fract. **10**(4), 525–536 (1974)
7. Park, C.H., Bobet, A.: Crack coalescence in specimens with open and closed flaws: a comparison. Int. J. Rock Mech. Min. Sci. **46**(5), 819–829 (2009)
8. Wong, R.H.C., Chau, K.T.: Crack coalescence in a rock-like material containing two cracks. Int. J. Rock Mech. Min. Sci. **35**(2), 147–164 (1998)
9. Bobet, A., Einstein, H.H.: Fracture coalescence in rock-type materials under uniaxial and biaxial compression. Int. J. Rock Mech. Min. Sci. **35**(7), 863–888 (1998)
10. Sagong, M., Bobet, A.: Coalescence of multiple flaws in a rock model material in uniaxial compression. Int. J. Rock Mech. Min. Sci. **39**(2), 229–241 (2002)
11. Wong, L.N.Y., Einstein, H.H.: Crack coalescence in molded gypsum and Carrara marble: part 1. Macroscopic observations and interpretation. Rock Mech. Rock Eng. **42**(3), 475–511 (2009)
12. Wong, L.N.Y., Einstein, H.H.: Crack coalescence in molded gypsum and Carrara marble: part 2—microscopic observations and interpretation. Rock Mech. Rock Eng. **42**(3), 513–545 (2009)

13. Mas Ivars, D., Pierce, M., Darcel, C., Reyes-Montes, J., Potyondy, D.O., Young, R.P., Cundall, P.A.: The synthetic rock mass approach for jointed rock mass modeling. Int. J. Rock Mech. Min. Sci. **48**, 219–244 (2011)
14. Cho, N., Martin, C.D., Sego, D.C.: A clumped particle model for rock. Int. J. Rock Mech. Min. Sci. **44**, 997–1010 (2007)
15. Hazzard, J.F., Young, R.P., Maxwell, S.C.: Micromechanical modeling of cracking and failure in brittle rocks. J. Geophys. Res. **105**(B7), 16683–16697 (2000)
16. Potyondy, D.O., Cundall, P.A.: A bonded-particle model for rock. Int. J. Rock Mech. Min. Sci. **41**(8), 1329–1364 (2004)
17. Itasca Consulting Group Inc. PFC^{3D} – Particle Flow Code in 3 Dimensions, Version 4.0. Minneapolis, MN (2008)
18. Chu, W.: The stability and structural safety assessment of tunnel surrounding rock under buried deep conditions. Postdoctoral Report (2009)

Research on Confining Pressure Effect on Mesoscopic Damage of Rock Salt Based on CT Scanning

Lu Wang[1], Jianfeng Liu[2](✉), Huining Xu[2], and Yangmengdi Xu[2]

[1] School of Civil Engineering, Architecture and Environment, Xihua University, Chengdu, China

[2] State Key Laboratory of Hydraulics and Mountain River Engineering, Sichuan University, Chengdu, China

ljfscu@163.com

Abstract. The Rock salt is the optimal surrounding rock of underground energy storage, so its damage is directly related to the impermeability and stability of the storage. In this study, CT scanning technique was used to analyze the microscopic damage characteristics of rock salt, which damaged under the three-dimensional stress state. The distribution and evolution characteristics of porosity, crack area, crack volume and crack quantity of damaged rock salt under different confining pressures were obtained. The results revealed the confining pressure effect of the mesoscopic damage of rock salt. With the increase of confining pressure, the quantity and size of cracks in rock salt gradually decreased. The confining pressure of 10 MPa is the turning point where the porosity decreased from fast to slow and the failure characteristics of rock salt changed from the apparent shear failure surface to the surface micro cracks.

Keywords: Rock salt · CT scanning · Microscopic damage
Confining pressure effect

1 Introduction

Rock salt is the optimal surrounding rock of underground energy storage, so its damage is directly related to the sealing and stability of the storage. Stormont et al. studied the internal damage characteristics of rock salt under the seepage state [1–3]. Hou and Lux [4] established a Hou/Lux model which could be used for the study of rock salt damage and damage recovery. Schulze [2] analyzed the damage evolution characteristics of rock salt during its deformation process under the stress-seepage coupling condition. In addition, Methods of Acoustic emission [5], unloading modulus [6] and microscopic statistics [7] were also applied in the study of rock salt damage characteristics and lots of valuable results have been achieved.

Compared with the conventional research methods of damage, such as the acoustic emission and ultrasonic wave tests, CT scanning is a more intuitive method to show the internal damage characteristics of rocks [8]. For rock salt, Zhou et al. [9] and Liu et al. [10] analyzed the characteristics of microscopic cracks and density damage and Liu et al. [11] studied the corrosion damage of salt rock through CT scanning. Teles et al. [12]

L. Zhang et al. (Eds.): GSIC 2018, *Proceedings of GeoShanghai 2018 International Conference: Rock Mechanics and Rock Engineering*, pp. 254–262, 2018.
https://doi.org/10.1007/978-981-13-0113-1_28

proposed a method to quantify the rock porosity based on the CT scanning. Thiemeyer [13–15] utilized the grayscale of CT images to realize the visualization and quantification of the pores in rock salt, and revealed the deformation mechanism of rock salt from the microscopic view. However, the damage characteristics of rock salt are influenced by lots of factors, such as the stress states, environmental conditions and impurity content. There are few studies on the influence of these factors on the microscopic damage characteristics of rock salt.

The confining pressure has a great influence on the mechanical and damage characteristics of rock salt [7, 16, 17]. Therefore, in this study, CT scanning technique was used to analyze the microscopic damage characteristics of rock salt, which damaged under the three-dimensional stress state. Based on the CT images, the statistical method was adopted to analyze the quantity and distribution of pores and cracks, and to reveal the confining pressure effect of microscopic damage of rock salt.

2 Test Method

The rock salt sample was pure rock salt which removed the influence of impurities on the damage characteristics. The sample was reprocessed into Φ50 mm × H100 mm cylinders according to a standard for engineering rock testing methods in the lab (The Professional Standards Compilation Group of People's Republic of China [18]). Then, the triaxial compression tests were carried out on these rock salt specimens under the different confining pressures of 3 MPa, 5 MPa, 10 MPa, 15 MPa and 20 MPa by the THMC rock triaxial test system of Sichuan University, and the damaged rock salt specimens were obtained (Fig. 1). The triaxial strengths of rock salt based on the logarithmic strain model under the confining pressure of 3 MPa, 5 MPa, 10 MPa, 15 MPa and 20 MPa were 53 MPa, 61 MPa, 73 MPa, 80 MPa and 86 MPa, respectively. The triaxial strength of rock salt increased with the increase of the confining pressure.

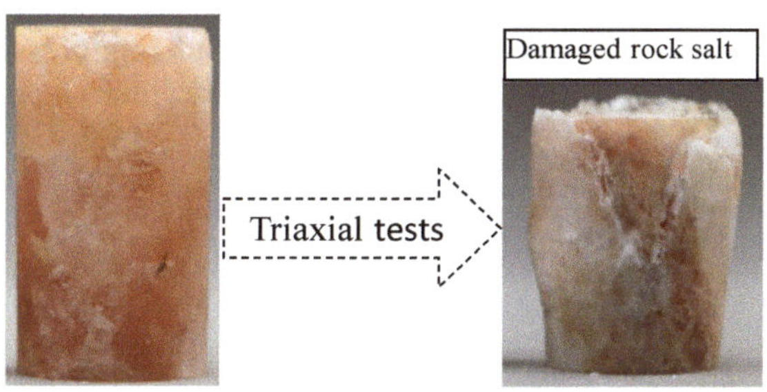

Fig. 1. Rock salt specimen figure

These damaged rock salt specimens were scanned through the GE Phoenix v|tome|x m300 scanning system, which adopts the surface array scan form with the maximum detection range of 400 mm × 400 mm. The CT images were analyzed through the VG

Studio MAX. In order to avoid the influence of the end effect on the damage characteristics, take the middle part of the specimen for section analysis and body analysis, as shown in Fig. 2.

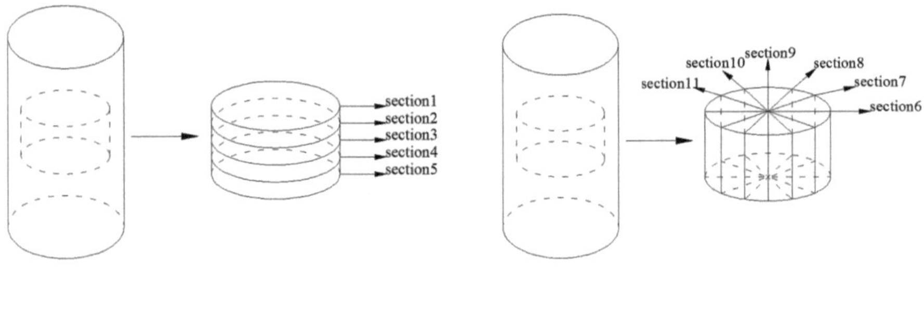

(a) Lateral section (b) Axial section

Fig. 2. Sections for analysis

3 Results and Analysis

3.1 Failure Feature

Rock salt is the optimal Based on the CT images of damaged rock salt specimens under different confining pressures, the axial section images and 3D images showed the distribution and features of the microscopic damage in the specimens (Fig. 3). When the rock salt specimen was constrained by the low confining pressure of 3 MPa (Fig. 3a), there were a large number of micro-cracks and a visible shear failure surface in the specimen. The rock salt under the confining pressure of 3 MPa exhibited the significant shear failures. In contrast, the damaged rock salt under the confining pressure of 10 MPa and 20 MPa had no obvious failure surfaces, which contained more spherical pores (Fig. 3b and c). As shown in the CT images, the confining pressure had a great influence on the development of cracks in the specimen. The confining pressure of 10 MPa is a turning point where the development of cracks and the failure feature of the salt rock changed significantly.

3.2 Microscopic Damage on Lateral Section

Rock salt is the optimal Based on the selection method of lateral sections in Fig. 2, the lateral sections CT images of damaged rock salt under different confining pressures were obtained. The specimens were homogeneous pure rock salt with the similar density, so the mesoscopic characteristics of the different specimens can reflect the confining pressure effect on the mesoscopic damage of rock salt. As shown in the CT images of section1 in Fig. 4, The quantity of cracks on the lateral section decreased as the confining pressure increased. When the confining pressures were 3 MPa and 5 MPa, there were several long cracks that surrounded by many small cracks. The cracks in the rock salt under the confining pressure of 3 MPa developed gradually during the deformation

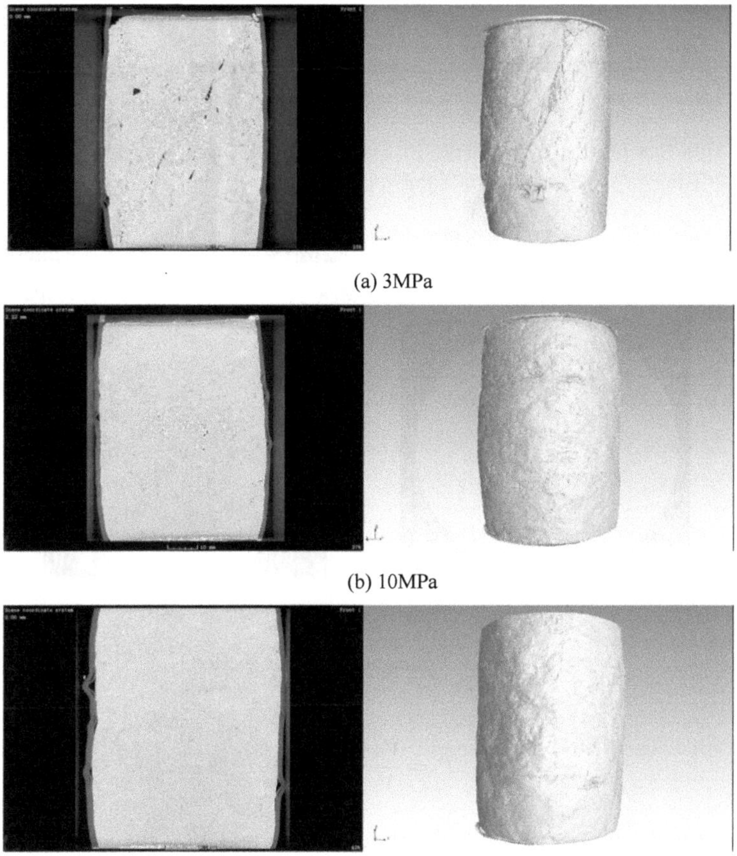

(a) 3MPa

(b) 10MPa

(c) 20MPa

Fig. 3. CT images of rock salt failure features

process and connected to a main fracture. With the increase of the confining pressure, there were no main fractures on the lateral section but the cracks with a complicated distribution under the confining pressure of 5 MPa. When the confining pressure increased to 10 MPa, the quantity and scale of cracks on the lateral section were significantly reduced (Fig. 4c), the CT images of cracks under the confining pressure of 15 MPa and 20 MPa showed the same characteristics as well. Therefore, the confining pressure has a significant effect on the germination and the radial expansion of cracks in the rock salt based on the mesoscopic damage characteristics on the lateral sections of rock salt, which causes the differences of rock salt deformation and failure under different confining pressures.

The change of the crack quantity and volume in the rock salt can be represented by the change of the porosity. In general, the porosity of the rock specimen can be calculated by the pore volume and the specimen volume

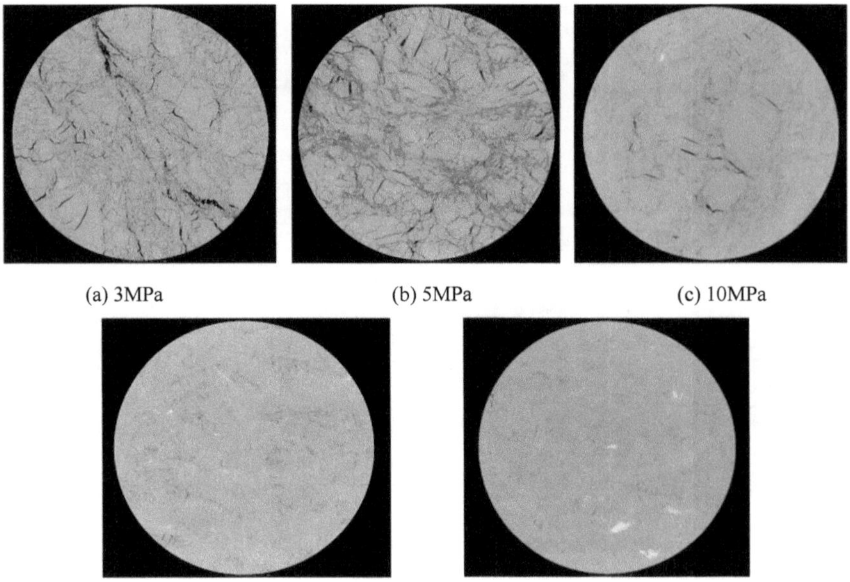

(a) 3MPa (b) 5MPa (c) 10MPa

(d) 15MPa (e) 20MPa

Fig. 4. CT images of lateral section1

$$P = \frac{V_{\omega}}{V} \times 100\% \tag{1}$$

Where P is the porosity, V_{ω} is the pore volume and V is the specimen volume.

In extreme cases, the volume of the specimen can be converted to the area multiplied by the height, so the porosity on the section can be calculated as

$$P = \frac{S_{\omega}}{S} \times 100\% \tag{2}$$

Where P is the porosity, S_{ω} is the pore area and S is the section area.

Based on the statistical Analysis of the cracks area on the sections through VG Studio MAX Data Analysis Software and the Eq. (2), the porosity on all lateral sections of rock salt specimens under different confining pressures were obtained (Fig. 5). As shown in Fig. 5, there were differences of the porosity among the different lateral sections under the same confining pressure, but the differences decreased gradually with the confining pressure increasing. The porosity of different rock salt specimens decreased with the increase of the confining pressure, and the relationship between the porosity on lateral sections and the confining pressure can be expressed by an exponential function

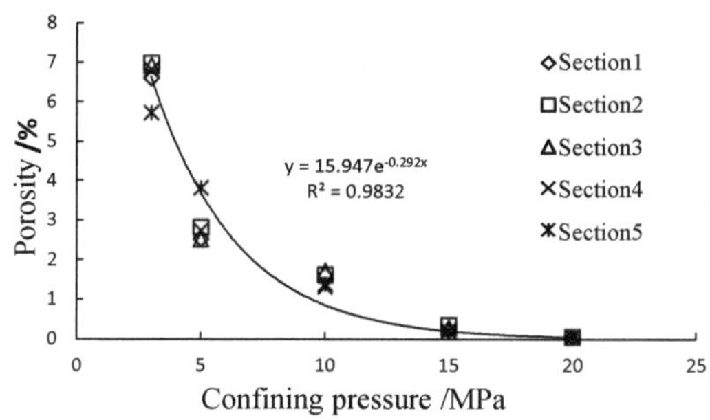

Fig. 5. Relationship between porosity on lateral sections and confining pressure

$$P = \mathrm{A}e^{B\sigma_3} \tag{3}$$

Where P is the porosity on the lateral section, σ_3 is the confining pressure, A and B are the relationship coefficients.

3.3 Microscopic Damage on Axial Section

Based on the selection method of axial sections in Fig. 2, the axial sections CT images of damaged rock salt under different confining pressures were obtained. Figure 6 showed the CT images on the axial section 6, in which the largest crack (Fig. 6b) in the upper section of the specimen under the confining pressure of 5 MPa was artificially caused after the triaxial compression test, so it is not considered. With the increase of the confining pressure, the quantity and scale of the crack on the axial section decreased, which were consistent with the mesoscopic damage characteristics on the lateral sections. There was a main through crack which was surrounded by many small cracks with the band-like distribution on the axial section of the specimen under the confining pressure of 3 MPa, but for the specimen under the confining pressure of 5 MPa, there was no an obvious through crack but a crack dense zone on the axial section (Fig. 6b). When the confining pressure is 10 MPa, the quantity and scale of cracks on the axial section decreased significantly, and the cracks distribution was scattered without the dense area and band-like characteristics (Fig. 6c). With the increase of the confining pressure, the cracks on the axial sections showed a homogeneous distribution with small pores (Fig. 6d and e). From the above, with the confining pressure increasing, the mesoscopic damage characteristics on the axial sections changed from the dense distribution of band-like cracks to the homogeneous distribution of pore-like cracks, and the confining pressure of 10 MPa is the turning point.

Based on the statistical Analysis of the cracks area on the axial sections and the Eq. (2), the porosities on the axial sections under different confining pressures were obtained. As shown in Fig. 7, the porosity on the axial sections decreased exponentially on the whole with the increase of the confining pressure. The porosities on the different sections showed obvious differences when the confining pressures were 3 MPa and 5 MPa, and that was caused by the main through crack and the dense

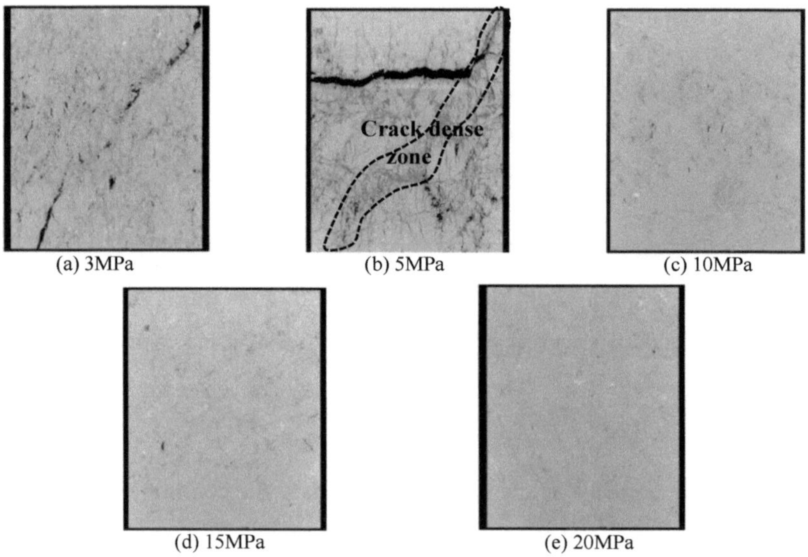

Fig. 6. CT images of axial section6

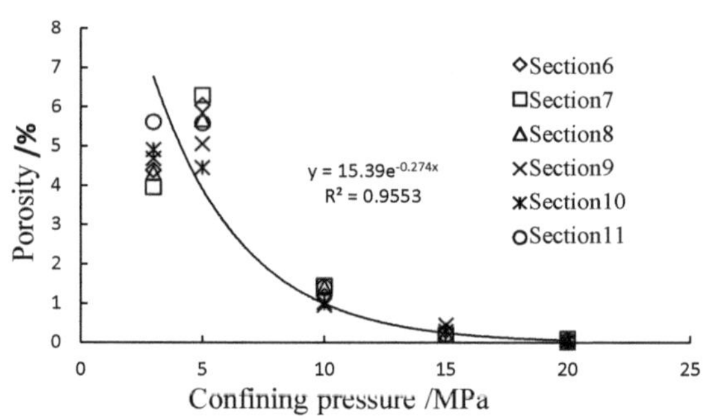

Fig. 7. Relationship between porosity on axial sections and confining pressure

distribution of cracks. With the confining pressure increasing, the distribution of cracks in rock salt tended to be homogeneous, so the differences of the porosity on the sections decreased.

3.4 Whole Microscopic Damage

Based on the above analysis of the microscopic damage characteristics on the sections, the confining pressure has a significant effect on the initiation and propagation of cracks in the rock salt. Because the CT images of sections can only reflect the distribution characteristics of the microscopic damage on a certain plane in rock salt, the statistical analysis of the total number and volume of cracks in the rock salt specimen were carried out to investigate the distribution characteristics of the damage in the whole specimen. The variation law of porosity of rock salt with the confining pressure was obtained, as shown in Fig. 8. For the porosity of the whole rock salt specimen, the porosity decreased exponentially with the increase of confining pressure. When the confining pressure was less than 10 MPa, the porosity increased sharply with the increase of confining pressure. With the confining pressure increasing to exceed 10 MPa, the porosity decline gradually tended to be gentle. It indicates that the porosity of rock salt has a significant sensitivity to the change of the confining pressure under low confining pressures, and the confining pressure of 10 MPa is the turning point where the porosity decline changed from fast to slow.

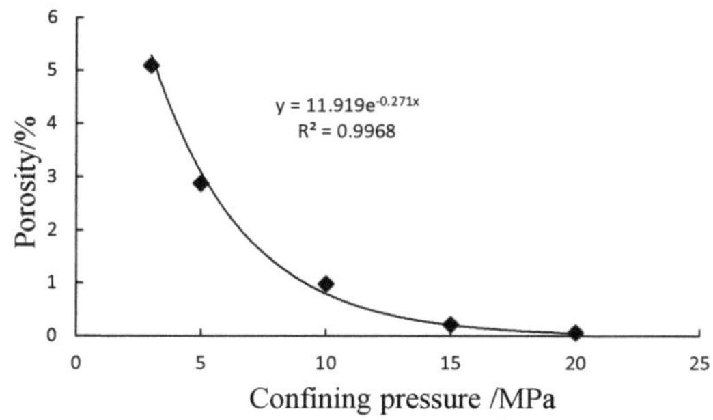

Fig. 8. Relationship between whole porosity and confining pressure

4 Conclusions

CT images of rock salt which damaged under the three-dimensional stress state were analyzed to obtain the microscopic damage characteristics of rock salt and reveal the confining pressure effect on rock salt damage. The results show that the confining pressure has a significant effect on the quantity and distribution of cracks in rock salt and the porosity. With the increase of confining pressure, the quantity and scale of cracks in rock salt gradually decrease, and the porosity decreases exponentially as well. The confining pressure of 10 MPa is the turning point where the porosity decreased from fast to slow and the failure characteristics of rock salt changed from the apparent shear failure surface to the surface micro cracks.

Acknowledgments. The authors are grateful for the financial support from the National Natural Science Foundation of China (Grant No. 51374148, No. 51641405), Sichuan Youth Science and Technology Fund (Grant No. 2017JQ0003). The authors wish to offer their gratitude and regards to the colleagues who contributed to this work.

References

1. Stormont, J.C.: In situ gas permeability measurements to delineate damage in rock salt. Int. J. Rock Mech. Mining Sci. **34**, 1055–1064 (1997)
2. Schulze, O., Popp, T., Kern, H.: Development of damage and permeability in deforming rock salt. Eng. Geol. **61**, 163–180 (2001)
3. Hou, Z.: Mechanical and hydraulic behavior of rock salt in the excavation disturbed zone around underground facilities. Int. J. Rock Mech. Mining Sci. **40**, 725–738 (2003)
4. Hou, Z., Wu, W.: Improvement of design of storage cavity in rock salt by using the Hou/Lux constitutive model with consideration of creep rupture criterion and damage. Chin. J. Geotech. Eng. **25**, 105–108 (2003)
5. Liu, J.F., Xu, J., Yang, C.H., Hou, Z.M.: Mechanical characteristics of tensile failure of salt rock. Chin. J. Geotech. Eng. **33**, 580–586 (2011)
6. Liu, J.F., Xu, J., Pei, J.L., Zhang, R.: Research on unloading modulus for salt rock damage testing. J. Sichuan Univ. (Eng. Sci. Ed.) **43**, 57–62 (2011)
7. Chen, J.W., Yang, C.H., Gao, X.P., Li, X.H., Jiang, D.Y.: Study on the coupled damage of temperature and mechanics for salt rock. Chin. J. Rock Mech. Eng. **24**, 1986–1991 (2005)
8. Ge, X.R., Ren, J.X., Pu, Y.B., Ma, W., Zhu, Y.L.: Primary study of CT real - time testing of fatigue meso-damage propagation law of rock. Chin. J. Geotechn. Eng. **23**, 191–195 (2001)
9. Zhou, H.W., He, J.M., Wu, Z.D.: Permeability and meso-structure characteristics of bedded salt rock. Chin. J. Rock Mech. Eng. **28**, 2068–2073 (2009)
10. Liu, Z.H., Xu, S.G., Hu, Y.Q., Feng, Z.C., Liang, W.G., Zhao, Y.S.: Experimental study of corrosion of glauberite salt rock. Chin. J. Rock Mech. Eng. **29**, 3616–3621 (2010)
11. Liu, Y.D., Yuan, F.L., Dong, J.H.: CT analysis of meso-structure changes in rock salt with Brine corrosion. Adv. Mater. Res. **960–961**, 82–86 (2014)
12. Teles, A.P., Lima, I., Lopes, R.T.: Rock porosity quantification by dual-energy X-ray computed microtomography. Micron **83**, 72 (2016)
13. Thiemeyer, N., Pusch, M., Hammer, J., et al.: Quantification and 3D visualisation of pore space in Gorleben rock salt: constraints from CT imaging and microfabrics. Zeitschrift der deutschen gesellschaft fur geowissenschaften **165**, 15–25 (2014)
14. Thiemeyer, N., Habersetzer, J., Peinl, M., et al.: The application of high resolution X-ray computed tomography on naturally deformed rock salt: multi-scale investigations of the structural inventory. J. Struct. Geol. **77**, 92–106 (2015)
15. Thiemeyer, N., Zulauf, G., Mertineit, M., et al.: Microfabrics and 3D grain shape of Gorleben rock salt: constraints on deformation mechanisms and paleodifferential stress. Tectonophysics **676**, 1–19 (2016)
16. Höfer, K.H., Thoma, K.: Triaxial tests on salt rocks. Int. J. Rock Mech. Mining Sci. **5**, 195–196 (1968)
17. Chen, J., Liu, J.X., Jiang, Y.D., Fan, J.Y., Ren, S.: An experimental study of strain and damage recovery of salt rock under confining pressures. Rock Soil Mech. **37**, 105–112 (2016)
18. The Professional Standards Compilation Group of People's Republic of China. GB/T 50 266-2013 Standard for Tests Method of Engineering Rock Massas. China Planning Press, Beijing (2013)

Shear Behavior of Rock Joints Under CNS Boundary Condition

Caichu Xia[1,2(✉)], Qiangfeng Yu[1,2], Yang Gui[1,2], Xin Qian[1,2], Xiaoqing Zhuang[1,2], and Songbo Yu[1,2]

[1] Department of Geotechnical Engineering, Tongji University, Shanghai 200092, China
tjxiaccb@126.com

[2] Key Laboratory of Geotechnical and Underground Engineering of Ministry of Education, Tongji University, Shanghai 200092, China

Abstract. To improve our understanding of the shear behavior of rock joints under constant normal stiffness (CNS) boundary condition, 48 groups of shear tests on plaster specimens replicated from three rough rock joints were conducted. The effects of joint roughness, normal stiffness, and initial normal stress on mechanical characteristics of joints were studied comprehensively. Test results indicate that compared to constant normal load (CNL) condition, CNS condition is more reasonable to describe the shear behavior of rock joints in underground engineering. The existence of normal stiffness restrains joint dilation, and this constraint effect is strengthened with the increase of normal stiffness. The shear strength increase with the increase of joint roughness and initial normal stress, while the influence of normal stiffness to shear strength is more obvious with the improvement of initial normal stress and the roughness of joint. Meanwhile, three typical shear curves were proposed under CNS boundary condition where both the strain softening and strain hardening behaviors are shown depending on the ratio of initial normal stress to normal stiffness. Moreover, 0.625 of σ_{n0}/k_n is a critical value for plaster mixture used in this study, there will certain be phenomenon of stress hardening when the value of σ_{n0}/k_n is lower than 0.625, and whether the phenomenon of stress hardening exists depends on the roughness and other conditions when the value of σ_{n0}/k_n is 0.625.

Keywords: Rock joint · Constant normal stiffness · Joint roughness
Initial normal stress

1 Introduction

Mechanical behaviors of rock masses are significantly affected by the presence of joints. Therefore, accurate evaluation of the shear strength of rock joints plays an important role in the design of rock structures in jointed rock masses such as stability analysis of rock slopes, deep underground openings, and risk assessment of underground waste disposal.

L. Zhang et al. (Eds.): GSIC 2018, *Proceedings of GeoShanghai 2018 International Conference: Rock Mechanics and Rock Engineering*, pp. 263–274, 2018.
https://doi.org/10.1007/978-981-13-0113-1_29

Conventionally, the shear behaviors of rock joints are investigated in the laboratory under constant normal load (CNL) boundary condition where the normal stress acting on the joint surface remains constant during shearing. Under the CNL condition, the deformation of surrounding rock mass is supposed to occur sufficiently during the shearing process. For example, the rock block slides down a slope where the normal stress remains unchanged because of the free surface. However, in confined environments, as in underground mining operations, rock deformation may be constrained. In such a case, the CNL condition is unrealistic in determining the shear behavior. Shrivastava et al. (2013) indicated shear testing under the CNL boundary condition was only beneficial for non-reinforced rock slopes or planar rock joints. However, in stability analysis and evaluation of roadway top, calculation of side resistance of rock socketed pile and stability analysis of potential dumping damaged joint layer, the stiffness of the surrounding rock mass has significant effects on the shear behavior. This particular mode of shearing is known as shearing under the CNS boundary conditions; the importance of this condition in accurately simulating rock joint shear behavior has been described by many researchers (Heuze et al. 1979; Anno et al. 1984; Johnston et al. 1987; Skinas et al. 1990; Indraratna et al. 1998; Jiang et al. 2004; Mirzaghorbanali et al. 2014). In underground construction, the structure is always in a certain surrounding rock, the dilatation of the shear surface was confined in the shear process, and the boundary condition is closer to the normal stiffness (CNS) boundary condition. The shear behaviors of rock joints under CNS boundary condition were experimentally investigated in the past decades. Indraratna et al. (1997) investigated the mechanical behaviors of rock joints and infilled rock joints with regular triangular asperities under CNL and CNS conditions, and Shrivastava et al. (2015) employed modelled rock joints with different asperity angle to investigate the effect of boundary conditions and the roughness of joints on shear behaviors, but the regular triangular asperities were not appropriate for representing the surface topography of real rock joints. Jiang et al. (2006) reported that the normal stiffness has a significant influence on the mechanical behavior of joints during the shear process. Lee et al. (2014) studied the effect of initial normal stress and roughness of joints on the shear behavior of rock joints. However, the effects of roughness, initial normal stiffness and normal stiffness on shear behaviors were usually investigated independently, and the corporate influence of these factors to shear behavior of rock joints are supposed to be further investigated.

The objective of the present study is to explore the difference of the mechanical behaviors of rock joints under the CNS and CNL boundary conditions. Moreover, the effects of joint roughness, normal stiffness, and initial normal stress on mechanical characteristics of rock joints were studied comprehensively. Typical shear curves were divided into three categories under CNS condition according to the test results, and the normal stiffness and initial normal stress were linked together to further study the shear behavior of rock joints under CNS condition in this work. The results contribute to an improved understanding of the shear behavior of rock joints under the CNS boundary condition.

2 Experiment

2.1 Testing Equipment

Two typical practical engineering cases under different boundary conditions are shown in Fig. 1. Achieving normal stiffness is a problem for the shear tests of rock joints under the CNS boundary condition, and it is usually simulated by a spring with a constant stiffness in several studies (Johnston et al. 1987; Ohnishi et al. 1990). These studies simulated the effect of the normal stiffness during the shear process to a certain extent. However, the joint surfaces were easily damaged when k was extremely strong. The main problem with this method is that springs should be changed according to the stiffness conditions of the experiment.

In this study, a new system for direct shear and coupled shear-flow test under the different boundary conditions was developed to study the shear behaviors of rock joints under the CNS boundary condition. In this system, the computer receives a signal of normal displacement monitored by displacement sensors and sends a signal to control the normal load acting on the sample through a servo motor. The basic principle of the shear test under the CNS boundary is shown in Fig. 2. The calculation of normal load is based on the following value of stiffness set in control software:

$$\Delta P_n = k_n \Delta \delta_n \tag{1}$$

$$P_n(t + \Delta t) = P_n + \Delta P_n, \tag{2}$$

where ΔP_n and $\Delta \delta_n$ are the changes in normal load and normal displacement.

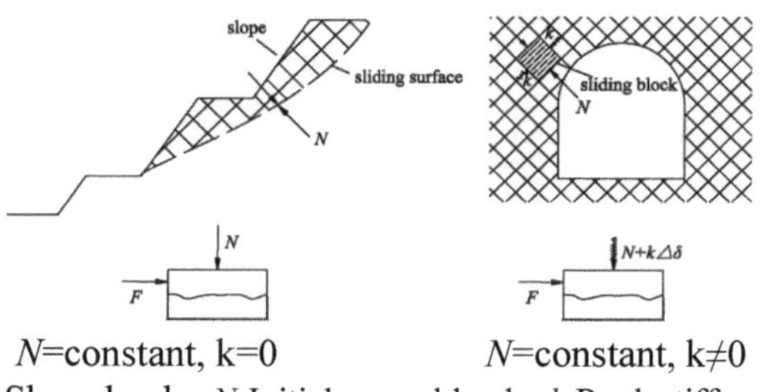

Fig. 1. Simulation of different boundary conditions in shear test.

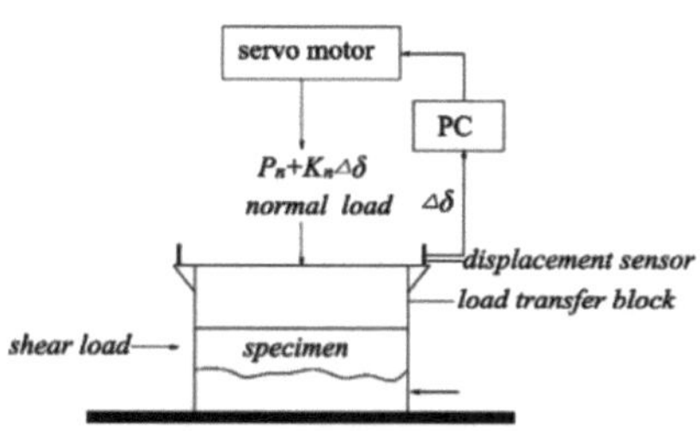

Fig. 2. Basic principle of shear test under CNS boundary condition

2.2 Specimen Preparation

JRC has been used to characterize the surface roughness of rock joints in previous studies and has proved to be a representative characterization method (Jiang et al. 2006). In the present study, a 3D laser machine was used to obtain the joint surface and the JRCs of profiles with a spacing of 0.5 mm were calculated using the regression equation between JRC and Z2, as proposed by Tse et al. (1979):

$$JRC = 32.2 + 32.47 \log Z_2 \tag{3}$$

$$Z_2 = \left(\frac{1}{L_n} \sum_{i=1}^{N_p - 1} \frac{(z_{i+1} - z_i)^2}{(x_{i+1} - x_i)} \right)^{1/2}, \tag{4}$$

where values (x_i, z_i) and (x_{i+1}, z_{i+1}) represent the adjacent digitized coordinates of the profile separated by the sampling interval, N_p is the number of digitized points along the profile, and L_n is the nominal length of the digitized joint profile. The estimated JRC values of RJ1, RJ2, and RJ3 were 6 to 8, 10 to 12, and 14 to 16, respectively.

Three rock-joint moulds made of silica gel with various degrees of roughness were replicated from split-granite joints through the Brazil splitting method, as shown in Fig. 3.

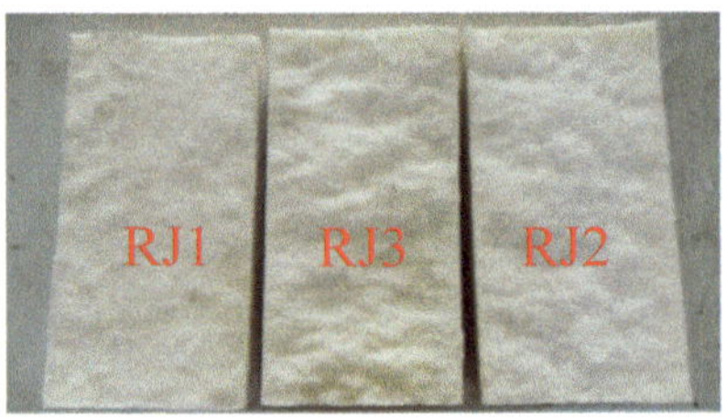

Fig. 3. Three rock-joint moulds made of silica gel

As plaster is universally available and easily moulded into any shape when mixed with water, it was adopted as the material for making artificial rock joints. The standard specimen was made by mixing plaster and water at a weight ratio of 1:0.28. The mixture was cured at room temperature (20 °C) for 7 days before testing. The mechanical properties of the specimens are shown in Table 1.

Table 1. Mechanical properties of specimens used in this study

Tensile strength/MPa	Compressive strength/MPa	Young's modulus/GPa	Basic friction angle/°
1.26 ± 0.10	24.53 ± 2.07	6.00 ± 0.54	38.03 ± 2.94

In this test, all specimens were 100 mm in width, 200 mm in length, and 100 mm in height. A shear displacement rate of 0.50 mm/min was applied for all the tests. The initial normal stresses were set to 1, 2, 3, and 4 MPa and the stiffness was set to 1.6, 3.2, and 4.8 GPa/m based on Eq. (5), as proposed by Skinas et al. (1990). The experimental cases are summarized in Table 2, in which the CNL boundary condition is identical to zero stiffness.

$$k_n = \frac{E_r \times L}{2c(1 - v^2)}, \tag{5}$$

where k_n is the boundary normal stiffness of rock mass, E_r is Young's modulus of rock mass, L is the length of a rectangular jointed block, v is Poisson's ratio, and c is a constant.

Table 2. Test cases under CNS boundary condition

Sample	Roughness	Stiffness (GPa/m)	Initial normal stress (MPa)	Total groups
RJ1	JRC6–8			
RJ2	JRC10–12	0, 1.6, 3.2, 4.8	1, 2, 3, 4	48
RJ3	JRC14–16			

3 Test Results and Discussion

3.1 Comparison of Shear Behaviors of Joints Under CNL and CNS Boundary Conditions

RJ3 specimen described in the previous section and the normal stiffness of the 3.2 GPa/m for the CNS condition is illustrated in Fig. 4. Figure 4(a) shows the relationship between shear stress and shear displacement under varying initial normal stress. Obviously, the shear strength and residual strength under the CNS condition are always higher than those under the CNL condition with the same initial normal stress, which means that the strength of rock joints is underestimated under the CNL condition compared to the CNS condition. Designing and constructing a practical engineering case may be more reasonable under the CNS boundary condition. As shown in Fig. 4 (a), a strain softening behavior is indicated under the CNL boundary condition, whereas both strain softening and hardening behaviors occurred under the CNS boundary condition according to normal stiffness.

The results of Fig. 4(b) indicate that the normal stress increases under the CNS condition and the growth of normal stress tends to be mild during the shear process. The relationship between normal displacement and shear displacement are shown in Fig. 4(c). The dilation of joints is confined under the CNS condition because of the increase of normal stress caused by normal stiffness. As shown in Figs. 4(b) and 4(c), the normal stress and dilation have the same trend, and the normal stress is reduced while the dilation decreases due to the breakage of joint asperity.

3.2 Shear Behavior of Joints Under CNS Boundary Condition

The shear behavior of rock joints is significantly influenced by joint roughness, normal stiffness, and initial normal stress. In this section, the effects of joint roughness and initial normal stress on joint shear characteristics are studied comprehensively.

For example, Fig. 5 shows the shear behavior of rock joints with different roughness under the CNS condition in the case of $\sigma_{n0} = 1\,MPa$ and $k_n = 1.6\,GPa/m$. As shown in Fig. 5(a), the peak shear strength of rock joints is significantly affected by the roughness of rock joints, and the shear strength is higher when the roughness increases.

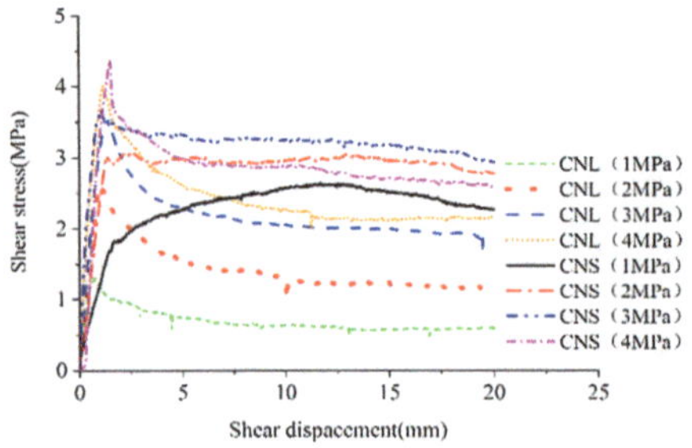

(a) Shear stress vs. shear displacement

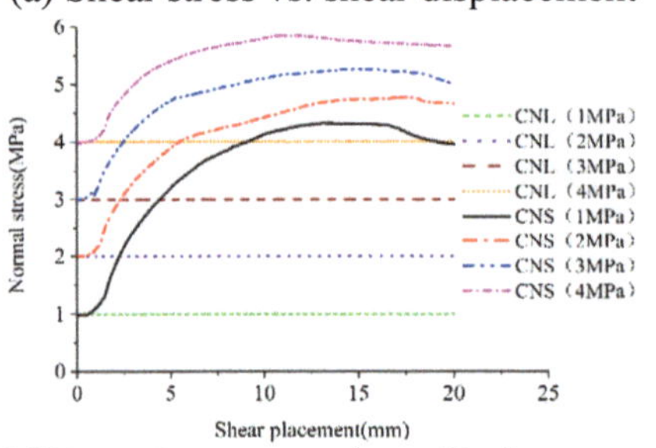

(b)Normal stress vs. shear displacement

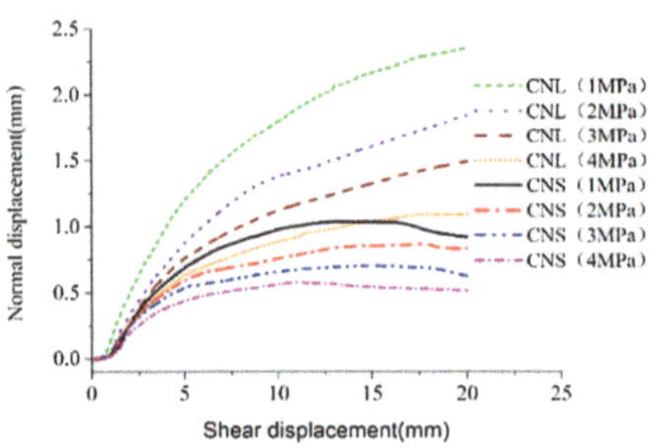

(c) Normal displacement vs. shear displacement

Fig. 4. Shear behavior of rock joints with different initial normal stress under CNL and CNS condition (RJ3, $k_n = 3.2\,GPa/m$)

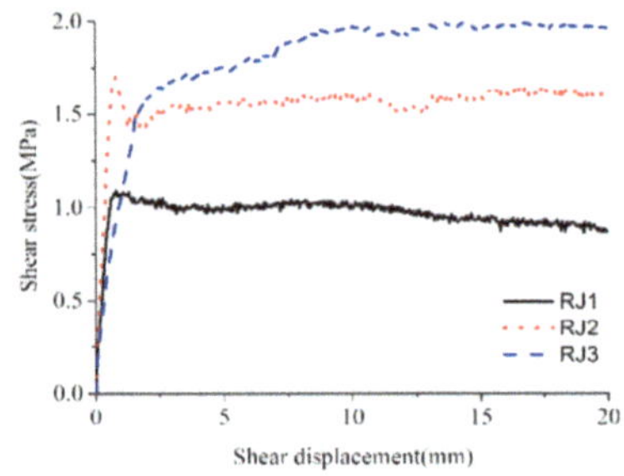

(a) Shear stress vs. shear displacement

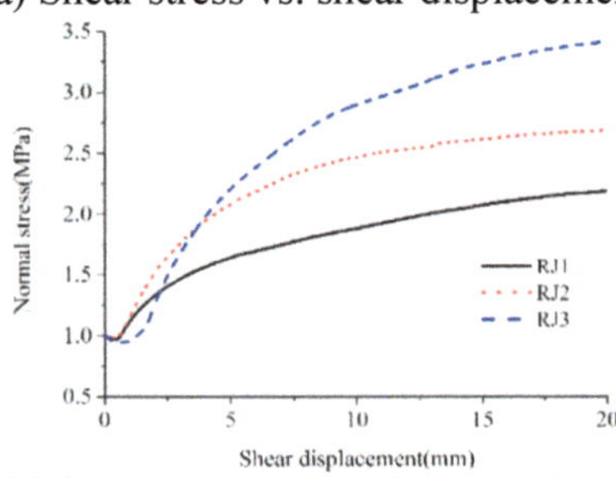

(b) Normal stress vs. shear displacement

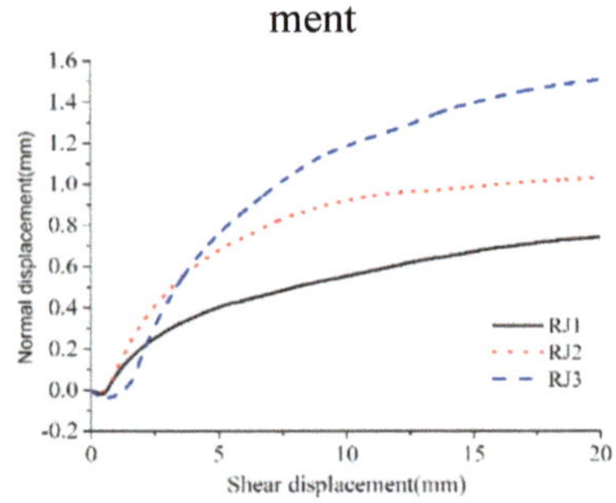

(c) Normal displacement vs. shear displacement

Fig. 5. Shear behavior of rock joints with different roughness under CNS condition ($\sigma_{n0} = 1\,MPa$, $k_n = 1.6\,GPa/m$)

The results of Fig. 5(c) indicates that the dilation of rock joints increases with the increase of roughness, this is attributed to the deformation of asperities. And a large shear displacement of peak stress is observed when joints are rougher. Meanwhile, the curve of normal stress in Fig. 5(b) seems to be the same as that in Fig. 5(c) because the normal stress on the surface of joints depends on the dilation of joints under the CNS condition and confines the dilation of joints at the same time.

Figure 6 shows the shear behavior of rock joints under different initial normal stress under the CNS condition in the case of the RJ1 specimen and $k_n = 1.6\,GPa/m$.

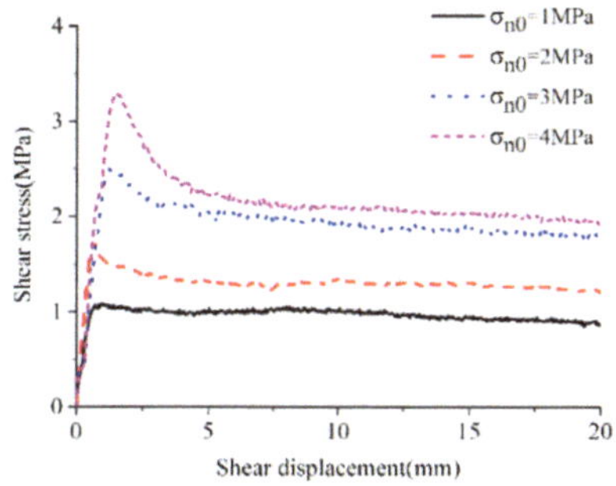

(a) Shear stress vs. shear displacement

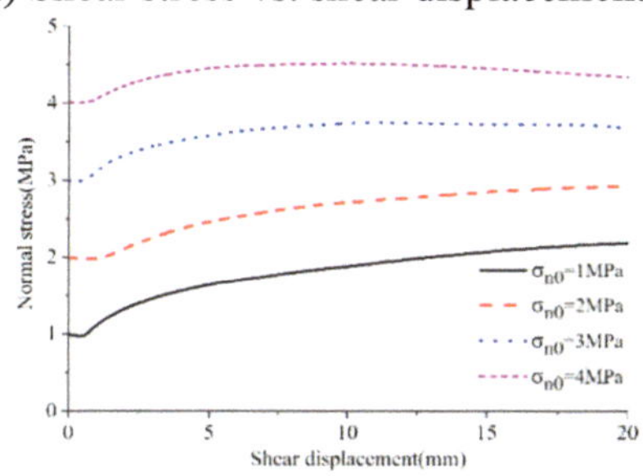

(b) Normal stress vs. shear displacement

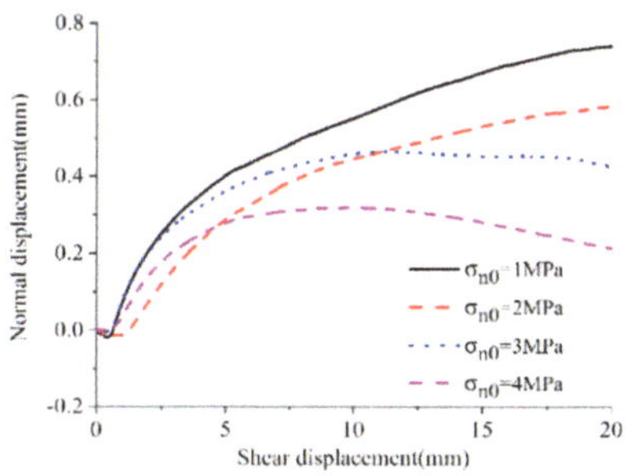

(c) Normal displacement vs shear displacement

Fig. 6. Shear behavior of rock joints with different initial normal stress under CNS condition (RJ1, $k_n = 1.6\,GPa/m$)

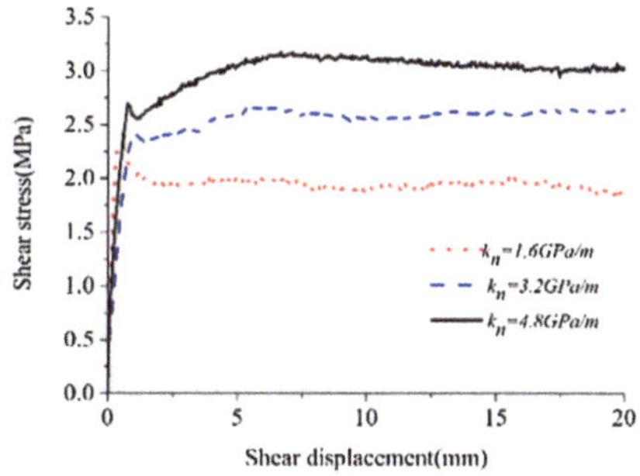

(a) Shear stress vs. shear displacement

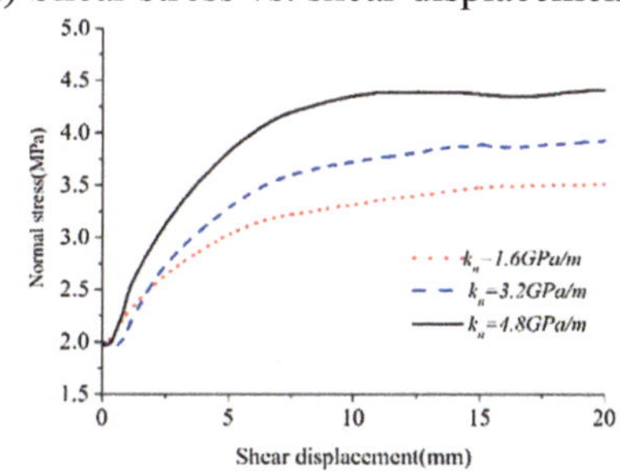

(b) Normal stress vs. shear displacement

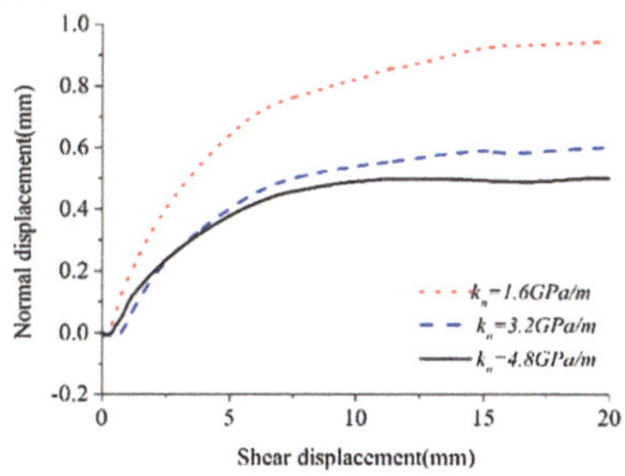

(c) Normal displacement vs. shear displacement

Fig. 7. Shear behavior of rock joints with different normal stiffness under CNS condition (RJ2, $\sigma_{n0} = 2\,MPa$)

Figure 6(a) shows the relationship between shear stress and shear displacement. Obviously, the shear stress increases with the increase of initial normal stress. However, the residual strength shows different characteristics under the CNS condition, the ratio of shear strength and residual strength decrease as the initial normal stress rises, which may be related to the ratio of initial normal stress to normal stiffness. Figures 6(b) and 6(c) show the relationship between normal stress and normal displacement in shear process. The dilation of rock joints exhibits different behaviors under lower normal stress and higher normal stress, and the dilation continues to grow under lower normal stress. However, under higher normal stress of 3 and 4 MPa, the dilation grows in the beginning and decreases in the last stage, which might be due to the destruction of asperity and compression of the specimen.

3.3 The Effect of Stiffness on Shear Behavior Under CNS Boundary Conditions

To investigate the effect of stiffness on shear behavior of rock joints under CNS condition. the test result for various stiffness under the CNS condition with initial normal stress of 2 MPa in RJ2 are plotted in Fig. 7. Figures 7(b) and 7(c) show the behavior of normal stress and dilation under various stiffness. It is clear that the normal stress increase with the growth of dilation. The dilation is confined under CNS boundary condition and this constraint effect is strengthened with the increase of normal stiffness. With balance between normal stress and dilation, the growth of dilation improves the normal stress while the increase of normal stress restricts the dilation. Therefore, both normal stress and dilation begin to grow faster and then level off.

It is clear in Figs. 4(a), 5(a) and 7(a) that the shear stress do not always show an obvious peak point throughout shear process under CNS boundary condition. The curve of shear stress shows different forms depending on the initial normal stress, normal stiffness and roughness. Thus three typical shear curves in Fig. 8 were proposed under CNS boundary condition according to the test results. The nearly linear stage can be observed in all the typical curves which can be defined as the first stage of shear process where shearing of the asperity becomes more predominant than the sliding and shear stress increases rapidly. In the end of this stage, the shear stress is up to the first

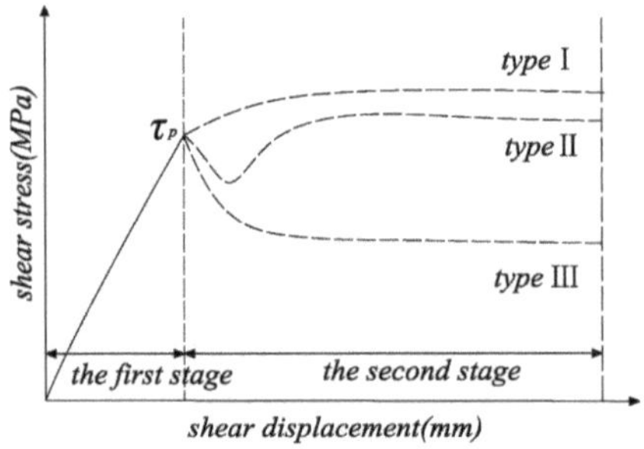

Fig. 8. Three typical shear curve under CNS boundary condition

peak shear stress when the joint asperity is broken. There can be find completely different performance for three typical curves in the second stage, the phenomenon of stress softening can be observed in curve of type III, while both of stress softening and stress hardening can be observed in curve of type II and only stress hardening in type I. The sliding of joint surface becomes predominant in this stage due to the nearly broken of asperity in the first stage, thus the change of normal stress may lead to these different curves under CNS boundary condition through influencing friction force. While the increment of normal stress is too larger to cause the increase of the shear stress, then the curves of type I and type II may appear in shear process as Figs. 7(a) and 7(b). Obviously, it is difficult to define the shear strength of rock joints for type I and type II, Shrivastava et al. (2017) defined the maximal shear stress in shear process as shear strength, while Jiang et al. (2004) defined the point of shear stress with sharp change in curvature as the shear strength. As mentioned above, the shearing and broken of asperity is mainly happened in the first stage and the increasing of shear stress in the second stage is mainly related to low growth of normal stress, thus the peak shear stress in the first stage are more appropriate to define as in Fig. 8.

As shown in Fig. 8, the phenomenon of stress hardening is observed in shear curves of type I and type II, and the factors contributed to this are complicated including normal stiffness, initial normal stress and roughness of rock joints. To deeply investigate the phenomenon of stress hardening in shear process and consider the normal stress closely related to this, normal stiffness and initial normal stress are linked together to further study the stress hardening in this paper. The ratio of τ_h to τ_p can be the basis to judge whether the shear stress-shear displacement curve shows phenomenon of stress hardening where τ_h is the maximal shear stress in shear process and τ_p is the peak shear strength of rock joints under CNS boundary condition and τ_h is equal to τ_p for shear curve of type III, While τ_h is higher than τ_p for shear curves of type I and type II. Table 3 showed the test results under CNS boundary condition, and it is clear from the Table 3 that 0.625 of σ_{n0}/k_n is a critical value for the test material in this study, there will certain be phenomenon of stress hardening when the value of σ_{n0}/k_n is lower than 0.625, and that whether the phenomenon of stress hardening exists depends on the roughness and other conditions when σ_{n0}/k_n is 0.625. The critical value of 0.625 is applicable for plaster mixture used in this study, and there may be another critical value for other material mixture.

Table 3. Test results under CNS boundary condition

Joint type	σ_{n0} (MPa)	k_n (MPa/mm)	τ_p (MPa)	τ_h (MPa)	Value of $\frac{\sigma_{n0}}{k_n}$	$\frac{\tau_h}{\tau_p}$	Phenomenon of stress
RJ1	1.00	1.60	1.09	1.09	0.625	1.00	Softening
	1.00	3.20	1.02	1.77	0.313	1.74	Hardening
	1.00	4.80	1.12	2.01	0.208	1.79	Hardening
	2.00	1.60	1.80	1.80	1.250	1.00	Softening
	2.00	3.20	1.86	1.86	0.625	1.00	Softening
	2.00	4.80	1.85	2.31	0.417	1.25	Hardening

(*continued*)

Table 3. (*continued*)

Joint type	σ_{n0} (MPa)	k_n (MPa/mm)	τ_p (MPa)	τ_h (MPa)	Value of $\frac{\sigma_{n0}}{k_n}$	$\frac{\tau_h}{\tau_p}$	Phenomenon of stress
	3.00	1.60	2.51	2.51	1.875	1.00	Softening
	3.00	3.20	2.48	2.48	0.938	1.00	Softening
	3.00	4.80	2.69	3.04	0.625	1.13	Hardening
	4.00	1.60	3.30	3.30	2.500	1.00	Softening
	4.00	3.20	3.41	3.41	1.250	1.00	Softening
	4.00	4.80	3.87	3.87	0.833	1.00	Softening
RJ2	1.00	1.60	1.70	1.70	0.625	1.00	Softening
	1.00	3.20	1.67	2.17	0.313	1.30	Hardening
	1.00	4.80	1.80	2.70	0.208	1.50	Hardening
	2.00	1.60	2.57	2.57	1.250	1.00	Softening
	2.00	3.20	2.60	2.67	0.625	1.03	Hardening
	2.00	4.80	2.72	3.20	0.417	1.18	Hardening
	3.00	1.60	2.86	2.86	1.875	1.00	Softening
	3.00	3.20	3.37	3.37	0.938	1.00	Softening
	3.00	4.80	3.58	3.58	0.625	1.00	Softening
	4.00	1.60	3.89	3.89	2.500	1.00	Softening
	4.00	3.20	3.98	3.98	1.250	1.00	Softening
	4.00	4.80	4.38	4.38	0.833	1.00	Softening
RJ3	1.00	1.60	1.54	2.01	0.625	1.31	Hardening
	1.00	3.20	1.84	2.64	0.313	1.44	Hardening
	1.00	4.80	2.23	3.24	0.208	1.45	Hardening
	2.00	1.60	2.80	2.80	1.250	1.00	Softening
	2.00	3.20	2.62	3.08	0.625	1.18	Hardening
	2.00	4.80	2.76	3.55	0.417	1.29	Hardening
	3.00	1.60	3.58	3.58	1.875	1.00	Softening
	3.00	3.20	3.67	3.67	0.938	1.00	Softening
	3.00	4.80	3.97	3.97	0.625	1.00	Softening
	4.00	1.60	4.31	4.31	2.500	1.00	Softening
	4.00	3.20	4.42	4.42	1.250	1.00	Softening
	4.00	4.80	4.63	4.63	0.833	1.00	Softening

It also can be observed from Table 3 that the influence of normal stiffness to shear strength is more obvious with the improvement of initial normal stress and the roughness of joint. There is nearly no difference in shear strength for RJ1 and RJ2 when the initial normal stress is 1 MPa and 2 MPa, but the shear strength improves with the improvement of normal stiffness when the initial normal stress is 3 MPa and 4 MPa for all the joints. Moreover, the effect of normal stiffness on shear strength is more obviously for RJ3 compared to RJ1. In general, τ_h is significantly improved when the value of σ_{n0}/k_n is smaller.

4 Conclusions

The shear behavior of rock joints under the CNS and CNL boundary conditions were analyzed comparatively. Then, the effects of joint roughness, normal stiffness, and initial normal stress on the joint shear characteristics were studied comprehensively. Based on the test results, the following conclusions are presented:

(1) The shear strength and residual strength obtained from the CNS boundary condition is always higher than that from the CNL boundary condition. Therefore, the strength of rock joints is underestimated under the CNL boundary condition and CNS boundary condition may be more reasonable for the design and construction of practical engineering.
(2) The shear behavior of rock joints is significantly influenced by joint roughness, normal stiffness, and initial normal stress under CNS boundary condition. With the increase of joint roughness and initial normal stress, the shear strength and residual strength also increase, while the influence of normal stiffness to shear strength is more obvious with the improvement of initial normal stress and the roughness of joint. Moreover, the dilation of rock joints in shear process are confined under this condition, and this constraint effect is strengthened with the increase of normal stiffness.
(3) The shear process can be separated into two stages and the peak shear stress in the first stage are more appropriate to define as peak shear strength τ_p. Meanwhile, three typical shear curves were proposed under CNS boundary condition according to the test results where the strain softening or strain hardening behaviors are shown depending on the ratio of initial normal stress to normal stiffness.
(4) The value of σ_{n0}/k_n is a major factor influence the shear behavior in the second stage, and 0.625 of σ_{n0}/k_n is a critical value for plaster mixture used in this study, there will certain be phenomenon of stress hardening when the value of σ_{n0}/k_n is lower than 0.625, and whether the phenomenon of stress hardening exists depends on the roughness and other conditions when the value of σ_{n0}/k_n is 0.625.

Acknowledgements. This research was supported by project of National Natural Science Foundation of China-development of coupled shear-flow tests technology and system for rock joints (41327001).

References

Shrivastava, A.K., Rao, K.S.: Shear behaviour of rock joints under CNS boundary conditions. In: Proceedings of the 18th International Conference on Soil Mechanics and Geotechnical Engineering, Paris, pp. 961–964 (2013)

Heuze, F.E.: Dilatant effects of rock joints. In: Proceedings of the 4th ISRM Congress, Montreux, p. 169e75 (1979)

Johnston, I.W., Lam, T.S.K.: Shear behavior of regular triangular concrete/rock joints — analysis. Geotech. Eng. Div. ASCE **115**(5), 711–727 (1989). Anno, T., Teruya, H.: J. Chem. Phys. **81**(1), 582–583 (1984)
Johnston, I.W., Lam, T.S.K., Williams, A.F.: Constant normal stiffness direct shear testing for socketed pile design in weak rock. Geotechnique **37**(1), 83–89 (1987)
Indraratna, B., Haque, A., Aziz, N.: Laboratory modelling of shear behaviour of soft joints under constant normal stiffness conditions. Geotech. Geol. Eng. **16**(1), 17–44 (1998)
Skinas, C.A, Bandis, S.C., Demiris, C.A.: Experimental investigations and modelling of rock joint behaviour under constant stiffness. In: Proceedings of the International Conference on Rock Joints, Loen, Norway, pp. 301–308 (1990)
Jiang, Y., Xiao, J., Tanabashi, Y., Mizokami, T.: Development of an automated servo controlled direct shear apparatus applying a constant normal stiffness condition. Int. J. Rock Mech. Min. Sci. **41**(2), 275–286 (2004)
Mirzaghorbanali, A., Nemcik, J., Aziz, N.: Effects of cyclic loading on the shear behavior of infilled rock joints under constant normal stiffness conditions. Rock Mech. Rock Eng. **47**(4), 1373–1391 (2014)
Indraratna, B., Haque, A.: Experimental study of shear behavior of rock joints under constant normal stiffness conditions. Int. J. Rock Mech. Min. Sci. **34**(3–4), 141.e1–141.e14 (1997)
Shrivastava, A.K., Rao, K.S.: Shear behaviour of rock joints under CNL and CNS boundary conditions. Geotech. Geol. Eng. **33**(5), 1205–1220 (2015)
Lee, Y.K., Park, J.W., Song, J.J.: Model for the shear behavior of rock joints under CNL and CNS conditions. Int. J. Rock Mech. Min. Sci. **70**(9), 252–263 (2014)
Ohnishi, Y., Dharmaratne, P.G.R.: Shear behavior of physical model of rock joints under constant normal stiffness condition. In: Barton, N., Stephannsson, O. (eds.) Proceedings of the International Conference on Rock Joints, Loen, Norway, pp. 267–273 (1990)
Jiang, Y., Li, B., Tanabashi, Y.: Estimating the relation between surface roughness and mechanical properties of rock joints. Int. J. Rock Mech. Min. Sci. **43**(6), 837–846 (2006)
Tse, R., Cruden, D.M.: Estimating joint roughness coefficients. Int. J. Rock Mech. Min. Sci. Geomech. Abstr. **16**, 303–307 (1979)
Shrivastava, A.K., Rao, K.S.: Physical modeling of shear behavior of infilled rock joints under CNL and CNS boundary conditions. Rock Mech. Rock Eng. **3**, 1–18 (2017)

Combined Tension-Shear Experimental Study of Rock Failure

Shouding Li[1,2,3(✉)], Linanan Liu[1,2,3], Xiao Li[1,2,3], Jingyun Guo[1], Jianming He[1,2,3], and Yanhui Liu[1,4]

[1] Key Laboratory of Shale Gas and Geo-engineering, Institute of Geology and Geophysics, Chinese Academy of Sciences, Beijing, China
lsdlyh@mail.iggcas.ac.cn
[2] Institutions of Earth Science, Chinese Academy of Sciences, Beijing, China
[3] University of Chinese Academy of Sciences, Beijing, China
[4] China Institute for Geo-Environment Monitoring, Beijing, China

Abstract. This paper deals with experimental study of rock failure under combined tension-shear condition. Combined tension-shear stress state was achieved using self-developed apparatus. Tension-shear tests were conducted on brittle granite sample to investigate the characteristics of tension-shear failure. The failure criterion for whole stress state was obtained, a modified Mohr-Coulomb criterion, with consideration of tension-shear stress state, was proposed by analyzing the experimental results. The obtained failure envelope presents continuous double-lined variation, with relatively steeper slope in tension-shear zone. To quantitatively characterize failure surfaces under tension-shear conditions, fractal dimension (FD) was introduced to represent the roughness of failure surfaces, the technique of scanning electron microscopy (SEM) and 3D laser scanning technique were utilized to observe failure surfaces morphology. Analytical results suggested the variation of FD values calculated from the two methods coincide with each other, and that the roughness of failure surfaces increase with tensile stress. Different from the situation in compression-shear stress condition, under combined tension-shear stress condition, when applying a constant tensile load, the strength decreases with the increment of shearing rates.

Keywords: Rock failure · Tension-shear · Failure criterion
Experimental study

1 Introduction

In traditional near-surface geological engineering, the yield criterion of geomaterials often presented on the basis of compression-shear failure type. However, in natural earth or engineering endeavors, induced tension stresses combined with shear stresses frequently occurs (e.g., slope failure due to unloading effects, bank slope failure in anchorage section, roof failure of deep tunnels), and tension-shear failure becomes the main failure type (Zhou et al. 2016).

Restricted to the difficulty in directly applying tensile load on specimens, research efforts have mainly been focused on the design of unique-shaped rock specimens and

L. Zhang et al. (Eds.): GSIC 2018, *Proceedings of GeoShanghai 2018 International Conference: Rock Mechanics and Rock Engineering*, pp. 275–283, 2018.
https://doi.org/10.1007/978-981-13-0113-1_30

innovative mechanical rock testing devices for applying tensile and shear loads simultaneously. For instance, Ramsey and Chester (2004) used notch-cut dog-bone-shaped samples of Carrara marbles to investigate hybrid fracture subjected to tension-shear loadings. Dunand and Mohr (2011) employed a butterfly-shaped specimen to test sheet stock failure under combined shear and axial loads. Aimone-Martin et al. (1997) designed a mechanical rock testing device and in use to study the response of rock to combined tension-shear stress state. Haeri et al. (2016) developed a device which can convert compression loading to tensile loading for the determination of tensile strengths with a hole in the middle of specimens. As devices for various stress states of rock specimens steadily improves, different failure criterions under the combined tension-shear condition were proposed (e.g., Tekoğlu et al. 2012; Haltom et al. 2013). At microscales, Barsoum and Faleskog (2007) developed a micromechanical model based on the attainment of a critical value of shear deformation. Yuan and Xiao (2015) investigated the microscopic damage mechanism and proposed a multiparameter elastoplastic damage constitutive model considering combined tension-shear condition. Sun et al. (2016) introduced SEM technique to evaluate the cross-sectional profiles and established a biaxial tension-shear failure envelope of fiber materials. However, literatures reviews indicate that despite the many experimental studies that concerns materials' behavior under combined tension-shear conditions, most of them focus on ductile materials (e.g., Guillet 2011; Graham et al. 2012), with less attention focusing on the brittle materials (Ayatollahi and Aliha 2007; Torabi 2013), let alone those of geomaterials.

In the work presented herein, systematic studies were designed to study failure mechanism of rock under combined tension-shear conditions. The experiments were organized as follows: Firstly, an innovative device was developed, which has the capability to apply either combined tension-shear loads or compression-shear loads on rock specimens. Then, granite samples were tested using the device. By analyzing the experimental data, a modified Mohr-Coulomb failure criterion was proposed to depict failure law for the whole stress state (combined tension-shear stress state and compression-shear state). Finally, the failure surfaces of granite samples under combined tension-shear condition were quantitatively characterized by SEM and 3D laser scanning respectively. The results of rock mechanical experiments under combined tension-shear conditions can be summarized as follows: (a) as a reference, rock mechanical tests under combined tension-shear condition was performed; (b) the whole failure envelope of granite was obtained and characterized; (c) the influence of tensile strengths and shearing rates on the roughness of failure surface was discussed.

2 Experimental Methodology

The combined tension-shear mechanical experiments reported were conducted employing self-developed DSC-800 electro-hydraulic servo testing system (Fig. 1). The critical part is a high stiffness main machine of tension-compression and shear system, it allows for axial loading of specimens along with simultaneous horizontal shearing. Axial stress is applied using a bidirectional actuator, while shearing force is applied via the horizontal power system, the rollers mounted between the chamber and actuator guarantee the chamber to move horizontally. A series of experimental tests were carried out under the combined tension-shear conditions for granite samples with an external size of 100 mm × 100 mm × 100 mm. Specimen was placed in the cubic chamber and then was glued with the chamber by polyurethane adhesive so that the tensile force can be applied when the bidirectional actuator moves upward. Figure 2 shows the specimen installed in the chamber. It should be noted the tensile properties of polyurethane adhesive materials and granites were pre-investigated. Brazilian test results suggested that the tensile strength of granite samples (50 mm in diameter with 25 mm in height) is averagely 8.95 MPa, which is lower than the bonding strength of polyurethane adhesive (approximately 20 MPa). Besides, during entire tests no signs of detachment between the end faces of samples and chamber surface were observed, proving that the polyurethane adhesive can provide sufficient tensile strength.

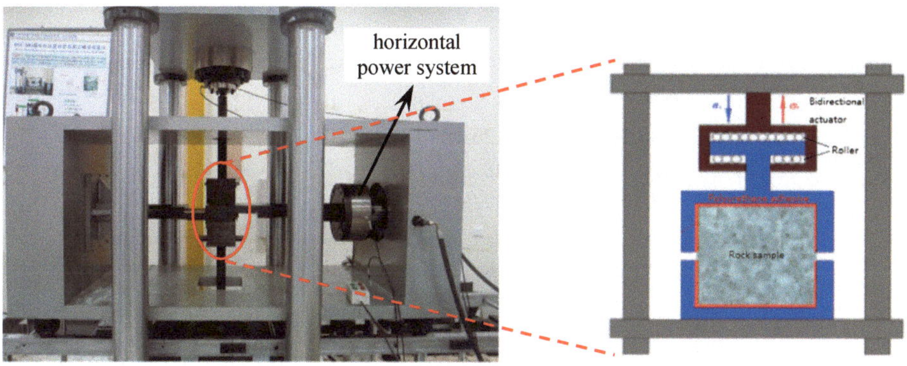

Fig. 1. Photo of experimental devices installed with schematic diagram showing the application of tensile strength simultaneously with shearing force.

Fig. 2. Illustration of granite samples, chamber and the polyurethane adhesive materials.

The procedure of the combined tension-shear experiment was set to two steps. Firstly, the tensile load was applied to the target level at a constant rate, then horizontal load was applied at a constant shearing rate until specimen failure. During the experimental procedure, all the data including axial stress, shearing stress, and axial displacement were recorded once every 0.1 s. The sample was enclosed in the chamber, and the axial loads were applied in sequence. Once the desired axial load was achieved, the axial stress was kept at constant value and the horizontal shearing force was set to a constant increasing rate. Because the increment of shearing force was held at a constant value, a sharp variation of shearing force can be a sign of sample failure.

3 Experimental Results

3.1 Failure Criterion

In the experiments, a series of 15 sequences of tensile loads were applied. The corresponding failure shearing forces S, was recorded. The results for mechanical strength parameters are shown in Table 1. Note that the values are the average values of three parallel tests.

Table 1. The applied tension sequence and corresponding failure shearing forces (unit: kN)

T	0	2	4	6	8	10	12	15
S	143.30	140.00	128.56	123.18	109.23	105.60	103.08	92.46
T	20	22	25	28	30	32	35	
S	72.86	67.49	60.70	47.78	43.47	34.18	24.16	

[a]T denotes tensile load, S denotes shear force when the specimen failure under the corresponding tensile load.

To obtain the whole failure envelope of granite, experiments under compression-shear condition were also conducted. The obtained data were then projected into the Cartesian coordinate, as shown in Fig. 3. It could be observed that under the combined tension-shear state, the tensile stress and shear stress can be correlated as linear variation (Fig. 3a). In the compression-shear regime, the failure envelope presents linear variation as well, yet its slope is smaller than that in the tension-shear regime (see in Fig. 3b). It is therefore, impracticable to extrapolate the empirical Mohr-Coulomb failure criteria to the combined tension-shear regime. The results also demonstrate that when subjected to tension-shear condition, the rock samples are more likely to breakdown when compared to the situation of compression-shear condition.

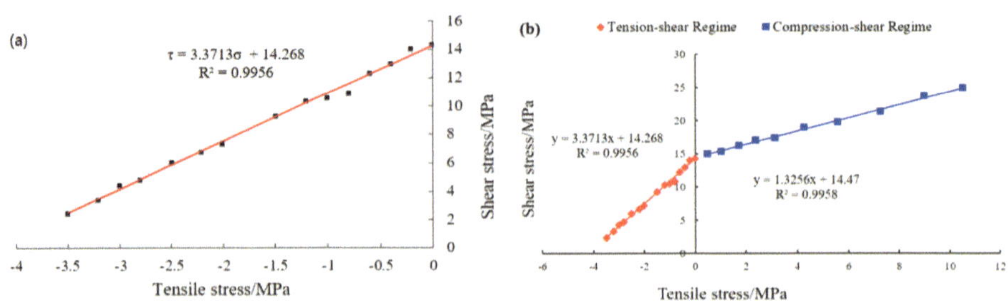

Fig. 3. Failure envelope of granite samples.

3.2 Fracture Surface Characterization

It should be noted that in the experiment, the potential failure surfaces were predetermined, so the ultimate failure surfaces were almost along the predetermined surface. However, under different tensile stress levels, the fracture patterns varied differently. To characterize these different fracture features under the combined tension-shear condition, 3D laser scanning technique and SEM technique were utilized. 3D laser scanning technique depicts the macroscopic topography of the failure surface (Fig. 4), while SEM characterizes failure surface morphology from microscopic aspect (Fig. 5).

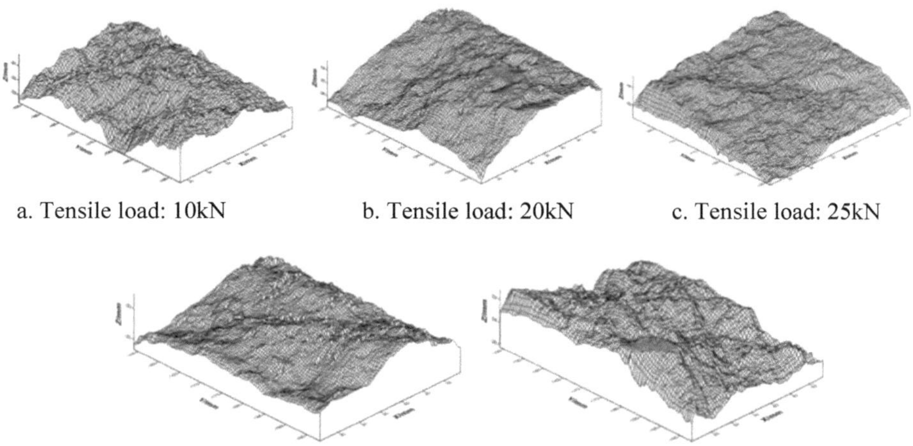

a. Tensile load: 10kN b. Tensile load: 20kN c. Tensile load: 25kN

d. Tensile load: 30kN e. Tensile load: 35kN

Fig. 4. Topography of fractures under combined tension-shear state.

Qualitative observations from Figs. 3 and 4 show that at higher tensile loads, the failure surfaces seem to be coarser. To quantitatively evaluate the fracture surface characteristics, fractal dimension (FD) was introduced to depict surface roughness. In the images obtained by 3D laser scanning, the cubic covering method was used to calculate the FD. In SEM scanning images, Box-counting method was applied to calculate the FD. Table 2 lists calculation results of each images. The FD represents the roughness of failure surface, it is universally acknowledged that the larger the FD values, the coarser the failure surface. FD values calculated from the two methods suggested that with the increase of tensile strength, the FD values increases corresponding, indicating that under tension-shear condition, higher tensile stress state results in coarser failure surface, this is mainly because that with the increase of external tensile force, the inner frictional force on the potential failure surface decreases, thus aggravating the concave-convex degree of the failure surface.

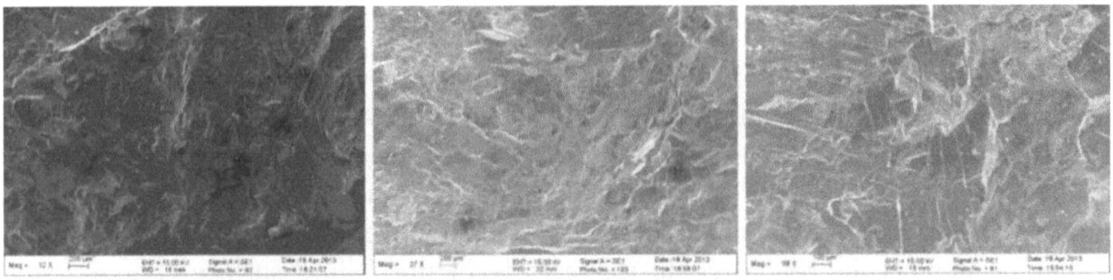

a. Tensile load: 10kN b. Tensile load: 20kN c. Tensile load: 25kN

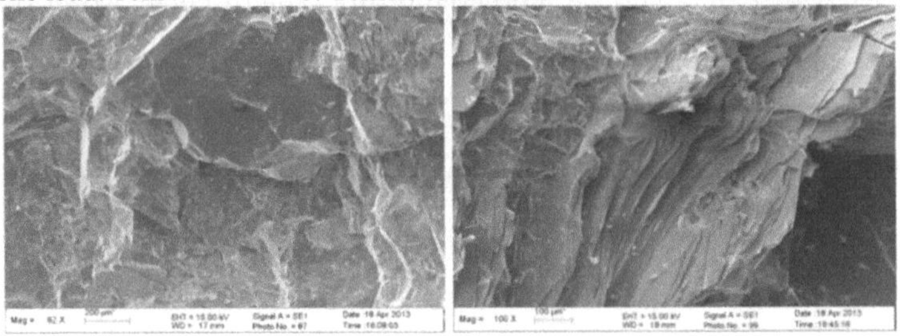

d. Tensile load: 30kN e. Tensile load: 35kN

Fig. 5. SEM scanning images of the fractures.

Table 2. Fractal dimension of failure surfaces under different tensile loads.

Tensile load/kN	Image type	Linear fitting results	FD	Correlation coefficient
10	3D laser scanning SEM	$y = 2.1248x + 6.9588$	2.1248	0.9999
		$y = 1.9145x + 13.3139$	1.9145	0.9971
20	3D laser scanning SEM	$y = 2.1273x + 7.3659$	2.1273	0.9998
		$y = 1.9433x + 13.2925$	1.9433	0.9976
25	3D laser scanning SEM	$y = 2.1544x + 6.5921$	2.1544	0.9999
		$y = 1.9667x + 13.2781$	1.9667	0.9982
30	3D laser scanning SEM	$y = 2.1623x + 7.1852$	2.1623	0.9997
		$y = 1.9699x + 13.1787$	1.9699	0.9982
35	3D laser scanning SEM	$y = 2.2083x + 7.2209$	2.2083	0.9998
		$y = 1.9758x + 13.0658$	1.9758	0.9973

In addition, the variation of FD values with different tensile loads was plotted in Fig. 6. It can be observed that higher tensile force corresponds to higher FD values, and the FD values derived from two techniques present similar variation, i.e., with the increase of tensile strength, the FD values increase respectively.

3.3 Effect of Loading Rate on Rock Strength

The shear strength determined from the experiment is closely related to the shearing rate. To explore the influence of shearing rate on shear strength under combined

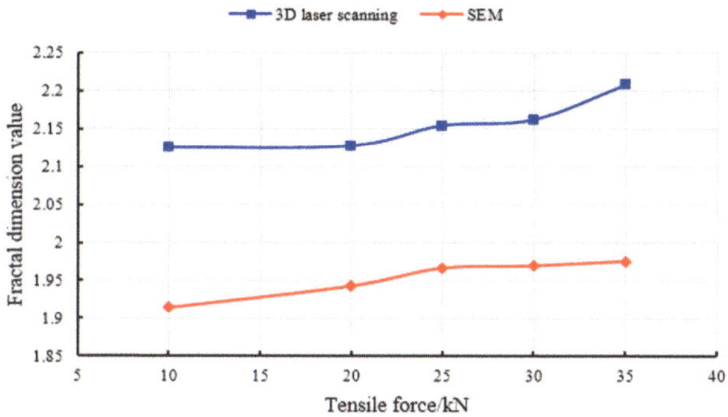

Fig. 6. Variation of FD values derived from 3D laser scanning and SEM.

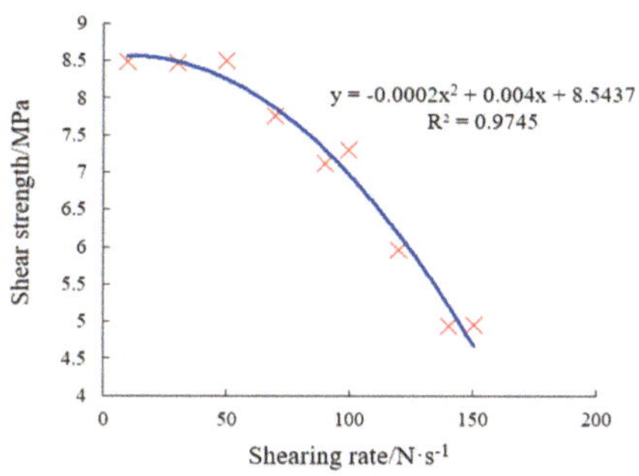

Fig. 7. Correlation between shear rate and failure strength.

tension-shear condition, a constant tensile force (T = 20kN) was applied firstly, then the horizontal shear force was applied at different rates until samples breakdown. The shear strengths under different shearing rates was shown in Fig. 7. According to the testing results, the shear strength of the specimen nearly kept constant when shearing rate is less than 50 $N \cdot s^{-1}$. When shearing rate exceeds 50 $N \cdot s^{-1}$, the shear strength drop dramatically with the increase of shearing rate. Under combined tension-shearing condition, the shear strength presents nonlinear inverse correlation with shearing rate, which is quite different from those under the compression-shear condition.

4 Conclusion

As a special stress state, the mechanical behavior of rocks under combined tension-shear state is of great concern yet poorly understood. In this article, a series of mechanical experiments with granite specimens were conducted. The effects of shearing rate/tensile stress on shear strength were presented. The following conclusions can be drawn:

(1) The failure criteria of granite in the whole stress regime can be represented by a turning line: in the combined tension-shear stress regime, the shear strength varies negatively with tensile stress. The slope of failure envelope in the combined tension-shear stress regime is steeper when compared with that in the compression-shear stress regime, it indicated that the rocks are more likely to fail under the combined tension-shear condition.
(2) At higher tensile load level, the failure surfaces under combined tension-shear condition becomes coarser, it was validated by the variation of FD values. In addition, the calculated FD values from 3D laser scanning and SEM show good coincidence with each other.

Acknowledgement. This study was finally supported by the Strategic Priority Research Program of Chinese Academy of Sciences (Grant No. XDA14040401 and No. XBD10030000), National Science and Technology Major Project (Grant No. 2016ZX05034003-005), Major Projects of National Natural Science Foundation (Grant No. 41227901) and National Natural Science Foundation (41272352). The authors appreciate the kind revision suggestions from the anonymous reviewers.

References

Zhou, H., Lu, J.J., Xu, R.C., Zhang, C.Q., Chen, J., Meng, F.Z.: Research on tension-shear failure characteristics and yield criterion of hard brittle marble. Rock Soil Mechanics **37**, 305–314, 349 (2016). https://doi.org/10.16285/j.rsm.2016.02.001

Ramsey, J.M., Chester, F.M.: Hybrid fracture and the transition from extension fracture to shear fracture. Nature **428**(6978), 63 (2004)

Dunand, M., Mohr, D.: Optimized butterfly specimen for the fracture testing of sheet materials under combined normal and shear loading. Eng. Fract. Mech. **78**(17), 2919–2934 (2011)

Aimone-Martin, C.T., Oravecz, K.I., Nytra, T.K.: A mechanical device for the measurement of combined shear and tension in rocks. Int. J. Rock Mech. Min. Sci. **34**(1), 147–151 (1997)

Haeri, H., Sarfarazi, V., Hedayat, A., Tabaroei, A.: Effect of tensile fracture toughness using experimental test and PFC2D simulation. J. Min. Sci. **52**(4), 647–661 (2016)

Tekoḡlu, C., Leblond, J.B., Pardoen, T.: A criterion for the onset of void coalescence under combined tension and shear. J. Mech. Phys. Solids **60**(70), 1363–1381 (2012)

Haltom, S.S., Kyriakides, S., Ravi-Chandar, K.: Ductile failure under combined shear and tension. Int. J. Solids Struct. **50**(10), 1507–1522 (2013)

Barsoum, I., Faleskog, J.: Rupture mechanisms in combined tension and shear-Experiments. Int. J. Solids Struct. **44**(6), 1768–1786 (2007)

Yuan, Y.H., Xiao, M.: A multiparameter damage constitutive model for rock based on separation of tension and shear. Math. Problems Eng. **2015**, 10 p. (2015). Article ID 821093. https://doi.org/10.1155/2015/821093

Sun, J., Hudspeth, M., Chen, W.: Biaxial shear/tension failure behavior of spectra single fibers. Compos. A Appl. Sci. Manuf. **88**, 286–294 (2016)

Guillet, T.: Behavior of metal anchors under combined tension and shear cycling loads. ACI Struct. J. **108**(3), 315 (2011)

Graham, S.M., Zhang, T., Gao, X., Hayden, M.: Development of a combined tension–torsion experiment for calibration of ductile fracture models under conditions of low triaxiality. Int. J. Mech. Sci. **54**(1), 172–181 (2012)

Ayatollahi, M.R., Aliha, M.R.M.: Fracture toughness study for a brittle rock subjected to mixed mode I/II loading. Int. J. Rock Mech. Min. Sci. **44**(4), 617–624 (2007)

Torabi, A.: Failure curves for predicting brittle fracture in V-notched structural components loaded under mixed tension/shear: an advanced engineering design package. Eng. Solid Mech. **1**(3), 99–118 (2013)

Study on Crack Repair of Tunnel Segment by Electrochemical Deposition Method

Hehua Zhu[1], Xin Wang[1], Zhengwu Jiang[2], and Qing Chen[2(✉)]

[1] Department of Geotechnical Engineering, Tongji University, Shanghai, China
[2] Department of Civil Engineering Materials, Tongji University, Shanghai, China
chenqing19831014@163.com

Abstract. In the complicated environment, such as groundwater and rock-soil medium, there are some limitations in traditional methods for repairing underground structure. Electrochemical deposition method is a novel method of repairing cracks of concretes in water environment and it is of great significance to explore its application in repairing damaged underground structure. According to the asymmetric characteristics of underground structure, indoor simulation devices adopting electrochemical deposition method to repair damaged underground structure are proposed and an experimental study on repairing different parts of damaged underground structure is carried out. Under several conditions of different solution concentrations, the healing effectiveness in different sizes and positions of underground structure is evaluated using some evaluation indexes, such as surface coating rate and crack closure rate. Repair mechanism at the micro scale level is revealed by adopting Xrd method and SEM. The results indicate that solution concentrations and crack depth distribution affect the healing effectiveness. In conclusion, this experimental study provides a more reasonable selection of test parameters for subsequent application. It is proved that electrochemical deposition method is feasible to repair damaged underground structure (tunnel structure).

Keywords: Damaged underground structure · Asymmetric electrochemical deposition method · Indoor simulation devices · Crack repair

1 Introduction

Subjected to complex internal and external environment, the degradation of concrete lining in urban metro tunnels seems inevitable [1]. Before the concrete lining comes into being, initial damage will generate because of thermal stress produced by hydration heat, autogenous shrinkage, unsuitable methods of maintenance and installation and so on. During the service period, the surrounding construction load, additional load caused by uneven settlement and overground constructions will lead to microcracks connecting with each other resulting in aggravated damage. Eventually, the concrete segment will inevitably crack causing a series of problems. In addition, underground structures are mostly non-removable or irreversible structures. Once damaged, it will be extremely difficult to replace them. Therefore, it is of great significance to explore an effective method in repairing damaged underground structure.

L. Zhang et al. (Eds.): GSIC 2018, *Proceedings of GeoShanghai 2018 International Conference: Rock Mechanics and Rock Engineering*, pp. 284–296, 2018.
https://doi.org/10.1007/978-981-13-0113-1_31

However, there are some limitations in traditional methods of repairing underground structure. Electrochemical deposition method (EDM) is a novel method of repairing cracks of concretes in water environment taking full advantage of concrete characteristics and water environment conditions. By applying a certain weak current, the harmful material (i.e., Cl^-) is removed, and at the same time the deposits are produced both on the surface of the concrete and in the cracks to repair them. EDM was first proposed in 1980s to repair offshore concrete structures by Japanese researchers [2]; Hilbertz conducted studies on repairing concrete structure using sea water by EDM [3, 4]. The repair objects can be various, i.e., reinforced concrete with dry shrinkage cracks [5], onshore concrete structures [6–8], and concrete beams [9]. In addition, related studies on EDM are conducted by other researchers [10–25].

It should be noted that the previous research on EDM is mainly focused on the level of materials. The objects are all concrete specimens of onshore or offshore structure and lacks consideration of the characteristics of underground structure. Therefore, in order to prove the practicability of EDM in the repair of damaged underground structure, an experimental study is conducted with indoor simulation devices for repairing different parts of damaged underground structure. Under several conditions of different solution concentrations, the healing effectiveness in different sizes and positions of underground structure is evaluated using several evaluation indexes, such as surface coating rate, crack closure rate and filling depth of crack. To reveal the repair mechanism at the micro scale, analysis of mineral compositions of deposits is conducted by Xrd method and SEM is adopted to analyze the sedimentary morphology.

2 Electrochemical Deposition Test Program

2.1 Raw Material and Mix Proportions

The cement used is Grade-42.5 ordinary Portland cement produced by Huaxin Cement Group. Fine aggregate is river sand and its fineness modulus is 2.6. Coarse aggregates with size of 5–15 mm are employed. Mix proportions are listed in Table 1.

Table 1. Mix proportions of concrete

Dosage per cubic meter (kg)			
Water	Cement	Sand	Gravel
175	350	755	1135

2.2 Simulation of Crack Repair of Tunnel Segment at the Bottom of Tunnel (Experiment One)

Geometric Sizes of Tunnel Segment Specimens

The length, width and height of tunnel segment are 300 mm, 300 mm and 100 mm, separately. There is a steel framework inside segment with the diameter of steel bar of 8 mm and steel bars are laid out in two dimensions. The steel framework and its location inside the segment are shown in Fig. 1. The size of transparent storage tank is a little smaller than that of segment.

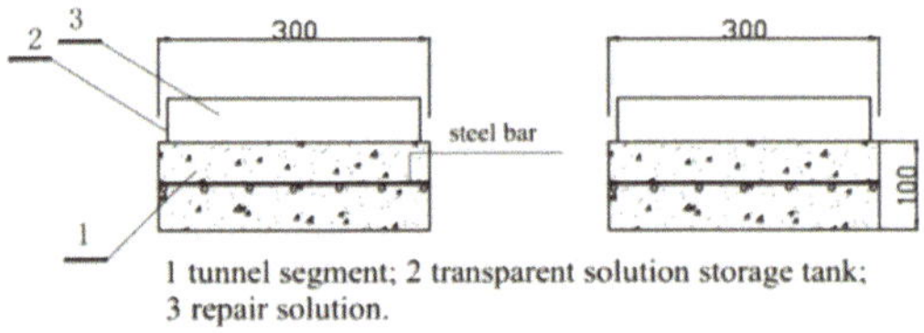

Fig. 1. Geometric sizes of tunnel segment specimens of experiment one

Crack Prefabrication

In order to simulate concrete cracking in practical environment, 12 cracks are prefabricated on the upper surface of each segment. The specific distribution and number of cracks in the specimen are indicated in Fig. 2.

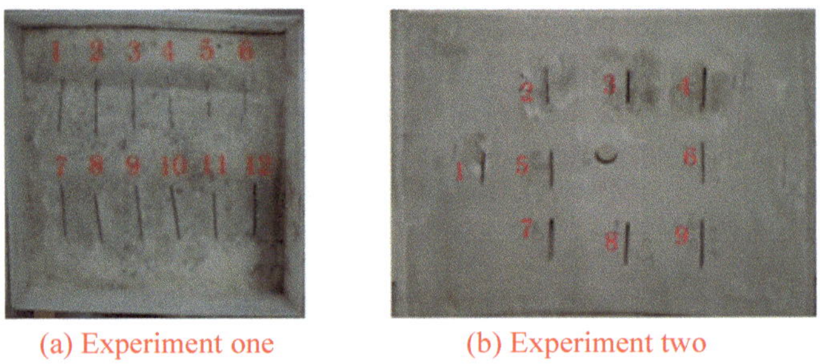

Fig. 2. Distribution and number of cracks in specimens of two experiments

The length and width of each crack are 50 mm and 3 mm, separately. Among 12 cracks, the prefabrication depth of crack 1, 2, 3 and 4 is 10 mm; 5 mm for crack 5, 6, 7 and 8; and 15 mm for crack 9, 10, 11 and 12. The actual size of each crack is slightly off balance due to uncertain factors of construction.

Repair Environment Settings

Experiment one is made up of two test groups, in which repair solution is $ZnSO_4$ solution, repair current is 0.2 A and the solution concentration is 0.1 mol/L and 0.2 mol/L, respectively.

2.3 Simulation of Crack Repair of Tunnel Segment at the Top and Side of Tunnel (Experiment Two)

Geometric Sizes of Tunnel Segment Specimens

The specimen is a reduced concrete tunnel segment model in proportion of 1:5. The length, width and height of tunnel segment are 600 mm, 540 mm and 800 mm, separately. The outer and inner radius is 650 mm and 550 mm respectively. The steel framework and its location inside the segment can be seen in Fig. 3. The size of transparent storage tank which has the function of containing repair solution is designed according to the size of tunnel segment.

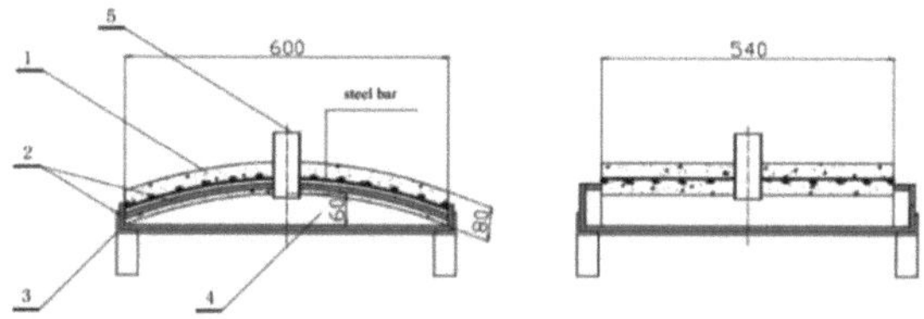

1 tunnel segment; 2 transparent solution storage tank; 3 support structure; 4 repair solution; 5 fluid filled tube.

Fig. 3. Geometric sizes of tunnel segment specimens of experiment two

Crack Prefabrication

In order to simulate concrete cracking in practical environment, 9 cracks are prefabricated on the lower surface of the test segment. The specific distribution and number of cracks in the specimen are indicated in Fig. 2. The length of each crack is 60 mm and the actual width and depth are listed in Table 2. It is remarkable that both crack 2 and 3 are through cracks.

Table 2. The actual width and depth of the precast concrete cracks of experiment two (mm)

Number	1	2	3	4	5	6	7	8	9
Depth	10.61	84.76	83.87	13.07	10.57	9.46	9.09	14.86	15.87
Width	3.35	2.57	3.16	5.18	3.85	3.97	8.33	5.10	4.31

Repair Environment Settings

In experiment two, repair solution is still $ZnSO_4$, the concentration is 0.4 mol/L and repair current is 0.2 A.

2.4 Simulation Devices and Its Principle Introduction

The difference between these two experiments is that cracks are prefabricated on different surfaces of test segment which means gravity function is distinct. However, they have common features in the aspect of experiment principle. In these two experiments, titanium plates are employed as assisting electrode. The cathode is composed of the steel framework inside tunnel segment while the anode is formed by titanium plate immersed in the repair solution. Finally, the steel bar, repair solution, titanium plate, the wire and the power supply form a complete loop. The device diagrammatic sketch of these two experiments is indicated in Fig. 4. Compared to experiment one, the joints between solution storage tank and test specimen are sealed with sealant in experiment two in case of leakage of repair solution. After the seal is done, repair solution is poured through fluid filled tube and thus a certain confining pressure to the inner surface of the segment is formed by repair solution, using the law of connected vessels in physics.

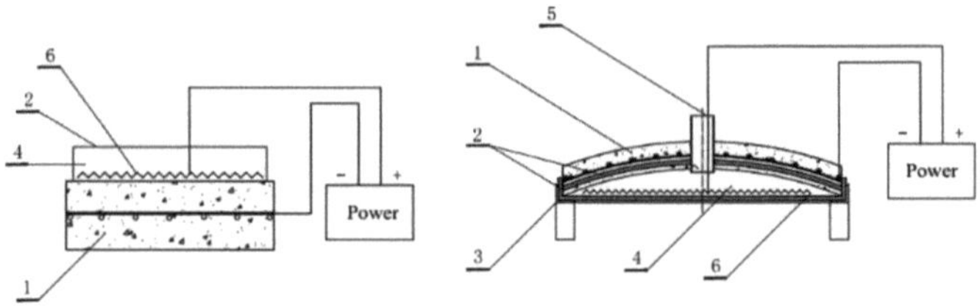

1 tunnel segment; 2 transparent solution storage tank; 3 support structure; 4 repair solution; 5 fluid filled tube; 6 titanium plate.

Fig. 4. Device diagrammatic sketch of the experiment

Since there are a large amount of ions inside repair solution, conduction of current can be accomplished by directional movement of ions when the current passes by. The anions move toward the anode (titanium plate) while the cations move toward the cathode (steel bar). If the anions and cations are close to the opposite electrode, the exchange of electrons occurs at the interface between the electrode and the solution. The cations that migrate to the cathode are combined with OH^- and thus form insoluble sediments.

2.5 Evaluation Methods and Evaluation Indexes

Choosing reliable evaluation methods and evaluation indexes is an essential key to evaluate the reliability and validity of EDM for crack repair in both laboratory and engineering [13]. In the aspect of macroscopic performance evaluation, the healing effectiveness is evaluated using several evaluation indexes, such as surface coating rate, crack closure rate and filling depth of crack. In the aspect of micro-analysis, Xrd method is used to conduct an analysis of mineral compositions of deposits and SEM is adopted to analyze the sedimentary morphology.

Surface Coating Rate
After the test specimen has been repaired, 3 digital photographs of each specimen will be taken. The corresponding coverage area is obtained by processing the photos and surface coating rate R_s is then calculated using Eq. (1). The arithmetic average of the 3 measured values is used as final surface coating rate.

$$R_s = (A_c/A) \times 100\% \tag{1}$$

where A_c is the coverage area of deposits (unit: mm^2); A is the total surface area of specimen needed to be repaired (unit: mm^2).

Crack Closure Rate

MapInfo system is used to measure the curve length of cracks before repair and those which are unable to be repaired after repair. Crack closure rate R_c is then achieved by Eq. (2).

$$R_c = (L_c/L) \times 100\% \tag{2}$$

where L_c is the curve length of crack repaired (unit: mm); L is the total curve length of crack before repair (unit: mm).

Filling Depth of Crack

After repair, the specimen is cut along the crack direction and 5 measuring points are chosen for equal interval of a crack. Filling depth of crack is measured at each measuring point using vernier caliper and the arithmetic average is calculated as final filling depth of crack.

Xrd Method

Xrd method is a technique which can be used to analyze mineral compositions of substance by diffraction effect. When substance specimen is exposed to X-ray owning sufficient energy, secondary fluorescence X-ray will be produced because substance has already been stimulated and then further diffraction data, which is an inevitable reflection of its crystal structure, will be obtained. In this experiment, X-ray diffractometer of D/max2550VB3+/PC type having graphite monochromator and Cu-$K_{\alpha 1}$ ray is employed to conduct the analysis. Scan mode is continuous scan, scan rate is 5°/min and scan angle is 5°–75°.

SEM Analysis Method

SEM is a type of electron microscope that produces images of a sample by scanning the surface using a focused beam of electrons. The electrons interact with atoms in the sample, producing various signals that contain information about the sample's surface topography and composition [26]. In this experiment, SEM of S-2360N type is adopted to analyze the sedimentary morphology of substance.

3 Test Results and Analysis

3.1 Simulation of Crack Repair of Tunnel Segment at the Bottom of Tunnel (Experiment One)

Test Results

When electrified, in 28 days, as shown in Fig. 5, cracks of both test groups are completely repaired basically. However, in 0.1 mol/L condition, some of cracks are not fully repaired; while in 0.2 mol/L, cracks are almost filled with deposits. It should be noted that there are obvious deposits on the surface of concrete and deposits tend to be gathered in both conditions.

(a) 0.1mol/L, $ZnSO_4$ (b) 0.2mol/L, $ZnSO_4$

Fig. 5. Healing effectiveness of tunnel segment of experiment one in 28 days

Evaluation Indexes

Indexes of surface coating rate and filling depth of crack are used to evaluate the healing effectiveness. The variation of the filling depth of crack with solution concentrations in different precast depth conditions is shown in Fig. 6 and comparison of surface coating rate of specimens is indicated in Table 3.

As shown in Fig. 6, under different precast depth conditions, filling depth of crack is greater with the increase of concentration of the solution; as indicated in Table 3,

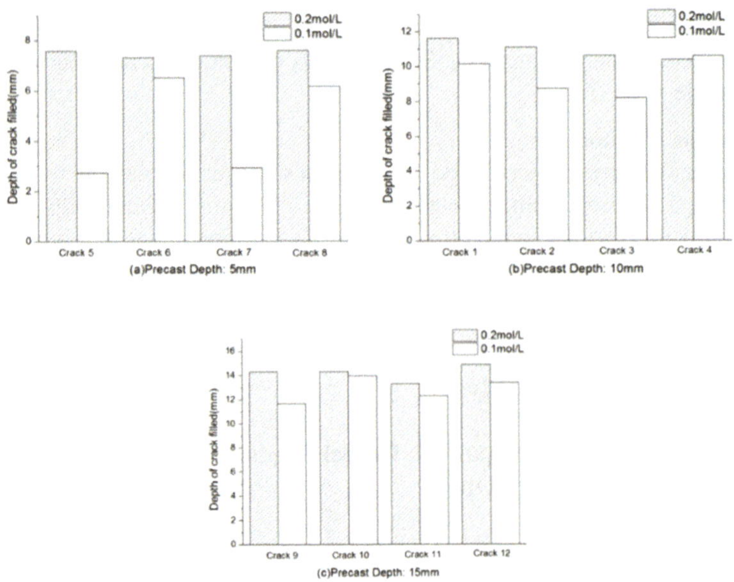

Fig. 6. The variation of the filling depth of crack with solution concentrations in different precast depth conditions

Table 3. Comparison of surface coating rate of specimens

Conditions	0.1 mol/L, $ZnSO_4$	0.2 mol/L, $ZnSO_4$
Surface coating rate (%)	33.10	41.77

surface coating rate is higher and thickness of the coating is larger with the rise of concentration. Since the number of ions in the solution increases as the concentration of the solution rises, and the ion concentration gradient in the different regions is greater. These lead to a faster migration of ions in solution when electrified and thus anions and cations combine to form deposits at a faster rate.

SEM Analysis Method

SEM is employed to scan the deposits inside cracks in both conditions and the micrographs obtained are shown in Fig. 7.

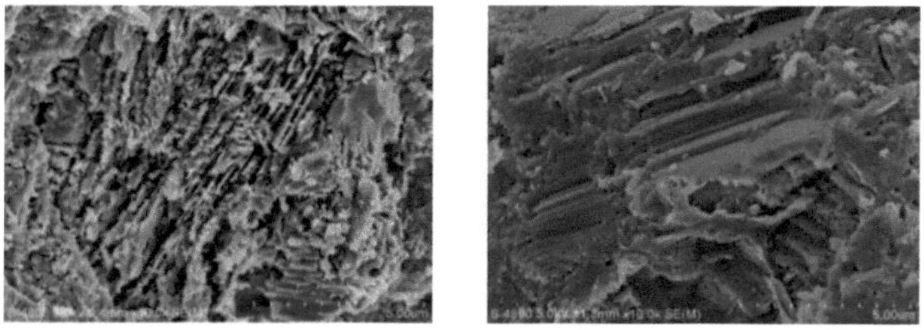

(a) 0.1mol/L, $ZnSO_4$ (b) 0.2mol/L, $ZnSO_4$

Fig. 7. Scanning electron micrograph of deposits in specimen crack

In 0.1 mol/L condition, most of the crystals are in the form of lamellar spindle structure, of which the size is small. And crystals are arranged closely but not gathered in a regular way; while in 0.2 mol/L condition, the crystals are in the form of plate-type structure, thickness of the coating is larger and degree of compactness of crystals is relatively high. There are no voids between different plate-type structures, and some crystals develop in a zigzag way due to the growth of other neighboring crystals.

When the concentration is low, ions immigration is slower, which is not conducive to the orderly and directional growth of crystals. On the contrary, if concentration is too high, the growth speed of the crystal unit is too fast, so that there is not enough time for orderly arrangement. That's why the growth of a small number of crystals near the edge will be squeezed by other crystals in one or two dimensional space, and the final crystal structure is changed, showing a distorted shape.

3.2 Simulation of Crack Repair of Tunnel Segment at the Top and Side of Tunnel (Experiment Two)

Test Results

Compared to experiment one, the repair process lasts much longer because of function of gravity. Healing effectiveness of tunnel segment in 50 days is presented in Fig. 8.

In contrast with Fig. 2, healing effects of crack 2 and 3 are remarkable while healing effects of the other cracks are not satisfying. However, there are still needle-like deposits produced inside the other cracks and crystals grow and accumulate attached to the side walls of crack, see details for Fig. 9.

Fig. 8. Healing effectiveness of tunnel segment of experiment two in 50 days

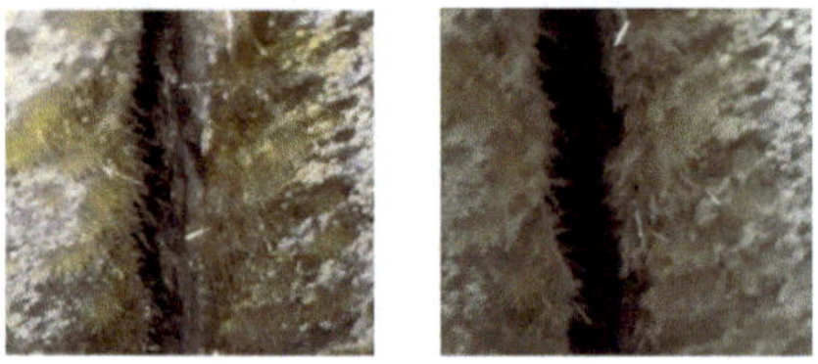

Fig. 9. The small deposits in No. 4 crack

Evaluation Indexes

Crack closure rate and filling depth of crack are employed as evaluation indexes. The variation of evaluation indexes with width or depth of cracks is shown in Fig. 10.

From the above figures, there are two main reasons that affect the healing effectiveness of EDM. One is the variation of width of cracks. The smaller the width of crack is, the faster the crack will be closed. If the width of crack exceeds a certain value, crack is hard to be fully repaired. That's because crystals grow and accumulate attached to the side walls of crack and eventually deposits will fill the crack with the growth of crystals [14]. But when the width of crack exceeds a certain value, the growing crystals attached to different side walls of crack cannot reach each other easily and then the growth is interrupted causing deposits cannot filling the crack.

The other is the distribution of depth of cracks. Since the depth of crack 2 and 3 are far larger than that of other cracks, concrete resistance at these two cracks are smaller; both crack 2 and 3 are through cracks which means the steel bar is exposed to repair solution directly, and the conductivity of the steel bar is much larger than that of concrete. These two lead to a fact that the current density at through cracks is several hundred times larger than that of other parts [10]. So the healing effectiveness of through cracks is the best.

However, the width of crack 1, 5 and 6 is a little larger than that of crack 3 but the healing effectiveness is not satisfying. That's because the steel bar is not exposed to solution at the location of crack 1, 5 and 6, which leads to less deposits.

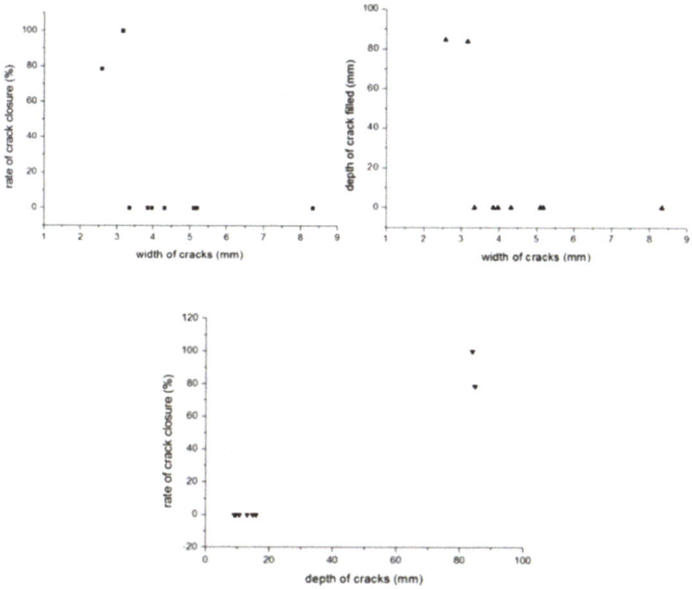

Fig. 10. The variation of evaluation indexes with width or depth of cracks

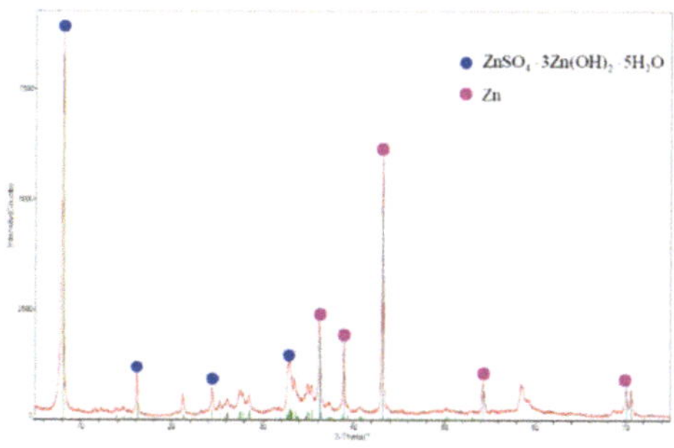

Fig. 11. X ray diffraction atlas of deposits in No. 2 crack

Xrd Method

In 50 days, the deposit samples are obtained from crack 2, and analysis of mineral compositions is conducted by Xrd method.

As indicated in Fig. 11, the main mineral compositions of deposits in crack 2 are crystals of $ZnSO_4 \cdot 3Zn(OH)_2 \cdot 5H_2O$. The above crystals are produced under the environment of $ZnSO_4$ solution, validating the feasibility of EDM. In this experiment, the chemical reaction equations are listed:

The anode: $2H_2O - 2e^- \rightarrow O_2 \uparrow + 2H^+$

The Cathode: $2H_2O + 2e^- \rightarrow 2OH^- + H_2 \uparrow$ (On the surface of steel bar)

$4Zn^{2+} + SO_4^{2-} + 6OH^- + 5H_2O \rightarrow ZnSO_4 \cdot 3Zn(OH)_2 \cdot 5H_2O \downarrow$
(On the surface of concrete)

The generation of Zn indicates that the listed chemical reaction at crack 2 also happens:

The cathode: $Zn^{2+} + 2e^- \rightarrow Zn \downarrow$ (On the surface of steel bar)

The steel bar is exposed to repair solution directly since crack 2 is through crack. Both hydrogen evolution reaction and reduction reaction of Zn^{2+} occur on the surface of steel bar.

4 Conclusions and Discussions

In this paper, simulation of crack repair of tunnel segment using EDM at the bottom and at the top and side of the tunnel, separately, is conducted. The healing effectiveness in different sizes and positions of underground structure is evaluated using some evaluation indexes and repair mechanism at the micro scale is revealed. The following conclusions can be obtained from our present work.

(1) It is proved that EDM is feasible to repair damaged underground structure (tunnel structure) at the level of concrete component.
(2) Both the variation of width of cracks and the distribution of depth of cracks can influence the effect of EDM. In experiment two, the distribution of depth of cracks is the main reason. At through cracks, concrete resistance is smaller and current density is larger leading to more deposits. Therefore, the healing effect is better.
(3) From the results of SEM analysis method, different solution concentrations will lead to distinct crystal microstructure of deposits. If the concentration is higher, degree of compactness of deposits is relatively high.
(4) Compared to 0.1 mol/L condition, the crack repair rate is faster and the healing effectiveness is better in the 0.2 mol/L condition at the macro scale, providing a more reasonable selection of test parameters for subsequent application.

Acknowledgements. This work is supported by the National Natural Science Foundation of China (51508404). This work is also supported by the Program of State Key Laboratory of High Performance Civil Engineering Materials (2015CEM008), Central University Basic Research Fund Project of Chang'an University (310821151113) and Innovative Training Program for College Students in Shanghai. We would also like to express our gratitude to Mr. Kukai Feng, Xin Zhou and Yan Wang for their help in the experiment process, Dr. Chen Li for the support of many literature and data about DEM, and Mr. YanBang Lou for the suggestions of simulation device.

References

1. Zhou, S., Zhu, H., Yan, Z., Ju, J.W.: Self-healing micromechanical model of microcapsule-enabled self-healing concrete in underground structures. J. Tongji Univ. **42** (10), 1467–1472 (2014)
2. Yokoda, M., Fukute, T.: Rehabilitation and protection of marine concrete structure using electrodeposition method. In: Proceedings of the International RILEM/CSIRO/ACRA Conference on Rehabilitation of Concrete Structures, RILEM, Melbourne, pp. 213–222 (1992)
3. Hilbertz, W.: Mineral accretion of large surface structure, building components and elements. US (1982)
4. Hilbertz, W.: Repair of reinforced concrete structures by mineral accretion. US (1984)
5. Otsuki, N., Ryu, J.S.: Use of electrodeposition for repair of concrete with shrinkage cracks. J. Mater. Civil Eng. **13**(2), 136–142 (2001)
6. Ryu, J.S.: An experimental study on the repair of concrete crack by electrochemical technique. Mater. Struct. **34**(241), 433–437 (2001)
7. Ryu, J.S., Otsuki, N.: Crack closure of reinforced concrete by electro deposition technique. Cem. Concr. Res. **32**(1), 159–264 (2002)
8. Ryu, J.S., Otsuki, N.: Experimental study on repair of concrete structural members by electrochemical method. Scripta Mater. **52**(11), 1123–1127 (2005)
9. Chang, J.J., Yeih, W.C., Hsu, H.M., et al.: Performance evaluation of using electrochemical deposition as a repair method for reinforced concrete beams. SL **1**(2), 75–93 (2009)
10. Yao, W., Zheng, X.: Experimental study on crack repair of reinforced concrete by electrodeposition technique. J. Tongji Univ. (Nat. Sci.) **34**(11), 1441–1444 (2006)
11. Chu, H., Jiang, L., Xu, Y.: Influence of current density in electrodeposition method for repair of concrete cracks. J. Build. Mater. **12**(6), 729–733 (2009)
12. Chu, H., Jiang, L., Hua, W.: Influence of accessorial electrode and electrode distance on electrodeposition effect. J. Build. Mater. **8**(4), 456–462 (2005)
13. Jiang, Z., Xing, F., Sun, Z., et al.: Preliminary study on crack repair of reinforced concrete by using electrodeposition method. Adv. Sci. Technol. Water Resourc. **27**(3), 5–9 (2007)
14. Jiang, Z., Sun, Z., Wang, P.: Mechanism on rehabilitation of cracks in reinforced concrete using electrodeposition technique. J. Tongji Univ. (Nat. Sci.) **32**(11), 1471–1475 (2004)
15. Jiang, Z., Sun, Z., Wang, P.: Healing effectiveness of crack rehabilitation in reinforced concrete using electrodeposition method. J. Southeast Univ. (Nat. Sci.) **36**(2), 129–135 (2006)
16. Zhu, H.H., Chen, Q., Yan, Z.G., et al.: Micromechanical model for saturated concrete repaired by electrochemical deposition method. Mater. Struct. **47**, 1067–1082 (2014)
17. Yan, Z.G., Chen, Q., Zhu, H.H., et al.: A multiphase micromechanical model for unsaturated concrete repaired by electrochemical deposition method. Int. J. Solids Struct. **50**(24), 3875–3885 (2013)
18. Chen, Q., Jiang, Z.W., Yang, Z.H., et al.: Differential-scheme based micromechanical framework for saturated concrete repaired by the electrochemical deposition method. Mater. Struct. **49**(12), 5183–5193 (2016)
19. Zhu, H.H., Chen, Q., Ju, J.W., et al.: Maximum entropy based stochastic micromechanical model for two-phase composite considering the inter-particle interaction effect. Acta Mech. **226**(9), 3069–3084 (2015)
20. Chen, Q., Zhu, H.H., Yan, Z.G., et al.: A multiphase micromechanical model for hybrid fiber reinforced concrete considering the aggregate and ITZ effects. Constr. Build. Mater. **114**, 839–850 (2016)

21. Chen, Q., Zhu, H.H., Ju, J.W., et al.: A Stochastic Micromechanical model for Multiphase Composite Containing Spherical Inhomogeneities. Acta Mech. **226**(6), 1861–1880 (2015)
22. Chen, Q., Mousavi Nezhad, M., Fisher, Q., et al.: Multi-scale approach for modeling the transversely isotropic elastic properties of shale considering multi-inclusions and interfacial transition zone. Int. J. Rock Mech. Min. Sci. **84**, 95–104 (2016)
23. Chen, Q., Jiang, Z.W., Yang, Z.H., et al.: An experimental study on the repair of deteriorated concrete by electrochemical deposition method. In: Environmental Sustainability in Transportation Infrastructure, pp. 87–94 (2015)
24. Chen, Q., Zhu, H.H., Yan, Z.G., et al.: Micro-scale description of saturated concrete repaired by electrochemical deposition method based on Mori-Tanaka method. J. Build. Struct. **36**(1), 98–103 (2015)
25. Chen, Q., Zhu, H.H., Yan, Z.G., et al.: Micro-scale description of the saturated concrete repaired by electrochemical deposition method based on self-consistent method. Chin. J. Theor. Appl. Mech. **47**(2), 367–371 (2015)
26. Stokes, D.J.: Principles and Practice of Variable Pressure Environmental Scanning Electron Microscopy (VP-ESEM). Wiley, Chichester (2008). ISBN 978-0470758748

Study on the Rock Damage Characteristics Based on SEM Test

Yimeng Zhou[1], Cheng Zhao[1,2(✉)], Chunfeng Zhao[1], and Junfei Xie[1]

[1] Key Laboratory of Geotechnical and Underground Engineering of Ministry of Education, Department of Geotechnical Engineering, Tongji University, Shanghai, China
zhaocheng@tongji.edu.cn

[2] Tibet University Institute of Technology, Lhasa, China

Abstract. The meso morphologies of the fracture surface of cracked rock mass indirectly reflects its internal damage evolution process. In this paper, the SEM scanning images of the specimens with different pre-existing cracks under uniaxial compression are post processed by MATLAB and IPP. Moreover, the damage evolution of meso structure in the failure process of the specimen is quantitatively studied, which establishes the relationship between macroscopic failure and meso mechanism. It is found that when the crack initiates, the damage degree of meso structure in the crack tips decreases with the increase of the inclination angle of pre-existing crack, and the weakening effect of tensile cracks on the strength of specimen is greater than that of shear cracks. Subsequently, when the crack propagates, the inclination angle of pre-existing crack has little influence on the damage rate of the meso structure.

Keywords: Damage evolution · SEM · Tensile and shear properties

1 Introduction

The meso structure of jointed rock mass will be damaged during the failure process under loading. With the accumulation of damage, the cracks will initiate, propagate, and coalesce the rock on the macro level. Thus, the study on the damage evolution law of meso structure inside the rock is of great importance. The conception of damage was first proposed by Kachanov [1] to study metal creep. On this basis, many scholars have studied the damage model of rock [2, 3], the meso morphology of facture surface [4] and the fracture mechanism of rock mass [5] by conducting theoretical, experimental, and numerical methods. By comprehensive use of the theories of structure mechanics of rock mass, geometrical damage mechanics and rock fluid mechanics, the model of seepage-damage coupling was established by Zhao et al. [6]. Yang et al. [7] experimentally studied the deformation failure and energy properties of marble specimens under conventional triaxial compression. Wong [8] numerically studied the coalescence of two coplanar cracks in rock under compression using AUTODYN. These studies have comprehensively promoted the understanding of crack propagation process and failure evolution of rock mass. However, most of the researches are based on the microscopic mechanic properties of rock mass or the macroscopic crack propagation

L. Zhang et al. (Eds.): GSIC 2018, *Proceedings of GeoShanghai 2018 International Conference: Rock Mechanics and Rock Engineering*, pp. 297–304, 2018.
https://doi.org/10.1007/978-981-13-0113-1_32

process, and there is little quantitative study on the damage evolution of meso structure. Moreover, little attention has been paid to the research on the relationship between damage evolution of rock interior and macro mechanical property.

According to the previous study, the geometrical morphologies of crack fracture surface can be obtained by SEM. By comparing with the morphologies of pure tensile and pure shear specimens, the evolution pattern of the distribution of tensile and shear stress during the crack propagation process is analyzed [9]. In this paper, the SEM images of different specimens are post processed by binaryzation based on MATLAB in advance. Subsequently, the image analysis software Image-Pro Plus (IPP) is used to extract the data from SEM images after binaryzation, and these meso damage data are quantitatively analyzed. Finally, the damage evolution law of meso structure is summarized, and the relationship between the damage of meso structure and macroscopic crack propagation is established.

2 SEM Test

SEM is a microscopic observation method, which can directly scan the fracture surface of a specimen, as shown in Fig. 1. In order to obtain the criterion for judging tensile and shear cracks, the fracture surfaces of pure tensile and pure shear are manufactured artificially by Brazil splitting test and compression shear test. As shown in Fig. 2, under the action of pure tension, the typical fracture surface is rough and the meso-surface presents granular feeling. Under the pure shear, the typical fracture surface is smooth and the meso-surface presents friction sliding feeling [10]. On this basis, a weight parameter D is proposed for the analysis of tensile and shear properties of the crack propagation process and is defined as follows:

$$D = (A_s - A_t)/A \tag{1}$$

where A is the total area of the scanning section, A_s is the area with shear characteristics, A_t is the area with tensile characteristics. The range of D is $-1 \leq D \leq 1$, when D = −1, the section is pure tensile stress region, the new crack is tensile crack. When D = 1, the section is pure shear stress region, the new crack is shear crack.

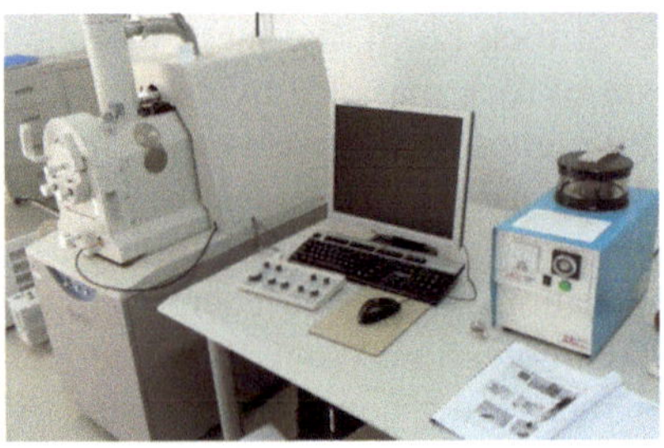

Fig. 1. The schematic of SEM.

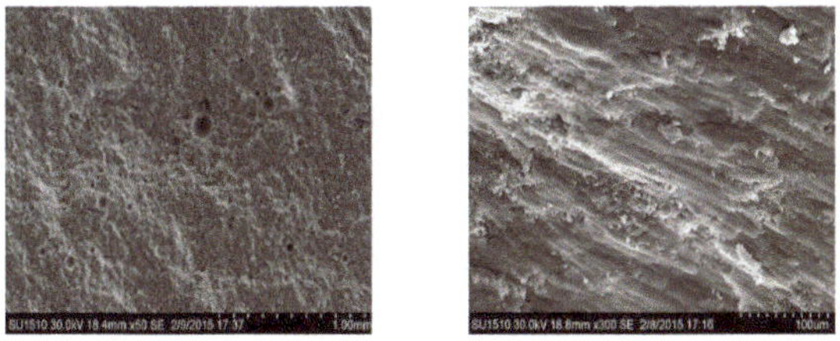

Fig. 2. Tensile (left) and shear (right) crack SEM images.

As shown in Fig. 3, the failure surface of the specimen was divided into three parts, and the SEM samples were selected from the corresponding areas [9]. Figure 4 shows the scanning images of initial stage, middle stage, and late stage of a 30° specimen. Observing the morphologies of fracture surface and the particle distribution at high magnification, the D values of the three stages are −0.8, −0.7, −0.4, respectively [10].

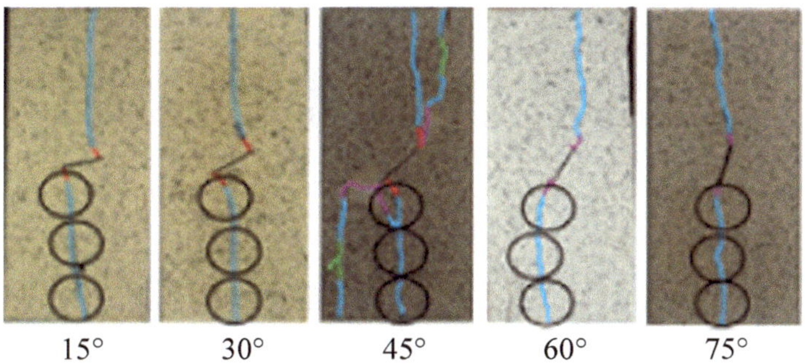

Fig. 3. Failure diagrams and sampling points of different specimens.

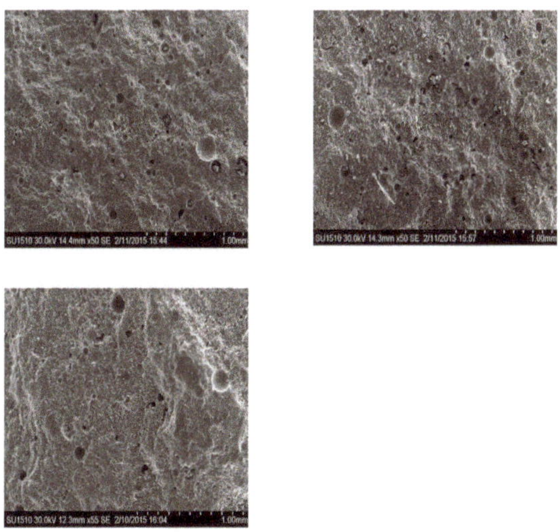

Fig. 4. SEM images of a 30° specimen.

It is found that tensile stress dominates the crack initiation and propagation of a 30° specimen. With the propagation of wing cracks, the weight of tensile stress decreases and the weight of shear stress increases gradually.

3 Post Processing of SEM Images

Based on the concept of validity [1], the ratio of the damaged area to the total area of the meso structure is taken as the meso damage variable. Therefore, the damage degree can be obtained by extracting damage area and total area from SEM images. The formula is as follows:

$$\Omega = \frac{S_e}{S_0} \tag{2}$$

where Ω is meso damage variable, S_e is the damage area, S_0 is the total area, the range of it is $0 \le \Omega \le 1$. When Ω is 0, there is no damage to the meso structure, when Ω is 1, the meso structure is completely damaged. According to the imaging principle of SEM, when the incident electron remains unchanged, the number of electrons reflected by rock crystal is more than that of the crystal void. On the macro level, the color of the undamaged area was shallow and the color of the damaged area was deeper. Using the threshold method, the SEM image with magnification of 300 times is transformed into binaryzation images. The images after binaryzation are shown in Table 1, the white parts represent the undamaged areas, and the black parts represent the damaged areas.

The graphic analysis software Image-Pro Plus(IPP)is used to analyze the binaryzation images, the amount of black and white pixels is extracted. Moreover, the value of Ω of crack initiation, crack propagation, and crack coalescence are obtained in Table 2.

4 Analysis of Meso Damage

Figure 5 presents the evolution curve of Ω at the initial stage with X axis representing inclination angles and Y axis the value of Ω. It is found that the damage degree Ω of the crack tips decreases gradually with the increase of the crack inclination angle. Zhao et al. [9] found that when the wing crack initiates, the proportion of tensile stress on the fracture surface decreases and the proportion of shear stress increases gradually with the increase of crack inclination angle. Based on the above analyses, a conclusion can be obtained as follows: With the increase of the shear stress on the crack surface, the damage degree Ω of the meso structure decreases gradually when crack initiates. That is to say, shear stress reduces the damage degree of meso structure when crack initiates.

Based on the previous study [11], Fig. 6 shows the initiation stress of different specimens in experimental tests and numerical simulation. Analyzing the curve of experimental tests, with the increase of inclination angle, the crack initiation stress increases gradually, indicating that the meso structure must have changed, which makes crack initiation more difficult to occur. This phenomenon also confirms the conclusion

Table 1. Binaryzation images of different specimens.

Inclination angle	Initial stage	Middle stage	Late stage
15°	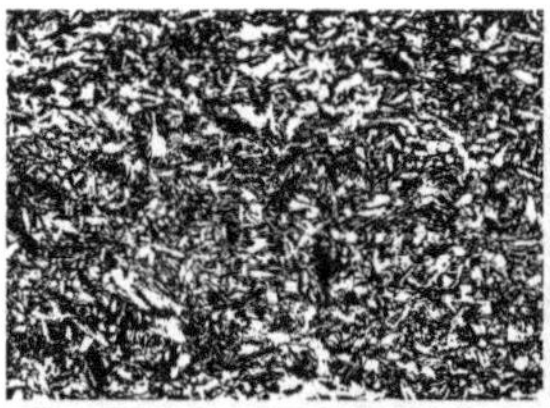	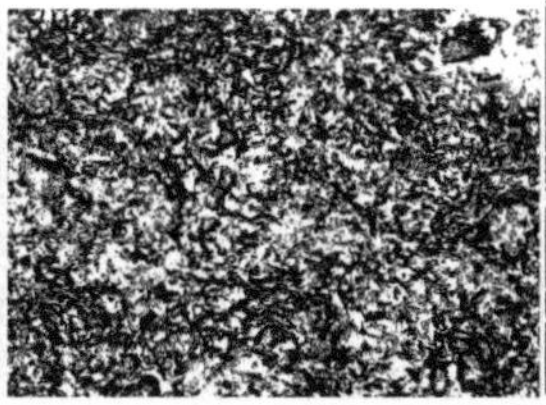	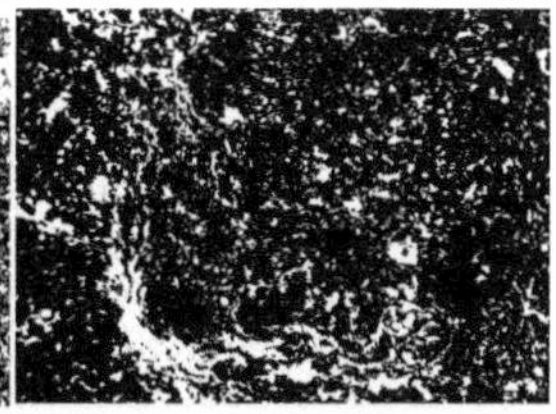
30°	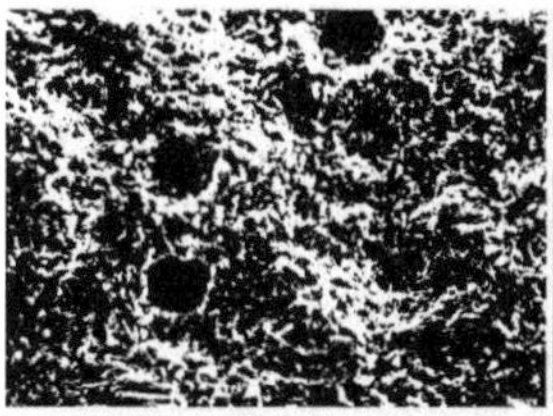	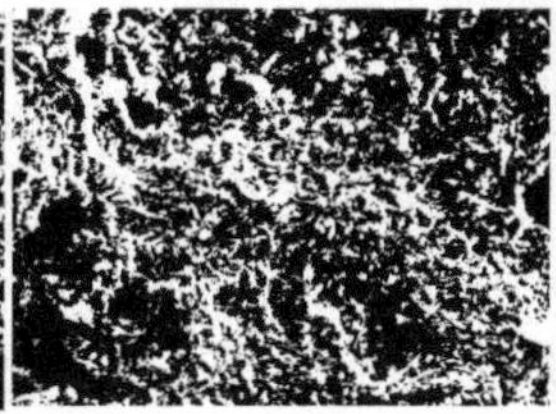	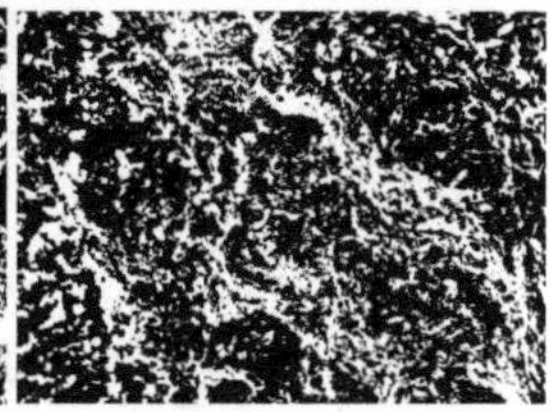
45°	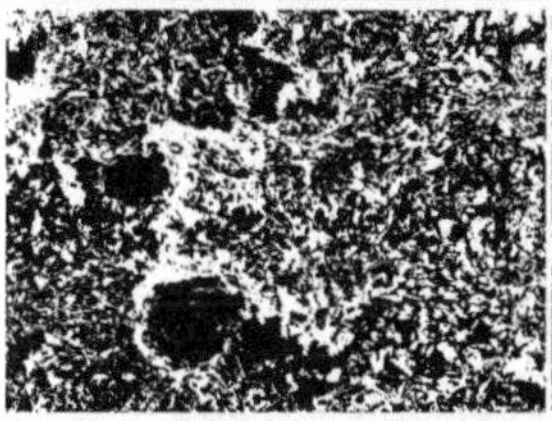	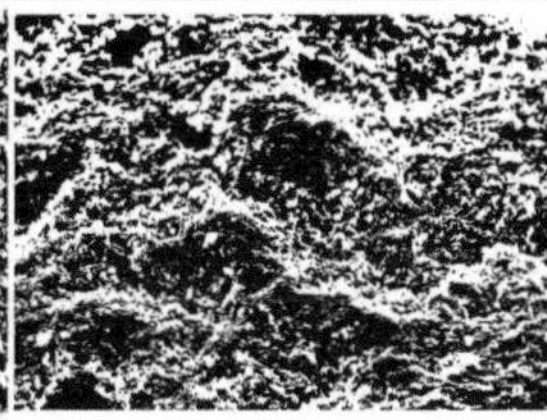	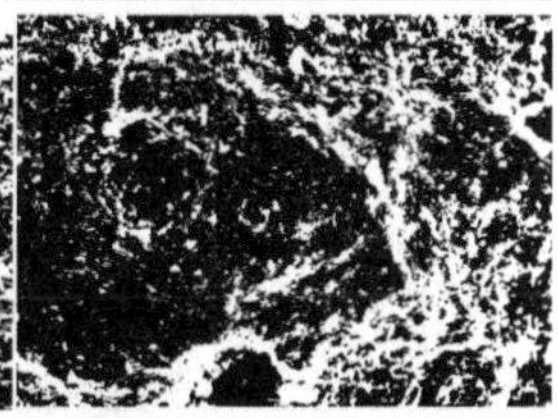
60°			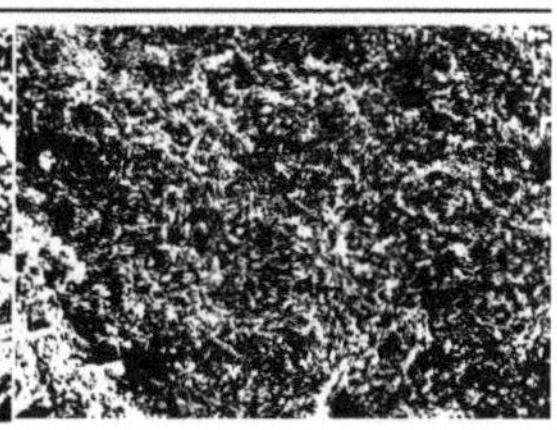
75°	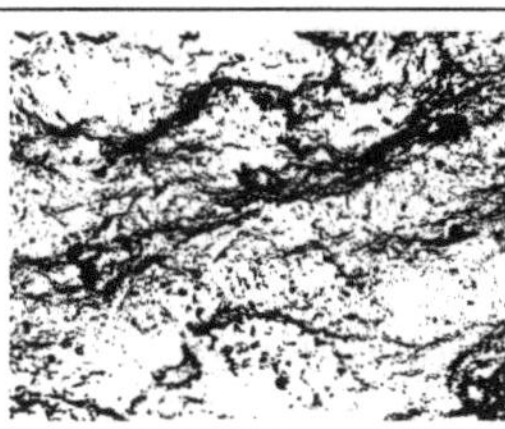	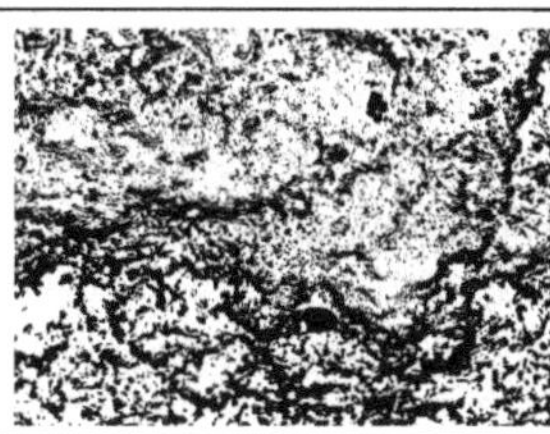	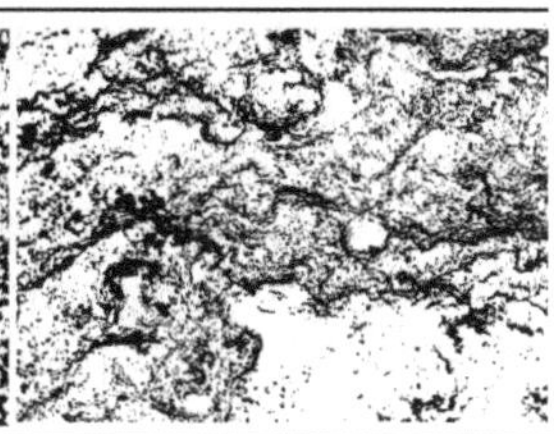

that damage proportion is getting smaller with the increase of inclination angle. However, when the inclination angles reach 60° and 75°, the crack initiation stress increases rapidly and has a multiple increase compared with other angles. The analysis of this particular case is as follows: In the case that the internal damage degree of the

Table 2. The value of Ω in different stages of different specimens.

Inclination angle	The value of Ω
15°	36.3%, 40.7%, 49.2% (initial stage, middle stage, late stage)
30°	36.1%, 38.4%, 47.2%
45°	34.8%, 41.6%, 45.5%
60°	32.5%, 36.0%, 43.7%
75°	28.3%, 32.0%, 41.0%

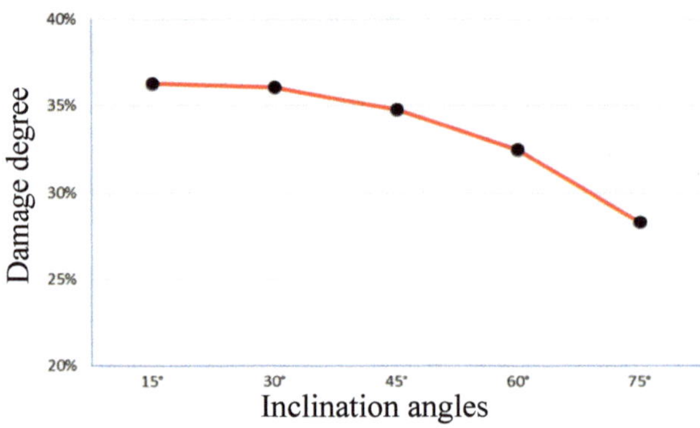

Fig. 5. The evolution curve of Ω at the initial stage.

specimens with different inclination angles is approximately uniform decreasing, the sudden change of crack initiation stress can only depends on the different weakening effect of tensile and shear cracks. When the inclination angles are 15°, 30°, and 45°, tensile crack dominates the crack initiation, when the inclination angles are 60° and 75°, shear crack dominates the crack initiation [9]. Therefore, it can be concluded that the weakening effect of tensile cracks on the strength of specimen is greater than that of shear cracks. Analyzing the curve of numerical simulation, it is found that there is no sudden change section in the stress curve, this is because numerical simulation does not take into account the different weakening effect between the tensile and shear cracks on the strength of the specimen.

Analyzing the variation of Ω from crack initiation to crack coalescence in Table 2, the five groups of data are 12.9%, 11.1%, 10.7%, 11.2%, and 12.7% respectively. It is found that although the damage degree of meso structure at crack initiation stage decreases gradually, the damage variation during the propagation process remains almost unchanged. Once the crack initiates, the tensile and shear stress on the fracture surface has little effect on the damage of meso structure.

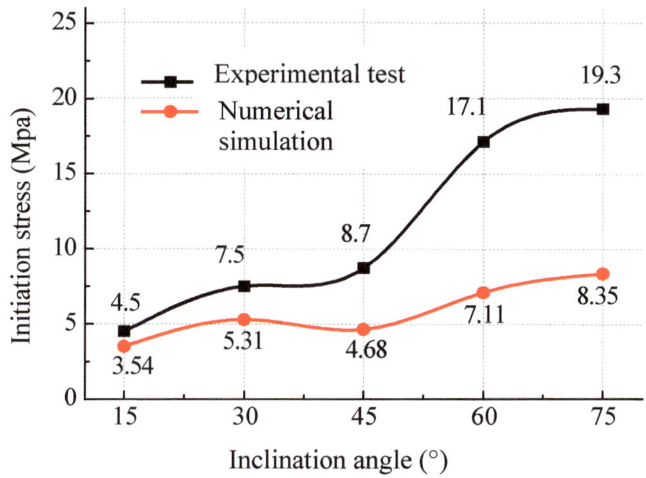

Fig. 6. Initiation stress of different specimens in experimental tests and numerical simulation.

5 Conclusion

In this paper, the SEM scanning images of the specimens with different pre-existing cracks under uniaxial compression are post processed by MATLAB and IPP. Moreover, the evolution of damage degree of meso structure in the failure process of the specimen is quantitatively studied, which establishes the relationship between macroscopic failure and meso mechanism. The conclusions are as follows:

(1) When the crack initiates, the damage degree Ω of meso structure in the crack tips decreases with the increase of the inclination angle. Moreover, the weakening effect of tensile cracks on the strength of specimen is greater than that of shear cracks, which leads to the rapid increase of initiation stress of 60° and 75° specimen.

(2) When the crack propagates, the inclination angle of pre-existing crack has little influence on the damage rate of the meso structure.

Disclosure statement

No potential conflict of interest was reported by the authors.

Acknowledgements. The authors would like to acknowledge the financial support of the National Natural Science Foundation of China (No. 41202193 and No. 41572262), Innovation Program of Shanghai Municipal Education Commission (No. 15ZZ016), and Shanghai Rising-Star Program (No. 17QC1400600).

References

1. Kachnov, M.: Effective elastic properties of cracked solids: critical review of some basic concepts. Appl. Mech. Rev. **45**(8), 304–355 (1992)
2. Jiang, W., Deng, J., Li, Y.: Study on constitutive model of rock damage based on lognormal distribution. Chin. J. Undergr. Space Eng. **6**(6), 1190–1194 (2010)
3. Zhang, M., Wang, F., Yang, Q.: Statistical damage constitutive model for rocks based on triaxial compression tests. Chin. J. Geotech. Eng. **35**(11), 1965–1971 (2013)
4. Lemaitre, J.: A course on damage mechanics, pp. 29–30. Springer, Berlin (1992)
5. Krajcinovic, D.: Statistical aspects of the continuous damage theory. Int. J. Solids Struct. **18**(7), 551–562 (1982)
6. Zhao, Y.L., Wang, W.J., Huang, Y.H., et al.: Coupling analysis of seepage-damage-fracture in fractured rock mass and engineering application. Chin. J. Geotech. Eng. **32**(1), 24–32 (2010)
7. Yang, S.Q., Dai, Y.H., Han, L.J., et al.: Uniaxial compression experimental research on deformation and failure properties of brittle marble specimen with pre-existing fissures. Chin. J. Rock Mech. Eng. **28**(12), 2391–2404 (2009)
8. Wong, L.N.Y., Li, H.Q.: Numerical study on coalescence of two coplanar pre-existing flaws in rock. Int. J. Solids Struct. **50**, 3685–3706 (2013)
9. Zhao, C., Yu, Z.M., Wang, W.D., et al.: Meso-experiment study on feature mechanism of rock based on uniaxial compression test. Chin. J. Rock Mech. Eng. **35**(12), 2490–2498 (2016)
10. Zhao, C., Ma, C., Zhao, C., et al.: Simulation on crack propagation of rock-like specimen using strain criterion (2017). https://doi.org/10.1080/19648189.2017.1359677
11. Zhao, C., Liu, F., Tian, J., et al.: Study on single crack propagation and damage evolution mechanism of rock-like materials under uniaxial compression. Chin. J. Rock Mech. Eng. **35**(2) (2016)

The Influence of Water Absorption Increment on Infrared Radiation Characteristics of Rammed Earth

Zhang Fang[1,2(✉)], Liu Wenfang[1,2], Wang Zhuo[3,4], Fu Yingyu[2], Wang Shiyu[2], Liu Runqi[2], and He Manchao[1]

[1] State Key Laboratory for GeoMechanics and Deep Underground Engineering, Beijing 100083, China
zhangf76@163.com
[2] School of Mechanics and Civil Engineering, China University of Mining and Technology (Beijing), Xuzhou 100083, China
[3] Chinese Society for Rock Mechanics and Engineering, Beijing 100029, China
[4] Beijing Zhongke Force Blasting Technology Co., LTD., Beijing 101318, China

Abstract. Targeting for the description of water occurrence status in the rammed earth of Mogao Grottoes, the infrared thermography was applied to study the relationship between water absorption increment and infrared radiation temperature at various water absorption stages. The result showed that (1) before the infiltrating of all rock samples, and due to the impact of the increasing volume of water passing, the overall relationship between the water absorption increment of unit mass of dry rocks and the infrared radiation temperature was positive, but no obvious regular difference existed in various water absorption stages; (2) by making use of the characteristic that *NMI* of adjacent thermal images was 0, the moment when water in rammed earth turned from partial infiltration to complete infiltration could be judged; (3) by making use of the characteristic that *NMI* of adjacent thermal images was 0, the moment when rammed earth turned from unsaturation to complete saturation could be judged. A non-destructive, convenient and real-time method was expected to be provided for the judgment of water occurrence conditions in rammed earth of Mogao Grottoes.

Keywords: Water occurrence conditions · Water absorption increment
Infrared radiation characteristics · Mutual information

1 Introduction

Water is one of the most active factors causing geotechnical engineering disasters and damages; it is also the direct reason causing the damages in grottoes, murals and ruins in stone and earthen historic relics protection field. Therefore, it is very important to dynamically monitor the water content in rock-soil medium [1–3]. Due to its non-destructive and real-time monitoring performance, the infrared thermography has been attracting more and more attention worldwide, and many scholars have started

L. Zhang et al. (Eds.): GSIC 2018, *Proceedings of GeoShanghai 2018 International Conference: Rock Mechanics and Rock Engineering*, pp. 305–315, 2018.
https://doi.org/10.1007/978-981-13-0113-1_33

relative studies [4–8]. Domestically, Mingde [9] was the first who pointed out that the water, to some extent, could absorb some certain intrinsic radiation of the rock sample and obviously promote the infrared radiation characteristics of stressed rocks by contrasting the infrared radiation temperature changing characteristics of water-bearing rock samples to the dry rock samples under the condition of stress variation. Huihui [10] contrasted the direction and extent of the infrared radiation temperature changes of the rock samples that were with different moisture contents under the same lighting condition in tests, and put forward the possibility of discrimination and recognition of moisture content and water percolation of rocks through infrared radiation variation; Yuhua [11] used the infrared thermal imaging technology to monitor the specific seepage positions on painting cliffs, clarify seepage grades, and judge seepage severity of different areas by taking the highest and lowest temperatures in the seepage area as the indexes, thus to put forward the feasibility of applying thermal imaging technology to monitor the seepage damages of cliff paintings; Avdelidis [12] came to the conclusion that the damper the rock was, the lower its surface temperature was, which indicated that rock's water content was in negative correlation to temperature according to the infrared radiation temperature variety rule in the ascent process of water in porous rock capillary; Furmański [13] established the mathematical model of heat and mass transfer of moisture evaporation of wet sample and put forward that when sudden change happened to the radiation temperature filed of wet sample surface, it would be the most proper moment for testing the water content of the sample; Gayo-Monco [14] applied different interference filters to eliminate impacts of interference factors on infrared radiation in water-bearing buildings under environments with high humidity or high CO^2 index, strengthening the display effect of infrared thermal imaging of building component humidity.

In a word, in the protection fields of rock, soil, and cultural relic at home and abroad, the utilization of infrared thermal imaging technology in detection and monitoring of water disasters has been studied and applied a lot, but most of these researches and application examples are limited to qualitative identification of static moisture in materials. Apparently, these methods could not describe time effect of moisture, or, they could not judge occurrence state of moisture in medium at a certain moment.

With infrared thermography, this paper studied the relationship between water absorption increment and infrared radiation temperature at various stages of rammed earth absorption in the hope to provide a non-destructive, convenient and real-time method for the judgement of water occurrence conditions in rammed earth of Mogao Grottoes.

2 Mutual Information Theory

Mutual Information (*MI*) can be applied to describe the similarity between two random signals or the correlation between statistical variables. Currently, it has been also widely studied on the measurement of similarity between images [15–17]. *MI* is usually used as judgment criterion; a higher value of *MI* indicates a higher similarity between two images. The computing method of *MI* is to take the two images that are to be registered (at moment t_i and t_{i+1} respectively) as two random events, According to the image of the complete gray information to calculate the probability of occurrence of each level of gray and joint probability of the combination of the two images. Based on this, information entropy $H(t_i)$ and $H(t_{i+1})$ of thermal images at moment t_i and t_{i+1} respectively and joint information entropy $H(t_i, t_{i+1})$ can be calculated and finally the *MI* value is obtained. The application of *MI* shows that the reduction of overlapping regions between different images will cause the incompleteness of mutual information, which generates local extremum that cannot accurately evaluate actual situation and may be leads to wrong match results. Normalized Mutual Information (*NMI*) is an extended concept in order to solving the problem of misjudgment. It is constrained by joint entropy and edge entropy at the same time, the image registration effect is more accurate. Definitions of information entropy and joint information entropy of infrared thermograms:

$$H(t_i) = -\sum_{n=1}^{N} P_n(t_i) \lg P_n(t_i) \tag{1}$$

$$H(t_{i+1}) = -\sum_{m=1}^{m} P_m(t_{i+1}) \lg P_m(t_{i+1}) \tag{2}$$

$$H(t_i, t_{i+1}) = -\sum_{k=1}^{K} P_k(t_i, t_{i+1}) \lg P_k(t_i, t_{i+1}) \tag{3}$$

$$MI(t_i, t_{i+1}) = H(t_i) + H(t_{i+1}) - H(t_i, t_{i+1}) \tag{4}$$

In the formula, t_i is the data acquisition time ($i = 1, 2, \ldots, l$); N is the distribution of N radiation temperature values in the thermal image at time t_i; $P_n(t_i)$ is the probability of the nth radiation temperature in the thermal image at t_i; M is the thermal image at time t_{i+1}, M is the probability of mth radiation temperature in the thermal image at t_{i+1}, $P_m(t_{i+1})$ is the probability that the mth radiation temperature appears in the thermal image at time t_{i+1}; $K = N \cdot M$, K means that there are K infrared radiation temperature combinations, in the two thermograms at the adjacent time t_i, t_{i+1}; $P_k(t_i, t_{i+1})$ is the probability of infrared radiation temperature combinations in the two thermograms at the adjacent time t_i, t_{i+1}.

In this paper, the normalized mutual information was applied to evaluate the similarity between adjacent thermal images with the same water content increment, and analyze the characteristics of the infrared radiation intensity change of the water-contained rocks under different water absorption status.

3 Experimental Method

The sample used for the experiment was rammed earth in the Mogao Grottoes of Dunhuang, China, which is a cylinder with a diameter of 50 mm and a height of 100 mm. The intelligent test system for water absorption of deep soft rock, which was independently researched and developed by State Key Laboratory of Deep Rock Mechanics and Underground Engineering, was used in the experiment to simulate the whole process in which dry rock absorbed water until they were saturated. Meanwhile, the infrared thermal imager was used in the experiment to get the infrared radiation information on rock surface throughout the absorption process. Test system of the experiment is shown in Fig. 1.

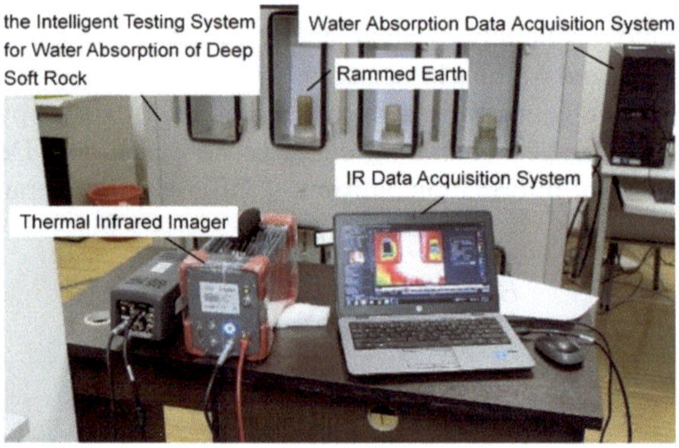

Fig. 1. Experimental system for water test of deep soft rocks

Before the experiment, proper lens was selected, distance and angle between the infrared thermal imager and rammed earth sample were set, and related instrument parameters were configured according to the experimental environment.

In the experiment, injecting water into a container first, discharge bubbles inside the conduit to prevent the bubbles from affecting the experiment; put the filter paper on the experiment table and collect data at the same time; then, start up the infrared thermal imager to collect the infrared radiation temperature on the sample surface; when the rammed earth was saturate, stop the experiment (note: The experiment in this study was terminated because the rammed earth sample fell down and broke).

4 Result Analysis

4.1 Absorption Stage and Its Infrared Radiation Characteristics

Figure 2 is the characteristic curve of water absorption of the rammed earth, in which, t is the water absorption time, Q is the water absorption. By observing rammed earth, the whole process of water absorption. The whole process of water absorption can be divided into four stages: AB, BC, CD and DE. the water absorption characteristics and the changes rules of temperature field of the four stages of absorption as follows:

Stage AB is the contact stage. The bottom of the rammed earth touched water, the stage lasted a very short period, slight changes occurred at the bottom of the infrared thermal image.

Stage BC is the initial stage of the water absorption. Water absorption went up rapidly with the absorption velocity. the stage lasted for a very short period, slight changes occurred at the bottom of the infrared thermal image.

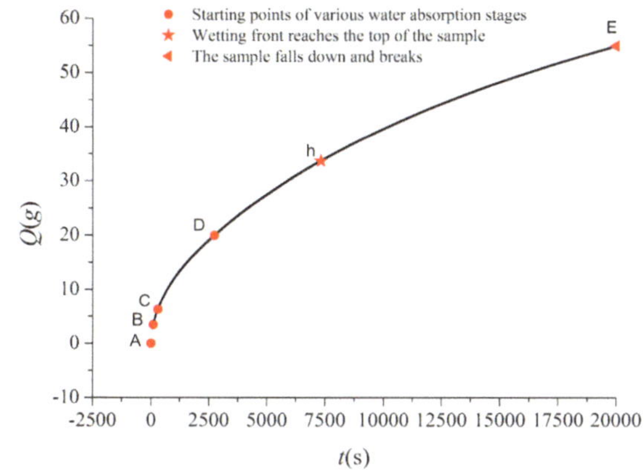

Fig. 2. The whole process of the water absorption curve of rammed earth

Stage CD is the water absorption stage with a changing velocity. In which, water absorption showed nonlinear growth and absorption velocity slowed down and gradually to 0. This stage is one of the main stages of water migration, accounting for about 15% of the whole process. Large changes took place in infrared thermal image in this stage. Temperature dropped drastically where the peak area passed.

Stage DE is the absorption stage with steady state and constant velocity. The water absorption showed linear increase and the absorption velocity did not change basically; the wetting front of rammed earth reached the top at Point h, when the rammed earth was still absorbing water; it broke at Point E and the absorption process came to an end. This stage is one of the main stages of water migration, accounting for about 85% of the whole process. Before Point h, the overall tendency of infrared thermal image temperature went down in this stage; temperature began to see consistency from the bottom between Point h and Point E, which meant that the whole sample tended to be saturate.

4.2 Analysis on Relation Between Water Absorption Increment and Infrared Radiation Characteristics

Many experiments had shown the close correlation between water absorption increment and temperature change of infrared radiation. Therefore, the temperature change rules of the infrared radiation at 1 g and 2 g were analyzed respectively to reveal the causal relationship between water absorption increment and infrared temperature characteristics.

The temperature changing rule on the surface of the rammed earth surface was analyzed when water absorption increment was 1 g. In the three stages of CD, Dh and hE, with rammed soil absorption increment 1 g as the time gradient, select the thermal image at different times, of which, $t10, t11, t12, t13, t14, t15$ were in Stage CD; $t20, t21, t22, t23, t24, t25, t26, t27, t28, t29$ in Stage Dh and $t30, t31, t32, t33, t34, t35, t36, t37, t38, t39$ in Stage hE. There were 26 moments totally, and $t10, t20, t30$ were the

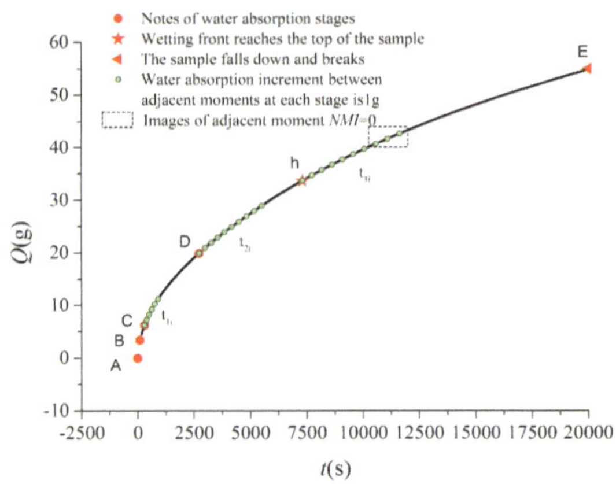

Fig. 3. Sketch map of water absorption characteristic curve and characteristic position

corresponding moments of Point C, D and h, which were the beginning points of Stage CD, Dh and hE (Fig. 3).

Infrared thermal images of the 6 moments of $t10, t11, t12, t13, t14, t15$ were extracted from Stage CD (as is shown in Fig. 4.). Based on the mutual information theory in Sect. 2, the entropy value of each thermal infrared image and the joint information entropy of the adjacent thermograms were calculated, and thereby the normalization of adjacent thermal images under the same increment of water absorption were obtained. The calculation results are shown in Table 1.

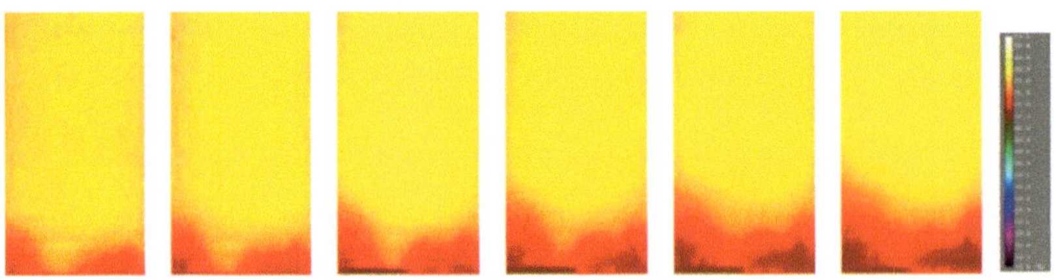

(a) $t_{10} = 300$ s (b) $t_{11} = 395$ s (c) $t_{12} = 500$ s (d) $t_{13} = 620$ s (e) $t_{14} = 750$ s (f) $t_{15} = 900$ s

Fig. 4. Infrared radiation temperature (CD stage)

Table 1. Calculation results of mutual information (CD stage)

Moment t_i(s)	300	395	500	620	750	900
Information entropy $H(t_i)$	5.371347	5.625987	5,685766	5.855466	5.778684	5.370109
Joint information entropy $H(t_i, t_{i+1})$	9.487079	9.748628	9.630516	9.687693	8.938239	/
$NMI(t_i, t_{i+1})$	0.274658	0.276372	0.331111	0.334611	0.396555	/

Note: $i = 10, 11, 12, 13, 14, 15$

It can be seen from Table 1 that in Stage CD, similarity between adjacent thermal images enlarges gradually, that is, distribution change of infrared radiation temperature of rock sample is getting smaller and smaller. This indicates that, with the time increasing of water absorption, for every increased 1 g of water absorption volume, the change in thermal images gets smaller and smaller gradually. By combining with the analysis on water adsorption curve characteristics, it can also be known that, the absorption velocity is very fast in Stage CD. With the pasture of time, volume of water passing of the rock sample grows up rapidly. Whenever water absorption increases 1 g, the dispersed volume in rock also grows bigger quickly. So the temperature distribution influence of moisture on adjacent thermal images becomes smaller and smaller.

Infrared thermal images of $t20, t21, \ldots, t29$, 10 moments in total were extracted from Stage Dh (as is shown in Fig. 5). Based on the same method as the CD segment is used to calculate the *NIM*. The calculation results are shown in Table 2.

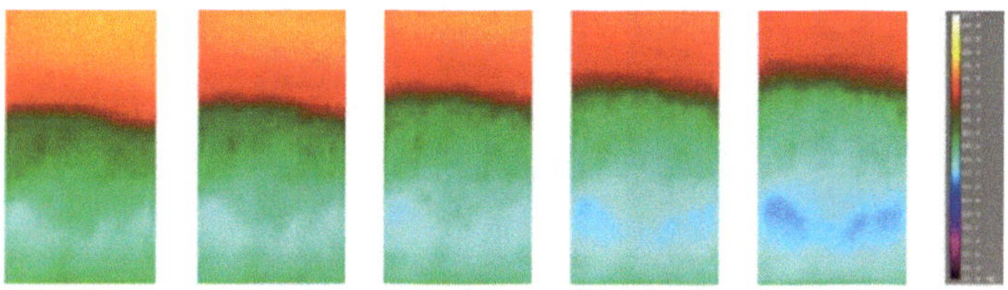

(a) $t_{20} = 2735$ s (b) $t_{21} = 3005$ s (c) $t_{22} = 3280$ s (d) $t_{23} = 3565$ s (e) $t_{24} = 3860$ s

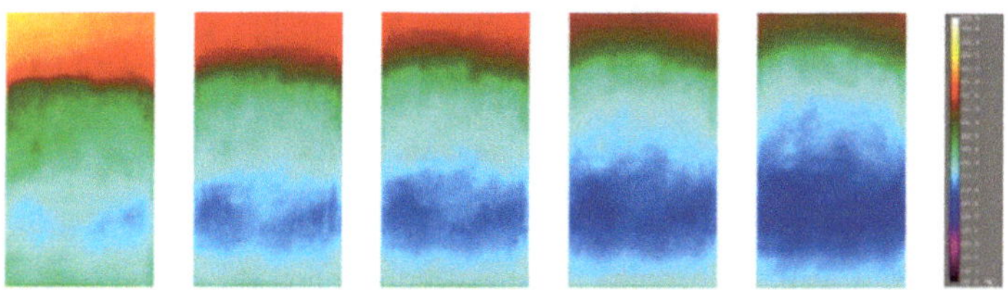

(f) $t_{25} = 4175$ s (g) $t_{26} = 4495$ s (h) $t_{27} = 4835$ s (i) $t_{28} = 5180$ s (j) $t_{29} = 5520$ s

Fig. 5. Infrared radiation temperature (Dh stage)

Table 2. Calculation results of mutual information (Dh stage)

Moment t_i(s)	2735	3005	3280	3565	3860	4175	4495	4835	5180	5520
Information entropy $H(t_i)$	6.552694	6.530062	6.540520	6.442925	6.347258	6.657930	6.400355	6.288786	6.458167	6.78907
Joint information entropy $H(t_i, t_{i+1})$	10.955929	11.080763	10.915058	10.819145	11.478292	11.699939	10.855203	10.800322	11.039712	/
$NMI(t_i, t_{i+1})$	0.325134	0.304473	0.318619	0.308211	0.234813	0.208044	0.289056	0.305427	0.333361	/

Note: $i = 20, 21, 22, \ldots, 29$.

By analyzing Table 2, it can be known that the similarity between adjacent thermal images decreases first and increases later, that is, the temperature drop of adjacent thermal images increases first and decrease later. Compared with conclusion of

Stage CD, it can be known that, after the initial period of rapid water absorption, the temperature of the rock sample dropped rapidly at early Stage Dh; the dropping slowed down at late Stage Dh, meaning that, with the pasture of time, volume of water passing of the rock sample grew up continuously. Temperature dropping velocity increased first and decreased later as total water absorption increased 1 g each time. The overall rule was not obvious and *MI* of adjacent thermal images was between 0.2 and 0.35.

Infrared thermal images of $t30, t31, \ldots, t39$, the 10 moments in total, were extracted from Stage hE (as is shown in Fig. 6). Based on the same method as the CD segment is used to calculate the *NIM*. The calculation results are shown in Table 3 and Fig. 7.

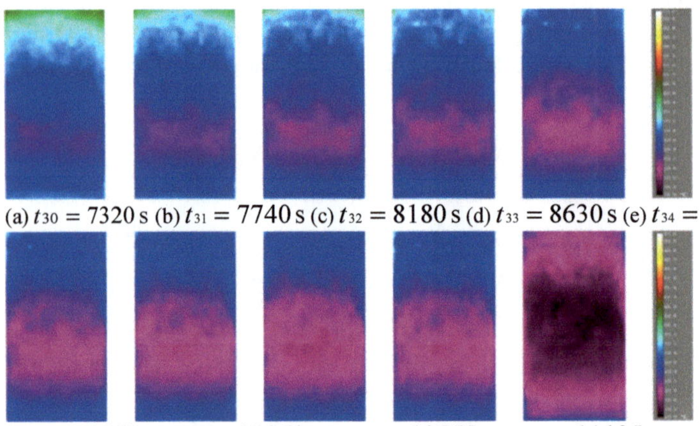

(a) $t_{30} = 7320$ s (b) $t_{31} = 7740$ s (c) $t_{32} = 8180$ s (d) $t_{33} = 8630$ s (e) $t_{34} = 9090$ s

(f) $t_{35} = 9565$ s (g) $t_{36} = 10060$ s (h) $t_{37} = 10575$ s (i) $t_{38} = 11105$ s (j) $t_{39} = 11650$ s

Fig. 6. Infrared radiation temperature (hE stage)

Table 3. Calculation results of mutual information (hE stage)

Moment t_i(s)	7320	7740	8180	8630	9090	9565	10060	10575	11105	11650
Information entropy $H(t_i)$	6.817765	6.682700	6.610790	6.508592	6.240700	6.203141	6.146832	6.072168	6.139289	6.503203
Joint information entropy $H(t_i, t_{i+1})$	11.333656	10.906165	10.552496	10.828472	10.399089	10.087253	10.013134	12.215039	12.642192	/
$NMI(t_i, t_{i+1})$	0.320998	0.359172	0.391312	0.301322	0.328637	0.366433	0.361055	0	0	/

Note: $i = 30, 31, 32, \ldots, 39$.

By analyzing Table 3 and Fig. 7, it can be known that, before moment t_{37}, similarity between adjacent images has no significant change. It indicates that, before t_{37}, for every 1 g of total water absorption increasing, the temperature drops almost by the same degree. On the other hand, it can be noticed that the wetting front has reached the top of the rock sample at Point h (moment t_{30}) in curve of water absorption, which means that the rock sample surface is totally infiltrated. It can be inferred that in period of $t_{30} \sim t_{37}$, volume of water passing is not changed. So whenever water absorption

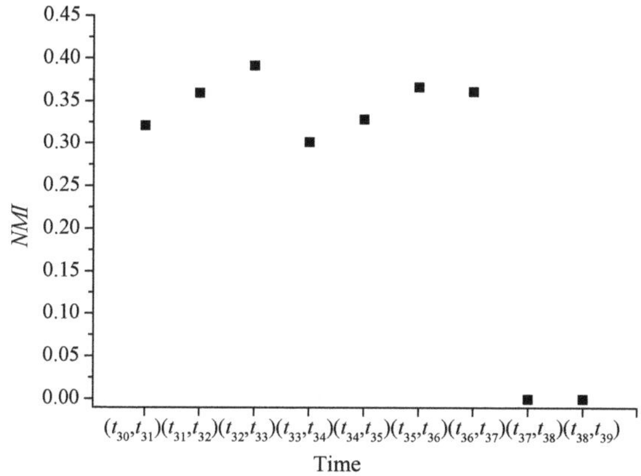

Fig. 7. *NMI* value of thermal image at adjacent time (hE stage)

increases 1 g, temperature drops almost by the same degree. That is, at moment t_{37}, water has reached the top of both the rock surface and its inside.

Especially, *NMI* between adjacent thermal images is 0 at the three moments of t_{37}, t_{38} and t_{39}. This means that adjacent thermal images are uncorrelated. Before and after moment t_{38}, the increase of every 1 g of water causes large temperature change. Though rock keeps on absorbing water after that moment, its physical essence has changed, predicting that the rock is converting from unsaturation to complete saturation at the time.

The same comparative analysis was made on the variation of the temperature field of infrared radiation under the condition of 2 g water absorption increment. The conclusion is consistent with that under the condition of 1 g water absorption increment, here no longer expatiatory.

Based on the above analysis, it can be known that in Stage CD $\sim$ Dh, due to the ever-increasing volume of water passing, temperature dropping range is closely linked to the increased volume of water passing; under the boundary conditions limited by the experiment, the rule of temperature drop and water absorption increment was not obvious under the influence of the ever-changing velocity of volume of water passing. At the stage from around Point h to Point E, it can be judged from the experimental phenomena that rock at the stage has been wet from inside to outside, which means that when volume of water passing does not change, *NMI* fluctuates at about 0.3 when water absorption increment is 1 g and 2 g and no obvious changes occur. All these tell us that water absorption increment is in positive correlation to temperature dropping, which is in consistency to infrared radiation physics law.

It is noteworthy that around a certain moment at Stage hE, *NMI* between adjacent thermal images was 0, predicting that a part of the water in rock has conversed from bound water to free water after absorption and the rock is transiting from unsaturation to complete saturation.

5 Conclusions

Targeting for the description of water occurrence status in the rammed earth of Mogao Grottoes, the infrared thermography was applied to study the relationship between water absorption increment and infrared radiation temperature at various water absorption stages. And the conclusions were obtained as below:

Under the mechanical boundary conditions limited by the experiment, before the infiltration of all rock samples, due to the impact of the increasing volume of water passing, the overall relationship between water absorption increment of unit mass of dry rocks and infrared radiation temperature was positive, but no obvious regular difference was shown in various water absorption stages;

In the process when the whole rock sample was completely infiltrated and kept absorbing water, the moment could be judged as the point when water inside of the rock sample transited from absorbed water to free water and the rock sample turned form unsaturated state to complete saturated state based on absorption process of water molecules in mineral if right before and after the very moment, mutual information of adjacent thermal images was 0. This meant that by using *NMI* of adjacent thermal images, the moment when porous medium turned from unsaturated state to complete saturated state could be judged.

Acknowledgements. This work was supported by the National Natural Science Foundation for Young Scientists of China [Grant Nos. 51604276 and 41502323].

References

1. Kuchitsu, N., Ishizaki, T., Nishiura, T.: Salt weathering of the brick monuments in Ayutthaya. Eng. Geol. **55**(1), 91–99 (1999)
2. Angeli, M., Bigas, J.P., Benavente, D., et al.: Salt crystallization in pores: quantification and estimation of damage. Environ. Geol. **52**(2), 205–213 (2007)
3. Lerma, J.L., Cabrelles, M., Portalés, C.: Multitemporal thermal analysis to detect moisture on a building facade. Constr. Build. Mater. **25**(5), 2190–2197 (2011)
4. Ye, D., Hongjun, G., Lei, Q., et al.: Application of infrared thermography in the water-proof and leakage detection of underground buildings. China Building Waterproofing **5**, 33–36 (2014)
5. Wu, L., Liu, S., Wu, Y.: Infrared imaging detection of water permeation on field large-scale rock relics. IEEE Trans. Geosci. Remote Sens. **49**(1), 581–590 (2010)
6. Grinzato, E., Ludwig, N., Cadelano, G., et al.: Infrared thermography for moisture detection: a laboratory study and in-situ test. Mater. Eval. **69**(1), 97–104 (2011)
7. Shakmak, B., Al-Habaibeh, A.: Detection of water leakage in buried pipes using infrared technology; a comparative study of using high and low resolution infrared cameras for evaluating distant remote detection. In: Applied Electrical Engineering & Computing Technologies, pp. 1–7 (2015)
8. Mercuri, F., Cicero, C., Orazi, N., et al.: Infrared thermography applied to the study of cultural heritage. Int. J. Thermophys. **36**(5–6), 1189–1194 (2015)
9. Mingde, D., Zongfei, F., Xiaohong, L., et al.: Research on the action of water in the infra-red radiation of rocks. Earthq. Res. China **13**(3), 288–296 (1997)

10. Huihui, Z.: The experimental study on the use of infrared thermal imaging for nondestructive detection of deterioration disease in stone relics. J. Liaoning Prov. College Commun. **15**(06), 20–22 (2013)
11. Yuhua, W., Shanjun, L.: Infrared imaging detection on water-permeation of rock paint-ings. Geotech. Invest. Surv. **38**(05), 31–35 (2010)
12. Avdelidis, N.P., Moropoulou, A., Theoulakis, P.: Detection of water deposits and movement in porous materials by infrared imaging. Infrared Phys. Technol. **44**(3), 183–190 (2003)
13. Furmański, P., Wiśniewski, T., Wyszyńska, E.: Detection of moisture in porous materials through infrared methods. Arch. Thermodyn. **29**(1), 19–40 (2008)
14. Gayo, E., Frutos, J.D.: Interference filters as an enhancement tool for infrared thermography in humidity studies of building elements. Infrared Phys. Technol. **38**(4), 251–258 (1997)
15. Yong, W.: Study of the machine learning methods based on mutual information and prior knowledge. Institute of Automation, Chinese Academy of Sciences (2014)
16. Yu, Z.: Scene matching techniques based on mutual information and spatial information. Xidian University (2013)
17. Zhengwei, X.: Medical image registration based on normalized mutual information. Kunming University of Science and Technology (2014)

Applications of Rock Mechanics and Case Studies

A Dynamic Rock Mass Classification Method for TBM Tunnel

Ya-dong Xue[1,2(✉)], Xing Li[1,2], Zhen-xing Diao[3], Feng Zhao[1,2], and Han-xiang Zhao[1,2]

[1] Key Laboratory of Geotechnical and Underground Engineering of Education Ministry, Tongji University, Shanghai 200092, China
yadongxue@126.com
[2] Department of Geotechnical Engineering, Tongji University, Shanghai 200092, China
[3] Shanghai Nuclear Engineering Research and Design Institute Co. Ltd., Shanghai 200233, China

Abstract. As the traditional rock mass classification is not suitable for TBM tunnel, the comprehensive rock mass classification for TBM is established by combing the boreability classification and adaptability classification, which are accomplished respectively by analyzing TBM performance. In order to evaluate surrounding rock continuously, Markov Chain method is used in this paper. Four geological parameters are chosen to describe the ground conditions. Based on YHJW project, we can get the state of each parameter through geological investigation. Counting the state transformation along the alignment at 50 m intervals, the state transformation probability matrix of each parameter is obtained. Comparing the parameter state in the same position between investigation data and revealed condition, the likelihood matrix of each parameter in Qinling Region is acquired. Probability of boreability classification of K29 + 900 ~ K31 + 150 is computed and the expectation of thrust is considered as the thrust in prediction. The high penetration rate appears when the field thrust is in the range of predicted thrust. Dynamic Rock Mass Classification Method is proved effective. Tunneling parameters can be predicted preliminary through the new classification.

Keywords: Rock mass classification · TBM · Boreability · Adaptability Markov chain · Performance prediction

1 Introduction

Rock mass classification is the determines of tunnel design and construction. Traditional rock classification methods usually focus on stability of surrounding rock. While using Tunnel Boring Machine (TBM), in order to achieve optimal advance rate (AR) and economic benefits, we should consider balancing the relationship between rock boreability and stability. Traditional rock mass classification methods can not reach the requirements in such situation [1].

Compared to other construction methods, TBM is more sensitive to the variation of ground condition, the engineers need to adjust tunneling parameters with the change of surrounding rock [2]. The present engineering geology investigation can't afford

L. Zhang et al. (Eds.): GSIC 2018, *Proceedings of GeoShanghai 2018 International Conference: Rock Mechanics and Rock Engineering*, pp. 319–328, 2018.
https://doi.org/10.1007/978-981-13-0113-1_34

successive and accurate information for design, except for several drilling holes. Geological uncertainty leads to frequent halting and even the damage to TBM. This makes a noticeable impact on TBM performance [3]. Though advanced geological prospecting can ameliorate this situation, its application is also limited by high time cost as well as economic profit loss. There are many soft methods proposed by researchers to complement the results of engineering geology investigation, such as the time series approach, neural networks approach, support vector machine networks and discrete-state continuous-space Markov process [4–7].

By analyzing the construction data supported by HANJIANG-TO-WEIHE RIVER VALLEY WATER DIVERSION PROJECT (YHJW Project) in China's Shaanxi Province, a rock mass classification system is proposed based on excavation mechanism and penetration performance of TBM. It can be used to evaluate the surrounding rock dynamically by Markov chain approach. Depending on the results of the classification, tunneling parameters and TBM performance could be predicted preliminarily, and further construction organization can be adjusted in time.

2 Comprehensive Rock Mass Classification Method Based on TBM Performance

2.1 Penetration Characteristics of TBM

The performance of TBM and benefit of the project are reflected by *AR*, which refers to the accumulated footage per day (m/day) and is calculated as:

$$AR = PR * U \tag{1}$$

PR is the penetration rate and expressed in m/h, which refers to linear excavation footage per unit time when TBM is in operation. And *U* is utilization rate (%) of the machine, representing the ratio of driving time to total construction time. *PR* is associated with boreability of rock mass. And from the point of the lowest energy, there are optimal tunneling parameters for their corresponding ground condition [8]. *U* reflects the geology adaptability for TBM and is often reduced by high quartz rock, fault fracture zone, squeezing deformation of soft rock, rock burst, etc.

Excavation of TBM tunnel is a result of the interaction between machine and surrounding rock. This process is affected by boreability of rock mass and TBM adaptability to the geology at the same time. And the above two factors are in opposition to each other. For example, low-strength rock is fairly easy to be broken and low thrust is required, while the time consumed by supporting is enormous. Therefore, in order to estimate the geology condition, boreability and adaptability should be analyzed respectively, and then the whole effection can be evaluated comprehensively.

2.2 Boreability Classification of Rock Mass

The boreability of rock mass refers to the degree of their resistance to the break of cutters, and the consumption of cutters. This ability is decided by character of a certain

Table 1. Boreability classification of rock mass and prediction of tunneling parameters

Boreability classification	FPI (kN/mm/rev)	Group of data	Prediction of thrust (kN)	Prediction of Penetration (mm/rev)
1	<50	19	4500～8500	14～24
2	50～100	163	7000～9500	12～16
3	100～150	105	9000～12000	9～13
4	150～300	79	10000～17000	7～10
5	300～600	82	16000～21000	4～7
6	600～1000	11	18000～22000	3～4

rock. Since *PR* relates to tunneling parameters in a given ground condition, *FPI* (field penetration index) is defined to describe this property. The analysis of field data indicates that FPI serves as a link between geological parameters and tunneling parameters, and can be set as the classification index [9].

The present engineering geology investigation in China complies with *Standard for engineering classification of rock mass*. When evaluating geological conditions for underground engineering, the basic geological parameters mainly includes standard values of rock-saturation uniaxial compressive strength (*Rc*), intactness index of rock mass (*Kv*), groundwater state (*W*), initial geo-stress state (*Gs*), etc. [10]. Before advanced geology investigation methods are adopted in national standard, the classification parameters are limited by survey methods.

We performed a regression analysis using field data of YHJW project. And a fitting formula was obtained as follows:

$$FPI = 12.35EXP(0.594 + 0.198Rc + 1.65Kv) \tag{2}$$

When $Rc > 90\ kV + 30$, $Rc = 90\ kV + 30$, *and when* $Kv > 0.04Rc + 0.4$, $Kv = 0.04 + 0.4$.

Statistics suggest that tunneling parameters are approximately follow normal distribution in FPI segments (at 50 m intervals). According to FPI segments, boreability classification of rock mass can be accomplished. And the corresponding tunneling parameters can be predicted preliminary as shown in Table 1.

2.3 Adaptability Classification of Rock Mass

The adaptability of TBM to geological conditions is embodied in TBM utilization. Many researches on TBM performance prediction indicate that it is quite difficult to estimate the utilization due to the numerous factors, as well as their uncertainty and complex relationship [2, 11]. It is necessary to judge the influence on utilization by each specific factor, mainly including support, wear of cutters, mucking, etc. The relative geological parameters are *Rc*, *Kv*, *W* and *Gs*.

According to the daily tunneling logs of YHJW project, and based on fault tree model, Xue [12] analyzed the time consumption of each construction event caused by the parameters listed above. The influence on utilization by different state of single

Table 2. Grades of each parameter condition.

Rc(MPa)	<30	30~60	60~100	100~150	>150	
Grades A1	2~4	4~6	6~8	4~6	2~4	
Kv	<0.35	0.35~0.45	0.45~0.75	0.75~0.85	>0.85	
Grades A2	2~4	4~6	6~8	4~6	2~4	
W	dry	wet~seepage	drip	linear flow	stream	water burst
Grades A3	8~9	10	6~8	4~6	2~4	1~2
GS(MPa)	0~5	5~30	>30			
Grades A4	10	4~6	1~2			

Table 3. Adaptability classification of rock mass and prediction of utilization.

Adaptability classification	1	2	3	4	5
Total grades (A1 + A2 + A3 + A4)	36 ~ 46	28 ~ 35	19 ~ 27	10 ~ 18	0 ~ 9
Prediction of utilization	32% ~ 40%	24% ~ 32%	16% ~ 24%	8% ~ 16%	0 ~ 8%

parameter is graded as shown in Table 2. Adaptability of rock mass is classified into 5 levels based on the sum of each grade. The results are listed in Table 3.

2.4 Comprehensive Classification of Rock Mass Based on TBM Performance

Another important purpose of rock classification is to predicate advance rate, which is calculated as follows:

$$AR = P * N * U * 60/1000 \tag{3}$$

with N (rev/min) is revolving speed of the cutter head, set as ~5 rev/min.

Comprehensive Rock Mass Classification Matrix for TBM Tunnel

Grades of adaptability	FPI	<50	50-100	100-150	150-300	300-600	>600	Prediction of Utilization
	Boreability / Adaptability	1	2	3	4	5	6	
36-46	1	I	I	I	II	II	III	32%-40%
28-35	2	I	I	II	II	III	III	24%-32%
19-27	3	I	II	II	III	III	IV	16%-24%
10-18	4	II	III	III	III	IV	IV	8%-16%
0-9	5	III	IV	IV	IV	IV	IV	0-8%
Prediction of thrust（kN）		4500-8500	7000-9500	9000-12000	10000-17000	16000-21000	18000-22000	
Prediction of penetration（mm/rev）		14-24	12-16	9-13	7-10	4-7	3-4	

（1）FPI=12.35exp(0.594+0.0198Rc+1.65Kv)
When Rc>90Kv+30，Rc=90Kv+30. When Kv>0.04Rc+0.4 时，Kv=0.04Rc+0.4.
（2）Grades of adaptability =A1+A2+A3+A4

Fig. 1. Comprehensive rock mass classification matrix.

According to Eq. 3, Tables 1 and 3, daily average *AR* (m/h) can be computed. A comprehensive classification matrix is concluded in Fig. 1.

3 Dynamic Rock Mass Classification Method for TBM Tunnel Based on Markov Chain

Usually, a geological parameter is defined to describe one kind of rock property. Single parameter state varies with the advance and is only influenced by the state in last location, i.e., it is independent of state in other previous locations. Therefore, the transformation of one single parameter can be regarded as a Markov process [7]. This random process is expressed as follows:

$$P[X(t_{i+1})|X(t_i) = x_i,\ X(t_{i-1}) = x_{i-1}, \ldots] = P[X(t_{i+1})|X(t_i) = x_i] \tag{4}$$

where t_i is the location in sequence of tunneling, and x_i is the state of parameter in t_i.

The state distribution of a parameter is called state space of a parameter, covering all status of this aspect's characters. In a certain location, a parameter might be in any state with corresponding probabilities. The whole condition of rock mass is decided by arrangement of all parameters' state, transforming with the changing of any of them. This arrangement can be expressed by geology vector (a row vector).

For each parameter, probability of each possible state is expressed by state distribution vector $\mathbf{s}(t_i)$, which can be looked up in likelihood matrix $\mathbf{L}$ of this parameter:

$$\mathbf{L} = \left[l_{jk}\right],\ where\ l_{jk} = P[Y(t_i) = k|X(t_i) = j] \tag{5}$$

l_{ab} refers to the probability that the investigation state $X(t_i)$ of a parameter is j in location t_i, but the real state $Y(t_i)$ revealed later is k. Likelihood matrix $\mathbf{L}$ reflects the accuracy of geology investigation. For a certain region and a certain geological investigation method, $\mathbf{L}$ is supposed as constant.

Suppose that parameter state is j in location t_i, it might transfer to any states within its state space in the next location t_{i+1}. Element Tr_{jk} of transition probability matrix $\mathbf{Tr}$ denotes the probability that state in t_i is j and in t_{i+1} is k, shown as follows:

$$\mathbf{Tr} = \left[Tr_{jk}\right],\ where\ Tr_{jk} = P[X(t_{i+1}) = k|X(t_i) = j] \tag{6}$$

Based on Bayes' theorem, when the parameter state revealed in location between t_i and t_{i+1} is q, $\mathbf{Tr}$ can be updated to $\mathbf{Tr'}$ dynamically by Eq. 7.

$$\mathbf{Tr}' = \left[Tr'_{jk}\right],\ where\ Tr'_{jk} = Tr_{jk}l_{kq}/\left(\sum Tr_{jk}l_{kq}\right) \tag{7}$$

The probability of parameter state in location t_{i+1} can be calculated as follows:

$$s(t_{i+1}) = s(t_i)\mathbf{Tr}'(t_i - t_{i+1}) \tag{8}$$

While some difficulties occur when applying this method in TBM tunnel. Firstly, since the machine is more sensitive to variation of surrounding conditions, the states of parameters should be divided more delicate. Further, TBM tunnel is always longer, therefore, more complex conditions could be encountered along the alignment. Geology data are more limited because the geology investigation is more difficult. Finally, the classification should be relevant to the tunneling parameters.

3.1 State Space Definition and Division of Geological Parameters

Four parameters are defined in Sect. 2 as standards of classification. On account of the influence on tunneling, state space of each single parameter is divided as presented in Table 4. The division criterion is based on field data and construction logs.

From the combination of different parameters' state, there are 300 different geology vectors. The FPI value and adaptability grades of each geology vector can be calculated according to Eq. 2 and Table 2. Then the geology vector can be classified into corresponding boreability, adaptability, and comprehensive classification based on Fig. 1.

Coring is conducted at 50 m intervals following the excavation in YHJW project. Thus, observation points are set at 50 m intervals as well. On the basis of field-acquired geological profile and longitudinal section of YHJW project (about an 82 km tunnel), we can get the state of each parameter in every observation point according to Table 4. Counting the state transformation along the alignment, the frequency of transformation among these states is acquired. For example, the state of Rc transforms 38 times from 1 to 1, 28 times from 1 to 2, and 0 times from 1 to other state. The corresponding transformation frequency of Rc from 1 to 1, 2, 3, 4, 5 is 0.5758, 0.4242, 0, 0 and 0 respectively. Supposing the state transformation frequency as the state transformation probability, the state transformation probability matrix of each parameter is obtained respectively, as listed in Table 5.

The daily tunneling logs (about 12 km excavated tunnel) record actual geological information. Comparing the parameter state in each observation point between investigation data and revealed condition, the likelihood matrix of each parameter in Qinling Region can be acquired (as shown in Table 6). In this 12 km section, according to geological investigation, state of W is 1 in 96 observation points. In these points, actual state of W revealed after excavation is 1 in 57 points, 2 in 29 points, 3 in 8 points and 4

Table 4. State definition of single parameter

State	1	2	3	4	5
Rc(MPa)	<30	30 ~ 60	60 ~ 100	100 ~ 150	>150
Kv	<0.35	0.35 ~ 0.45	0.45 ~ 0.75	0.75 ~ 0.85	>0.85
W	dry	wet ~ drip	linear flow	water burst	–
Gs(MPa)	0 ~ 5	5 ~ 30	>30	–	–

Table 5. State transformation probability matrix of each parameter.

Tr-Rc	0.5758	0.4242	0	0	0	Tr-Kv	0.5758	0.4242	0	0	0
	0.0841	0.7855	0.1275	0.0029	0		0.0841	0.7855	0.1275	0.0029	0
	0	0.0616	0.9260	0.0123	0		0	0.0616	0.9260	0.0123	0
	0	0.0023	0.0208	0.9746	0.0023		0	0.0023	0.0208	0.9746	0.0023
	0	0	0	0.0164	0.9836		0	0	0	0.0164	0.9836
Tr-W	0.9716	0.0258	0.0026	0		Tr-GS	0.8235	0.1765	0		
	0.0299	0.9179	0.0507	0.0015			0.0137	0.9817	0.0046		
	0.0145	0.2464	0.5870	0.1521			0	0.0083	0.9917		
	0	0.0196	0.4118	0.5686							

in 2 points. This means when the state of W in a location indicated by geological investigation is 1, the real state might be 2 with a probability of 0.3.

3.2 Application in YHJW Project K29 + 900 ~ K31 + 150 Section

There is a borehole CZK-4 in K29 + 900, and the geology condition judged by investigation can be parameterized by geology vector (4, 4, 1, 2). The boreability classification is 6 and the adaptability classification is 3 in this location. According to Table 6, the corresponding state probability vector $\mathbf{s}(t_1) = (0.75, 0.75, 0.6, 0.7)$. Then we can list the state distribution probability of each parameter according to Table 6. The state distribution probability vector of each parameter in posterior location can be computed as Eq. 8.

Table 6. Likelihood matrix of each parameter.

L-Rc	0.80	0.15	0.05	0	0	L-Kv	0.80	0.15	0.05	0	0
	0.10	0.75	0.10	0.05	0		0.10	0.75	0.10	0.05	0
	0.05	0.10	0.70	0.10	0.05		0.05	0.10	0.70	0.10	0.05
	0	0.05	0.10	0.75	0.10		0	0.05	0.10	0.75	0.10
	0	0	0.05	0.15	0.80		0	0	0.05	0.15	0.80
L-W	0.60	0.30	0.08	0.02		L-GS	0.80	0.15	0.05		
	0.10	0.75	0.10	0.05			0.15	0.70	0.15		
	0.05	0.10	0.75	0.10			0.02	0.08	0.90		
	0.01	0.04	0.15	0.80							

The state with the highest probability of each parameter forms the geology vector in a location. The probability of each geology vector is the product of the state probabilities of the four parameters related to it. The probability of a classification is the sum of probabilities of all geology vectors belong to this classification. And when the TBM arrived at location t_n, the revealed geology condition is different from the information supplied by investigation, state distribution probability vector should be updated dynamically by Eq. 7.

From K29 + 900 to K31 + 150, the probability of each classification is computed based on the above method at 50 m intervals. The results are shown in Fig. 2. The boreability classification can be considered as 6. The probability of classification 6 is decreasing along the alignment. The rock mass condition of boreability is in transition from classification 6 to classification 5. Intuitively, the rock mass will be easier to excavate and lower thrust is needed to break rock.

According to Figs. 1 and 2, the expectation of thrust $E(T)$ can be calculated as follows:

$$E(T) = \sum p(b_i)\, Th\,(b_i) \tag{9}$$

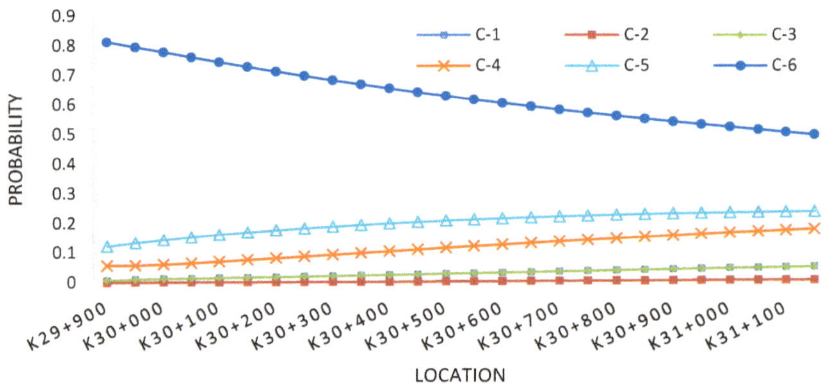

Fig. 2. Probability classification of borabability.

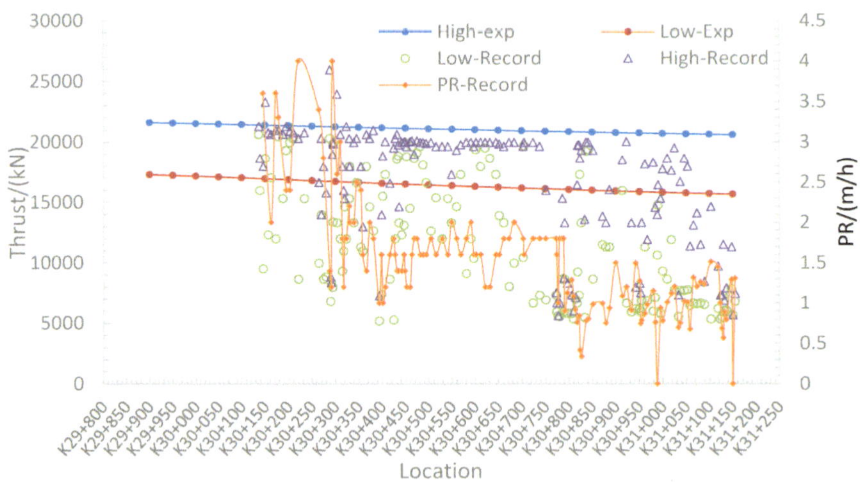

Fig. 3. Expectation of thrust and analyzation field data.

where $p(b_i)$ is the probability of each boreability classification in this location, $Th(b_i)$ is the thrust range offered by Fig. 1. The result of $E(T)$ and the thrust recorded by TBM are listed in Fig. 3. When actual thrust is within the expectation thrust range in a high density, the penetration rate is higher. And when the most actual thrust is out of the range, the penetration rate is quite low.

4 Conclusion

Based on field data of geology and TBM performance, an innovative comprehensive rock mass classification method is formed. After parameterizing the geological information, Markov Chain method is used to computing the state distribution of parameters at 50 m intervals. The probability of each classification can be calculated. The expectation of thrust is considered as the thrust in prediction. Comparing the thrust in prediction with the field thrust, the high penetration rate appears when the field thrust is in the range of predicted thrust. This method is proved effective.

All calculations can be computed by program completely. When the program is applied in construction site, technicians only need to input the new geology information revealed by excavation, the results will be given by program or apps automatically.

Acknowledgement. The authors acknowledge the support of National Natural-Science Foundation of China (Grant No.41072206), Fundamental Research Funds for the Central Universities (Grant No.0200219209) and the support of Science and Technology Plan of Department of Communication of Zhejiang province (Grant No.2015J05).

References

1. Liu, Q.S., Liu, J.P., Pan, Y.C., Kong, X.X., Hong, K.R.: A case study of TBM performance prediction using a Chinese rock mass classification system – hydropower classification (HC) method. Tunn. Undergr. Space Technol. **65**, 140–154 (2017)
2. Rostami, J.: Performance prediction of hard rock tunnel boring machines (TBMs) in difficult ground. Tunn. Undergr. Space Technol. **57**(57), 173–182 (2016)
3. Frough, O., Torabi, S.R., Yagiz, S., Tajik, M.: Effect of rockmass conditions on TBM utilization factor in Karaj–Tehran water conveyance tunnel. In: World Tunneling Conference (2012)
4. Jeon, J.S., Martin, C.D., Chan, D.H., Kim, J.S.: Predicting ground conditions ahead of the tunnel face by vector orientation analysis. Tunn. Undergr. Space Technol. **20**(4), 344–355 (2005)
5. Alimoradi, A., Moradzadeh, A., Naderi, R., Salehi, M.Z., Etemadi, A.: Prediction of geological hazardous zones in front of a tunnel face using TSP-203 and artificial neural networks. Tunn. Undergr. Space Technol. **23**(6), 711–717 (2008)
6. Lau, S., Lu, M., Ariaratnam, S.T.: Applying radial basis function neural networks to estimate next-cycle production rates in tunnelling construction. Tunn. Undergr. Space Technol. **25**(4), 357–365 (2010)
7. Guan, Z.C., Deng, T., Du, S., Li, B., Jiang, Y.: Markovian geology prediction approach and its application in mountain tunnels. Tunn. Undergr. Space Technol. **31**(5), 61–67 (2012). Incorporating Trenchless Technology Research

8. Wang, L.H., Kang, Y.L., Zhao, X.J., Zhang, Q.: Disc cutter wear prediction for a hard rock TBM cutterhead based on energy analysis. Tunn. Undergr. Space Technol. **50**, 324–333 (2015)
9. Hassanpour, J., Rostami, J., Zhao, J.: A new hard rock TBM performance prediction model for project planning. Tunn. Undergr. Space Technol. **26**(5), 595–603 (2011)
10. The Ministry of Water Resources of P.R. China.: Standard for engineering classification of rock mass GB/T 50218-2014. 1st edn. China Planning Press, Beijing (2015). (in Chinese)
11. Delisio, A., Zhao, J.: A new model for TBM performance prediction in blocky rock conditions. Tunn. Undergr. Space Technol. **43**(7), 440–452 (2014). Incorporating Trenchless Technology Research
12. Xue, Y.D., Diao, Z.X., Zhao, F.: Analysis of TBM performance and disc cutter consumption in Yinhanjiwei water conveyance tunnel project. In: World Tunnel Conference 2016, San Francisco, USA 22–29 April 2016

Karst Development and Hydrogeological Characteristics of Typical Limestone Formations in Guizhou, China

Zhong-Mei Wang[1(✉)] and Liang-Bo Hu[2]

[1] Guizhou University, Guiyang 550025, Guizhou, China
re.zmwang@gzu.edu.cn
[2] University of Toledo, Toledo, OH 43606, USA
liangbo.hu@utoledo.edu

Abstract. Guizhou Province of southwest China is known for its extensive karst development, where carbonate rocks are strongly featured. The geological properties and hydrogeological features of the widely distributed carbonate rocks are particularly relevant to evolution and exploration of underground water resources. This paper is primarily focused on a field study of various dissolution features developed in limestone formations which originated from three different geological periods. It was found that the most mature karst development occurred in the Permian Qixia-Maokou Formation, indicating the strongest potential for water resource exploitation among various geological formations. The hydrogeological characteristics of these formations were also studied, focusing on their subterranean river discharge rate, runoff module number and specific capacity. The results also indicate the large water-bearing capacity of Permian Formation of carbonate rocks in Guizhou.

Keywords: Karst · Water resources · Rock properties · Hydro-geology Guizhou

1 Introduction

Guizhou is a southern province of China and 73.6% of its area is formed by karst landscapes (Zhou et al. 2012), featuring an abundance of subterranean drainage systems in the karstic limestone and dolomite. The potential large capacity of geological formations associated with underground cavities, conduits and fissures is a main characteristic of karst (e.g., Ford and Williams 2007; Worthington and Ford 2009; Qian and Wang 2011). For example, approximately 25% of the world's population rely on karst groundwater for drinking water resources (e.g., Kolliarits et al. 2006; Ford and Williams 2007).

In Guizhou Province, karst groundwater is a major source for water supply; it provides around 21.5 billion m^3 of water, over 80% of the overall groundwater resources of Guizhou. Karstic rocks in Guizhou mainly feature carbonates with various lithofacies which are widely distributed across the province. Evaluation of the geological properties of the carbonate rocks in the karst terrain can have strong economic and social impacts in the water resources management in Guizhou. This topic has

L. Zhang et al. (Eds.): GSIC 2018, *Proceedings of GeoShanghai 2018 International Conference: Rock Mechanics and Rock Engineering*, pp. 329–337, 2018.
https://doi.org/10.1007/978-981-13-0113-1_35

received tremendous interest among the research communities in recent years. Gao (1983) investigated the effects of Neotectonic movement in Guizhou's karst development. (Jin and Feng 1994) studied the causes of Permian limestone in Yunnan-Guizhou areas. Luo and Song (2004) surveyed various methods of water resources identification on the deep river valley in Guizhou. Deng (2015) examined karst development characteristics and controlling factors in Bijie County. The relationship between limestone dissolution characteristics and karst groundwater in Guizhou was also recently studied by Wang (2017) and Wang et al. (2017).

The present study is focused on three major geological formations widely distributed in Guizhou and their relevant geological properties related to the karst development and water capacity potential. Each formation is characterized by its geological age. The results of field sampling and testing revolving around the fissures, solution pits, grooves and caves in the limestone rocks, characteristic of karst development, are reported and analyzed. Subsequently the water-bearing capacities of the studied geological formations are studied based on the geo-hydrological records collected in the last few years. The correlation between karst development and water-bearing capacity is then examined among the three major geological formations to identify the potential implications on water resources exploration.

2 Dissolution Features in the Karst Development

Three most widely distributed limestone strata in Guizhou were studied in this investigation: the Triassic Maocaopu Limestone Formation (T_1m), the Permian Qixia-Maokou Limestone Formation (P_2q_+m) and the Carboniferous Huanglong-Maping Limestone Formation ($C_{2-3}h_+m$), each of which is denoted with the symbol of its geological period and conventionally named in association with its specific location. The Triassic Maocaopu Limestone Formation features the interbeded limestone/dolomite; its depth ranges from 378–643 m, mainly distributed over Zunyi and Xifeng Counties. The Permian Qixia-Maokou Limestone Formation is primarily composed of a combination of thick, massive, crystalline, biogenic limestone and thin to medium limestone; its depth ranges from 500–870 m, widely distributed over the entire studied area. The Carboniferous Huanglong-Maping Limestone Formation features crystalline, micritic, bioclastic carbonate and thick-layered structure of fine-grained limestone; its depth ranges from 50–1240 m, located in the south of Bijie-Fuquan area.

The primary goal of the study was to determine the amount and the density of fissures, as well as locations of dissolution pits, grooves and caves, in order to identify the maturity and extent of the karst development in these formations. In total 21 boreholes of the three strata were drilled and investigated, 10 in the Triassic Maocaopu Formation, 7 in the Permian Qixia-Maokou Formation and 4 in the Carboniferous Huanglong-Maping Formation. It should be noted that the choice of boreholes numbers and locations in each formation was largely dictated by conditions and environments suitable for drilling at the time of field study. Typical depth of boreholes ranges from 150–200 m. Underground borehole television videos were employed in the field study. Results are summarized in Table 1 and Fig. 1 shows the distribution of the limestone dissolution for each formation.

2.1 Characteristics of Dissolution Features Among the Studied Limestone Formations

Table 1 summarizes the total number of dissolution features and in particular, the density of dissolution fissures at each borehole. The dissolution features resulting from limestone dissolution include dissolution pits (or so-called holes or pores) which are characterized by small dissolution holes of a few centimeters, dissolution fissures, dissolution grooves and dissolution caves which are reflective of advanced dissolution and karst development. The density of features is measured in terms of the number of features per meter depth of the formation.

Table 1. Summary of dissolution features in different limestone strata.

Stratum	Borehole	Number of features	Density (Number of features per meter)
Triassic Maocaopu formation (T_1m)	Xifeng #1	15	0.094
	Xifeng #2	13	0.065
	Xifeng #3	18	0.100
	Xifeng #4	17	0.089
	Xifeng #5	17	0.100
	Xifeng #6	13	0.081
	Xifeng #7	12	0.08
	Xifeng #8	14	0.088
	Xifeng #9	14	0.082
	Xifeng #10	21	0.117
	Mean	15.4	0.090
Permian Qixia-Maokou formation (P_2q_+m)	Dusdhan #1	9	0.113
	Dusdhan #2	6	0.060
	Changshun #1	11	0.085
	Changshun #2	31	0.258
	Changshun #3	5	0.042
	Luodian #1	14	0.100
	Luodian #2	5	0.050
	Mean	10.9	0.097
Carboniferous Huanglong-Maping formation ($C_{2-3}h_+m$)	Changshun #4	5	0.039
	Dushan #3	2	0.022
	Dushan #4	4	0.033
	Dushan #5	19	0.146
	Mean	7.5	0.061

Table 1 suggests that the Triassic Maocaopu Formation (T_1m) may have developed more mature system of dissolution features than the other formations investigated. All boreholes were drilled at the same county (Xifeng County, Borehole #1–10). The strips of dissolution features were found to range from 12–21 per hole with the average being

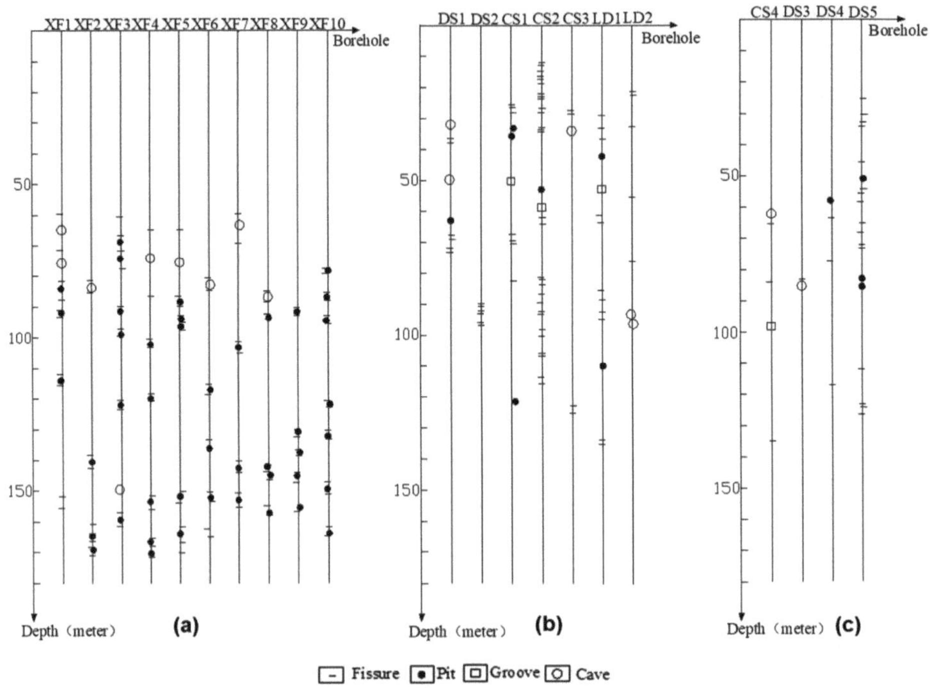

Fig. 1. Distribution of dissolution features in the karst development in (a) the Triassic Maocaopu Formation (T_1m); (b) the Permian Qixia-Maokou Formation (P_2q_+m); and (c) the Carboniferous Huanglong-Maping Formation ($C_{2-3}h_+m$).

15.4. The dissolution feature density, indicated by the number of dissolution features per meter, ranges from 0.065/m to 0.117/m. The highest density appears in Xifeng #10 while the smallest density in Xifeng #2. Compared with other two formations, the features seem to develop more uniformly.

The Permian Qixia-Maokou Formation (P_2q_+m) has the most widely distributed results across the boreholes among the studied formations. The boreholes were drilled at three counties (Dusdhan County, Changshun County and Luodian County). Their average number of dissolution features found per hole is moderately fewer than that from the Triassic Maocaopu Formation but the density is quite high, slightly exceeding the Triassic Maocaopu Formation. However, it is evident that the results vary tremendously across the boreholes. The dissolution features were found to range from 5–31 per hole with the average being 10.9. The density ranges from 0.042/m to 0.258/m with the average being 0.097/m. Changshun County apparently has some very large amount of dissolution features and density, among which Changshun #2 was identified with the highest density. It can be assessed that overall the Permian Qixia-Maokou Formation has a very reasonably mature development of karst and, but its extent varies significantly and the inhomogeneity is quite strong across the stratum. The considerable difference may be attributed to rock composition, mineralogical structures and sedimentary history; the underlying mechanisms is worthy of further investigation.

The Carboniferous Huanglong-Maping Formation ($C_{2-3}h_{+}m$) clearly features the lowest quantities of dissolution features and the lowest density. The dissolution features range from 2–19 per hole with the average being 7.5. Dushan #3 stands out with a large number of dissolution features and a very high dissolution feature density, but the rest were found to feature very few dissolution features. It can be concluded that the karst of the Carboniferous Huanglong-Maping Formation has not fully developed yet and its extent varies considerably as well.

2.2 Dissolution Features Across the Depth of the Geological Formations

Each type of dissolution features and their distribution across the depth are detailed in Fig. 1.

The karst development in the Triassic Maocaopu Formation (T_1m) clearly occurs primarily below 60 m; almost no dissolution features were found above 60 m. Especially from 60 m to 100 m, significant amounts of various types of dissolution features including pits, fissures and caves were identified. Below 100 m, the limestone features almost exclusively pits and fissures only which are generally of small-scale structures resulting from dissolution, showing a very modest karst development. In addition, most of the investigated boreholes share such similar trends of distribution, thus demonstrating a more uniform karst development across this formation.

In the Permian Qixia-Maokou Formation ($P_2q_{+}m$), fissures seem to dominate over other types of dissolution features identified. Some fissures start to appear in shallow underground (within 30 m); main dissolution features are concentrated within 30–100 m and manifest in all four different types. Only small dissolution pits/holes and fissures exist below 100 m. Although overall the distribution suggests a fairly advanced karst development, but it varies considerably across the investigated boreholes.

The results on the Carboniferous Huanglong-Maping Formation ($C_{2-3}h_{+}m$) evidently show that it undergoes a less advanced karst progression. There were only three caves found in two boreholes; the rest features either pits or fissures, concentrated within 30–130 m.

The presented results thus far indicate that overall the Triassic Maocaopu Formation has the most advanced karst development, closely followed by the Permian Qixia-Maokou Formation, while the Carboniferous Huanglong-Maping Formation is far less advanced. This order coincides with that of the geological age of these formations. It may indicate that the karst progression in the studied areas is still primarily dictated by natural processes. However, further studies are much needed to substantiate this implication.

3 Hydro-geological Data and Water-Bearing Capacities in the Studied Limestone Formations

A principal consequence of the dissolution features and karst development is the formation and evolution of the subterranean drainage system, as networks of caves and conduits potentially provide a significant capacity for water resources.

3.1 Subterranean Rivers and Karst Springs in the Studied Limestone Formations

Subterranean rivers and karst springs are two major typical features of water flows that exist and run among the cave and conduit networks in the karst formation, and have a strong presence in the three studied geological formations. Table 2 summarizes the total length and quantity of the subterranean rivers in each formation, based on the regional hydrogeological records. The percentage of the number of the rivers in the total number of the rivers of all formations is also listed in Table 2. The subterranean rivers in the Permian Qixia-Maokou Formation (P_2q_+m) almost are 10 times that in the Triassic Maocaopu Formation (T_1m) in terms of the number of the rivers, showing a significant lead in the development of subterranean drainage network.

Table 2. Summary of subterranean rivers in the three formations.

Stratum	T_1m	P_2q_+m	$C_{2\text{-}3}h_+m$	Total
Length (km)	115	1,794	516	2,425
Number	29	281	75	385
Percentage (%)	7.5	73.0	19.5	100

The record of karst springs in these formations seems to also confirm the more developed stage in the Permian Qixia-Maokou Formation (P_2q_+m), as shown in Table 3. Interestingly the percentages of springs in three formations in Table 3 are very close to those of subterranean rivers in Table 2.

Table 3. Summary of karst springs in the three formations.

Stratum	T_1m	P_2q_+m	$C_{2-3}h_+m$	Total
Number	48	352	69	469
Percentage (%)	10.2	75.1	14.7	100

3.2 Assessment of the Water-Bearing Capacity of the Studied Limestone Formations

Table 4 summarizes a comprehensive series of hydrogeological records regarding the three studied formations, based on historical records (for Year 2007–2011) collected by the Bureau of Geology and Mineral Resources of Guizhou Province and recent field investigations.

Table 4. Summary of hydrogeological records of the three formations and overall ratings of their water-bearing capacities.

Stratum	T_1m	P_2q_+m	$C_{2-3}h_+m$
Subterranean river density (km/100 km^2)	5.6	10.0	25.0
Mean discharge of subterranean rivers (L/s)	346.4	516.8	527.0
Dry season mean flow (L/s)	189.1	215.4	158.6
Dry season runoff module number (L/s/km^2)	3.2–10	4.0–12.0	4.9–9.1
Mean specific borehole capacity (m^3/d)	398.6	649.7	404.9
Overall rating	M	H	H

Three major characteristics listed in Table 4, subterranean river discharge rate (L/s), runoff module number in dry seasons (L/s/km^2) and specific capacity from drilling borehole tests (m^3/day) are usually used to assess the water storage of karst aquifers among the common practices in Guizhou. The runoff module number refers to the runoff discharge per unit area of the aquifer. In practice, not all three parameters are always available; however, any of these three may be used to render a general assessment of the water-bearing capacity, rated as high (H), moderate (M) or low (L). The ratings of these formations are also offered in Table 4. Both the subterranean river discharge rate (>100 L/s) and specific borehole capacity (>500 m^3/d) of the Permian Qixia-Maokou Formation are very high and easily qualify it in "H" category. The mean specific borehole capacities of the Carboniferous Huanglong-Maping Formation and the Triassic Maocaopu Formation are considerably lower than the Permian Qixia-Maokou Formation, but the mean discharge of subterranean rivers of the Carboniferous Huanglong-Maping Formation is very high, even greater than that of the Permian Qixia-Maokou Formation. All three formations have strong dry season runoff module numbers. Overall the capacity of the Triassic Maocaopu Formation is behind the other two but a moderate rating seems a reasonable assessment.

It is of interest to note that the order of the water-bearing capacity of the three formations does not coincide with that of the assessment of the karst development discussed in the preceding section. This demonstrates the complex dynamics of the processes involved in the karst progression and subterranean drainage systems, a direct correlation between limestone dissolution and water-bearing capacity needs to be treated with caution. In addition, the hydrogeological data reported herein are averaged numbers across large amounts of testing locations; therefore their variability over the studied areas cannot be underestimated. The limited, different amount of boreholes among all three studied formations, primarily due to constraints of drilling conditions, may also potentially affect the observations or conclusions resulting from the presented data.

4 Conclusions

Karst groundwater is a major source for water supply in Guizhou; effective assessment of the karst characteristics and water capacities can have significant impact on the exploitation and management of regional water resources. The karst development is typically manifested in the evolution of various dissolution features in the geological formations. The present study shows that mature and strong dissolution features were found in the Triassic Maocaopu Formation and the Permian Qixia-Maokou Formation; but the former develops more uniformly while the development of the latter varies significantly across the formation. The water-bearing capacities of the studied formations are evaluated based on the hydrogeological data, showing reasonably strong potential for large water capacities from all three limestone formations. However, a discrepancy between the order of the water-bearing capacity and that of the karst development among the three studied formations is observed; it demonstrates the complex dynamics of the processes involved in the karst development and subterranean drainage systems. Further study is needed to substantiate their correlation or lack thereof in typical limestone formations in Guizhou.

Acknowledgements. The first author would like to acknowledge the financial support provided by Guizhou University and the Division of Science and Technology of Guizhou Province, Grant #LH[2017]7284.

References

Deng, G.: Characteristics of Karst development and control factors in Haizi Street area of Bijie County of Guizhou. Master thesis, Chengdu University of Technology (2015). (in Chinese)

Ford, D., Williams, P.: Karst Hydrogeology and Geomorphology. Wiley, West Sussex (2007)

Gao, P.: A study on the importance of neotectonic movement in the Karst areas of Guizhou. J. Guizhou Sci. **4**, 48–51 (1983). (in Chinese)

Jin, Z., Feng, Z.: Genesis of the Permian limestone in Yunnan-Guizhou areas. J. Petrol. Mineral. **23**(2), 133–137 (1994). (in Chinese)

Kolliarits, S., Kuschnig, G., Veselic, M., Pavicic, A., Soccorso, C., Aurighi, M.: Decision-support systems for groundwater protection: innovative tools for resource management. Environ. Geol. **49**(6), 840–848 (2006)

Luo, L., Song, W.: A study of water resources identification in deep valley area of Guizhou Province. J. Eng. Surv. **10**, 39–44 (2004). (in Chinese)

Qian, H.T., Wang, S.J.: Interconnection of Karst systems and flow piracy through Karst collapse in layered carbonate rocks. Environ. Earth Sci. **64**(6), 600–608 (2011)

Zhou, J., Tang, Y., Yang, P., Zhang, X., Zhou, N., Wang, J.: Inference of creep mechanism in underground soil loss of karst conduits I. Conceptual model. Nat. Hazards **62**(3), 1191–1215 (2012)

Wang, Z.M.: Distribution characteristics of carbonate rocks and their control over karst groundwater in Guizhou. Geol. Explor. **53**(2), 242–248 (2017). (in Chinese)

Wang, Z.M., Yang, G.L., Yang, R.D., Rawal, K., Hu, L.B.: Evaluating the factors influencing limestone dissolution characteristics in Karst regions of Quizhou, China. J. Test. Eval. **45**(1), 220–229 (2017)

Worthington, S.R.H., Ford, D.C.: Self-organized permeability in carbonate aquifers. Ground Water **47**(3), 326–336 (2009)

Stability of Large-Scale Rock Cavern Complex at the 800-Year Feifengyan Quarry

Xiaolong Deng[1(✉)], Lihui Li[2,3], Yufang Tan[2,3], Beixiu Huang[2,3], Zhifa Yang[2,3], Gaime Guo[4], and Tianbin Li[5]

[1] China Highway Engineering Consulting Corporation, Beijing 100089, People's Republic of China
lyttney@foxmail.com
[2] University of Chinese Academy of Sciences, Beijing 100049, People's Republic of China
[3] Key Laboratory of Shale Gas and Geoengineering, Institute of Geology and Geophysics, Chinese Academy of Sciences, Beijing 100029, People's Republic of China
[4] Geological Construction Engineering Group Corporation of Guandong Province, Guangzhou 510080, People's Republic of China
[5] State Key Laboratory of Geohazard Prevention and Geoenvironment Protection, Chengdu University of Technology, Chengdu 610059, People's Republic of China

Abstract. This paper deals with long-term stability of an ancient large underground rock cavern complex excavated some 800 years ago in Zhejiang Province, SE China. The rock cavern complex is characterized by complicated geometric aligns, unique structure designs, large spans with shallow depths. Its rock mass characteristics were firstly investigated and the favorable rock mass quality was verified using the collected in-situ data. Distinctive quarrying method-Digging Holes for Quarrying Vertical Flagstone (DQF) was introduced the ancients to ensure safe excavation, it proved much smarter than modern blast-casting method. The favorable rock mass quality, smart structure designs, and ingenious quarrying technique were concluded to collectively lead to the noticeable long-term stability of the cavern complex. These ancient achievements mark high level of quarrying in complicated geo-environments with manpower, the results are positive to extend our knowledge in category design of cavern scale and stability assessments.

Keywords: Ancient cavern complex · Stability analysis
Rock mechanics · DQF

1 Introduction

Underground openings are often excavated in certain geological environment for engineering design, construction, operation and maintenance. Among them, ancient rock caverns are of special values in multidisciplinary researches such as archaeology, engineering geology and rock mechanics. In general, ancient quarrying caverns are

L. Zhang et al. (Eds.): GSIC 2018, *Proceedings of GeoShanghai 2018 International Conference: Rock Mechanics and Rock Engineering*, pp. 338–345, 2018.
https://doi.org/10.1007/978-981-13-0113-1_36

mostly found to be abandoned and in turn form the integral parts of stone relics, some of which have been acknowledged as world heritages, and their cultural relic values have also been explored by several authors [1, 2].

As the witness of intermittent quarrying activities, the ancient rock caverns have been experienced long-term interaction with surrounding geo-environment, thereby resulting in some instability problems. Correlations between rock mechanics and slope instabilities have been discussed from extensive studies of ancient rock caverns [3–5]. It should be noted that recent years, numerous ancient quarry caverns of large scale have been reported in Zhejiang Province of China [6], these cavern complexes challenge us with long-term stabilities of large spans, which are beyond the ultimate span expectation on current knowledge [7, 8]. The engineering scientific highlights, relating to rock mechanics [9], structure designs [10], long-term weathering [11] and stability mechanisms [12] have been systematically studied, these achievements are not only complemented to ancient rock caverns inventory, but also contribute to the current modification of specifications or guidelines for risk engineering design.

As one distinctive example, the Feifengyan ancient quarry was found in Xianju County of Zhejiang Province, SE China. The quarry consists of numbers of interconnected single caverns forming a huge underground space, the caverns exhibit special structure characteristics (e.g., arc-shaped support, bell-shaped roof and ribbed inverted arch). Except for some local unstable mechanical behaviors, these caverns maintain preferable stabilities, even with the long-term unsupported span over 40 m. This contribution hereby presents comprehensive investigations of the cavern complex to interpret its stability mechanism, and provide as analogical model for conservation and protection of ancient quarry engineering.

2 The Feifengyan Rock Cavern Complex

Feifengyan rock cavern complex is named after the Feifengyan Hill in which it situates, it consists of more than 30 secondary caverns, covers 30 000 m^2 approximately (Fig. 1).

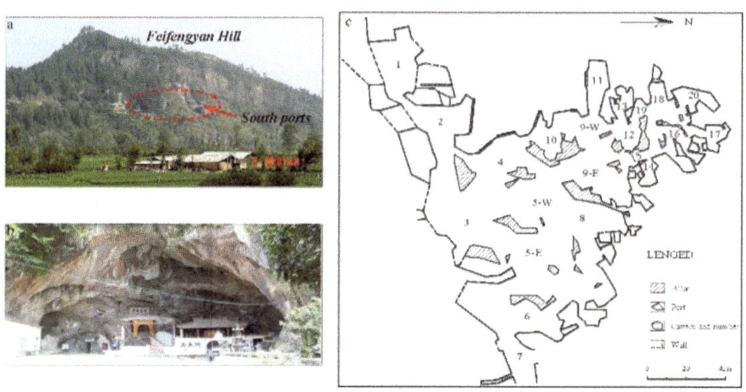

Fig. 1. Layout of the rock cavern complex. a. Northeast viewing outside of the quarry gate for the subdued Feifengyan Hill. b. Northward viewing outside No. 1 cavern. c. Plane viewing map.

2.1 Time of Original Quarry

Near the cavern complex, there is a village (Fig. 1a) owning a mutual family-preserved ancient book named "A genealogy of surname Guo at Xianju County". In the book, there is a sentence recording "In A.D. 1210, the Feifeng Temple (situated in No. 1 cavern, see in Fig. 2b) were prosperous and attractive for tourists". Besides, at Feifeng Temple there exists a tablet writing "the Feifengyan rock caverns were initially excavated during Song Dynasty (960–1279 A.D.) for flagstone materials". Therefore, it is reasonable to infer that the quarry was excavated no later than 800 years ago.

2.2 Geometry and Structure

The Feifengyan quarry is very huge as a cavern complex, and the individual caverns forming the cavern complex are generally featured by large spans. Taking No. 5 cavern for example, this cavern is divided into eastern zone and western zone due to its large spans, the overall span of No. 5 cavern reaches to 79.8 m, making it the largest cavern among the caverns (Fig. 2).

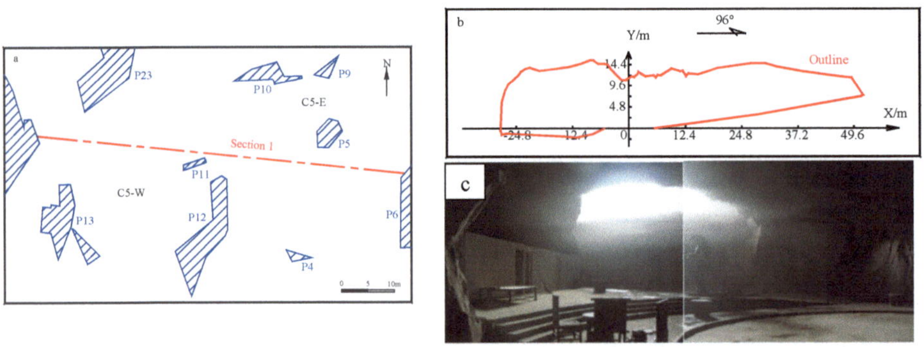

Fig. 2. Diagram showing the dimension of No. 5 cavern. a. Plan view of No. 5 cavern. b. Outline of No. 5 cavern derived from Sect. 1. c. South viewing inside No. 5 cavern.

To further gain the characteristics of the cavern complex's dimensions, we conducted extensive mappings among these caverns by the aid of laser total station (Leica 2003) and laser-gauge-detecting device (BJSD-2B, precision: 0.1 m). The mapping results are set out in Fig. 1c and Table 1 respectively.

It should be noted that given the considerable global size, large individual cavern spans and shallow buried depth, the cavern complex has maintained its stability for over 800 years. To explore its long-term stability mechanism, it is necessary to evaluate its rock mass characterization and quarrying technique in the followings.

Table 1. Dimensions of some main caverns in the cavern complex

Number	Surface elevation/m	Roof elevation/m	Floor elevation/m	Height/m	Buried depth/m
No. 3	142	97–107	94.7	2.3–12.3	35–45
No. 5-W	142	87–105	79	8–26	37–55
No. 5-E	145	82–85	74	8–11	60–63
No. 6	140	82	73	9	58
No. 9-E	125	84–87	76	8–11	38–41
No. 11	125	89	82	7	36
No. 12–17	110–120	84–87	77	7–10	25–32
No. 18	115–120	81–86	76	5-10	31–36
No. 19	115–120	81–84	76	5–8	34–37

3 Rock Mass Characterization

3.1 Mechanical Strength of Intact Tuff Specimens

Triaxial compression tests were performed on NX size samples prepared according to ISRM standard. Table 2 lists tests results of specimens retrieved from intact rocks, according to related standards and specifications, the tuff belongs to hard rocks.

Table 2. Mechanical properties of laboratory test for intact rocks

Lithology	UCS (MPa)	E (GPa)	μ	C (MPa)	φ (°)	R_t (MPa)	K_w
Tuff	90.29	20.66	0.258	14.2	42.69	9.852	0.927

3.2 Rock Mass Structure

Discontinuities are relatively well developed in the cavern complex. For instance, in No. 2 cavern, there expose 25 joints in a sidewall of 15 m length (Fig. 3a), and the joints present steeper angles with preferable extension.

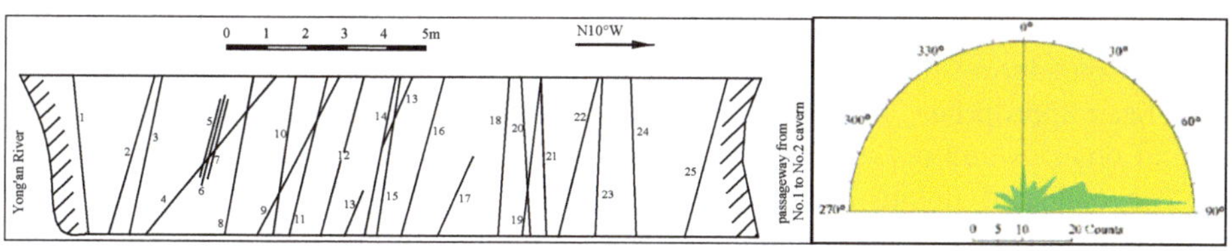

Fig. 3. Distribution features of the discontinuities. a. Sketch of discontinuities (location: western sidewall, No. 2 cavern). b. Rose diagram of discontinuities strikes.

Discontinuities in the cavern complex are mostly clean joints, only several joints are open (1–5 mm) with little argillaceous infilling materials. Almost all the observed joints are persistent and narrowly spaced. Bedding planes are absent on the rock mass scale, the exposed rocks in the cavern complex suffered slight weathering. A total of 143 discontinuities were measured and mapped in the cavern complex (Fig. 2b), many joints transect the caverns and exhibit trace lengths of tens of meters. The primary discontinuities developed with the occurrence of N50°–90°E, NW∠70°–90°, and most of the strikes are N50°–90°E, dominated by N70°–80°E.

3.3 Prediction of Rock Mass Quality

Based on the results of site investigation, mechanical test data and rock structure characteristics, the Q [13] methods are used to evaluate rock mass quality of the cavern

Table 3. RMR and Q classification for Feifengyan rock mass

Item	Index	Rating
RQD	Excellent	90
Number of joint sets (J_n)	One plus random	3
Roughness (J_r)	Rough/irregular, undulating	3
Degree of alteration (J_a)	Unaltered	1
Water inflow (J_w)	Dry excavation	1
Stress reduction factor (SRF) (σ_c/σ_1)	90/1.0 = 90	1.0
Q (grade)	90 (Very good)	

complex, the results are set out in Table 3.

The response of Q system to strength can be represented by the SRF parameter which incorporates the strength/stress ratio (σ_c/σ_1). If we consider a typical cavern (No. 5 cavern) with an average buried depth of 50 m (see in Table 1), the maximum tangential stress is 1.0 MPa approximately, and the SRF value can be returned as 90 roughly. Taking the Excavation Support Ratio (ESR) into consideration, the values of ESR are 3–5, 1.6 for temporary and permanent mining openings, respectively [14]. The corresponding maximum unsupported spans are 30–50 m, and 16 m. It shows that the predicted unsupported spans by Q system are not far from the unsupported spans that are measured in the cavern complex (Table 1), assuming an ESR suitable for temporary mines. Considering the original purpose of the cavern complex, an ESR of 5 would have been appropriate, for these caverns were initially designed as temporary underground quarries and not as long-term material disposals or dwelling facilities.

The collected in-situ data suggest that the Q system provides relative rational prediction in this case. Many singular caverns, with the observed unsupported spans of more than 10 m, are free of signs of deterioration, and they have been remarkably self-stabilizing for over 800 years.

4 Quarry Mechanism

During the in-situ survey, we found lots of quarrying traces in the cavern complex, these are small holes left in waste rocks and rows of holes regularly arranged on sidewalls (Fig. 4a). Inquiries of local craftsmen indicates that the ancient craftsmen used short chisel and hammer to manually quarry flagstones, just like splitting blocky sheets from vertical sidewalls. This quarrying technique was called Digging Holes for Quarrying Vertical Flagstone (DQF), it follows the procedures as: (1) determine the shape and dimension of the flagstones to be quarried, these flagstones are usually rectangular, and about 1.0 m wide, 3.0 m high, 80–100 mm thick (Fig. 4c). (2) Designate its position in vertical sidewall. (3) Groove along the bottom, left and right boundaries of the flagstone, the depth of these grooves should be deeper than that of the flagstone, usually 100–120 mm deep. (4) Dig holes along the top boundary of the flagstone, these digging holes are regularly arranged at an interval of 30 mm, and upward incline at 20°–35° so as to easily get the flagstone with uneven margins (Fig. 4b). (5) Penetrate the chisel into these holes, strike in sequence by the hammer to produce fracture around singular holes until these fractures connected with each other dissecting the left, right and reaching the bottom boundary groove. Finally, the stone decoupled from the sidewall as a flagstone.

In some cases, the mostly vertical joints outcropped on the sidewall are used instead of lateral grooves to quarry flagstone. At some quarrying sites, we measured the orientation of flagstones and the occurrence of virgin joints developed in their lateral side, the results are set out in Table 4. This replace of grooves by joints can save the grooving time, and increase the rate of stripping to improve quarrying efficiency.

The DQF technique has also been reported to use in the other ancient quarry not far from the Feifengyan quarry, He et al. [15] have pointed out that this quarrying technique is concordant with the principle of modern fracture mechanics. And the

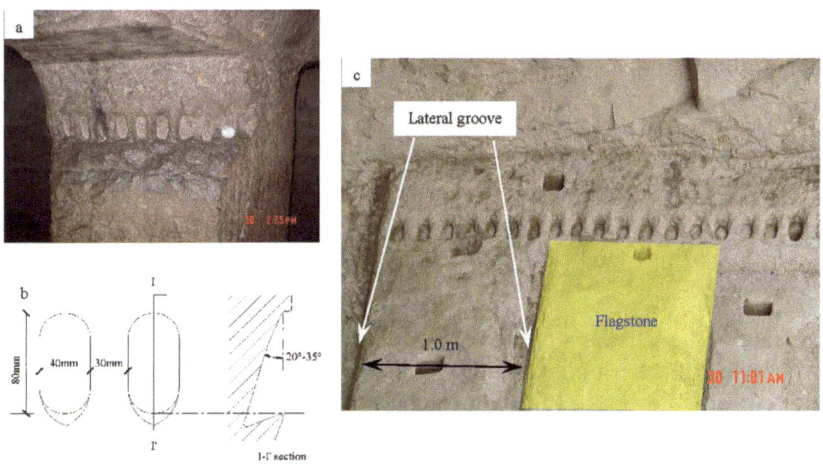

Fig. 4. Plates showing the traces of flagstone quarrying. a. A row of small digging holes left in rock pillar. b. Plane and section showing the small digging holes. c. Discard flagstone quarrying site.

Table 4. Cases showing the use of joints to replace lateral grooves for flagstone extraction

Quarry sites	Orientation of the flagstone's surface	Occurrence of joints developed in lateral sides of the flagstone
North wall of No. 18 cavern	N10°W	N85°E, SE∠85°
No. 24 pillar of No. 9E cavern	SN, E∠80°	N84°W, ⊥; N85°W, ⊥
No. 12 pillar of No. 9E cavern	N7°W, SW∠78°	N82°, NE87°
No. 15 cavern	N30°E, ⊥	N58°W, SW75°

fracturing mechanism revealed by 2-D numerical simulation (FEM) also explained that the cracks produced by striking chisel in the holes originated at the top boundary, propagated to grooves to spall the flagstone from surrounding rocks [16].

The DQF provides state of the art quarrying technique in ancient times, the excavation speed is slow, for the equipment employed is limited at that time. But it is a safe method because the induced excavation damage zones are much smaller compared with modern blasting and casting technique, which is faster yet subsequent effective measures must be taken so as to eliminate negative impacts on surrounding rocks. This comparison reminds us how to strike a balance among efficiency, safety and geoenvironment-friendly of the underground opening.

5 Summary and Conclusion

The Feifengyan ancient quarry with more than 30 interconnected singular caverns forms a huge underground rock cavern complex in an area of 30 000 m^2. Given its considerable global size, large individual cavern spans and shallow buried depth, the rock cavern complex have been standing unsupported for more than 800 years.

- There primarily develops one set of discontinuities, their traces are commonly observed in the cavern complex. Rock mass quality assessed by Q system returns realistic unsupported span prediction for the cavern complex.
- The DQF is characterized by the delimitation and breaking process, equipment is a limiting factor for the state of the art quarrying technique at ancient times. Compared with modern quarrying technique, it is a slow and long process, but it is safe and environment-friendly. It marks a high level in complicated geo-environment with manpower.

In conclusion, the relatively favorable rock mass quality, smart quarrying technique collectively lead to the remarkable long-term stability of the cavern complex. These results are positive to extend our knowledge on category design of cavern scale, and provide a distinct analogical and extrapolative model for stability assessment.

Acknowledgement. The work presented in this paper is financially supported by the National Natural Science Foundation of China (41172269, 41372322), the authors would also like to express their sincerest gratitude to the anonymous reviewers for their valuable and constructive comments.

References

1. Grenne, T., Heldal, T., Meyer, G.B., Bloxam, E.G.: From Hyllestad to Selbu: Norwegian millstone quarrying through 1300 years. In: Slagstad, T. (ed.) Geology for Society, Geological Survey of Norway, Trondheim, Special Publication 11, pp. 47–66 (2008)
2. Bolxam, E.G.: New directions in identifying the significance of ancient quarry landscapes: four concepts of landscape. In: Abu-Jaber, N., Bloxam, E.G., Degryse, P., Heldal, T. (eds.) QuarryScapes: Ancient Stone Quarry Landscapes in the Eastern Mediterranean, Geological Survey of Norway, Trondheim, Special Publication 12, pp. 165–183 (2009)
3. Hatzor, Y.H., Talesnick, M., Tsesarsky, M.: Continuous and discontinuous stability analysis of the bell-shaped caverns at Bet Guvrin, Israel. Int. J. Rock Mech. Min. Sci. **39**, 867–886 (2002)
4. Talesnick, M.L., Haztor, Y.H., Tsesarsky, M.: The elastic deformability and strength of a high porosity, anisotropic chalk. Int. J. Rock Mech. Min. Sci. **38**, 543–555 (2001)
5. Tsesarsky, M., Gal, E., Machlav, E.: 3-D global-local finite element analysis of shallow underground caverns in soft sedimentary rock. Int. J. Rock Mech. Min. Sci. **57**, 89–99 (2013)
6. Yang, Z.F., Li, L.H., Pan, W., Lu, M., Zheng, J.: Discussion about large-scale ancient underground engineerings. Sci. Technol. Eng. **3**(5), 464–466 (2003)
7. Barton, N.: Some new Q-value correlations to assist in site characterization and tunnel design. Inte. J. Rock Mech. Min. Sci. **39**, 617–641 (2002)
8. Hatzor, Y.H., Wainshtein, I., Mazor, D.B.: Stability of shallow karstic caverns in blocky rock masses. Inte. J. Rock Mech. Min. Sci. **47**, 1289–1303 (2010)
9. Zhang, Z.J., Yang, Z.F., Li, L.H., Zhang, L.Q., Yang, T.Y.: Back analysis of long-term shear strength of structural plane group in Qingfeng Ancient Quarry of Shepan Island. Rock Soil Mech. **32**(7), 2123–2127 (2011)
10. Zhang, Z.J., Li, L.H., Xu, W.J., Fu, Y., Feng, J.: Flat-plate roof collapse of shallow caverns and protective measures: a case study of Longyou ancient siltstone caverns. Nat. Hazards **76** (1), 191–213 (2015)
11. Guo, G.M., Li, L.H., Yang, Z.F., Tao, B., Qu, Y.X., Zheng, J.: Weathering mechanism of the cretaceous argillaceous siltstone caverns. Bull. Eng. Geol. Environ. **64**(4), 397–407 (2005)
12. Li, L.H., Yang, Z.F., Yue, Z.Q., Zhang, L.Q.: Engineering geological characteristics, failure modes and protective measures of Longyou rock caverns of 2000 years old. Tunn. Undergr. Space Technol. **24**, 190–207 (2009)
13. Bieniawski, Z.T.: Engineering classification of jointed rock masses. Trans. S. Afr. Inst. Min. Metall. **74**, 335–344 (1976)
14. Barton, N.R., Lien, R., Lunde, J.: Engineering classification of rock masses for the design of tunnel support. Rock Mech. **6**, 189–236 (1974)
15. He, W.T., Shang, Y.J., Sun, Y.L., Li, L.H., Yang, Z.F.: Insight of the environmental awareness on waste rock disposal at Heidong Quarry dated 1000 years ago in SE China. Environ. Earth Sci. **75**, 163 (2016)
16. Yang, Z.F., Shang, Y.J., Li, L.H., Zhang, L.Q., Zhang, Z.J., Li, T.B., Zhang, W., Xu, S.S.: Long-term stability of large span caverns at the 1400-year Heidong quarry. Episodes **36**(1), 39–46 (2013)

Mechanical Analysis of Removable Blocks During TBM Tunnelling in Blocky Rock Masses in View of TBM-Block Interaction

Shuaifeng Wang[1,2], Zixin Zhang[1,2(✉)], and Xin Huang[1,2]

[1] Department of Geotechnical Engineering, College of Civil Engineering, Tongji University, Shanghai 200092, People's Republic of China
zxzhang@tongji.edu.cn
[2] Key Laboratory of Geotechnical Engineering, Tongji University, Shanghai 200092, People's Republic of China

Abstract. Interaction between TBM and the surrounding rock masses is consist of the cutterhead-block interaction and shield-block interaction, leading to the stability consideration of TBM tunnels. The equilibrium equation of removable blocks is derived by taking into consideration of the interaction forces and the net sliding forces of moving modes is also deduced. For different types of blocks in terms of the new classification of blocks in TBM tunnels, sliding and falling will happen for rear blocks, while front blocks and corner blocks will only experience sliding motion.

Keywords: Block theory · Tunnel · TBM-block interaction
Mechanical analysis

1 Introduction

Rock blocks together with the geological planes constitute a complex discontinuous system, of which the geometrical and mechanical properties are dominated by the discontinuous planes, such as faults, beddings and joints. When tunnelling in the blocky rock masses, the tunnel boring machine (TBM) excavates the surrounding rock and therefore the interaction forces between them come into action. The interaction includes normal contact force along the tunnelling direction, tangential friction force parallel to the excavation face and the friction force between the shield and surrounding rock. The normal and tangential forces reflect the interaction of excavation face and cutterhead. However, the interactional friction force between the TBM shield and surrounding rock is often considered to distribute only in the lower part of the circular tunnel [1]. Stability of TBM tunnels in blocky rock mass is a synthetic issue of the combination of rock blocks and the interaction forces.

The paper analyzes the normal and tangential interaction forces from the cutterhead. The net sliding forces of moving modes are obtained by substitute the above interaction forces into the equilibrium equation, which is derived from traditional block theory. In the view of new classification of blocks in contact with TBM, mechanical features and moving modes of different types of blocks are illuminated.

L. Zhang et al. (Eds.): GSIC 2018, *Proceedings of GeoShanghai 2018 International Conference: Rock Mechanics and Rock Engineering*, pp. 346–351, 2018.
https://doi.org/10.1007/978-981-13-0113-1_37

2 Interaction Between TBM and Rock Mass

During TBM advancement, the cutterhead and the rock blocks on the excavation face form a complex mechanical system where they are coupled and interactive. The interaction forces include: (a) the cutterhead-block interaction induced by the cutterhead progression and rotation and (b) the shield-block interaction due to the contact of shields and the surrounding rocks.

2.1 Cutterhead-Block Interaction

Tunnels are excavated by cutterhead progression and rotation causing the formation of concentric circular grooves on the working face (Fig. 1(a)). The interacting forces between rock mass and TBM at the cutterhead include: (a) normal contact force by the cutterhead progression and (b) tangential forces induced by annually-distributed frictional forces due to cutterhead rotation (Fig. 1(b)).

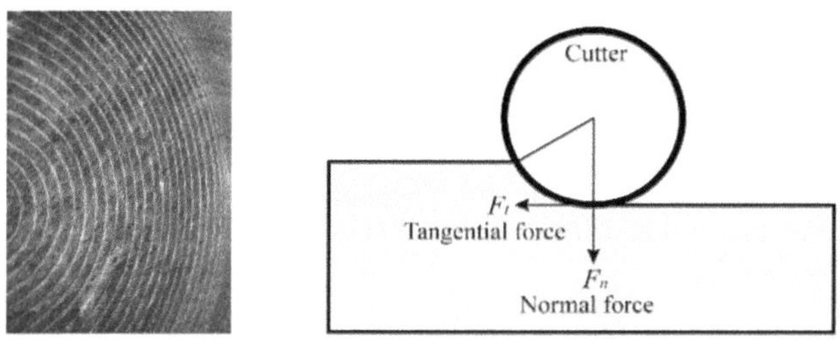

(a) (b)

Fig. 1. Schematic diagram of cutterhead-block interaction: (a) grooves by cutters (from Benato, Oreste 2015 [4]); (b) cutterhead-block interaction forces

Assuming that the ith surface of block B is formed by excavation face, then the normal force is derived as follow:

$$F_n^B = F_n \cdot N_c \cdot \frac{A^i}{S_c} \tag{1}$$

where F_n^B is normal force induced by one cutter, N_c is the number of cutters on the cutterhead, A^i is the area of the ith surface and S_c is the area of the cutterhead.

The tangential force is expressed as:

$$F_t^B = F_t \cdot N_c \cdot \frac{A^i}{S_c} \tag{2}$$

For hard rock TBMs, F_n is in a range of 0–400 kN and F_t lies between 0 and 35 kN [2, 3].

2.2 Shield-Block Interaction

The interaction forces between surrounding rock and the shield is dependent on the overboring magnitude and ground convergence.

In blocky rock masses, there is free space between shield of TBM and the surrounding rocks before convergence [5]. Therefore, only the bottom of the shield will be in contact with the rock mass. The compressing region is considered to be a 1/4 invert of the tunnel perimeter [1, 6]. This part of bedrock lies the shield and bears a portion of the TBM weight, causing radial pressures on the rock blocks which direct inside the blocks.

Blocks in this region are stable due to the downward weight and the pressure forces, which are supported by bedrocks. So the shield-block interaction can be ignored for its positive effect on the stability of tunnels.

3 Mechanical Analysis of Removable Blocks

This part mainly deals with the mechanical features and sliding modes of removable blocks in view of the new classification of blocks in contact with TBM.

3.1 Shield-Block Interaction

According to the traditional block theory [7] in Fig. 2, the equilibrium equation of block B is expressed as

$$\vec{N} + \vec{T} + \vec{F_r} = 0 \tag{3}$$

where $\vec{N}$ is the reaction force on the sliding plane represented by Eq. (4), $\vec{T}$ is the tangential frictional forces along the sliding plane as expressed in Eq. (5) and $\vec{F_r}$ is the resultant of all other forces acting on block B. In this paper, $\vec{F_r}$ mainly includes self-weight of block B and interactive forces from TBM as shown in Eq. (6).

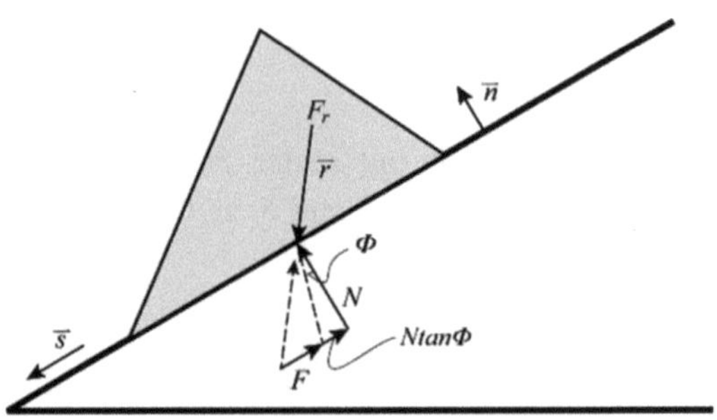

Fig. 2. Schematic diagram of forces on a removable block

$$\vec{N} = \sum N_i \vec{n_i} \tag{4}$$

where $\vec{n_i}$ is the normal unit vector of plane i, directing into the block B.

$$\vec{T} = \sum -N_i \tan\phi_i \vec{s} - F\vec{s} \tag{5}$$

where $\vec{T}$ is the tangential frictional forces, in the opposite direction of $\vec{s_i}$. $\vec{s_i}$ is the sliding direction. ϕ_i is the friction angle of plane i. F is the fictitious force, also called net sliding force.

$$\vec{F_r} = \vec{G} + \vec{F_n^B} + \vec{F_t^B} \tag{6}$$

where $\vec{G}$ is the self-weight of block B.

Due to the existence of the net sliding force F, this equation is feasible. The sign of F represents the equilibrium state of block B. When F is positive, the block B tends to slide along some plane(s) without essential support. On the contrary, a negative F indicates that the block B is stable. Obviously, if F equals to 0 the block B is at limit equilibrium status.

3.2 Moving Modes of Contacting Blocks

For three traditional moving modes, i.e. falling, sliding along one single plane and sliding along the intersection line of two planes, the net sliding force can be derived in Eq. (7), respectively.

$$\begin{cases} F = \left|\vec{F_r}\right| \\ F = \vec{F_r} \cdot \vec{s} - N_i \tan\phi_i \\ F = \vec{F_r} \cdot \vec{s} - \left(N_i \tan\phi_i + N_j \tan\phi_j\right) \end{cases} \tag{7}$$

In terms of the spatial location with respect to cutterhead, blocks contact with TBM can be categorized into three types: Front Block, Rear Block and Corner Block [8]. As shown in Fig. 3, Front Block is defined as blocks formed by geological discontinuities and the excavation face and is located in front of the excavation face and next to the cutterhead. Rear Block is defined as blocks determined by geological discontinuities plus the tunnel perimeter and is located at the rear of the cutterhead and next to the shield. Corner Block is those blocks located at the face-wall corner of the tunnel. The outer boundaries of Corner Blocks simultaneously include joint planes, the excavation face and the tunnel circumferential periphery. According to spatial positions relative to the cutterhead, front blocks and corner blocks will experience both normal and tangential forces, while the rear blocks is free of interaction forces.

Confined by the outer boundaries, different types of contacting blocks have different kinematic modes. Sliding and falling will happen for rear blocks, while front blocks and corner blocks will only experience sliding motion due to the deviation of

Block type	Location	Moving modes			Mechanical features
		Sliding		Falling	
		Single-plane	Double-plane		
Front Block		√	√	×	
Corner Block		√	√	×	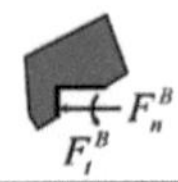
Rear Block		√	√	√	

Fig. 3. Summary of moving modes and mechanical features for the contacting blocks

gravity from the normal of excavation boundaries. Figure 3 summaries the moving modes of three types of contacting blocks.

4 Conclusion

This paper discusses the interaction forces between the rock mass and TBM, including the cutterhead-block interaction and shield-block interaction. Cutterhead-block interaction is consist of normal contact force and tangential frictional force, which can be determined from forces of single cutter on the cutterhead. Due to its positive effect on the stability of TBM tunnels, the shield-block interaction is ignored in this study.

Afterwards, the sliding forces of three types of moving modes are derived from the ultimate equilibrium equation by applying the interaction forces. Classifying contacting blocks into Front Block, Corner Block and Rear Block, the paper indicates: sliding and falling will happen for rear blocks, while front blocks and corner blocks will only experience sliding motion.

Future research will analyze the relationship between the stability of TBM tunnels and the magnitude of interaction forces. Furthermore, a computer-aided program will be developed.

Acknowledgements. The research was conducted with funding provided by the National Basic Research Program of China (973 Program, No. 2014CB046905), the National Science Foundation of China (Grant No. 41172249), and the State Key Laboratory for Geomechanics and Deep Underground Engineering (No. SKLGDUEK1303).

References

1. Hasanpour, R., Rostami, J., Ünver, B.: 3D finite difference model for simulation of double shield TBM tunneling in squeezing grounds. Tunn. Undergr. Space Technol. **40**, 109–126 (2014)

2. Gertsch, R., Gertsch, L., Rostami, J.: Disc cutting tests in Colorado Red Granite: Implications for TBM performance prediction. Int. J. Rock Mech. Min. Sci. **44**(2), 238–246 (2007)
3. Cho, J., Jeon, S., Jeong, H., Chang, S.: Evaluation of cutting efficiency during TBM disc cutter excavation within a Korean granitic rock using linear-cutting-machine testing and photogrammetric measurement. Tunn. Undergr. Space Technol. **35**, 37–54 (2013)
4. Benato, A., Oreste, P.: Prediction of penetration per revolution in TBM tunneling as a function of intact rock and rock mass characteristics. Int. J. Rock Mech. Min. Sci. **74**, 119–127 (2015)
5. Shang, Y., Xue, J., Wang, S., Yang, Z., Yang, J.: A case history of Tunnel Boring Machine jamming in an inter-layer shear zone at the Yellow River Diversion Project in China. Eng. Geol. **71**(3–4), 199–211 (2004)
6. Zhao, K., Janutolo, M., Barla, G.: A completely 3D model for the simulation of mechanized tunnel excavation. Rock Mech. Rock Eng. **45**(4), 475–497 (2012)
7. Goodman, R., Shi, G.: Block Theory and Its Application to Rock Engineering. Prentice Hall, Upper Saddle River (1985)
8. Zhang, Z., Wang, S., Huang, X., Kwok, C.Y.: TBM–block interaction during TBM tunneling in rock masses: block classification and identification. Int. J. Geomech. **17**(5), E4016001 (2017)

Modeling and Inversion of Complex 3D Geologic Field Stress and Its Application in Rockburst Evaluation

Quanjie Zhu[1,2(✉)], Jinhai Liu[1,2], Linsheng Gao[1], Zhihua Chen[3], and Xiaohui Liu[1]

[1] School of Safety Engineering, North China Institute of Science and Technology, Beijing 101601, China
youyicun2008@163.com
[2] Hebei State Key Laboratory of Mine Disaster Prevention, North China Institute of Science and Technology, Beijing 101601, China
[3] School of Environment, Henan Normal University, Xinxiang 453007, China

Abstract. Geostress is the basis for mining and geotechnical engineering designs, and dynamic disaster preventions. And the geostress inversion is a mathematical procedure that to calculate and extend the geostress field distribution base on a handful of measurement results. In order to make use of fine distribution characteristics of geostress field to evaluate the rock burst risk in a target area, this study carried out corresponding research contents for a deep mine in Shandong Province: (1) Conducted site measurements of geostress by means of hollow inclusion method; (2) Developed a complex 3D geological model based on multi-software to inverse and reconstruct the complex geostress fields coupled with multiple linear regression (MLR); (3) Compared the microseismic monitoring results with 3D geological model nephogram and provide theoretical basis for rock burst risk evaluation and regionalization in target area. Results showed that the reconstitution of 3D geological model gave well agreement with experimental results. The geostress evaluated results of 3D model of were consistent with the actual rock burst risk situation. The rapid constructed 3D geological model construction was expected to provide a certain reference meanings for geostress analysis to similar mines.

Keywords: In-situ stress · Modeling and inversion · Hollow inclusion Evaluation · Microseismic

1 Introduction

Currently, rock burst is one of the most threatened dynamic disasters in deep coal mines. Plenty of practices indicated that the regional geostress field distribution was closely related to unstable geological factors, such as roadway surrounding rock instability, ground pressure, and coal-gas outburst, etc. (Kang et al. 2007; Manouchehrian et al. 2012). Meanwhile, geostress measurement and inversion also is considered as a key factor to prevent and control the impact of ground pressure and consequently ensure the safe and orderly production of mine. For a mining engineering

L. Zhang et al. (Eds.): GSIC 2018, *Proceedings of GeoShanghai 2018 International Conference: Rock Mechanics and Rock Engineering*, pp. 352–363, 2018.
https://doi.org/10.1007/978-981-13-0113-1_38

area, corresponding studies on its stress field distribution analysis and inversion are of important significances on development of underground production activities and prejudgments on unknown disasters. These advantages of are beneficial to provides a theoretical basis for efficient, orderly and safe production of mine (Cai et al. 2010; Monjezi et al. 2011; Matsuku et al. 2004; Chen et al. 2013).

In order to investigated the geostress field distribution and inversion of coal mines, 3D geological models had been used by several researchers (Yuan et al. 2011; Pei et al. 2014; Xiang et al. 2015). Yuan et al. (2011) presented a method to carry out 3D geological model by using AutoCAD-ANSYS-FLAC3D simulation. Attributed to the inaccuracy caused by AutoCAD geometrical model, a relatively larger deviation between simulations and experiments was generated. To minor the deviation before input to ANSYS, the AutoCAD data was extended by MICROMINE, GOCAD or SURPAC (Pei et al. 2014), in which the data was optimally extracted and reconstructed. Benefited from the data optimization procedure, the mesh of geometric models was optimized and simulated results gave better agreement with experimental data using FLAC. Ji et al. (Xiang et al. 2015) developed a 3D geological model for the evolution of coal mine rock burst based on Visual C++ designing, in which the complex geological form was automatically meshed with more refinement partition. To a certain extent, those methods above compensated the time-consuming and tedious drawbacks of numerical simulations corresponding to a complex geological mine form.

The geostress inversion of was a subsequently inversed step of the initial stress field based on actual measurement data. Many researchers investigated the geostress field inversion in allusion to different mine bodies. The surrounding rock's stress field distribution of deep roadway was obtained by Luo et al. (2010) through directly measured the geostress in roadway. Zhang et al. (2010) carried out geostress field inversion in an exploration area by measuring the geostress via hydraulic fracturing technique in the deep hole. To receive an accurate calculation and expansion on geostress inversion distribution, accelerate optimization algorithm was utilized by Wang and Li (2015). However, owing to the anomalous features (such as fault and fold etc.) of a coal measure strata, it is difficult to develop a 3D model to descript the development and distribution of geostress field. A common method is to simplify the complex geological model, which was regulated and numerical calculated by using numerical simulation software (such as FLAC). But the calculation results obtained by this method are incapable to describe the actual stress field distribution accurately.

Based on the analysis above, the relevant research about "Modeling and Inversion of Complex 3D Geologic Field Stress and its Application in Rockburst Evaluation" was performed in this paper. Our studied target was a coal mine (located in Yuncheng, Shandong Province) that included three main productive coal seam (seam thickness: 4.75–7.80 m, and seam depth: 530 m–1200 m). In 2014–2015, there were 6 recorded accidences caused from rock burst disasters, therein four were gave rise to the tail-entry of 1301 working face which was induced by mining advance of 1300 work face. Figure 1 shows a typical rock burst happened in October 9, 2015, which located in 1300 workface. These accidences caused huge economic losses and seriously threatened the safe production of mine.

Therefore, the geostress field distribution and inversion of three mining areas of this coal mine were investigated by comparing with the results measured from hollow inclusion stress relief method. Based on these results, the regional geostress parameters as well as variation features were accurately provided for coal roadway stability evaluation, mining field geostress control, and rock burst prevention.

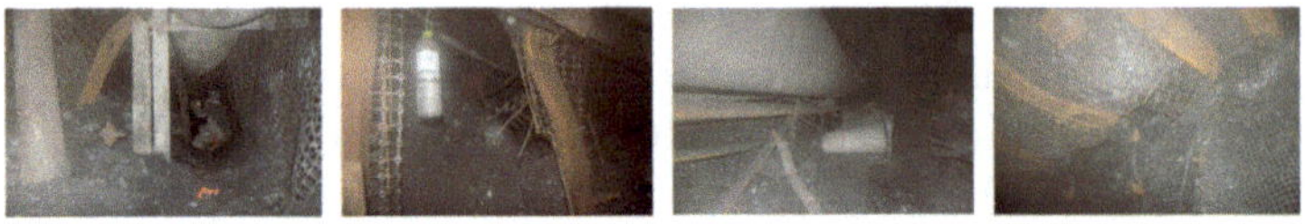

Fig. 1. Rock burst disaster in a coal mine

2 Hollow Inclusion Measurement and Field Application

2.1 Hollow Inclusion Method and Filed Testing

Considering the integrated mine existing conditions, measurement precision and technology maturity, we chose the hollow inclusion borehole stress relief method for 3D geological model construction and corresponding geostress inversions of the target of deep coal seam. Geostress measurement target areas and corresponding drilling borehole were depicted in Fig. 2. The approximately horizontal borehole was drilled on surrounding rock of roadway in 133 mm diameter, which was obtained by using a small drill bit hollow bag hole (in 370 mm length and 40 mm diameter) and drawing out the drill pipe (5 cm) after the formation of hole.

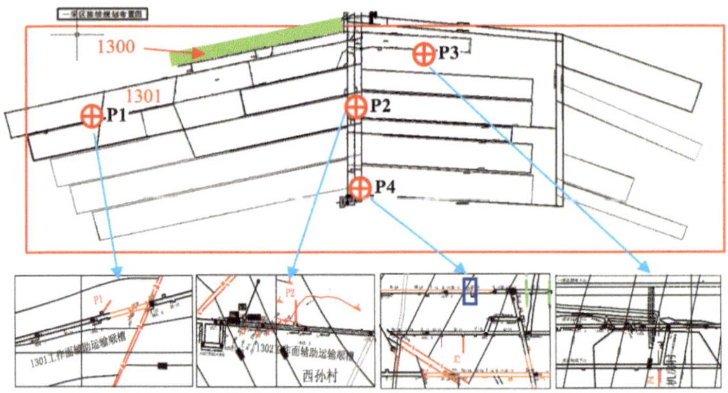

Fig. 2. Target area of geostress inversions and boreholes' arrangement

The surrounding rock strain collecting system in this study was a KJ327-f mine pressure monitoring system with a KX-81 hollow package strain sensor, as shown in Fig. 3. The main geostress measurement process could be successively summarized as four steps: geological borehole drilling (GBD), the borehole cleaning, hollow package installing, taking the guide bar, and geostress relieving. The borehole in GBD step was located in a roadway with favorable lithology area. The piercer was aforehand cleaned and dried with acetone impregnated gauze two times.

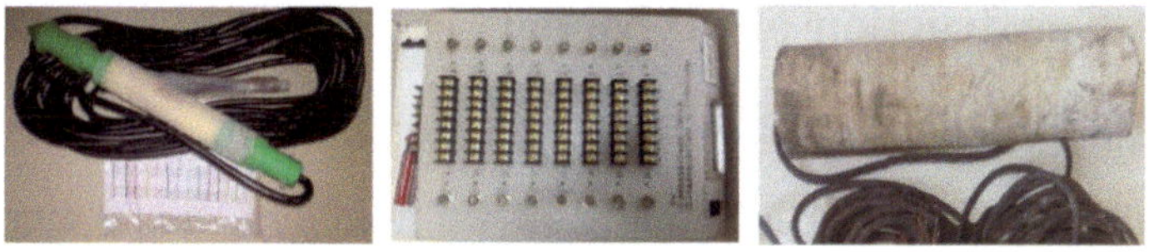

(a) hollow inclusion cell (b) data acquisition instrument (c) rock core in the site

Fig. 3. Testing system of stress relief by borehole overcoring technique

2.2 Data Processing and Analysis

According to the field measurement (collected strain value, elastic modulus, and Poisson's ratio), the geostress magnitude and direction, maximum and minimum geostress value and so on (P1–P4 in Fig. 4) were obtained and summarized in Table 1 (MV is the measured values).

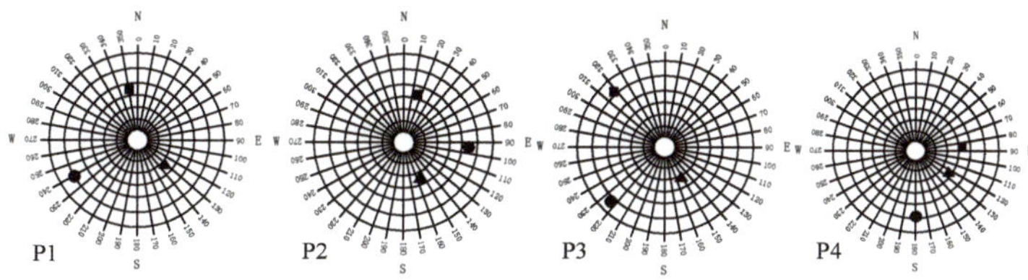

Fig. 4. Network diagram of the main stress distribution

Table 1. Result of in-situ rock measurement

No	Depth/m	σ_1			σ_2			σ_3		
		Azimuth/(°)	Inclination/(°)	MV/MPa	Azimuth/(°)	Inclination/(°)	MV/MPa	Azimuth/(°)	Inclination/(°)	MV/MPa
P1	−894	239	19.75	34.5	−9.31	45.09	24.6	133.1	38.31	17.7
P2	−880	94	18.61	30.2	15.49	−28.87	22.9	156.46	−54.63	18.7
P3	−810	230	−5.82	38.7	−42.17	23.47	35.5	153.44	−65.73	17.6
P4	−1030	180	15.01	33.5	84.66	20.48	22.8	124.09	−64.19	20.1

It could be found that the azimuth angle ranged from 90°–240°, and included angle between the maximum measured geostress and horizontal plane are fluctuated nearby 20°. The maximum geostress value was 1.09–1.95 and 1.40–2.20 times to the value of minimum geostress and vertical geostress. It indicated that the maximum principal geostress direction of the mining area was in accordance with the horizontal direction, which inferred the regional geostress field was dominated by horizontal tectonic stress field. Thus, the corresponding gravity stress and tectonic stress ought to be taken into major consideration.

3 Rapid Construction of 3D Model

3.1 Modeling Details and Its Implementations

To reduce the geostress inversion error generated from imprecise of numerical simulation, it was important to construct a geometric model which approximates the actual topography of coal mine before 3D modeling. In our present study, a new method for large-scale complex 3D geological model was presented based on coalesce of 3DMINE, GOCAD, SURFER, RHIRO, ANSYS and FLAC^{3D}. Such method was further utilized to inverse the geostress field of a deep mine area in Yuncheng Coal Mine (Shandong Province, China). It should be pointed out that the construction process of 3D geological model can be divided into five parts: contour lines generation, discrete lattice generation, data grid processing, irregular surface generation, and geometric model construction. Figure 5 depicted the modeling process which was based on the gradually constructed idea of point-line-surface-volume.

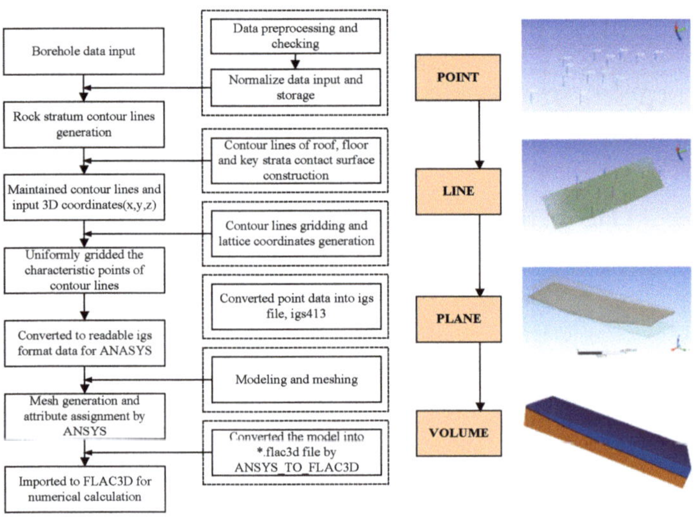

Fig. 5. Process of 3D numerical modeling

The regular modeling method was common used in conventional numerical simulations. Attributed to its inflexibility and software dependency, it is difficult to fully express the fluctuation of quality body, fault structure, and other information. In order to precisely describe the complex formation structure, we used the drilling information to generate a rock interfacial contour, which was further developed to complex geological of line structure and surface structure to improve the accuracy and reliability of numerical simulation.

3.2 3D Numerical Simulation

It is believed that FLAC^{3D} software is a powerful software in mining and geotechnical engineering attributed to its attractive advantages of numerial calculation and analysis. Nevertheless, the modeling capabilities of FLAC^{3D} are restrained by the manual modeling method using command flow, which is difficult to makes the modeling. The ANSYS finite element program has powerful pretreatment function on data processing. Therefore, we used the ANSYS software to grid the 3D geometric model which was subsequently imported into FLAC^{3D} to perform numerical calculation. The modeling process was significantly simplified and then improved the efficiency and precision of numerical simulation. Based on field measured information, the various working conditions formed by the initial geostress field were set as undetermined factors, which were further used as boundary conditions and loads and applied to the computational model. Figure 6 showed the numerical results of the above six factors in 3D finite element model.

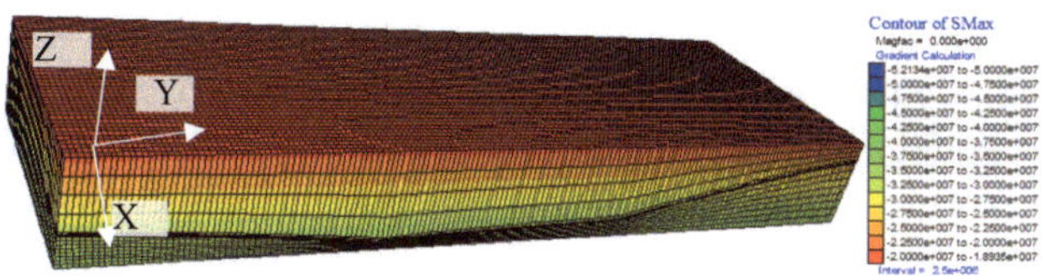

Fig. 6. Main stress distribution of 3D numerical modeling

4 Inversion and Reconstruction of 3D Geostress Field

4.1 3D Geostress Field Modeling and Inversion

The multivariate regression analysis was calculated based on the assumption that the tectonic stress obeyed a given distribution basis. Generally, the geostress value was obtained through field measurement (Sect. 2), which was further used to perform regression analysis to obtain corresponding regression coefficients. In this study, the 3D geologic field model was constructed and calculated by using numerical simulation coupled with multivariate regression analysis. The procedure could be briefly described as: (1) A large 3D model was constructed based on the geologic exploration data in studied area. (2) According to the geostress theory, affecting factors on geostress distribution was studied through numerical simulation, such as weight factor (OWF), etc.. (3) The geostress multi-regression equation of measured points was derived according to the stress linear superposition principle, in which the factors (i.e. OWF, etc.) being the independent variables. (4) The coefficients of the multi-regression equation were estimated via least square method by minimizing residual sum of squares (RSS) objective function between measured data and multi-regression equation. (5) Once coefficients were obtained, the regressive equation was substituted into numerical model to carry out iterative computations via FLAC^{3D} with known boundary conditions.

The multi-regressive calculation was coded in MATLAB® with vector component data measured in FEM model. According to Cholesky decomposition method, we can calculate any point initial stress σ in the given region (Qiu et al. 2003). The initial stress σ_{jp} can be decomposed as follows:

$$\sigma_{jp} = \sum_{i=1}^{n} L_i \hat{\sigma}_{jp} \qquad (j = 1, 2, \ldots, 6) \tag{1}$$

in which $\hat{\sigma}_{jp}$ is the calculated results of geostress at measurement point p, L is regression coefficient of the model. The corresponding regressive result led to an equation using field measurement data detailing in Sect. 2 and Eq. (1):

$$\begin{aligned} \sigma &= 3.2906 + 0.3195\sigma_{\text{sw}} + 0.7883\sigma_{\text{wc1}} + 0.4198\sigma_{\text{wc2}} + 0.7579\sigma_{\text{wc3}} \\ &\quad - 0.3314\sigma_{\text{wc4}} + 0.6667\sigma_{\text{wc5}} \end{aligned} \tag{2}$$

where σ denotes self-weight, tectonic stress corresponded to the six working conditions.

The satisfied high correlation coefficient (R^2 = 0.8826) indicated the result of multi-regression could be described the inversion results well. Theoretically, as for any points of 3D geologic model, corresponding values of geostress vector component could be obtained once their coordinate was ascertained. Thus, the corresponding geostress values were calculated through the procedure of: (1) Regressive coefficients of Eq. (2) were substituted into corresponding working condition model and multiplied by initial condition; (2) The working conditions were subsequently imported into geometric model of FLAC^{3D} via stepwise loading method, i.e. seriatim loaded and overlapped the six working conditions; (3) The initial geostress field of the studied region was obtained by the proceeding of elastic-plastic calculations.

4.2 Comparisons Between Inversion and Measurements

Table 2 detailed the multi-regression results of geostress inversions, MV is the field measured values, CV is the calculated values. It could be found that errors (ε) of shear stress referred to the normal stress were relative high, mainly since shear stress itself is relatively minor. The mean square errors (Δ) of shear stress were also higher than that of normal stress (the lowest 2.21%), which considered the total fitting goodness of positive stress (relevancy) was the highest one.

Thus, the following conclusions could be drawed: (1) The geostress inversion of 3D geologic model gave a relatively satisfied consistency to experimental data with mean errors of 24.18% (for normal stress) and 155.51% (for shear stress), respectively. (2) Most of the regressive values were closed to measurements except stress vector component values of several points.

Table 2. Comparison between MV and CV of field stress.

Stress components	Result	Number of Testing Point (unit: MPa)				Average error	Mean-square error
		P1	P2	P3	P4		
σ_x	MV	28.82	29.2	36.7	22.46	33.43%	16.32%
	CV	22.42	18.53	24.68	31.94		
σ_y	MV	24.92	21.77	34.46	32.63	8.26%	5.20%
	CV	27.33	21.05	29.86	30.44		
σ_z	MV	23.06	20.83	20.64	21.31	30.86%	2.21%
	CV	14.56	23.69	14.66	30.66		
τ_{xy}	MV	5.98	−0.02	2.77	0.27	367.84%	38.38%
	CV	3.24	−0.28	2.37	0.57		
τ_{yz}	MV	0.69	−1.99	18.01	−3.26	82.06%	11.91%
	CV	0.18	1.37	16.15	−0.81		
τ_{zx}	MV	5.14	2.98	−8.66	0.86	16.62%	14.01%
	CV	6.05	2.39	−6.25	0.85		

5 Field Application and Comparative Analysis

5.1 Application in Mining Field

In order to show and observe the regional geostress distribution of Yuncheng coal mine, the geostress inversion and reconstruction result was projected onto AutoCAD diagram to obtain the distribution of the maximum principal stress (SMax). Figure 7 showed the mapping of geostress of No. 1 mining area was simulated according to the drilling information. It indicated that the distribution of vertical stress was highly related to coal seam yield, burial depth, and spatial form. By comparing to the contour lines, an obvious upturned "scalloped" area with blue rendering was found with a high vertical stress value of ~35 MPa. Referring to the above drilling information, the height of such regional coal seam was ~1200 m, while the calculated elevation was ~800–1000 m as depicted in cylindrical diagram of outer ellipse shell (Fig. 7). It was interesting that the points of P2 and P4 (Table 1) also directed to the elliptic region. This phenomenon was in accord with the measurements and suggested that being fold (buried deep of coal seam) was an important reason accounted for the occurrence of "stress concentration zone" in studied region. While the vertical stress in coal rock seam presented as a rising trend with trapezoidal shape. Such results illustrated that the vertical stress was related to its depth.

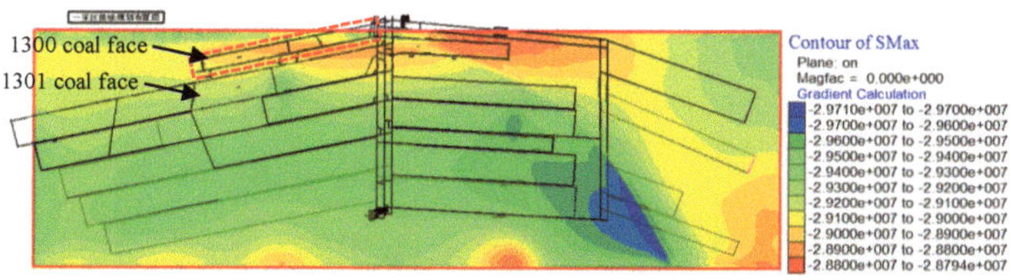

Fig. 7. Target area of geostress inversions and boreholes' arrangement

From the Fig. 7, we can find that SMax of 1300 workface is up to 29 MPa which is greater than the uniaxial compressive strength (UCS) of coal. With the increase of buried depth, the SMax becomes higher gradually. So the SMax in 1301 workface is higher than 1300 workface. Therefore, along with the working face 1300 being advanced, the area adjacent the 1301 workface will influenced by the dynamic pressure from 1300 workface advancing and high static stress of field.

5.2 Evaluation and Verification of Rock Burst Risk

It should be noted that the 1300 working face had a possibility of rock burst with coal seam thickness of 7 m and mining depth >800 m. Based on aforementioned geostress inversion and microseismic monitoring results, the multi-factors coupling evaluation method was used in rock burst risk evaluation in allusion to 1301 working surface. Risky distribution was delineated in Fig. 8 where red, pink and white denoted the high-risk, general-risk, and non-risk area, respectively. By means of this division, risk level in coal mine can be determined quantitatively and corresponding prevention measures are expected to be carried out efficiently in mining operations.

At the same time of geostress inversion, we introduced the microseismic monitoring system for the aims of rock burst monitoring and warning. Figure 9 showed the

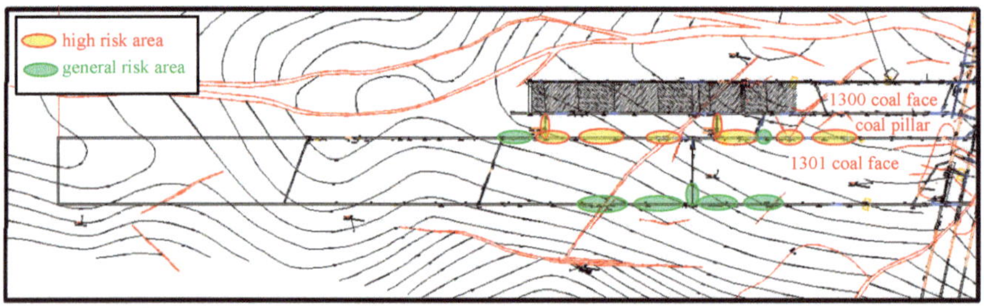

Fig. 8. Rock burst risk evaluation of 1301 working face

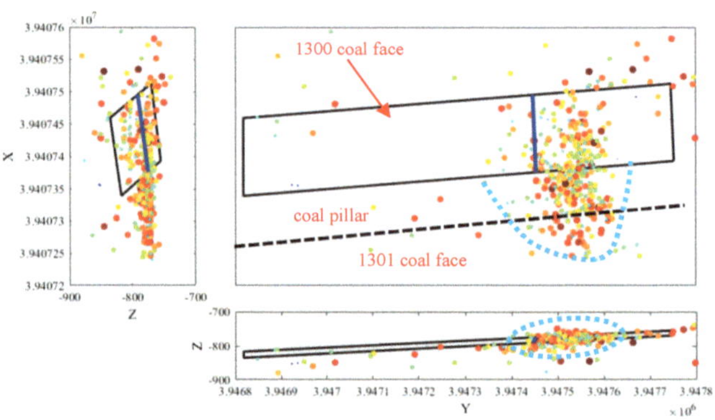

Fig. 9. Microseismic event distribution in the 1300 and 1301 work face

microseismic measurement results of the studied coal mine. This figure showed lots of microseismic events appeared around the working face, especially at recovery process of 1300 working face. Possible reasons for microseismic events appearance could be accounted for two factors: on one hand, the microseismic events were generated by the working face mining, which led to a broken of initial rock stress and subsequently induced a supported pressure peak; on the other hand, the microseismic actually was related to the high geostress value state in studied region.

The geostress distribution nephogram (Fig. 7) showed a maximum principal geostress in 1300 working face was 30–40 MPa, which was quite greater than uniaxial compressive strength of coal seam. It hinted that microseismic was induced by geostress concentration during recovery process on working surface (Fig. 9). The geostress concentration phenomenon was in accordance with the massive microseismic events as well as the high energy microseismic events. Therefore, the geostress distribution in studied region could be indirectly reflected and verified by microseismic monitoring.

It is believed that inversion and analysis works for mining area initial geostress field are complicated and overloaded. In this study, the mulit- regressive value of most measurement points were similar to field measured data, excepted geostress vector component value several points which were caused by factors of borehole number, regression calculation, numerical simulation results, measurement error, etc.. To obtain more real and reliable fitting results, it is considered that the advance multifarious measurements and the guarantees on accuracy and precision of field data are necessary. In addition, the test and optimization of inversion methods are also important in future work.

6 Conclusion

With the gradual increase of mining depth, the factors of geostress, temperature, and flood etc. will seriously affect the mine safety and bring new challenges to safe and efficient mining task. This study developed a rapid method for large, complex 3D geological model construction. The constructed 3D geological model was further used in geostress inversion and reconstitution of a coal mine based comparisons between simulation and measurements. Corresponding the above studies, following conclusions were obtained:

(1) The drilling stress contact method was used to measure the geostress of No. 1 mining area of Yuncheng coal mine by setting four monitoring point. Results showed that the first major stress in original rock was horizontal stress that 1.09–1.95 and 1.40–2.20 times to the minimum value and vertical stress respectively with a direction range of 94°–239.58°. The geostress orientation has certain relevance to regional geological structure.

(2) 3D geologic model was rapid constructed by combining with multiple softwares for geostress analysis. Results indicated that the noval method were fast and accurate for 3D geological model constructions, and could effectively solve difficult problems about low efficiency and grid partitioning difficulties.

(3) Regional geostress field model of No. 1 mining area was obtained by inversion and reconstruction using multi-regression method. The mathematical expression model of geostress field in the target inversion region was obtained using regression coefficients. Comparative analysis showed that mean error of the positive stress and shear stress was 24.18% and 155.51% respectively. The calculation value reflected the distribution characteristics of geostress field to a certain extent, which could provide a basis for later mining design and safety analysis.
(4) The distributed regularity of geostress in No. 1 mining area was obtained from the reconstruction of horizontal stress field in −1000 m−−800 m depth. Based on geostress inversion and microseismic monitoring results, the risky distribution of 1301 coal face was obtained and verified by multi-factors evaluation method.

Acknowledgment. The authors gratefully acknowledge financial support from the National Natural Science Foundation of China (No. 51604115, 51504096), the Fundamental Research Funds for the Central Universities (3142017002), the Technology Project and Self-financing Foundation of Hebei Province (No. 16275512) and the Project of Langfang Key Technology Research and Development Program, China (Grant No. 2016013116).

References

Cai, M., Liu, W., Li, Y.: In-situ stress measurement at deep position of Linglong gold mine and distribution law of in-situ stress field in mine area. Chin. J. Rock Mech. Eng. **29**(2), 227–233 (2010)

Chen, Z., Chen, L., Ji, H.: Geostress field fitting based on the partial least square method. J. Univ. Sci. Technol. Beijing **35**(1), 1–7 (2013)

Kang, H., Lin, J., Zhang, X.: Research and application of in-situ stress measurement in deep mines. Chin. J. Rock Mech. Eng. **26**(5), 929–933 (2007)

Luo, C., Li, H., Liu, Y.: Study of distributing characteristics of stress in surrounding rock masses and in situ stress measurement for deeply buried tunnels. Chin. J. Rock Mech. Eng. **29**(7), 1418–1423 (2010)

Monjezi, M., Hesami, S.M., Khandelwal, M.: Superiority of neural networks for pillar stress prediction in board and pillar method. Arab. J. Geosci. **4**(5), 845–853 (2011)

Matsuki, K., Kaga, N., Yokoyama, T., et al.: Determination of three-dimensional in-situ stress from core discing based on analysis of principal tensile stress. Int. J. Rock Mech. Min. Sci. **41** (7), 1167–1190 (2004)

Manouchehrian, A., Sharifzadeh, M., Moghadam, R.H.: Application of artificial neural networks and multivariate statistics to estimate UCS using textural characteristics. Int. J. Min. Sci. Technol. **22**(2), 229–236 (2012)

Pei, Q., Li, H., Liu, Y., et al.: Two-stage back analysis of initial geostress field of dam areas under complex geological conditions. Chin. J. Rock Mech. Eng. **33**(S1), 2779–2785 (2014)

Qiu, X., Li, S., Li, S.: 3D geostress regression analysis method and its application. Chin. J. Rock Mech. Eng. **22**(10), 1613–1617 (2003)

Wang, J., Li, F.: Review of inverse optimal algorithm of in-situ stress and new achievement. J. China Univ. Min. Technol. **44**(2), 189–205 (2015)
Xiang, P., Ji, H., Zou, J., et al.: An automation platform for generating the 3D numerical grids of mining fields with complex geomorphology. Rock Soil Mech. **36**(4), 1211–1216 (2015)
Yuan, H., Wang, J., Cai, M.: Geometric cross-scale back-analysis and reconstruction method for complicated geo-stress field. J. Min. Saf. Eng. **28**(4), 589–595 (2011)
Zhang, Y., Song, C., Cai, M., et al.: Geostress measurements by hydraulic fracturing method at great depth of boreholes and numerical modelling predictions of stress field. Chin. J. Rock Mech. Eng. **29**(4), 778–786 (2010)

Numerical Simulation Analysis of Layered Slope Landslide Mechanism in Open Pit Mine

Zhigang Tao[1,2(✉)], Wenshuai Han[1,2], Mengnan Li[1,2], and Xiaohui Zheng[1,2]

[1] State Key Laboratory for Geomechanics and Deep Underground Engineering, Beijing 100083, China
1003653730@qq.com
[2] School of Mechanics and Civil Engineering, China University of Mining and Technology, Beijing 100083, China

Abstract. There have been 7 local landslides accumulatively in Nanfen Open-pit Iron Mine since 2010, all of which issued early warning information prior to 4–8 h and avoided danger successfully. However, traditional anchor materials are usually broken off because they don't adapt to the large deformation of slope rock in the monitoring process, leading to the result that the numerical simulation method cannot reach the goal of analyzing force, deformation, sliding surface linking up and the whole process of failure of slope body. Therefore, firstly reveals the characteristics of the whole failure process of slope rock mass from cracks, expansion, penetration, disintegration and slippage, according to the evolution law of the monitoring curve of large-deformation cable with constant resistance (NPR cable) in "landslide 16-1101"; then, builds the large-deformation numerical analysis mechanics model of NPR cable using 3DEC, and carries out the numerical modeling calculation on the whole process that the deposits load of "13-0223 landslide" induces "16-1101 landslide". Finally, carries out the contrastive analysis on the measured result and simulation result, finding that there is a significant consistency between them. It is proved that the large-deformation numerical analysis model of NPR cable can meet the large deformation of rock mass without being pulled off and broken.

Keywords: Landslides
Large-deformation cable with constant resistance (NPR Cable)
Monitoring and warning · Numerical simulation

1 Introduction

The instability failure of slope is a progressive accumulative process [1]. With the accumulation of inner stress of slope, the partial region reaches yielding, and gradually cracks, expands and penetrates, and thus, the sliding surface is formed, and the structural failure occurs. Currently, many scholars develop the monitoring and warning research work with many methods in regard to the instability failure of slope [2–4]. However, the conventional landslide monitoring material belongs to the small-deformation material. When the slope deformation reaches 500 mm, the monitoring material has been snapped for failure, and thus, cannot measure the later deformation characteristics.

L. Zhang et al. (Eds.): GSIC 2018, *Proceedings of GeoShanghai 2018 International Conference: Rock Mechanics and Rock Engineering*, pp. 364–374, 2018.
https://doi.org/10.1007/978-981-13-0113-1_39

With regard to the above problems, He Manchao developed a high constant resistance and large-deformation monitoring cable (NPR) [5–7], which can deform for 2000 mm without being snapped under a condition of 850 kN constant resistance [8], thereby achieving the whole process monitoring and warning goal of landslide disaster. However, limited by the traditional numerical simulation method, it not only cannot directly display the development process of sliding surface, but also cannot quantitatively display the stress, deformation and failure process of sliding surface. In recent years, with the continuous development of computer technology, the calculation and analysis of slope stability with finite element limit analysis method [9, 10] and discrete element method [11, 12] have become the new trend.

This paper builds the large-deformation numerical analysis mechanics model of NPR cable with 3DEC procedure, carries out the numerical modeling calculation on the whole process that the deposits load of "13-0223 landslide" in Nanfen Open-pit Iron Mine induces "16-1101 landslide", and carries out the contrastive analysis on the measured result and simulation result of failure characteristics of NPR cable and slope, striving to explore mechanics model capable of meeting the calculation requirement for whole-process numerical simulation of large-deformation disaster, and providing a theoretical basis for large-deformation numerical simulation calculation of other similar landslides.

2 Regional Geological Characteristics and Analysis of Monitoring Results

2.1 Project Profile

Nanfen Open-pit Iron Mine is one of main production bases of iron ore of Benxi Iron & Steel (Group) Co., Ltd. Affected by the comprehensive internal and external factors, such as geometrical characteristics of slope, lithology, texture and structure, the slope of footwall has occurred about 60 landslide disasters successively since 1964, and the old landslide mass with area of about 11×10^5 m^3 is formed [13, 14], which seriously threatens the safety mining of mine. In recent years, the Mine has been guaranteed with two measures of "slope safety management" and "deep mechanics monitoring and warning system", and the landslide has been effectively controlled.

2.2 Landform and Formation Lithology

Nanfen Open-pit Iron Mine belongs to the Presinian Anshan-type sedimentary metamorphic iron deposit. The mining area has a north-south length of 3 km and east-west width of 2 km, which consists of footwall, hanging wall, silica stone mountain and north mountain. The topographic map of mining area is shown in Fig. 1. The slope of hanging wall belongs to the toppling rock slope, and relies mainly on gneissic migmatite and mica-quartzose schist. The slope of footwall of mining area belongs to the bedding rock slope, and relies mainly on two-mica-quartzose schist and epidotite-hornblende schist.

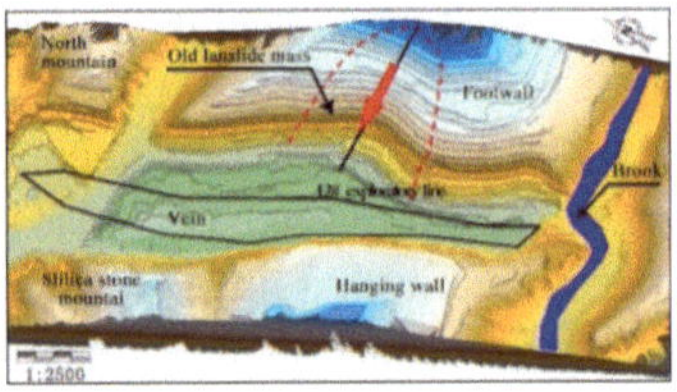

Fig. 1. Topographic map of mining area

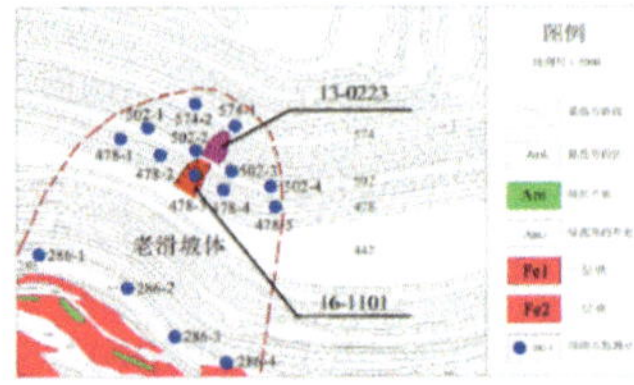

Fig. 2. Distribution of monitoring points

2.3 Layout Characteristics of Monitoring Point

Nine deep sliding force monitoring points were added when designing the footwall (478–502 m bench) of mining area in June 2015 [15], with the buried depth of 55–84 m. And all devices adopt the NPR cable, as shown in Fig. 2.

2.4 Analysis on Monitoring Results of Deep Monitoring Results

In October–November 2016, the long-range forecast was made to “16-1101 landslide” in Nanfen Open-pit Iron Mine at NO. 478-3 monitoring point prior to 45 days according to the warning criteria and threshold of landslide [15], the medium-range forecast was made prior to 30 days, the disaster warning before sliding was made prior to 4 h within the scope of whole mine. The NO. 478-3 monitoring curve is shown in Fig. 3.

As the timely warning is made to “16-1101 landslide” and the hedge response measure is taken decisively, no casualties and property loss are caused. The failure characteristics of top of slope after the landside are shown in Fig. 4.

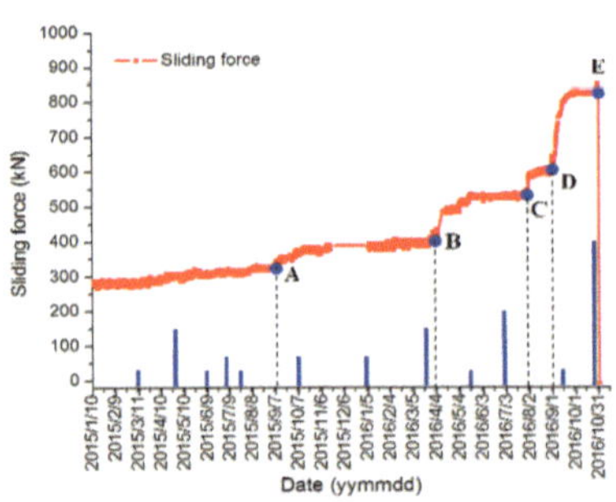

Fig. 3. Coupling monitoring curve

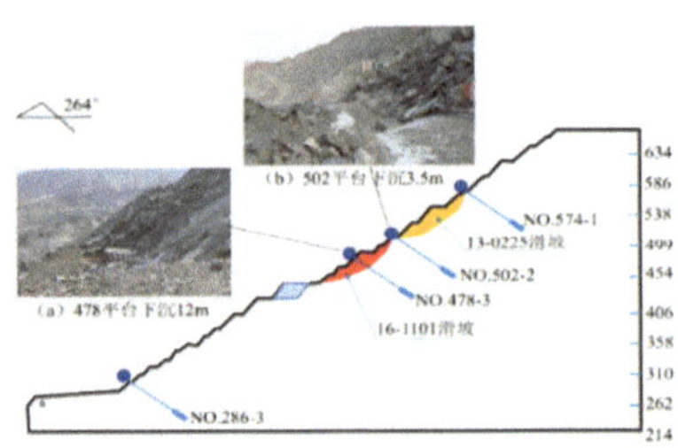

Fig. 4. Characteristics of “16-1101” landslide

3 Characteristics of Slip Mass and Analysis of Inducement Mechanism

3.1 Characteristics of Slip Mass

(1) Overall characteristics of landslide mass

"16-1101 landslide" is distributed in the invert U shaped. The slip mass is leaned for 264°, has a length of about 60 m along the leaning and a width of about 50 m along the strike. This landslide belongs to the bedding stone wedge landslide, and has the maximum thickness of 4.5 m.

(2) Failure characteristics of structural planes of slip mass

The outcrop of structural plane (A surface) on the south side is blurry, the structural plane is rough, it relies mainly on the epidotite-hornblende schist; and the outcrop of structural plane (B surface) on the north side is clear, the structural plane is smooth, it relies mainly on two-mica-quartzose schist. The angle between the A surface and B surface is 50.54°, as shown in Fig. 5. The occurrence information of structural plane and line of intersection is shown in Table 1.

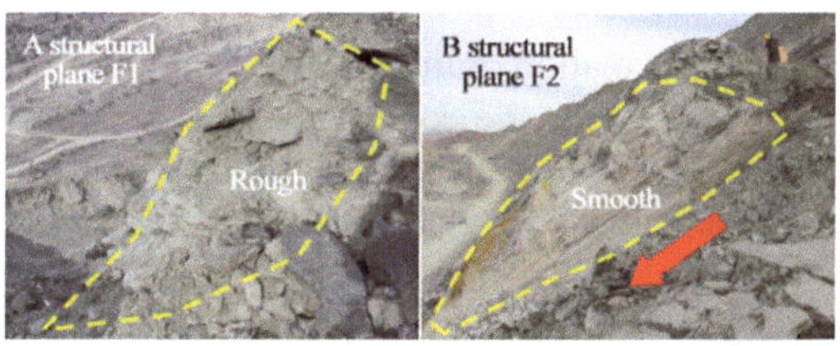

Fig. 5. Characteristics of wedge failure plane AB

Table 1. Parameters of the wedge failure plane

Structural plane	Dip angle/°	Dip angle/°	Formation complex	Physical and mechanical property
F3 slope surface	264	56	Green mudstone-hornblende formation Am (rock mass)	φface = 38°, cface = 414 kPa
F1 (A surface)	300	45	Epidotite-hornblende schist formation Aml (joint surface)	φA = 27.6°, cA = 77.4 kPa
F2 (B surface)	228	48	Two-mica-quartzose schist formation GPel (joint surface)	φB = 27°, cB = 69 kPa
Intersecting line of structural plane	263.13	40.34	/	

3.2 Landslide Mechanism and Cause Analysis

After the occurrence of 13-0223 landslide, about 11.4 × 10^4t loosened rock is accumulated on 502 m bench all the year round and not yet cleared. On October 30, 2016, shovel located on 430 m bench excavates about 20 m below NO. 478-3 monitoring point from north to south. Therefore, 16-1101 landslide occurs due to the disturbances of unloading of slope toe and deposits load of top of slope.

4 Numerical Simulation Analysis of 16-1101 Landslide

4.1 Simulation Stage Design

Based on the above analysis, the landslide simulation is analyzed by two stages:

(1) Deposits load influence stage of 13-0223 slip mass: it simulates that a lot of loose deposits are loaded at the top of 16-1101 slip mass, which accelerates the expansion and penetration of sliding surface after the occurrence of 13-0223 landslide;
(2) Excavation and unloading influence stage of slope toe of 16-1101 slip mass: it simulates that the unloading rebound is generated by carrying out the mechanical excavation disturbance at the slope top on 430 m bench.

To realize the large-deformation numerical simulation calculation of landslide, NPR cable mechanics model having the constant-resistance large-deformation function is firstly built with Fish language in discrete element 3DEC5.2 software, and the numerical modeling analysis is carried out on “16-1101 landslide”.

4.2 Large Deformable Design Principle of NPR Cable

In general, the cable reinforcement unit in 3DEC5.2 is defined by the geometrical parameter, material parameter and characteristics of anchoring agent.

The “tensile yield strength Ft” and “compressive strength Fc” of cable can be designated in 3DEC5.2, but the parameter set in the constitutive model cannot exceed the two limits. Therefore, the elements controlling the cable member only can leave two kinds of key words, i.e. “tensile strength (tension)” and “parameter (gr_coh, gr_fric, gr_k, gr_per) of anchoring agent”, once the constitutive model of cable is determined. The conventional small-deformation cable failure mechanism does not have the constant-resistance large-deformation characteristics, therefore, the small deformation and failure or anchoring agent failure phenomenon is inevitably generated after exceeding the tensile strength.

Conversely, with regard to the NPR cable having the constant-resistance large-deformation characteristics, the anchoring strength which has the same performance as that of common cable is set at the end of anchoring section of NPR cable by Fish language in 3DEC5.2 modeling process; the rigid contact with the slope rock mass is set at the end of free section of NPR cable, i.e. anchor node, to simulate the actual anchor effect, and its large deformation work principle is as follows: when the axial force applied to NPR cable reaches the design constant-resistance value P_0, NPR cable generates the axial tensile deformation immediately; and when the deformation reaches the design elongation, it judges that the anchoring gent is failed, and the cable element is released, so as to reach the large-deformation effect. The parameters are set with Fish language during the work of cable to judge the elongation of NPR cable, and meanwhile, monitor its axial change characteristics. The tensile mechanical characteristics of the NPR cable model and its static force are shown in Fig. 6.

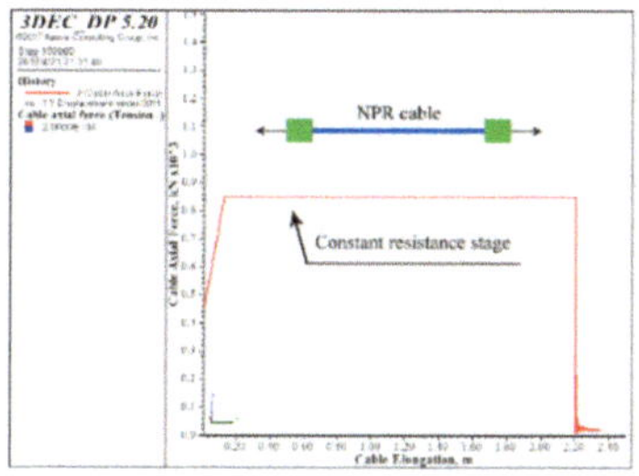

Fig. 6. Numerical anchor model and its static mechanical feature

4.3 Model Building and Grid Division

The maximum width of both 13-0223 landslide mass and 16-1101 landslide mass is 50 m, the elevation limit of the two landslide masses is 502 m–550 m and 430 m–502 m respectively. Considering that the width of model boundary should be more than 2.5 times of width of target body, the elevation of model is taken 277 m–634 m, and the width of strike is taken 200 m. The contact surface element is set between the slip mass and the rock foundation to simulate the spatial relationship of the upper and lower slip masses, thereby simulating the interface characteristics of slip mass in the starting and sliding process. The models of the two slip masses are shown in Fig. 7a, and the setting of main mechanical parameter, which derived from Table 1, is shown in Table 2.

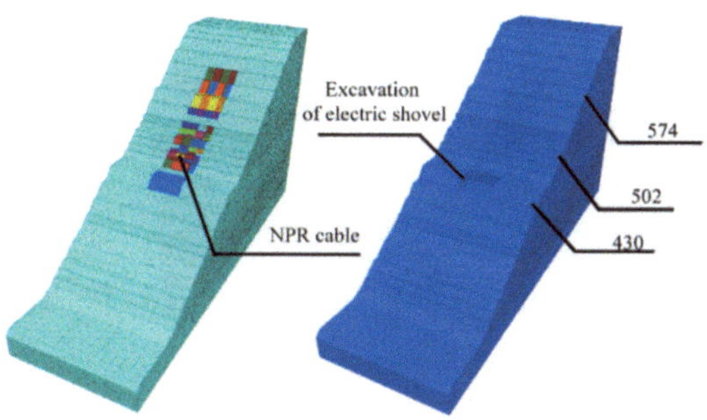

(a) Contact surface model of slip mass (b) Excavation model of slope toe

Fig. 7. Double sliding body model establishment

Table 2. Parameters of rock mechanics

Name of rock mass	Volume weight (kg/m³)	Bulk modulus (GPa)	Shear modulus (GPa)	Cohesion (kPa)	Friction angle (°)	Tensile strength (MPa)
Basement	2873	20	10.9	420	36	1.5
Slip mass	2580	5.2	2.10	70	27	0.2

NPR cable module is built according to the point coordinates of site deep sliding force monitoring point and parameter of buried depth of NPR cable, and its parameters are as follows: the dip angle is 30°, the length is 75 m, and the pretightening force is 40 t. In addition, the excavation length of strike of electric shovel on 430 m platform is set as 70 m, as shown in Fig. 7b.

4.4 Analysis of Simulation Result

(1) Analysis for numerical simulation result of 13-0223 landslide

In this simulation, the displacement nephogram of sliding surface, observation curve of settlement and axial force monitoring curve of cable are observed with the elastic-plastic model based on Moore coulomb yield condition. When the settlement curve is not changed, it judges that the slope has reached the stable state, otherwise, continues to calculate. According to the above principle, the maximum displacement after stabilization of "13-0223 landslide" obtained through numerical calculation is 1.20 m. The slip mass is basically located at the position set on the model contact surface, and when the landslide starts, the strength of contact surface is set as the actual strength of the sliding surface, and then, slipped until to be in stable state. "16-1101 slip mass" is not slid and disintegrated integrally due to reinforcement of a plurality of constant-resistance large-deformation cables, and only presents the crushing failure at the tongue and presents the crack failure at the middle and posterior edge, as shown in Fig. 8a.

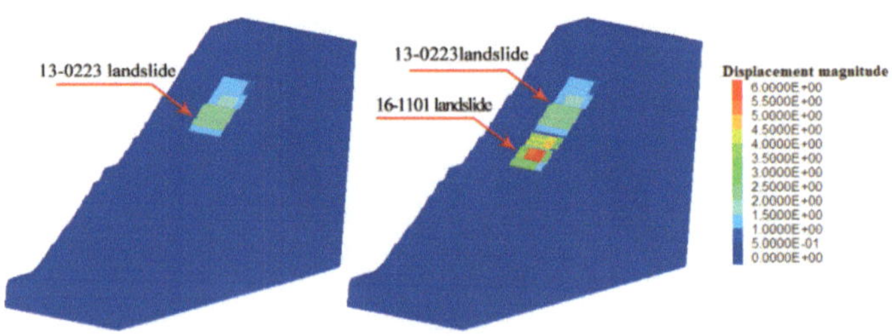

(a) Displacement contour of 13-0223 landslide (b) Displacement contour of 16-1101 landslide

Fig. 8. Displacement contour of 13-0223 and 16-1101 landslide

(2) Analysis for numerical simulation result of 16-1101 landslide

Before 16-1101 slip mass is subjected to the excavation and unloading calculation at the slope toe, the deformation and speed of 13-0223 landside are reset, the influence of deposits load to 16-1101 landslide is fully considered, and then, the operational program is started. The calculation result is shown in Fig. 8a. Figure 8b shows the final displacement nephogram of 16-1101 landslide.

A process of 16-1101 slip mass from being stable to completely unstable is a slow staged strength degradation process. Therefore, the processing of the strength of contact surface (sliding surface) in the actual calculation also should be from the staged weakening to instability of rock mass, and the staged parameters of strength of contact surface from the stable state to the final sliding state are shown in Table 3.

Table 3. Strength degradation stages criteria

Stage classification	Normal stiffness/GPa	Ttangential stiffness/GPa	Cohesion strength/kpa	Friction angle/°
The first stage	10	10	148	21
The second stage	10	10	36	20
The third stage	10	10	25	19
The fourth stage	10	10	18	18

After the starting of "16-1101 landslide", the distribution characteristics of two settlement monitoring points arranged on the surface of slip mass are shown in Fig. 9. From the key point A to key point F, the displacement nephogram of their slices, which can be seen from Fig. 10, appears the whole sliding process of "16-1101 landslide" completely again.

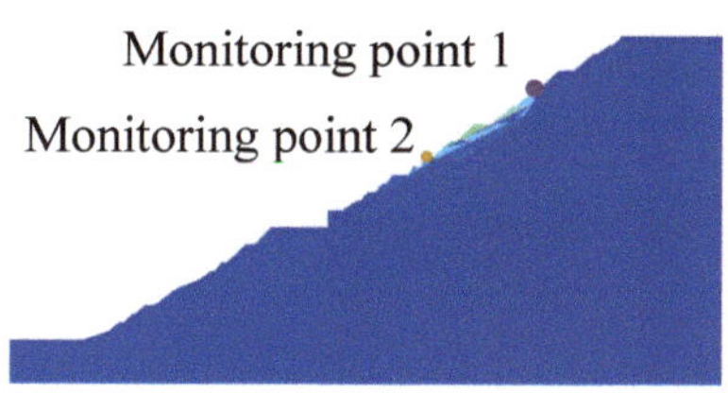

Fig. 9. Distribution of monitoring point for settlement

The change characteristics of axial force of NPR cable arranged on 478 m bench in the slip mass is shown in Fig. 11. The following change characteristics can be seen therefrom:

(1) 13-0223 slip mass reactivation starting stage (stage 1): the change of axial force of NPR cable at this stage is to increase from the initial force of 400 kN to 530 kN, the cable elongation reaches 0.3 mm, and as the reactivation activity of slip mass has the gradual increase trend, 0.6 m relative displacement is generated by sliding;

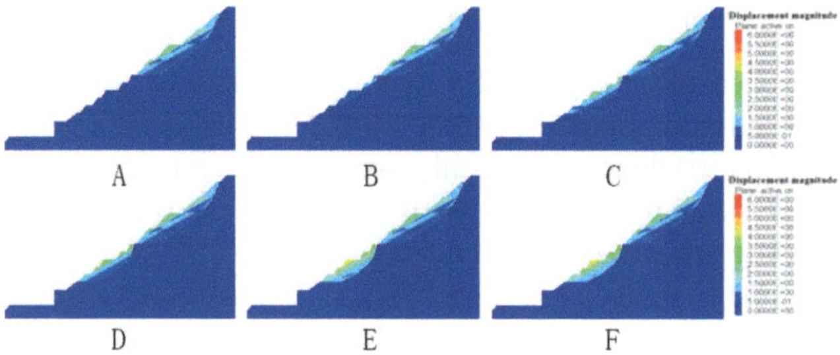

Fig. 10. Displacement slices nephogram of time key point

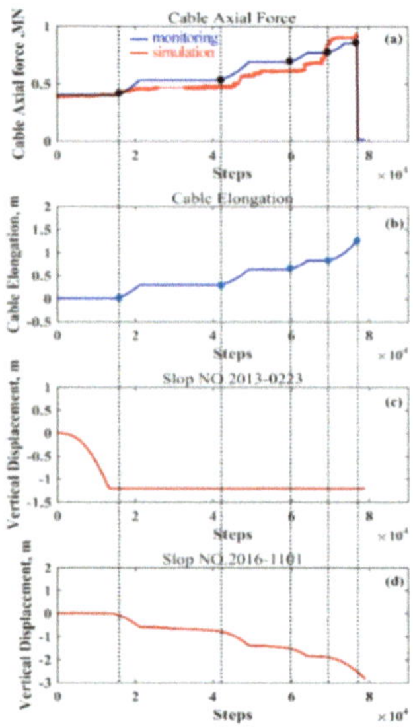

(a) Cable axial force of the measured and simulation result; (b) Cable elongation; (c) Vertical relative displacement (monitoring point 1);(d) Vertical relative displacement (monitoring point 2)

Fig. 11. Numerical simulation monitoring curve of 161101 landslide

(2) Excavation stage of slope toe of 16-1101 slip mass (stage2): Due to the excavation and unlading disturbance of electric shovel on 430 m bench at this stage, the strength of potential sliding surface is gradually decreased from being strong, the axial force monitoring curve is rapidly increased to 680 kN, the cable elongation reaches 0.64 m, and 1.54 m relative displacement is generated by the sliding surface;

(3) Relative movement stage of 16-1101 sliding surface (stage 3): The influence of slope expansion and excavation at 430 m bench at this stage has reaches the safety threshold, the axial fore of NPR cable is rapidly increased to 770 kN, the cable elongation reaches 0.84 m, and 1.87 m relative displacement is generated by the slip mass along the potential sliding surface;

(4) "16-1101 landslide" warning stage (stage 4): This stage undertakes the staged accelerating trending of stage 3, the sliding surface loses stability, and when the relative department of sliding surface reaches 2.54 m, the axial force is suddenly declined to 18 kN, the cable elongation reaches 1.45 m, and thus, the warning information is made.

It is found through the contrastive analysis from Fig. 11(a), that the actual measured result and simulation result have the remarkable consistency, thereby providing theoretical and practical basis for analysis and numerical simulation calculation of landslide mechanism of other similar landslides.

5 Conclusion

(1) The large-deformation numerical analysis mechanics model of NPR cable is firstly built using Fish language in 3DEC, which realized the numerical expression of large deformation theoretical model of NPR cable.
(2) With the staged weakening of contact strength, a reappearance of slip process of landslide, which is induced by disturbance of unloading of slope toe and deposits load of slope top, has been made.
(3) The actual monitoring result and simulation result are subjected to the contrastive analysis, and it is found that the two have the remarkable consistency, and the four stages have the same failure characteristics. Thereby providing a theoretical basis for large-deformation numerical simulation calculation of other similar landslides.

References

1. Tang, X., Zheng, Y., Tang, H.: Numerical analysis on the evolutionary features of deformation and failure modes of slopes. J. Chongqing Univ. **36**(10), 101–113 (2013)
2. Xu, L., Li, S., Liu, X., et al.: Application of real-time telemetry technology to landslide in Tianchi Fengjie of three gorges reservoir region. Chin. J. Rock Mech. Eng. **26**(Supp. 2), 4477–4484 (2007)
3. Wu, K., Sheng, Q., Zhang, Y., et al.: Development of real-time remote monitoring and forecasting system for geological disasters at subgrade slopes of mountainous highways and its application. Rock Soil Mech. **31**(11), 3683–3688 (2010)
4. Tang, R., Wang, J., Fan, X.: Application of TDR technology to landslides monitoring. J. Geol. Hazards Environ. Preserv. **18**(1), 105–110 (2007)
5. He, M.: Research on the double-block mechanics based on Newton force measurement. Chin. J. Rock Mech. Eng. **35**(11), 2161–2173 (2016)
6. He, M.: Real-time remote monitoring and forecasting system for geological disasters of landslides and its engineering application. Chin. J. Rock Mech. Eng. **28**(6), 1081–1090 (2009)
7. Tao, Z., Li, H., Sun, G., et al.: Development of monitoring and early warning system for landslides based on constant resistance and large deformation anchor cable and its application. Rock Soil Mech. **36**(10), 3032–3040 (2015)

8. He, M., Gong, W., Wang, J., et al.: Development of novel energy-absorbing bolt with extraordinarily large elongation and constant resistance. Int. J. Rock Mech. Min. Sci. **67**(67), 29–42 (2014)
9. Zheng, Y.R., Zhao, S.: Limit state finite element method for geotechnical engineering analysis and its applications. China Civ. Eng. J. **38**(1), 91–98 (2005)
10. Zheng, Y.R., Deng, C.J., Zhao, S.Y., et al.: Development of finite element limit analysis method and its applications in geotechnical engineering. Eng. Sci. **9**(3), 10–36 (2007)
11. Ge, Y., Tang, H., Xiong, C., et al.: Effect of sliding plane mechanical parameters on landslide stability-a case study of jiweishan rockslide in Wulong, Chongqing. Chin. J. Rock Mech. Eng. **33**(Supp. 2), 3873–3884 (2014)
12. Zhang, Y., Zhang, M., Hu, X., et al.: Deformation and failure mode and stability analysis of the slope under multi-stage dike intrusions. J. Eng. Geol. **23**(2), 227–232 (2015)
13. Fan, L., Zeng, P., Sun, D.: Discussion on haulage way transition by stages boundary in Nanfen open pit Iron mine. Metal Mine **42**(6), 37–40 (2012)
14. Sun, D., Tao, Z., Fan, L., et al.: Stability assessment and sensitivity analysis of slope at Nanfen open pit iron mine. Metal Mine **40**(5), 57–60 (2011)
15. Tao, Z., Zhang, H., Peng, Y., et al.: Frame structure and engineering applications of multi-source system cloud service platform of landslide monitoring. Rock Soil Mech. **36**(10), 1–10 (2017)

The Stability of Open-Pit Slope in Permafrost Regions Under Global Warming

Guofeng Li[1(✉)], Ning Li[1,2], and Naifei Liu[1]

[1] Xi'an University of Technology, Xi'an 710048, China
liguofeng205@126.com
[2] Lanzhou Jiao Tong University, Lanzhou 730070, China

Abstract. In the global warming trend, the permafrost area is decreasing. And the change of temperature seriously affects the safety and stability on Open-pit slope under the alternation of freezing and thawing. Based on FLAC^{3D}, the simplified algorithm of THM coupling with phase change is developed and upgraded again. The change law of failure area in one time freezing and thawing and various stability influence factors of the permafrost slope were discussed by yielding approach index. The results show that the frozen slope is local failure, from April to July is danger every year, and that the freezing temperature increment and the height of slope are still the main influence factors and that the water content, the times of freeze-thaw cycles and the temperature increment of surrounding boundary are severely affect the stability of the slope. And the research can be a significance reference for further understanding the slope stability under freeze-thaw cycles.

Keywords: Global warming · Phase transition · THM coupling
Slope stability

1 Introduction

Intergovernmental Panel on Climate Change (IPCC) estimated that the global average temperature in the 21st century will increase 1.4–5.8 °C, and the Qinghai-Tibet Plateau may rise 2.2–2.6 °C in the next 50 years [1]. In the global warming trend, the plateau shows a consistent warming trend with difference [2–5]. In the past 50 years, the temperature gradient has reached 0.037 °C/a and the four seasons has been generally increased, nevertheless the spring and summer were obvious but autumn and winter were not and the lowest is the highest 1–3 times. After 50 years and 100 years, when the temperature increment is 0.02 °C/a, the permafrost area respectively decreases by about 8.8% and 13.4%, and when the temperature increment is 0.052 °C/a, the permafrost area respectively decreases by about 13.5% and 46% [6]. Under the action of freeze-thaw cycle, global warming will aggravate the influence of frost heaving and thaw settling even exacerbate disasters occurring for Plateau permafrost region.

The freezing and thawing of rock and soil involves THM coupling and water-ice phase transition. Liu [7–10], considered the water migration and water-ice phase transition on the single joint, simulated the frost heave load with the equivalent thermal expansion coefficient, and analyzed the stress field distribution near the joint by

L. Zhang et al. (Eds.): GSIC 2018, *Proceedings of GeoShanghai 2018 International Conference: Rock Mechanics and Rock Engineering*, pp. 375–387, 2018.
https://doi.org/10.1007/978-981-13-0113-1_40

fracturing and frost heaving thermodynamic coupling model. He [11, 12], established a coupled model of moisture migration and developed a finite element program that could reflect various extrinsic effects, and then put forward a new FEM with transient heat transfer theory combining the temperature change characteristics on freezing-thawing. Kang [13], established rock quasi-creep frosting constitutive model, developed the frost heave program by FISH function and studied the frost heave fracture expansion algorithm with topological theory. Xu [14], established the coupling model of low temperature based on the convective heat transfer theory of porosity media. Zhang [15], introduced the pressure dissolution model of granular aggregates in the THM coupling finite element program to study the effect of pressure dissolution on porosity, permeability coefficient and pore pressure. Bai [16], established the frozen soil HT coupling differential equation based on unsaturated soil seepage and heat conduction theory, and achieved temperature and water full coupling by secondary development of the COMSOL software. Xue [17], established a composite unit model of seepage-heat transfer coupling for fractured rock mass taking into account the influence of fissure temperature and fluid. Shen [18], explored rock with crack internal temperature field and frost heave load evolution process from many aspects such as the heat transfer theory, phase transition theory, elastic-plastic mechanics and fracture mechanics theory.

Scholars have done a lot of numerical simulation researches on THM coupling and water-ice phase transition, however, the theory and simulation method were complex for engineers or beginner. Therefore, this paper improves and upgrades the simplified THM coupling algorithm proposed by author [19]. Based on the ideal slope model and yielding approach index, the change law of failure area in one time freezing and thawing and various stability influence factors of the permafrost slope were analyzed, especially under the global warming trend.

2 Upgrading the Simplified Algorithm

2.1 Water-Ice Phase Transition Simplified Algorithm

The phase change latent heat term is not included in the THM coupling energy balance equation of FLAC^{3D}. Although scholars have used some methods to equivalently treat, the calculation theory is relatively complex for beginners or engineers. So, in this section, the previous simplified algorithm [19] is improved and upgraded. The phase change energy term is replaced by the energy change from system temperature difference. The flow chart of water-ice phase transition is shown in Fig. 1, and default order is freezing before thawing. Ideas are as follows:

When the outside environment temperature is lower than the freezing point (T0), the model enters the 'freezing' detection state of the water-ice phase change. At this time, the temperature of each unit is traversed. When the cell temperature (T1) is lower than the freezing point (T0), the unit temperature is reset to the freezing point (T1 = T0) in the Δt time. And the cumulative energy ($\Sigma Q1 = \Sigma \Delta T \cdot Cv$) that corresponds to cumulative temperature difference ($\Sigma \Delta T = \Sigma |T1 - T0|$) is not less than the latent heat of unit phase transition, the water-ice phase transition is completed.

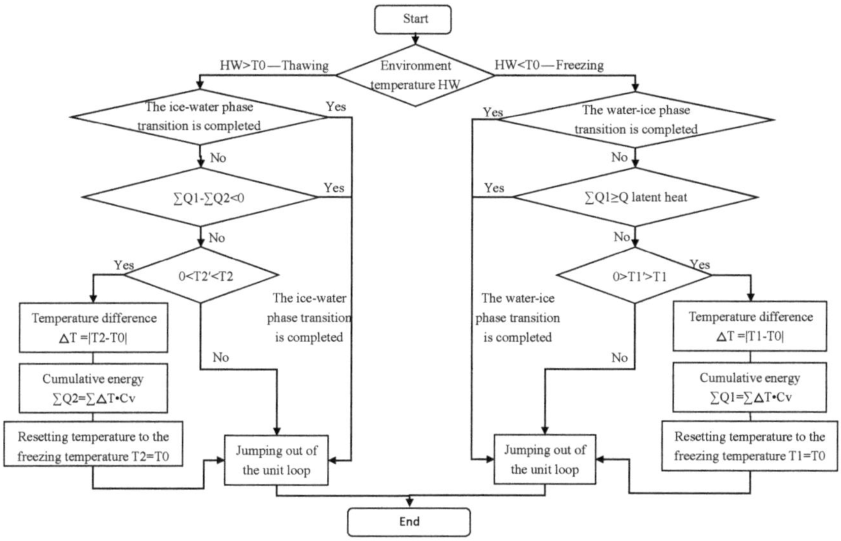

Fig. 1. Simplified flow chart of phase transition.

When the outside environment temperature is higher than the freezing point (T0), the model enters the "thawing" detection state of the ice-water phase change. At this time, the temperature of each unit is traversed. When the cell temperature (T2) is higher than the freezing point (T0), the unit temperature is reset to the freezing point ($T2 = T0$) in the Δt time. And the cumulative energy ($\Sigma Q2 = \Sigma \Delta T \cdot Cv$) that corresponds to cumulative temperature difference ($\Sigma \Delta T = \Sigma |T2 - T0|$) is not less than the latent heat of unit phase transition, the ice-water phase transition is completed. If considering the continuity of the unit bi-directional phase transition and freezing is before thawing, the cumulative energy will be canceled by ($\Sigma Q1 - \Sigma Q2 < 0$) and the ice-water phase transition is completed.

2.2 THM Coupling Simplified Algorithm

On the basis of mechanical, hydrological and thermal respective algorithm, the 'coupled loop core' algorithm is obtained by each other parallel-series connection. After the testing, the way of three fields successively acting that equivalent to series from thermal-hydrological-mechanical is relatively simple and practical. The flow chart of the coupling loop algorithm is shown in Fig. 2. The temperature detection is required, before and after the coupling cycle core, as well as before the phase change detection. In order to ensure the balance state, the time sub-step must be the same or in a multiple relation so that the total time in each 'coupled loop core' is equal.

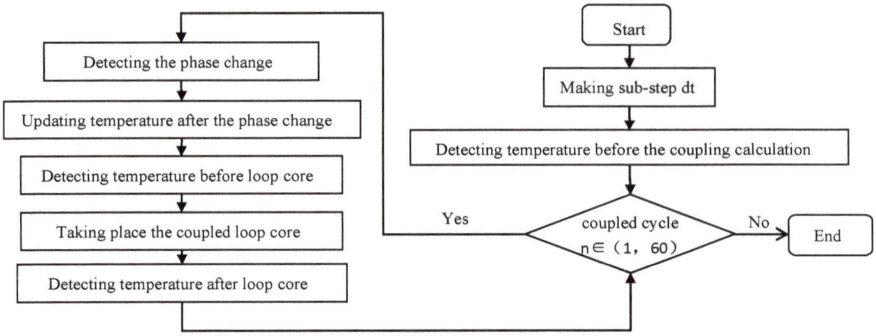

Fig. 2. Flow chart of coupling cycle.

3 Models and Parameters

3.1 Project Overview

Qinghai MU-LI Coal Mine that located in the permafrost zone of the Altun Mountains and Qilian Mountains in the Qinghai-Tibet Plateau is one of the largest coal resources zone in Qinghai Province. The thickness of seasonal frozen soil is about 5–6 m but the thickness of permafrost is about 58–60 m. The annual minimum temperature is −34 °C, the annual maximum temperature is 19.8 °C, the annual average temperature is −4.2–−5.1 °C and the annual average soil temperature is −1.0–−3.5 °C. The surface of permafrost forms a seasonal melting layer in the warm season for seasonal temperature changing. Excavating open-pit slope not only changes the stress field, but also changes the temperature field and seepage field. And the slope has appeared many creep and collapse under freezing and thawing cycle, especially under the global warming trend.

3.2 Model Establishment

Investigating the stability and stability influence factors of the MU-LI open-pit mine, taking into account the timeliness of the coupling calculation and the project requirement, the analysis model is selected as shown in Fig. 3. The constitutive model, boundary condition and initial condition are adopted as follows: The M-C model, the isotropic conduction-convection model and the isotropic flow model are used to calculate the THM coupling. The initial stress field is generated by the dead load of rock

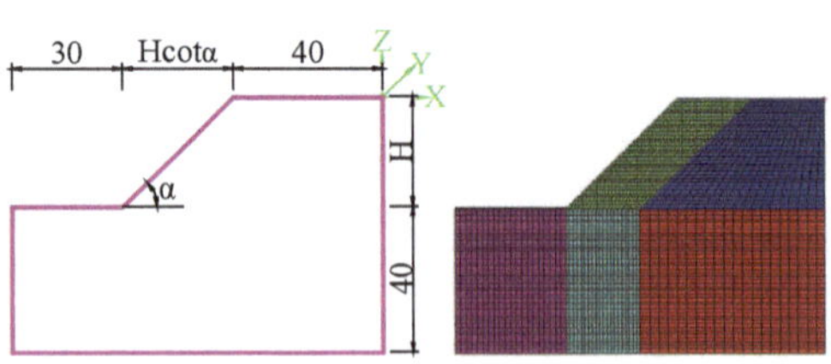

Fig. 3. Test model.

and the initial temperature is imposed by the real rock temperature, at the same time, the saturation of rock is assumed. The bottom of model is the XYZ direction displacement fix and sides are normal direction displacement fix. Assuming that there is no fluid and temperature exchange at sides and the bottom, so the impermeable adiabatic boundary condition is adopt, however, the slope surface has temperature exchange with the air, so the permeation convection temperature boundary is used.

3.3 Parameters Discussion

Assuming that the parameters change laws of rock mass with temperature and freeze-thaw cycles are the same as the rock, so that the initial values just need to be multiplied by the corresponding reduction coefficients. The year freeze-thaw cycle just is considered but the circadian cycle is neglected. The initial mechanical parameters of slope rock mass are shown in Table 1 and assuming the relationship of E and C with the temperature (*T*) and cycle times (*n*) [14, 20, 21] is as follows:

$$\begin{aligned} E_{(T)} &= 0.0086T^2 - 0.2721T + 9.8477 \\ C_{(T)} &= 0.0007T^2 - 0.0730T + 7.6967 \end{aligned} \tag{1}$$

$$\begin{aligned} K_E &= 2.0e{-}5n^3 + 0.0018n^2 - 0.0560n + 1 \\ K_C &= 2.0e{-}5n^3 + 0.0018n^2 - 0.0583n + 1 \end{aligned} \tag{2}$$

Table 1. Initial physical and mechanical parameters of rock.

	Density $\rho/(\mathrm{Kg/m^3})$	Internal friction angle $\Phi/(°)$	Internal cohesion C/(MPa)	Bulk modulus K/(GPa)	Shear modulus G/(GPa)
Rock	2650	50.0	1.5	13.3	8.0

It can be considered that specific heat (C_v) and thermal conductivity coefficient (λ) of rock are unchanged for the temperature range is relatively small. When the temperature is higher or lower than the freezing point, the water or ice parameters are respectively used. When the water-ice phase change or ice-water phase change, the water or ice parameters are respectively used. Based on the results [14, 22], thermal expansion coefficient (β) of rock matrix in the water-ice phase change is $-1.5\text{e-}5\ °\mathrm{C}^{-1}$ and in the ice-water phase change is $-1.2\text{e-}5\ °\mathrm{C}^{-1}$. The convective heat transfer coefficient (*h*) is respectively 5–25 $\mathrm{w/m^2}$ °C and 20–100 $\mathrm{w/m^2}$ °C under the natural convection and forced convection, so *h* may be 5 $\mathrm{w/m^2}$ °C for the stationary air. In $\mathrm{FLAC^{3D}}$ software, the water content can be replaced by porosity. When the saturation is 1.0, water density is 1000 $\mathrm{kg/m^3}$, substrate density is 2650 $\mathrm{kg/m^3}$, the relationship is as follows:

$$w = \frac{\rho_w S_r n}{\rho_d} = 0.38n \tag{3}$$

Table 2. Thermal-Hydraulic parameters of rock and water.

Materials		Specific heat C_v/(J/Kg·°C)	Thermal conductivity coefficient λ/(w/m·°C)	Permeability coefficient k/(cm/s)	Thermal expansion coefficient β/(°C^{-1})
Rock	Above the ice point	816	2.7	1.0e-7	3.0e-6
	Below the ice point	816	2.7	1.0e-10	−1.5e-5
Water-ice	Above the ice point	4190	0.52	/	2.1e-4
	Below the ice point	1880	2.21	/	−1.5e-3

The thermal and hydrological field parameters [13, 14, 16, 22, 23] are shown in Table 2. The initial temperature of the slope surface is shown in Fig. 4 and the initial temperature of the soil slope is shown in Fig. 5.

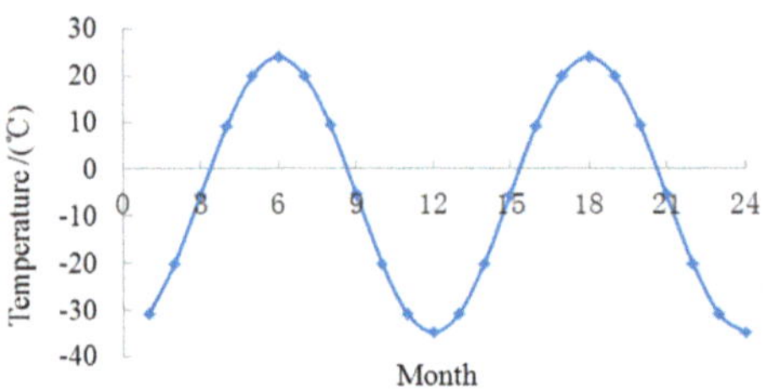

Fig. 4. The temperature changes with time.

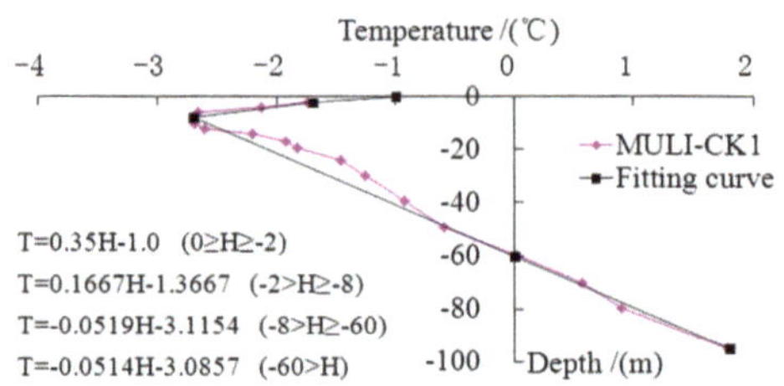

Fig. 5. The temperature changes with the depth of rock.

4 Slope Stability Analysis

4.1 Analysis Scheme

The stability affecting factors of slope under the freezing-thawing conditions are mainly occurrence state, lithology, temperature and water. Especially under global warming, the climate temperature, the water content and the migration water are the most serious. Therefore, this section, based on the simple algorithm and the basic scheme, designs the comparison analysis scheme as shown in Table 3.

4.2 Yielding Approach Index (YAI)

The froze-thaw slop mainly occurs sliding in the shallow surface and collapsing at the local block or sheet. Therefore, the general slope stability evaluation method will no longer be applicable but the indirectly local method may be used to evaluate, such as the withstanding the number of freeze-thaw cycles, local deformation, or the local damage failure area. Zhou [24] proposed the YAI and Gao [25] redefined and amended. The YAI is the ratio of the distance from the point to the yield boundary line and the

Table 3. Test scheme of slope stability under freeze-thaw cycle.

Affecting factors			Analysis scheme	
			Basic scheme	Comparison scheme
Shape	①	The height of slope/(m)	**30**	10/20/40/50/60/70
	②	The angle of slope/(°)	**45**	20/30/40/45/50/60
	③	The width of road/(m)	**0**	1/2/3/4
Global warming	①	The boundary temperature increment/(°C)	**0**	1/2/3/4/5
	②	The water content (The porosity)	**0.1**	0.05/0.2/0.3
	③	The water migration increment (The porosity increment)	**0**	0.01/0.02/0.03/0.04
Other	①	The freezing temperature reduction/(°C)	**0**	2/4/6
	②	The convective heat transfer coefficient/ (w/m^2°C)	**5**	10/15/20
	③	The coefficient of thermal expansion/(°C^{-1})	**1.5**	0.9/1.2/1.8
	④	The number of freeze-thaw cycles/(times)	**5**	10/15/20

distance from the coordinate origin of the π plane to the yield boundary line, along the direction from the coordinate origin of the π plane to the stress point. For M-C strength theory, it can be expressed:

$$Y = f(I_1, J_2, \theta_\sigma) = 1 - \frac{\sqrt{J_2}}{k - \alpha I_1} \quad (4)$$

When

$$\frac{\sigma_1 + \sigma_3}{2} < \frac{\sigma_t}{2}, k = \frac{3c\cos\varphi}{3\cos\theta_\sigma - \sqrt{3}\sin\varphi\sin\theta_\sigma}, \alpha = \frac{\sin\varphi}{3\cos\theta_\sigma - \sqrt{3}\sin\varphi\sin\theta_\sigma} \quad (4.1)$$

When

$$\sigma_t > \frac{\sigma_1 + \sigma_3}{2} \geq \frac{\sigma_t}{2}, k = \frac{3\sigma_t + \sqrt{3J_2}\sin\theta_\sigma}{3\cos\theta_\sigma}, \alpha = \frac{1}{3\cos\theta_\sigma} \quad (4.2)$$

Among them, I_1 is the first principal stress invariant, J_2 is the second partial stress invariant, θ_σ is the stress lode angle, c is the internal cohesion, φ is the internal friction angle.

Because of the limitation of initial numerical stage, based on the basic scheme, the YAI volume percentage on the 12-month of the fifth freeze-thaw cycle was analyzed as shown in Fig. 6.

It can be seen from Fig. 6 that the volume percentage of YAI subsection each month is basically the same. The part of YAI $\leq$ 0.6 actually is the seasonal frozen soil layer affected by the external environment. Since the YAI just is degree of yield or damage on one unit, so the local volume (YAI $\leq$ 0.6) weighting average may be to

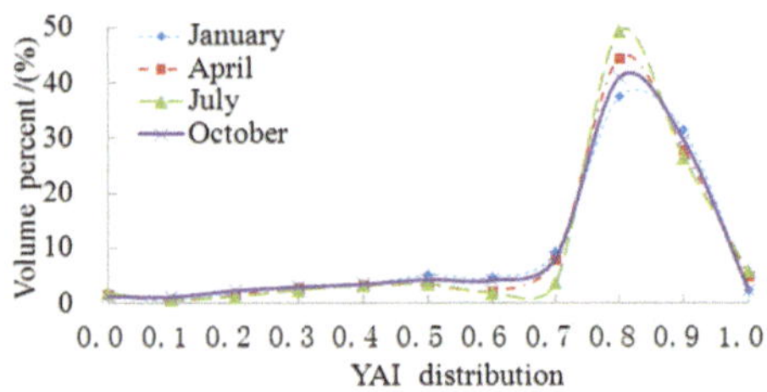

Fig. 6. The volume percentage changes with YAI distribution.

represent the degree of local security of the freeze-thaw slope. Thus, local yielding approach index (Y_I) can be defined as:

$$Y_L = \sum_{i=0}^{0.6} Y_i V_i \Big/ \sum_{i=0}^{0.6} V_i \tag{5}$$

4.3 The Most Dangerous Month

The Y_I of each month from 4th to 5th times freeze-thaw cycles is analyzed, as shown in Fig. 7, to find the most dangerous month under the freezing-thawing conditions.

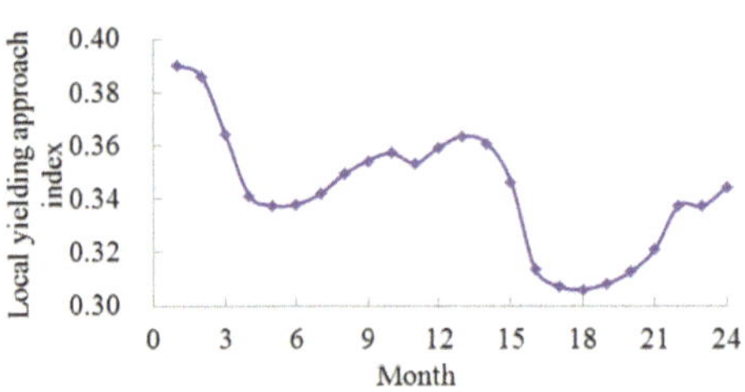

Fig. 7. The YI changes with month.

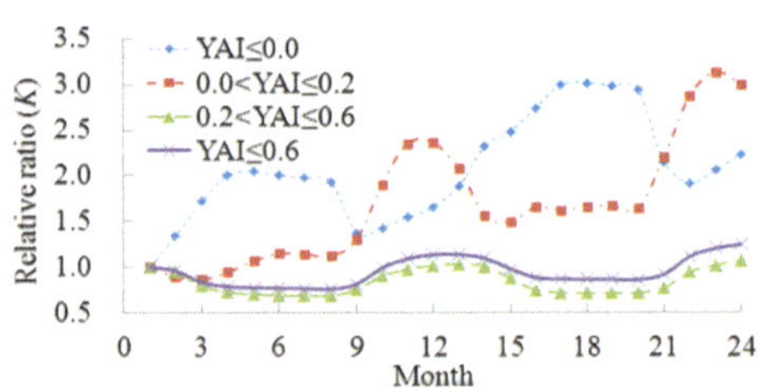

Fig. 8. The K changes with month.

It can be seen from Fig. 7 that as the freezing-thawing continuing, the Y_I decreases rapidly in March and the lowest month is from April to July, but then the Y_I rises with water frozen. That is, the dangerous month is from April to July every year, and this moment the seasonal frozen soil is most likely to have local collapse or slip damage. Therefore, follow-up stability analysis is based on the results of the annual June.

Gao pointed out that YAI $\geq$ 0.2 represents structural safety. The above analysis shows that YAI $\leq$ 0.6 is the effective part for the freezing-thawing rock slope. Therefore, the damage range is 0.2 < YAI $\leq$ 0.6, the near failure range is 0 < YAI 0.2 and the already failure range is YAI $\leq$ 0.0.

The relative ratio *K* are defined, which represents the range ratio of the damage failure on comparison scheme and basic scheme. The *K* of each month from fourth times to fifth times freeze-thaw cycles are shown in Fig. 8.

It can be seen from Fig. 8 that the affective volume (YAI $\leq$ 0.6) increases as the freeze-thaw proceeds continuously. The already failure volume (YAI $\leq$ 0.0) gradually increases that reaches the extreme value from April to August and that reduces after freezing and that is about doubled after one time freeze-thaw cycle as the increase of the temperature. However, the near failure volume (0.0 < YAI $\leq$ 0.2) and the damage volume (0.2 < YAI $\leq$ 0.6) increase with the freezing-thawing proceeds continuously and reach the extreme value when the water is frozen.

4.4 Stability Influence Factors Analysis

Under the action of freeze-thaw cycle, the slope stability influence factors mainly includes the height, the angle, the width of road, especially under global warming, the boundary temperature increment, the water content, the water migration increment, the freezing temperature, the convective heat transfer coefficient, the thermal expansion coefficient and the freeze-thaw cycles times. When changing these factors, based on the basic scheme, the ratio K is shown Fig. 9(a–j).

As can be seen from Fig. 9(a–c), with the increase of the height of slope, the already failure volume and the near failure increases and reaches the extreme at 40 m then gradually decreases, and the damage volume that is the same as the affective volume gradually increases. With the increase of the slope angle, the already failure volume that is the same as the affective volume is gradually increasing, and the near failure volume gradually increases and reaches the extreme at 50°, but the damage volume decreases. With the increase of the width of the road, the already failure volume firstly reduces to 2 m reaching the extreme value and then rises, and the near failure volume is fluctuant, and the damage volume that is the same as the affective volume gradually decreases. The existence on failure volume extreme values of slope height and angle may be related to the temperature field or the thickness of permafrost. And adding the width of road is equivalent to reducing the slope height and increasing the slope angle. That is, the slope with road endures the combined action from the slope angle and the slope height, and the action of the reduction on slope height is more obvious. The affective volume averagely respectively increases about 0.074, 0.013 and 0.006, when assuming that the height of slope (10–70 m), the angle of slope (20–60°) and the width of road (0–4 m) increase respectively 10%.

As can be seen from Fig. 9(d–f), with the increase of the boundary temperature, the already failure volume gradually decreases, but the near failure volume that is the same as the damage volume and the affective volume is gradually increasing. With the increase of the porosity, the already failure volume increases and reaches the extreme at 0.2 then gradually decreases, the near failure volume is fluctuant, and the damage volume that is the same as the affective volume is gradually increasing. With the increase of the porosity increment, the already failure volume and the near failure volume gradually reduces, and the damage volume that is the same as the affective volume is gradually increasing. The affective volume averagely respectively increases about 0.017, 0.059 and 0.004 when assuming that the boundary temperature increment (0–5 °C), the porosity (0.05–0.30) and the porosity increment (0.00–0.04) increase respectively 10%.

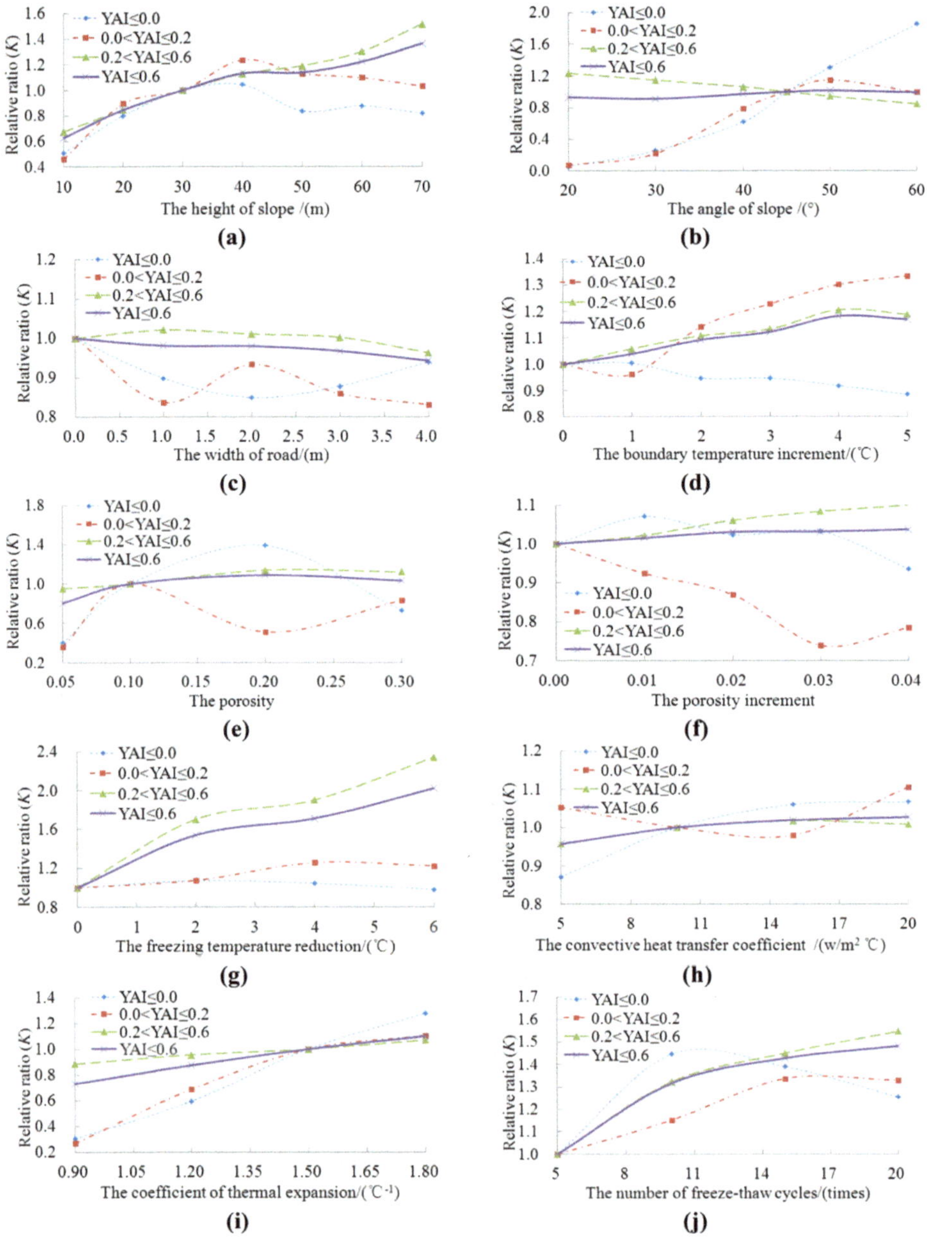

Fig. 9. The *K* changes respectively with the height of slope (a), the angle of slope (b), the width of road (c), the boundary temperature increment (d), the porosity (e), the porosity increment (f), the freezing temperature reduction (g), the convective heat transfer coefficient (h), the thermal expansion coefficient (i), the number of freeze-thaw cycles (j).

As can be seen from Fig. 9(g–j), with the increase of the freezing temperature reduction, the already failure volume has decreased tendency, the near failure volume that is the same as the damage volume and the affective volume is gradually increasing. With the increase of the convective heat transfer coefficient, the already failure volume is the same as the damage volume and the local yield volume is gradually increasing, but the near failure volume firstly reduces to 15 reaching the extreme value and then rises. With the increase of the coefficient of thermal expansion, the already failure volume, the near failure volume, the damage volume and the affective volume are all gradually increasing. With the increase of the number of cycles reflecting the damage history of freezing and thawing, the already failure volume firstly increases to 10 times reaching the extreme value and then reduces gradually, the near failure volume, the damage volume and the affective volume all gradually increasing. The affective volume averagely respectively increases about 0.102, 0.007, 0.037 and 0.048 when assuming that the reduction of freezing temperature (0–6 °C), the convective heat transfer coefficient (5–20 w/m^2 °C), the coefficient of thermal expansion (0.9–1.8 $°C^{-1}$) and the number of cycles (5–20 times) increase respectively 10%.

Overall, under the basic scheme, when each influence factors increase 10%, the degree of effect on the affective volume is: freezing temperature reduction > slope height > porosity > the number of freeze-thaw cycles > thermal expansion Coefficient > boundary temperature increment.

5 Conclusions

Under the trend of global warming, the temperature of permafrost is increasing year by year and the slope stands in the temperature of freezing-thawing alternation. The change of temperature affects the stability and safety of the project. The simplified algorithm of THM coupling with phase change is developed and upgraded again and the yield approach index is resorted. Based on $FLAC^{3D}$ and the slope ideal model, the change law on the region of damage or failure in a freezing and thawing cycle and the stability influence factors are discussed. The result shows that the effective range of the frozen rock slope is YAI $\leq$ 0.6, and the dangerous month is every year from April to July. The effect degree and the change trend of influence factors for the slope failure and damage region is different, especially the failure region on the water content and the width of road exist extreme. The freezing temperature reduction and the slope height are still the main influencing factor. The porosity, the number of freeze-thaw cycles and boundary temperature increment still seriously affect the slope stability.

References

1. Jiao, S.H., Wang, L.Y., Liu, G.N.: Prediction of Tibetan Plateau permafrost distribution in global warming. Acta Scientiarum Naturalium Universitatis Pekinensis **52**(02), 249–256 (2016)
2. Song, C., Pei, T., Zhou, C.H.: Research progresses of surfacc temperature characteristic change over Tibetan Plateau since 1960. Prog. Geogr. **31**(11), 1503–1509 (2012)

3. Zhou, N.F., Qin, N.S., Tu, Q.P., Li, D.L.: Analyses on regional characteristics of temperature changes over Qinghai Xizang Plateau in recent 50 years. Plateau Meteorol. (03), 344–349 (2005)
4. Liu, G.F., Lu, H.L.: Basic characteristics of major climatic factors on Qinghai Tibet Plateau in recent 45 years. Geograph. Res. **29**(12), 2281–2288 (2010)
5. Li, L., Chen, X.G., Wang, Z.Y., Xu, W.X., Tang, H.Y.: Climate change and its regional differences over the Tibetan Plateau. Adv. Clim. Chang. Res. **6**(03), 181–186 (2010)
6. Nan, Z.T., Li, S.X., Chen, G.D.: Research prediction of permafrost characteristic change over Tibetan Plateau in 50 and 100 years. Sci. China Ser. D Earth Sci. (06), 528–534 (2004)
7. Liu, Q.S., Kang, Y.S., Huang, X., Xu, C.Z.: Near problems of freeze-thaw damage in fractured rock and their research status. Rock Soil Mech. **33**(04), 971–978 (2012)
8. Liu, Q.S., Huang, S.B., Kang, Y.S., Cui, X.Z.: Advance and review freezing-thawing damage of fractured rock. Chin. J. Rock Mech. Eng. **34**(03), 452–471 (2015)
9. Liu, Q.S., Huang, S.B., Kang, Y.S., Pan, Y., Cui, X.Z.: Numerical and theoretical studies on frost heaving pressure in a single fracture of frozen rock mass under low temperature. Chin. J. Geotech. Eng. **37**(09), 1572–1580 (2015)
10. Liu, Q.S., Kang, Y.S., Liu, X.Y.: Analyses of stress field and coupled thermo-mechanical simulation of single-fracture frozen rock masses. Chin. J. Rock Mech. Eng. **30**(02), 217–223 (2011)
11. He, M., Li, N., Lui, N.F.: Analysis and validation of coupled heat-moisture-deformation model for saturated frozen soils. Chin. J. Geotech. Eng. **34**(10), 1858–1865 (2012)
12. He, M., Li, N., Gao, H.H., Liu, N.F.: Extended finite element method analysis for the transient temperature filed with phase change. J. Glaciol. Geocryol. **38**(04), 1044–1051 (2016)
13. Kang, Y.S.: Study on mechanical characteristics of damage due to freeze-thraw action in fractured rock mass and multi-field coupling in freezing rock. Institute of Rock and Soil Mechanics, Chinese Academy of Sciences, Wuhan (2012)
14. Xu, G.M.: Study on mechanical characteristics of rock at low temperature, damage due to freezing and thawing and multiphysical coupling problems of rock in cold regions. Institute of Rock and Soil Mechanics, Chinese Academy of Sciences, Wuhan (2012)
15. Zhang, Y.J., Yang, C.S., Xu, G.: Numerical simulation for influences of pressure solution on thermal-hydro-mechanical coupling in granule aggregate rock. Rock Soil Mech. **35**(05), 1461–1469 (2014)
16. Bai, Q.B., Li, X., Tian, Y.H., Fang, J.H.: Equations and numerical simulation for coupled water and heat transfer in frozen soil. Chin. J. Geotech. Eng. **37**(S2), 131–136 (2015)
17. Xue, L.L.: A composite element model for coupled seepage-heat transfer of fractured rock mass. Rock Soil Mech. **37**(01), 263–268+278 (2016)
18. Shen, Y.J., Yang, G.S., Rong, T.L., Liu, H.: Analysis of evolution of temperature field and frost heaving in hard rock with surface cracks under low temperature environment. Rock Soil Mech. **37**(S1), 521–529 (2016)
19. Li, G.F.: Research on slope stability of open-pit coal mine under freezing-thawing cycles. Xi'an University of Technology, Shaanxi (2016)
20. Zhang, H.M., Yang, G.S.: On the influence of moisture and freeze-thaw cycle on physical and mechanical properties of red sand stone. J. Exp. Mech. **28**(05), 635–641 (2013)
21. Fang, Y., Qiao, L., Chen, X., Yan, S.J., Zhai, G.L., Liang, Y.W.: Experimental study of freezing-thawing cycles on sandstone in Yungang Grottos. Rock Soil Mech. **35**(09), 2433–2442 (2014)
22. Ge, Q.: Research on the soil slope stability in seasonal frozen areas based on strength deterioration of the freeze-thaw surface. Jilin University, Jilin (2010)

23. Zhou, J.Z., Tan, L., Wei, C.F., Wei, H.Z.: Experimental research on freezing temperature and super-cooling temperature of soil. Rock Soil Mech. **36**(03), 777–785 (2015)
24. Zhou, H., Zhang, C.Q., Feng, X.T.: Analysis of rock mass stability in tunnel and underground engineering based on yield approach index. Chin. J. Rock Mech. Eng. **17**, 3083–3087 (2005)
25. Gao, L.Y., Yu, G.M., Zhao, J.F., et al.: Analysis and application of yielding approach based on material strength criteria. J. Chongqing Univ. **39**(05), 73–81 (2016)

Thermal Stress Effect on Fracture Integrity in Enhanced Geothermal Systems

Chao Zeng, Wen Deng(✉), Chenglin Wu, and Matt Insall

Missouri University of Science and Technology, Rolla, MO 65401, USA
wendeng@mst.edu

Abstract. In an enhanced geothermal system (EGS), fluid is injected into pre-existing fractures to be heated up and then pumped out for the electricity generation; injected fluid is cold as compared to surrounding bedrock. The rock-fluid temperature difference induces thermal stress along the fracture wall, and the large thermal stress could damage some of the self-propping asperities and result in a change of the topography and lifespan of the fractures. Although fracture sustainability has been extensively studied, the mechanism of asperity damage due to rock-fluid temperature difference remains unknown. We have constructed a finite-element based three-dimensional model, which uses a hemisphere contact pair to resemble a single self-propping asperity, to investigate the effect of temperature difference on the asperity damage. In the model, the rock mechanical properties are coupled with temperature and stress state of the bedrock. Two trends of asperity deformation with temperature effect are identified: opening zone and closure zone. Closure squeezes asperity further and induces more element damage at bottom. Higher temperature difference damages elements on asperity top whereas has negligible impact on elements at asperity bottom. In other aspect, a higher temperature expands closure zone and degrades elements at the asperity bottom. Accordingly, two potential mechanisms of asperity damage are qualitatively characterized.

Keywords: Fracture integrity · Thermal effect · Asperity failure

1 Introduction

Enhanced geothermal systems (EGS) offer a promising clean-energy resource in the world with enormous potential for baseload electricity generation [1, 2]. Unlike hydrothermal energy in shallow soil, EGS uses hot dry rock in deep earth formation: at depths of approximately 3 km to 10 km, and temperatures of up to 200 °C, the rock formations are ripe for energy extraction. An EGS injects cold water into such an environment and the heated water is brought back to the earth surface in the form of water vapor for electricity generation. In the EGS, the pre-existing fracture networks inside the hot rock play a crucial role, providing fluid (liquid water and steam) transport conduits and interfaces for heat exchange. The pre-existing fractures sustain themselves by propping due to asperities provided by contact roughness, which is different from corresponding mechanisms in the hydraulic fracture industry, which involve proppants that help maintain fracture integrity. The topography of fractures in EGS significantly

L. Zhang et al. (Eds.): GSIC 2018, *Proceedings of GeoShanghai 2018 International Conference: Rock Mechanics and Rock Engineering*, pp. 388–396, 2018.
https://doi.org/10.1007/978-981-13-0113-1_41

influences the geothermal energy extraction. Guo et al. [3] developed a thermos-hydro-mechanical coupled numerical model to analyze the effect of fracture aperture distribution on flow pattern evolution and temperature propagation. In their work, a generic fracture aperture of fracture was represented by a spatially spherical variogram model to consider the roughness effect. They observed that a reservoir tends to degrade flow channeling and endure heat production if the initial aperture field enables tortuous flow paths. Through a fracture with initially specified tortuosity, thermally-induced stress shrinks the rock matrix in cooled zones and deteriorates flow channeling, and accordingly shortens heat production life in contrast to its counterpart (which includes no consideration of thermal stress). Luo et al. [4] also pointed out that the distribution of fracture surface roughness, which is widely used to calculate fracture transmissivity, is of central importance to mechanical-hydraulic aperture correlations, because it leads to flow channeling and a steep temperature breakthrough curve.

When cold water is pumped into these fractures, the large temperature difference between rock and water induces extremely large tensile thermal stress which could deform the rock matrix. Pandey et al. [5] specifically analyzed the extent to which the thermos-elastic effect alters aperture. The cooling of the reservoir could cause contraction of the rock matrix and result in enlargement of fracture aperture in the vicinity of the injection well and conversely depresses regions outside of cooling zones. In light of increased aperture in direction between injection well and production wells, cold water directly flows from injection well to production well with least resistance and enables flow channeling and rapid drop in production temperature. Furthermore, the evolution of aperture field could reduce fluid pressure significantly in the long term, which influence aperture alteration in turn. Field circulation tests in enhanced fractured geothermal pilot program i.e. Soultz-sous-Forêts validated the occurrence of thermal contraction zone and temperature evolution [6]. In other aspect, the influence of thermal processes on new fracture opening had been addressed. Tarasovs et al. [7] studied the thermally-induced tensile cracking on primary fracture surface. The existing dominant fracture was simplified as half-plane surface and was subjected to sudden cooling effect. The multiple secondary thermal fractures could be created and propagate perpendicular to main fracture with different velocities and final lengths for distinctive temperature difference. This fracture initiation by thermal shock of reservoir rock were demonstrated as well in experiments [8] and analyzed theoretically [9, 10]. The effect of thermal-induced stress on aperture dilation between injection/production wells in long term and instant fracture surface damage by generation of new thermal cracks had also been investigated. They demonstrate the importance of fracture aperture variation on fluid flow and energy extraction by thermal effect. Nevertheless, none of previous studies had paid attention to the instant effect of large thermal stress on integrity of fracture asperities. Appreciable thermal stress due to the rock/water temperature difference would likely damage fracture asperities and result in notable aperture decrease and even the closure of fractures. This study aims to investigate such thermal stress effect on fracture asperity integrity. Through the finite element analysis, the damage of idealized fracture asperity, which is a pair of semi-spheres, due to the rock/water temperature difference has been discussed and demonstrated in this paper.

2 Problem Description and Modeling Methodology

Discrete fracture network constitutes the flow paths for fluid in EGS. To simplify the problem and keep focused, one arbitrary fracture is considered without loss of generality. Discrete fracture asperities bolster fracture surfaces and protrude into fluid pathway. Its deformation and integrity are essentially important to fluid transport and heat recovery. Generally, fracture asperities contact with distinct geometries and pairing patterns [11]. To make it tractable, individual asperity is idealized as semi-sphere contact with well mated pattern. Figure 1 depicts the geometry considered in this model at different scales.

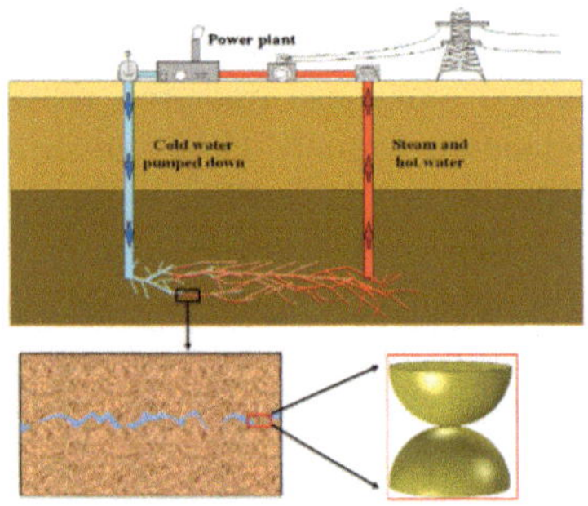

Fig. 1. Model geometry at different scales

In this model, both fluid and heat have impact on stress distribution on asperity. The difference of fluid pressure and initial overburden pressure is effective stress loading on asperities. With the exposure to high temperature difference, the fluid pressure change is negligible compared to large thermal stress [12]. Accordingly, the asperity stays at admissible effective stress on top and is subjected to large tensile thermal stress in a short time. Alteration of stress distribution would deteriorate asperity integrity. A quantitative model has been developed to describe the mechanical response of the asperity, which consider both a damage-based constitutive relationship and the temperature field from the heat conduction process.

2.1 Constitutive Relations

The constitutive relations for the rock including the thermal strain can be expressed in Eq. (1) [13].

$$\sigma_{ij} = \overline{\lambda}\varepsilon_{kk}\delta_{ij} + \overline{\mu}\varepsilon_{ij} - \left(3\overline{\lambda} + \overline{\mu}\right)\alpha\Delta T \tag{1}$$

Where σ_{ij} and ε_{ij} are the stress and strain tensor, separately. δ_{ij} is the Kronecker delta. α is the coefficient of thermal expansion in material. $\overline{\lambda}$ and $\overline{\mu}$ are damage dependent material constants, which follows the form of Lame's constants and related to damage based elastic modulus, $\overline{\lambda} = \frac{Ev}{(1+v)(1-2v)}$ and $\overline{\mu} = \frac{E}{2(1+v)}$. E is the damage based elastic modulus analyzed in following Eq. (2), v is the Poisson's ratio. In Eq. (1), ε_{ij} is sum of mechanical strain and thermal strain in materials, σ_{ij} is the mechanical stress induced only by mechanical strain. This thermal elastic model is suitable for brittle materials with negligible plastic deformation.

The failure of quasi-brittle heterogeneous materials such as rocks is mostly due to propagation and intersection of pre-existing micro-cracks. The micro-cracks connect and eventually form macro-cracks which lead to the failure of the materials [14, 15]. This macroscopic representation of microcrack development can be qualitatively described by continuum damage mechanics [16]. Damage variable is generally introduced to characterize surface density of intersection of micro-cracks in thermodynamic principle [17]. In this study, an isotropic damage variable (D) is used to model deterioration of elastic modulus:

$$E = E_0(1 - D) \tag{2}$$

Where E is the degraded elastic modulus. E_0 is the initial Young's modulus. D is an isotropic damage variable, $0 \leq D \leq 1$. The value of 0 means element is intact without any modulus degradation. The element loses its loading sustainability and fails when the value is equal to 1.

Referring to laboratory uniaxial experiments of granites [18–20], the uniaxial compression experimental curves has good representation of failure micro-mechanism of granite [21]. Whereas the uniaxial tension loading (usually by indirect Brazilian test) presents distinct features in terms of peak strength and failure modes. In fact, it is the macroscopic realization of different micro-cracking growth and proliferation [22]. In this sense, asymmetric constitutive model is suited for deformation and failure characterization for granites. The constitutive relation is depicted in Fig. 2.

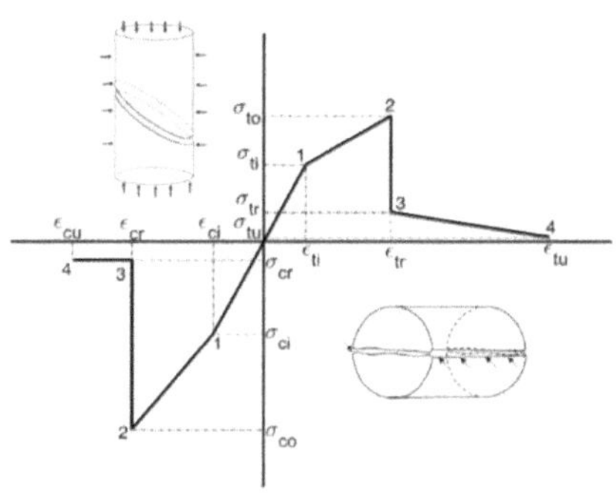

Fig. 2. Asymmetric constitutive model for granites and its failure modes under Brazilian test and triaxial compression test.

The definition of the sign in this paper follows that tensile stress is positive and compressive stress is negative. The first quadrant is under tension, while the third quadrant is under compression. The initial stage is elastic deformation. Once the strain threshold (i.e., corresponding to the peak stress) is exceeded, the Young's modulus will drop dramatically to a small residual value. That indicates the occurrence of element damage in simulation. A maintenance of small residual value (0.01% of initial value) is needed to avoid computational instability. From Fig. 2, the damage variable for tension and compression can be calculated respectively as:

$$D_t = \begin{cases} 0 & 0 < \varepsilon < \varepsilon_{ti} \\ \left(1 - \frac{\varepsilon_{ti}}{\varepsilon}\right)\left[1 - \frac{\frac{\sigma_{t0}}{\sigma_{ti}} - 1}{\frac{\varepsilon_{tr}}{\varepsilon_{ti}} - 1}\right] & \varepsilon_{ti} < \varepsilon < \varepsilon_{tr} \\ 1 - \frac{\sigma_{tr}}{\sigma_{ti}}\frac{\varepsilon_{ti}}{\varepsilon} - \frac{\frac{\sigma_{tu}}{\sigma_{ti}} - \frac{\sigma_{tr}}{\sigma_{ti}}}{\frac{\varepsilon_{tu}}{\varepsilon_{ti}} - \frac{\varepsilon_{tr}}{\varepsilon_{ti}}}\left(1 - \frac{\varepsilon_{ti}}{\varepsilon}\right) & \varepsilon_{tr} < \varepsilon < \varepsilon_{tu} \\ 1 & \varepsilon > \varepsilon_{tu} \end{cases} \tag{3}$$

$$D_c = \begin{cases} 0 & \varepsilon_{ci} < \varepsilon < 0 \\ \left(1 - \frac{\varepsilon_{ci}}{\varepsilon}\right)\left[1 - \frac{\frac{\sigma_{c0}}{\sigma_{ci}} - 1}{\frac{\varepsilon_{cr}}{\varepsilon_{ci}} - 1}\right] & \varepsilon_{cr} < \varepsilon < \varepsilon_{ci} \\ 1 - \frac{\sigma_{cr}}{\sigma_{ci}}\frac{\varepsilon_{ci}}{\varepsilon} & \varepsilon_{cu} < \varepsilon < \varepsilon_{cr} \\ 1 & \varepsilon < \varepsilon_{cu} \end{cases} \tag{4}$$

Where subscript t and c in D is for tension and compression respectively. Other stress and strain parameters signifies stress and strain at inflection points of constitutive curve in Fig. 2.

$$\overline{\sigma} = \begin{cases} |\sigma_1| & \sigma_1 > \sigma_2 > \sigma_3 > 0,\ or\ \sigma_1 > \sigma_2 > 0 > \sigma_3\ \&\ \left|\frac{\sigma_1}{\sigma_3}\right| < 0.1 \\ |\sigma_2| & 0 > \sigma_1 > \sigma_2 > \sigma_3,\ or\ \sigma_1 > 0 > \sigma_2 > \sigma_3\ \&\ \left|\frac{\sigma_1}{\sigma_3}\right| < 0.1 \end{cases} \tag{5}$$

Where $\{\sigma_1, \sigma_2, \sigma_3\}$ are three principal stress. In the principal stress system, when material is in tension-tension-tension mode $(\sigma_1 > \sigma_2 > \sigma_3 > 0)$ or tension-tension-compression mode $(\sigma_1 > \sigma_2 > 0 > \sigma_3)$, the failure is classified to tension failure. While material is subject to compression-compression-compression mode $(0 > \sigma_1 > \sigma_2 > \sigma_3)$ or tension-compression-compression mode $(\sigma_1 > 0 > \sigma_2 > \sigma_3)$, the failure is classified to compression failure. The restriction $\left|\frac{\sigma_1}{\sigma_3}\right| < 0.1$ comes from the empirical relationship between compressive and tensile strength of rocks with negligible cohesion forces [23]. Tension failure and compression failure modes can be depicted by failure surface. At here, $\overline{\sigma}$ is used to depict the failure surface in terms of principal stress.

2.2 Temperature Field and Boundary Conditions

The reservoir rock has high initial temperature, denoted as T_r. When the cold water pumped in with temperature T_w, rock temperature would drop forward the far field by cooling. This heat transfer follows the Fourier's Law. Assuming the isotropic thermal property of asperity, the energy balance equation is following:

$$k\nabla^2 T + q = \rho c \frac{\partial T}{\partial t} \tag{6}$$

where q denotes the heat source generated inside the medium in W/m^3, k is the apparent thermal conductivity of the medium in $W/m \cdot^{\circ} C$. ρ is the bulk density of medium in kg/m^3, c is the specific heat or heat capacity of the medium in $J/kg \cdot^{\circ} C$, and t is the elapsed time. Equation (6) is on the premise that thermal conductivity k is constant for temperature change and is a scalar. The top thermal boundary is set to constant temperature T_r. The curvy surface of hemispherical asperity is set to constant water temperature T_w. The above assumption is made based on the constant heat source from the adjacent rocks of fracture aperture and the relative small radius of the asperity (radius about 2.5 mm at here) [24]. The parameters in this study is presented as well in Table 1.

Table 1. Model data used in simulation

Young's modulus, E_0	18.56 GPa	Initial tensile strain, ε_{ti}	2.2×10^{-4}
Poisson's ratio, ν	0.25	Residual tensile strain, ε_{tr}	6.65×10^{-4}
Granite density, ρ	2750 kg/m^3	Maximum tensile strain, ε_{tu}	1.5×10^{-3}
Heat capacity, c	790 J/kg · K	Initial tensile stress, σ_{ti}	4.09 MPa
Thermal conductivity, k	10.7 W/m · K	Tensile strength, σ_{to}	6.7 MPa
Thermal expansion, α	$8 \times 10^{-6} 1/K$	Residual tensile stress, σ_{tr}	1.57 MPa
Rock temperature, T_r	300 °C	Ultimate tensile stress, σ_{tu}	0.25 MPa
Hemisphere radius, R	2.5 mm	Initial compressive strain, ε_{ci}	2.64×10^{-3}
Compressive strength, σ_{co}	100 MPa	Residual compressive strain, ε_{cr}	6.87×10^{-3}
Residual compressive stress, σ_{cr}	9.7 MPa	Maximum compressive strain, ε_{cu}	1.0×10^{-2}
		Initial compressive stress, σ_{ci}	49 MPa

3 Results and Discussion

In the deep fracture network of EGS, the asperities are initially subjected to over-burden pressure with initial compressive strain. The overburden pressure increases with the depth of EGS. When cold water is pumped into fracture network, temperature difference between water and rock decreases from the injection well to the production well. Therefore, thermal stress effect on the fracture deformation also varies. Because of the thermal stress, stress on asperities will redistribute, which would lead to asperity deformation and even the damage of asperities. In our study, we took different temperature differences into consideration. Figure 4 shows displacement of asperity subjected to different overburden pressure with and without considering temperature difference effects. It is clear the temperature difference in the EGS could cause considerable asperity deformation. Moreover, different overburden pressure can cause different thermal stress effects. Take ΔT = 50 K for example (Fig. 3(a)), with the overburden pressure increase, two curves intersect at different points. By considering

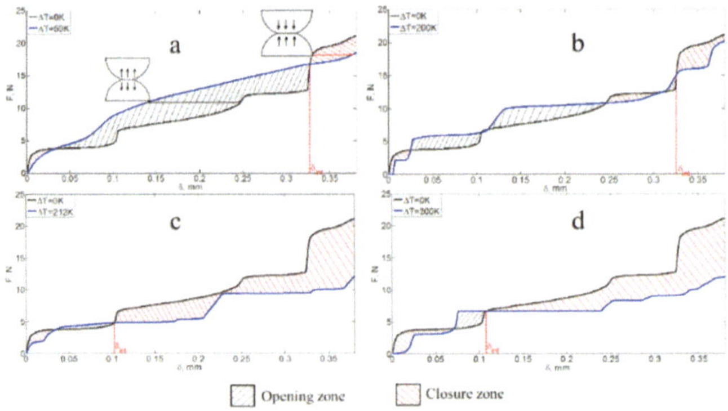

Fig. 3. The thermal effect on asperity displacement. With increasing temperature difference, opening zone shrinks and closure zone expands.

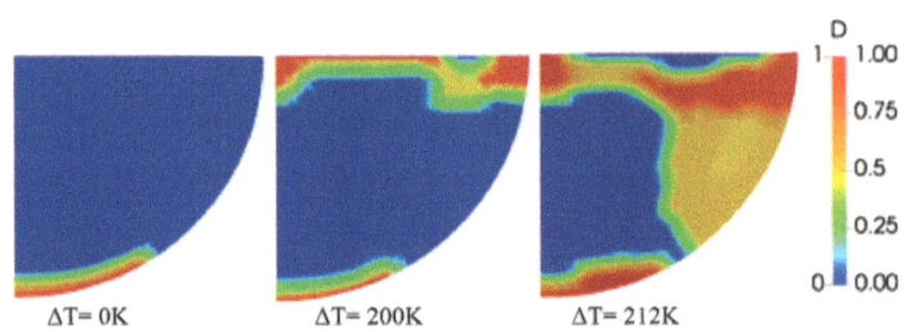

Fig. 4. The damaged elements in asperity with increasing temperature.

that the overburden pressure is constant at certain depth in EGS, the introduction of 50 K temperature could induce either the decrease of asperity displacement as indicated by the black arrow in Fig. 3(a) or the increase of asperity displacement as indicated by the red arrow in Fig. 3(a). This means temperature difference can cause either the opening or the closure of fracture. Although it is observed that the closure and opening of fracture occur alternately with varying overburden pressure. Generally speaking, the closure of fracture is more likely to occur for larger overburden pressure. By considering the temperature difference effects, the larger temperature difference can more likely cause the closure of fracture. As mentioned above, the fractures, which are the conduits for water transport and exchange of heat, are critical to the efficiency of EGS. Our results indicate that the temperature difference between water and rock could cause the opening and closure of the fracture, and therefore affect the pre-existing fracture network integrity, which is worthy of more concerns when studying the EGS fracture network.

Figure 4 presents the damage variable for above three temperature difference cases. Because of axisymmetric property of simulating models, only one quarter circle geometry is presented. From evolution of damage variable distribution, more elements damage with increasing temperature difference. Elements near top boundary starts to damage with temperature difference and exacerbate this damage with increasing temperature difference. After ΔT = 200 K, the damaged elements on top aggregate together to form a damaged band. However, the influence of temperature difference on evolution of damage at bottom is negligible. This alludes to potential damage on top by temperature effect and potential damage at bottom by displacement closure induced by temperature difference.

4 Conclusion

When cold water is pumped into fracture network in enhanced geothermal systems, the huge temperature difference of water and hot rock induces substantial tensile thermal stress. It alters the stress distribution in fracture asperity and changes its deformation. Numerical simulation is employed to capture the effect of temperature difference and intensity of its impact. Damage mechanics considering thermal effect is used in this model to analyze asperity failure. In pure mechanical loading, it presents two modes of deformation: closure and opening. Closure mode jeopardizes integrity of asperity and is of main concern in this analysis. Additionally, with thermal effect, two potential failure mechanisms interplay: (1) bottom failure by mechanical loading; (2) top failure induced by thermal stress. With high temperature difference, top failure becomes significant. High temperature difference also induces closure of asperity and prompts bottom failure. The interplay between these two mechanisms will be analyzed quantitatively. Furthermore, the critical temperature difference to bound closure and opening of asperity is quantified to be 207 K and it's also the boundary to distinguish top failure and bottom failure. In EGS field, zone close to injection well has higher temperature difference and more likely to close and induce asperity failure. While zone close to production well is less prone to damage asperity and more stable. Further analysis will be performed to investigate the relation of asperity failure and flowing length of fluid and reservoir depth.

References

1. Tester, J.W., Anderson, B.J., Batchelor, A.S., Blackwell, D.D., DiPippo, R., Drake, E., Garnish, J., Livesay, B., Moore, M.C., Nichols, K., Petty, S.: The future of geothermal energy. In: impact of enhanced geothermal systems (EGS) on the United States in the 21st Century, p. 372. Massachusetts Institute of Technology, Cambridge, MA (2006)
2. Bertani, R.: Geothermal power generation in the work 2010–2014 update report. Geothermics **60**, 31–43 (2016)
3. Guo, B., Fu, P., Hao, Y., Peters, C.A., Carrigan, C.R.: Thermal drawdown-induced flow channeling in a single fracture in EGS. Geothermics **61**, 46–62 (2016)
4. Luo, S., Zhao, Z., Peng, H., Pu, H.: The role of fracture surface roughness in macroscopic fluid flow and heat transfer in fractured rocks. Int. J. Rock Mech. Min. Sci. **87**, 29–38 (2016)
5. Pandey, S.N., Chaudhuri, A., Kelkar, S.: A coupled thermo-hydro-mechanical modeling of fracture aperture alteration and reservoir deformation during heat extraction from a geothermal reservoir. Geothermics **65**, 17–31 (2017)

6. Bruel, D.: Impact of induced thermal stresses during circulation tests in an engineered fractured geothermal reservoir: example of the Soultz-sous-Forets European hot fractured rock geothermal project, Rhine Graben, France. Oil Gas Sci. Technol. **57**(5), 459–470 (2002)
7. Tarasovs, S., Ghassemi, A.: Propagation of a system of cracks under thermal stress. In: 45th US Rock Mechanics/Geomechanics Symposium. American Rock Mechanics Association (2011)
8. Finnie, I., Cooper, G.A., Berlie, J.: Fracture propagation in rock by transient cooling. Int. J. Rock Mech. Min. Sci. Geomech. Abs. **16**(1) (1979)
9. Murphy, H.D.: Thermal Stress Cracking and the Enhancement of Heat Extraction From Fractured Geothermal Reservoirs. No. LA-7235-MS. Los Alamos Scientific Lab., N. Mex. (USA) (1978)
10. Barr, D.T.: Thermal Cracking in Nonporous Geothermal Reservoirs. Diss. Massachusetts Institute of Technology (1980)
11. Grasselli, G.: Shear Strength of Rock Joints Based on Quantified Surface Description (2001)
12. Ghassemi, A., Kumar, G.S.: Changes in fracture aperture and fluid pressure due to thermal stress and silica dissolution/precipitation induced by heat extraction from subsurface rocks. Geothermics **36**(2), 115–140 (2007)
13. Sadd, M.H.: Elasticity: Theory, Applications, and Numerics. Academic Press, Boston (2009)
14. Ashby, M.F., Sammis, C.G.: The damage mechanics of brittle solids in compression. Pure. appl. Geophys. **133**(3), 489–521 (1990)
15. Lockner, D.: The role of acoustic emission in the study of rock fracture. Int. J. Rock Mech. Min. Sci. Geomech. Abs. **30**(7) (1993)
16. Tang, C.A., Liu, H., Lee, P.K.K., Tsui, Y., Tham, L.G.: Numerical studies of the influence of microstructure on rock failure in uniaxial compression—part I: effect of heterogeneity. Int. J. Rock Mech. Min. Sci. **37**(4), 555–569 (2000)
17. Lemaitre, J.: A continuous damage mechanics model for ductile fracture. Transactions of the ASME. J. Eng. Mater. Technol. **107**(1), 83–89 (1985)
18. Okubo, S., Fukui, K.: Complete stress-strain curves for various rock types in uniaxial tension. Int. J. Rock Mech. Min. Sci. Geomech. Abs. **33**(6) (1996)
19. Hawkes, I., Malcolm, M., Stephen, G.: Deformation of rocks under uniaxial tension. Int. J. Rock Mech. Min. Sci. Geomech. Abs. **10**(6) (1973)
20. Schock, R.N., Louis, H.: Strain behavior of a granite and a graywacke sandstone in tension. J. Geophys. Res. Solid Earth **87**(B9), 7817–7823 (1982)
21. Tapponnier, P., Brace, W.F.: Development of stress-induced microcracks in Westerly granite. Int. J. Rock Mech. Min. Sci. Geomech. Abs. **13**(4) (1976)
22. Lajtai, E.Z., Bielus, L.P.: Stress corrosion cracking of Lac du Bonnet granite in tension and compression. Rock Mech. Rock Eng. **19**(2), 71–87 (1986)
23. Jaeger, J.C., Cook, N.G., Zimmerman, R.: Fundamentals of Rock Mechanics. Wiley, New York (2009)
24. Santucci, S., Grob, M., Toussaint, R., Schmittbuhl, J., Hansen, A., Maløy, K.J.: Fracture roughness scaling: A case study on planar cracks. EPL (Europhysics Letters) **92**(4), 44001 (2010)

Mechanical Characteristics and Control Technology of Floor in Gob-Side Entry Retaining

Ying Xu(✉) and Le Shen

China University of Mining and Technology, Xuzhou 221116, China
kdxuying@163.com

Abstract. The effect of abutment pressure caused by the mining influence of two working faces can easily lead to floor heave in gob-side entry retaining. According to the distribution law of abutment pressure loading on the floor of gob-side entry retaining, a mechanical calculation model is established based on the elastoplastic theory. The calculation results based on the production geology of track entry of 1024 working face in Zhongxin coal mine shows that the vertical stress and plastic zone in the floor increase with the advancing of working face at the stage of leading influence during first mining, increase first and then decrease at the stage of entry retaining during first mining, and increase at the stage of leading influence during second mining. Especially, the vertical stress distribution and plastic zone appear asymmetrical characteristic. In addition, the similarity model test of gob-side entry retaining demonstrates the forming process and shape of the main fracture plane of floor, that also proves the asymmetrical characteristics of floor heave. Finally, based on idea that high strength of roof and two sides can reduce the abutment pressure loading on the floor, the technology of strengthening support of roof and two sides is proposed to control floor heave in gob-side entry retaining, and the technology is proved by the calculation results based on the above mechanical calculation model. The research fruits can enrich theory of gob-side entry retaining technology to improve its application value.

Keywords: Gob-side entry retaining · Floor heave · Abutment pressure

1 Introduction

In underground coal mining, layout and control of a entry need to consider safety, efficiency and economy. Gob-side entry retaining technique is widely used in China for the last several decades. Comparing with the conventional layout of the entry with a wide coal pillar, the gob-side entry retaining technology has some safe, efficient and economic advantages, such as implementation of Y-shaped ventilation system, reduction of excavation work, improvement of coal recovery ratio and so on [1, 2]. However, the deformation of gob-side entry retaining is large, Especially severe floor heave. So far there is no the most effective technology to control floor heave in gob-side entry retaining. So the weakness of floor heave restrict the popularization and application of the technology.

L. Zhang et al. (Eds.): GSIC 2018, *Proceedings of GeoShanghai 2018 International Conference: Rock Mechanics and Rock Engineering*, pp. 397–408, 2018.
https://doi.org/10.1007/978-981-13-0113-1_42

Some researchers have done some work for mechanism and control of floor heave in gob-side entry retaining. For instance, Wang W.J. divided the floor heave process into five stages: the stage of entry driving, the stage of floor rheology after entry driving, the stage of first mining, the stage of floor rheology after first mining and the stage of second mining [3]. Wei hold that the high horizontal stress and pressure effect of filling body caused floor heave in gob-side entry retaining, and the floor heave appears zoning and skew characteristics [4]. Hua established a mechanical model of the floor in gob-side entry retaining based on beam theory. The left side of the beam (outside of backfilling body) was simplified as a fixed hinge support, while the right side of the beam (under the limit equilibrium zone of the coal side) was simplified as fixed end. Then the deformation of the floor beam in gob-side entry retaining was solved by using the principle of minimum potential energy [5]. Xu proposed a new approach named as steel pile method to control severe floor heaves in gob-side entry retaining [6]. From that, the characteristics, mechanism and control technology of floor heave in gob-side entry retaining have been preliminarily studied, still need study ever more to overcome the weakness of gob-side entry retaining.

This paper aims to present the effect of mining abutment pressure on floor of gob-side entry retaining, and then the mechanical characteristics and control technology of floor heave are obtained. The research fruits can enrich theory of gob-side entry retaining technology to improve its application value.

2 Effect of Mining Abutment Pressure on Floor of Gob-Side Entry Retaining

2.1 Mechanical Calculation Model

Force conditions of floor determine degree of floor heave in gob-side entry retaining. The forces loading on floor are mainly abutment pressures caused by movement of entry roof and transmitted by two entry sides. The abutment pressures are different at different stages of mining. A typical curve about distribution and change of abutment pressure in a practical gob-side entry retaining is shown in Fig. 1. Taking the point with distance −5 m from vertical midline entry center (coal side) as example, the value of abutment pressure increases from 12 MPa at the stage of leading influence during first mining to 22 MPa at the stage of entry retaining during first mining.

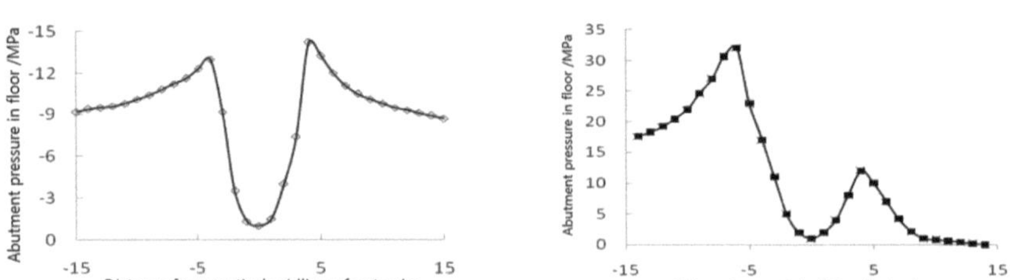

(1) Stage of leading influence during first mining (2) Stage of entry retaining during first mining

Fig. 1. Abutment pressure distribution in floor at different mining stage

According to the force conditions of floor in gob-side entry retaining, the mechanical calculation model of floor is established based on the following assumptions:

(1) The floor is taken as semi-infinite plane.
(2) The distribution of abutment pressures are linear (see Fig. 2).

In this model, the width of gob-side entry retaining is b_4, the mining influence width at both sides of entry are $b_2 + b_3$ and $b_5 + b_6$ respectively, the initial stress width at both sides are b_1 and b_7 respectively, γH is initial stress of entry, k_2 and k_3 are stress concentration factors.

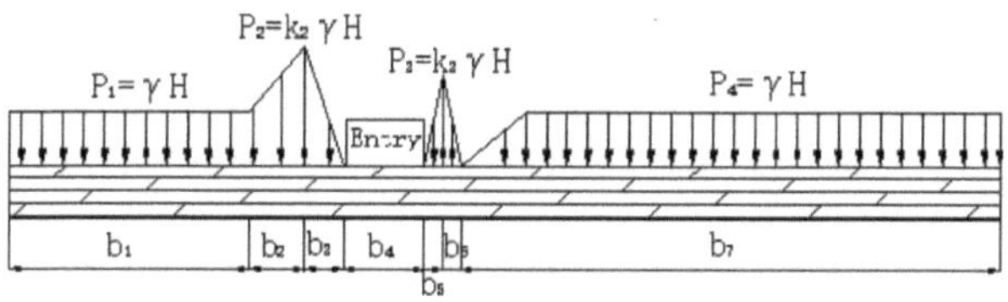

Fig. 2. Sketch of mechanical calculation model

Based on the elasticity theory [7], the stress component in floor caused by the abutment pressures are follows.

$$\begin{aligned}
\sigma_x &= \sigma_{x1} + \sigma_{x2} + \sigma_{x3} + \sigma_{x5} + \sigma_{x6} + \sigma_{x7} \\
\sigma_y &= \sigma_{y1} + \sigma_{y2} + \sigma_{y3} + \sigma_{y5} + \sigma_{y6} + \sigma_{y7} \\
\tau_{xy} &= \tau_{xy1} + \tau_{xy2} + \tau_{xy3} + \tau_{xy5} + \tau_{xy6} + \tau_{xy7}
\end{aligned} \quad (1)$$

Each stress component in the formula (1) is

$$\sigma_{x1} = -\frac{P_1}{\pi}\left(\arctan\frac{x+\sum_1^3 b_i}{y} - \arctan\frac{x+b_3}{y} + \frac{(x+\sum_1^3 b_i)y}{(x+\sum_1^3 b_i)^2+y^2} - \frac{(x+b_3)y}{(x+b_3)^2+y^2}\right)$$

$$\sigma_{y1} = -\frac{P_1}{\pi}\left(\arctan\frac{x+\sum_1^3 b_i}{y} - \arctan\frac{x+b_3}{y} - \frac{(x+\sum_1^3 b_i)y}{(x+\sum_1^3 b_i)^2+y^2} + \frac{(x+b_3)y}{(x+b_3)^2+y^2}\right)$$

$$\tau_{xy1} = \frac{P_1}{\pi}\left(\frac{y^2}{(x+\sum_1^2 b_i)^2+y^2} - \frac{y^2}{(x+b_3)^2+y^2}\right)$$

$$\begin{aligned}
\sigma_{x2} = &-\frac{p_2-p_1}{\pi b_2}(x+b_2+b_3)\left[\arctan\frac{x+b_2+b_3}{y} - \arctan\frac{x+b_3}{y} + \frac{(x+b_2+b_3)y}{(x+b_2+b_3)^2+y^2}\right. \\
&\left.-\frac{(x+b_3)y}{(x+b_3)^2+y^2}\right] - \frac{(p_2-p_1)y}{2\pi b_2}\left[\frac{y^2}{(x+b_2+b_3)^2+y^2} - \frac{y^2}{(x+b_3)^2+y^2}\right. \\
&\left.-\frac{(x+b_2+b_3)^2}{(x+b_2+b_3)^2+y^2} + \frac{(x+b_3)^2}{(x+b_3)^2+y^2}\right]
\end{aligned}$$

$$\sigma_{y2} = -\frac{p_2 - p_1}{\pi b_2}(x + b_2 + b_3)[\arctan\frac{x + b_2 + b_3}{y} - \arctan\frac{x + b_3}{y} - \frac{(x + b_2 + b_3)y}{(x + b_2 + b_3)^2 + y^2}$$
$$+ \frac{(x + b_3)y}{(x + b_3)^2 + y^2}] - \frac{2(p_2 - p_1)y}{\pi b_2}(\ln\frac{y}{\sqrt{(x + b_2 + b_3)^2 + y^2}} - \ln\frac{y}{\sqrt{(x + b_3)^2 + y^2}})$$
$$+ \frac{(p_2 - p_1)y}{\pi b_2}(\frac{y^2}{(x + b_2 + b_3)^2 + y^2} - \frac{y^2}{(x + b_3)^2 + y^2})$$

$$\tau_{xy2} = \frac{p_2 - p_1}{2\pi b_2}(x + b_2 + b_3)[\frac{y^2}{(x + b_2 + b_3)^2 + y^2} - \frac{y^2}{(x + b_3)^2 + y^2} - \frac{(x + b_2 + b_3)^2}{(x + b_2 + b_3)^2 + y^2}$$
$$+ \frac{(x + b_3)^2}{(x + b_3)^2 + y^2}] + \frac{(p_2 - p_1)y}{\pi b_2}[\arctan\frac{x + b_2 + b_3}{y} - \arctan\frac{x + b_3}{y}$$
$$- \frac{(x + b_2 + b_3)y}{(x + b_2 + b_3)^2 + y^2} + \frac{(x + b_3)y}{(x + b_3)^2 + y^2}]$$

$$\sigma_{x3} = \frac{p_2}{\pi b_3}x[\arctan\frac{x + b_3}{y} - \arctan\frac{x}{y} + \frac{(x + b_3)y}{(x + b_3)^2 + y^2} - \frac{xy}{x^2 + y^2}] - \frac{p_2 y}{2\pi b_3}[\frac{y^2}{(x + b_3)^2 + y^2}$$
$$- \frac{y^2}{x^2 + y^2} - \frac{(x + b_3)^2}{(x + b_3)^2 + y^2} + \frac{x^2}{x^2 + y^2}]$$
$$\sigma_{y3} = \frac{p_2}{\pi b_3}x[\arctan\frac{x + b_3}{y} - \arctan\frac{x}{y} - \frac{(x + b_3)y}{(x + b_3)^2 + y^2} + \frac{xy}{x^2 + y^2}]$$
$$+ \frac{2p_2 y}{\pi b_3}(\ln\frac{y}{\sqrt{(x + b_3)^2 + y^2}} - \ln\frac{y}{\sqrt{x^2 + y^2}}) - \frac{p_2 y}{\pi b_3}(\frac{y^2}{(x + b_3)^2 + y^2} - \frac{y^2}{x^2 + y^2})$$
$$\tau_{xy3} = -\frac{p_2}{2\pi b_3}x[\frac{y^2}{(x + b_3)^2 + y^2} - \frac{y^2}{x^2 + y^2} - \frac{(x + b_3)^2}{(x + b_3)^2 + y^2} + \frac{x^2}{x^2 + y^2}]$$
$$- \frac{p_2 y}{\pi b_3}[\arctan\frac{x + b_3}{y} - \arctan\frac{x}{y} - \frac{(x + b_3)y}{(x + b_3)^2 + y^?} + \frac{xy}{x^2 + y^2}]$$

$$\sigma_{x5} = -\frac{p_3}{\pi b_5}(x - b_4)[\arctan\frac{x - b_4}{y} - \arctan\frac{x - b_4 - b_5}{y} + \frac{(x - b_4)y}{(x - b_4)^2 + y^2}$$
$$- \frac{(x - b_4 - b_5)y}{(x - b_4 - b_5)^2 + y^2}] - \frac{p_3 y}{2\pi b_5}[\frac{y^2}{(x - b_4)^2 + y^2} - \frac{y^2}{(x - b_4 - b_5)^2 + y^2} - \frac{(x - b_4)^2}{(x - b_4)^2 + y^2}$$
$$+ \frac{(x - b_4 - b_5)^2}{(x - b_4 - b_5)^2 + y^2}]$$
$$\sigma_{y5} = -\frac{p_3}{\pi b_5}(x - b_4)[\arctan\frac{x - b_4}{y} - \arctan\frac{x - b_4 - b_5}{y} - \frac{(x - b_4)y}{(x - b_4)^2 + y^2}$$
$$+ \frac{(x - b_4 - b_5)y}{(x - b_4 - b_5)^2 + y^2}] - \frac{2p_3 y}{\pi b_5}(\ln\frac{y}{\sqrt{(x - b_4)^2 + y^2}} - \ln\frac{y}{\sqrt{(x - b_4 - b_5)^2 + y^2}})$$
$$+ \frac{p_3 y}{\pi b_5}(\frac{y^2}{(x - b_4)^2 + y^2} - \frac{y^2}{(x - b_4 - b_5)^2 + y^2})$$

$$\tau_{xy5} = \frac{p_3}{2\pi b_5}(x - b_4)[\frac{y^2}{(x - b_4)^2 + y^2} - \frac{y^2}{(x - b_4 - b_5)^2 + y^2} - \frac{(x - b_4)^2}{(x - b_4)^2 + y^2}$$
$$+ \frac{(x - b_4 - b_5)^2}{(x - b_4 - b_5)^2 + y^2}] + \frac{p_3 y}{\pi b_5}[\arctan\frac{x - b_4}{y} - \arctan\frac{x - b_4 - b_5}{y} - \frac{(x - b_4)y}{(x - b_4)^2 + y^2}$$
$$+ \frac{(x - b_4 - b_5)y}{(x - b_4 - b_5)^2 + y^2}]$$

$$\sigma_{x6} = \frac{p_3}{\pi b_6}(x - \sum\nolimits_4^6 b_i)[\arctan\frac{x - b_4 - b_5}{y} - \arctan\frac{x - \sum_4^6 b_i}{y} + \frac{(x - b_4 - b_5)y}{(x - b_4 - b_5)^2 + y^2}$$
$$- \frac{(x - \sum_4^6 b_i)y}{(x - \sum_4^6 b_i)^2 + y^2}] + \frac{p_3 y}{2\pi b_6}[\frac{y^2}{(x - b_4 - b_5)^2 + y^2} - \frac{y^2}{(x - \sum_4^6 b_i)^2 + y^2}$$
$$- \frac{(x - b_4 - b_5)^2}{(x - b_4 - b_5)^2 + y^2} + \frac{(x - \sum_4^6 b_i)^2}{(x - \sum_4^6 b_i)^2 + y^2}]$$

$$\sigma_{y6} = \frac{p_3}{\pi b_6}(x - \sum\nolimits_4^6 b_i)[\arctan\frac{x - b_4 - b_5}{y} - \arctan\frac{x - \sum_4^6 b_i}{y} - \frac{(x - b_4 - b_5)y}{(x - b_4 - b_5)^2 + y^2}$$
$$+ \frac{(x - \sum_4^6 b_i)y}{(x - \sum_4^6 b_i)^2 + y^2}] + \frac{2p_3 y}{\pi b_6}(\ln\frac{y}{\sqrt{(x - b_4 - b_5)^2 + y^2}} - \ln\frac{y}{\sqrt{(x - \sum_4^6 b_i)^2 + y^2}})$$
$$- \frac{p_3 y}{\pi b_6}(\frac{y^2}{(x - b_4 - b_5)^2 + y^2} - \frac{y^2}{(x - \sum_4^6 b_i)^2 + y^2})$$

$$\tau_{xy6} = -\frac{p_3}{2\pi b_6}(x - \sum\nolimits_4^6 b_i)[\frac{y^2}{(x - b_4 - b_5)^2 + y^2} - \frac{y^2}{(x - \sum_4^6 b_i)^2 + y^2}$$
$$- \frac{(x - b_4 - b_5)^2}{(x - b_4 - b_5)^2 + y^2} + \frac{(x - \sum_4^6 b_i)^2}{(x - \sum_4^6 b_i)^2 + y^2}] - \frac{p_3 y}{\pi b_6}[\arctan\frac{x - b_4 - b_5}{y}$$
$$- \arctan\frac{x - \sum_4^6 b_i}{y} - \frac{(x - b_4 - b_5)y}{(x - b_4 - b_5)^2 + y^2} + \frac{(x - \sum_4^6 b_i)y}{(x - \sum_4^6 b_i)^2 + y^2}]$$

$$\sigma_{x7} = -\frac{P_4}{\pi}(\arctan\frac{x - \sum_4^6 b_i}{y} - \arctan\frac{x - \sum_4^7 b_i}{y} + \frac{(x - \sum_4^6 b_i)y}{(x - \sum_4^6 b_i)^2 + y^2} - \frac{(x - \sum_4^7 b_i)y}{(x - \sum_4^7 b_i)^2 + y^2})$$
$$\sigma_{y7} = -\frac{P_4}{\pi}(\arctan\frac{x - \sum_4^6 b_i}{y} - \arctan\frac{x - \sum_4^7 b_i}{y} - \frac{(x - \sum_4^6 b_i)y}{(x - \sum_4^6 b_i)^2 + y^2} + \frac{(x - \sum_4^7 b_i)y}{(x - \sum_4^7 b_i)^2 + y^2})$$
$$\tau_{xy7} = \frac{P_4}{\pi}(\frac{y^2}{(x - \sum_4^6 b_i)^2 + y^2} - \frac{y^2}{(x - \sum_4^7 b_i)^2 + y^2})$$

In addition, in order to demonstrate the comprehensive response of stress distribution caused by mining abutment pressure, the floor is assumed to conform to D-P yielding criterion [8].

$$f = \alpha I + \sqrt{J_2} - K = 0 \tag{2}$$

2.2 Evolution Law of Stress and Plastic Zone in Floor

The track entry (Fig. 3) of 1024 working face in Zhongxin coal mine was taken as an example to analysis the evolution law of stress and plastic zone in floor of gob-side entry retaining. The track entry serves the mining of No. 2 coal seam which is of about 400 m depth and 2 m thickness. The width of entry and gob-side filling are 4.5 m and 2.5 m respectively [9]. The surrounding rocks are list in Table 1.

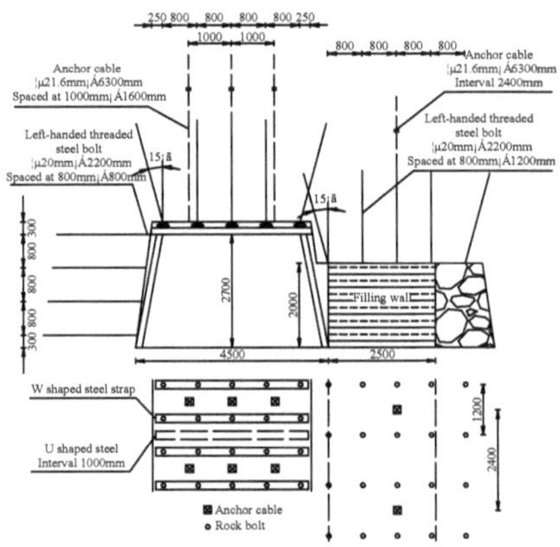

Fig. 3. The track entry of 1024 working face in Zhongxin coal mine

Table 1. Stress concentration factors at the stage of leading influence during first mining

Strata	Lithology	Thickness (m)	Density (kg/m^3)	Bulk modulus (GPa)	Shear modulus (GPa)	Cohesion (MPa)	Friction angle (°)
Main roof	Sand stone	8.0	2660	7.8	5.7	4.2	35
Immediate roof	Shale stone	1.3	1950	3.2	2.0	1.8	27
Coal seam	2# coal	2.0	1430	2.8	1.5	1.3	24
Floor	Sandy mudstone	0.5	2340	5.9	3.1	2.6	31
	Sandy shale	1.5	2170	4.1	2.5	2.1	29

With the advancing of 1024 working faces and 1025 working face, the abutment pressure on floor changes. So the stress concentration factors k_2 and k_3 are set to different value in the above mechanical calculation model (see Tables 2, 3 and 4). Other parameters of the track entry are substituted in the formula (1) and (2). Some results about vertical stress and plastic zone are illustrated in Figs. 4, 5 and 6.

Table 2. Stress concentration factors at the stage of leading influence during first mining

Distance ahead of 1024 working face/m	40	20	10	5
k_2	1.26	1.46	1.76	1.87
k_3	1.26	1.56	1.92	2.28

Table 3. Stress concentration factors at the stage of entry retaining during first mining

Distance behind of 1024 working face/m	−10	−20	−40	−80
k_2	2.24	2.54	2.40	2.30
k_3	0.67	1.38	1.06	0.89

Table 4. Stress concentration factors at the stage of leading influence during second mining

Distance ahead of 1025 working face/m	60	30	10
k_2	2.48	2.70	3.17
k_3	0.84	0.96	1.14

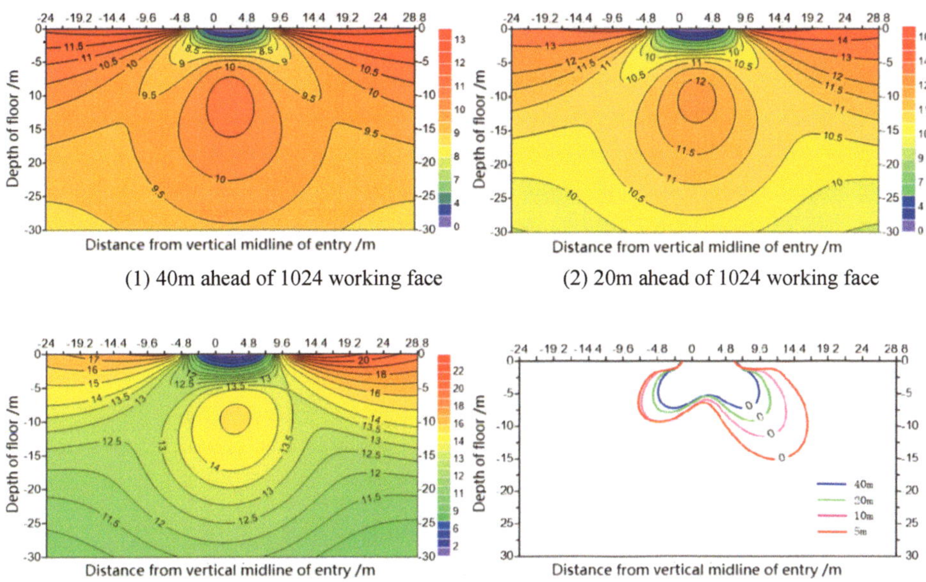

(1) 40m ahead of 1024 working face
(2) 20m ahead of 1024 working face
(3) 5m ahead of 1024 working face
(4) Plastic zone

Fig. 4. Distribution of vertical stress and plastic zone in floor at the stage of leading influence during first mining (MPa)

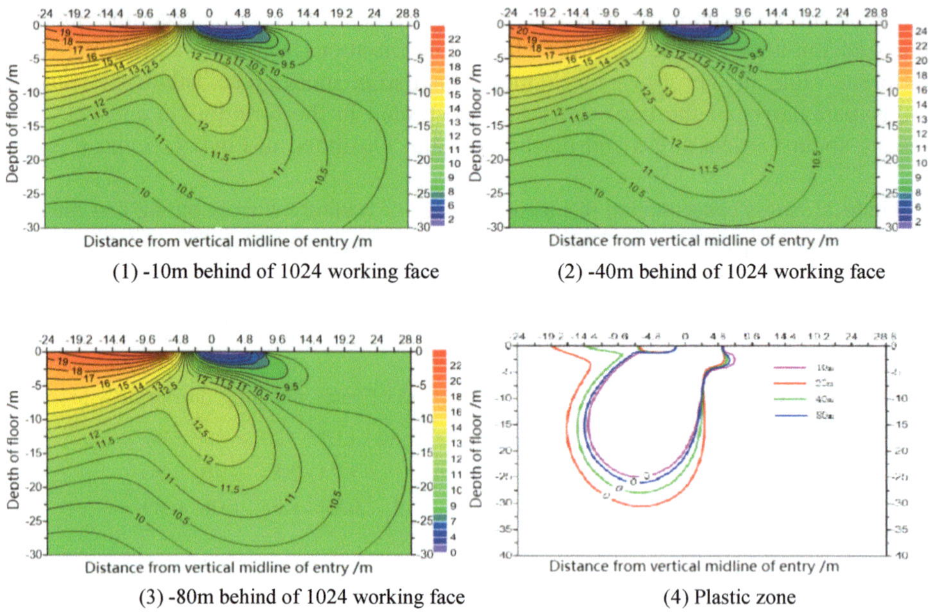

Fig. 5. Distribution of vertical stress and plastic zone in floor at the stage of entry retaining during first mining (MPa)

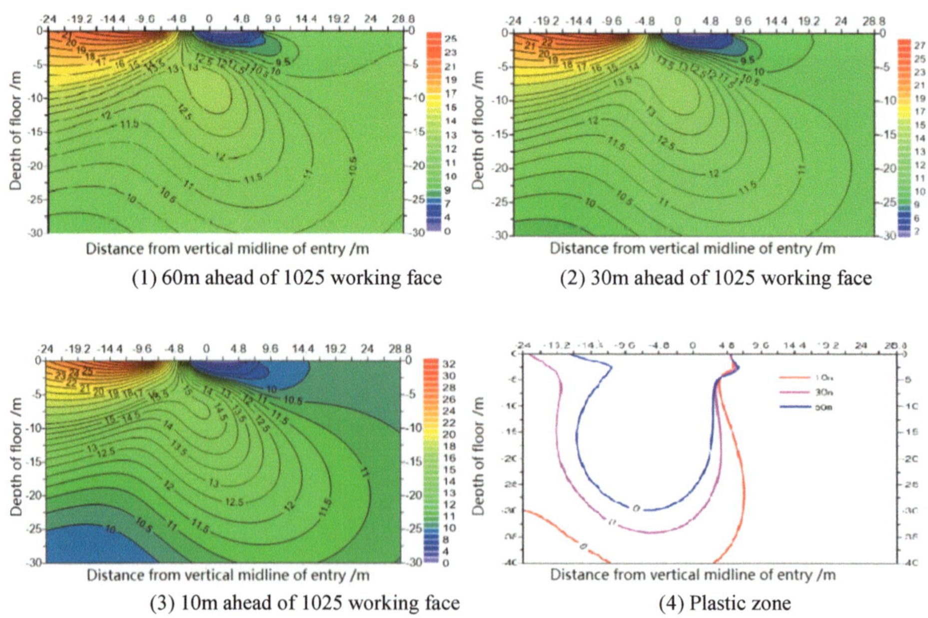

Fig. 6. Distribution of vertical stress and plastic zone in floor at the stage of leading influence during second mining (MPa)

As we can see from the Figures, the vertical stress and plastic zone in the floor change with the advancing of two working faces. (1) At the stage of leading influence during first mining, the vertical stress in floor increases with the advancing of the working face, and the increasing range of vertical stress at the mining side is larger than that at the coal side. Thereby, the distribution of stress and plastic zone shows a asymmetrical characteristic. (2) At the stage of entry retaining during first mining, the working face mining and roof collapse cause rise and fall of the abutment pressure on the floor. Thus the vertical stress and plastic zone increase first and then decrease. Particular care should be taken that, the vertical stress of floor at the coal side is larger than that at the gob side. thus the plastic zone slant to coal side. (3) At the stage of leading influence during second mining, basic roof instability gives rise to larger lead abutment pressure. That make floor bear the larger abutment pressure, and the vertical stress and plastic zone increase sharply.

2.3 Evolution Law of Rapture in Floor

In the effect process of abutment pressure on the floor, some floor rock may rapture when the stress stations exceed its strength. In order to find out the rapture process, a similarity model test was carried out. The test platform is composed of steel structure and toughened glass (see Fig. 7). The floor is set up based on the production geological condition of the track entry of 1024 working face in Zhongxin coal mine and similarity theory. The abutment pressure is loaded by lifting jack according to the load path (see Fig. 8). A1–A5 and B1–B5 represent abutment pressure at the stage of leading influence during first mining. A5–A9 and B5–B9 represent abutment pressure at the stage of entry retaining during first mining.

Fig. 7. Test platform

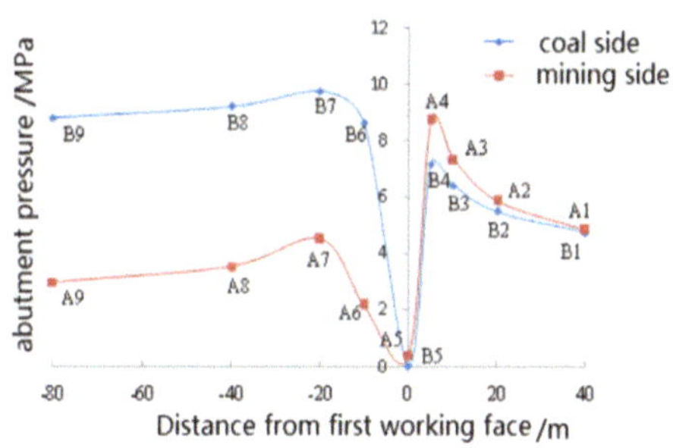

Fig. 8. Load path

The similarity model test of gob-side entry retaining demonstrates the forming process and shape of the main rapture plane of floor (see Figs. 9 and 10). (1) At the stage of leading influence during first mining, some small raptures formed from the two bottom corners, and then develop along about 45°. But the rapture degree at the coal side is larger than that at the mining side because the abutment pressure at the coal side is bigger. That also proves the asymmetrical characteristic of floor heave. (2) At the stage of leading influence during first mining, the main rapture gradually formed.

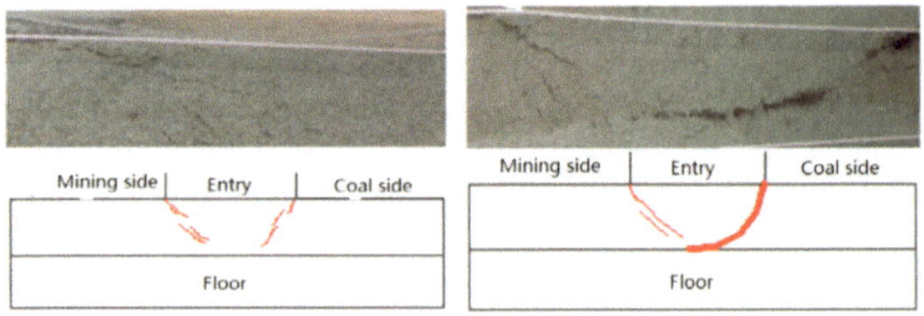

Fig. 9. Evolution law of rapture in floor at the stage of leading influence during first mining

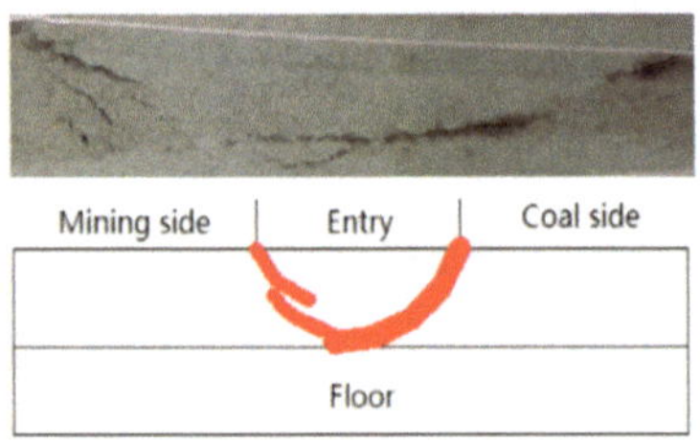

Fig. 10. Evolution law of rapture in floor at the stage of leading influence during first mining

It mainly occurred in B6–B8 period, especially the forming speed at the B7 is maximum. The floor rock slide along the arc-shape main rapture, and that gives rise to floor heave [10].

3 Control Technology of Floor Heave

The abutment pressure on the floor mainly originates from entry roof and transmits by two sides. So the property change of both roof and sides can influence the magnitude of abutment pressure. According to the idea that rigid support can bear more loading [11], the higher bearing capacity of roof and two sides can reduce the abutment pressure loading on the floor. The anchor bolts with high strength and high pre-tightening force can improve the bearing capacity of roof and two sides.

As for the entry roof, the anchor bolts can not only improve the bearing capacity of surrounding rock in anchorage scope but also reduce separation between the main roof and immediate roof. In addition, the anchor cables can transfer strata pressure into the deeper surrounding rock [12]. Thus the magnitude of abutment pressure originated from entry roof will decrease.

As for the two entry sides, The anchor bolts can made the state of entry sides change from biaxial compression state to triaxial compression state, which can reduce the deformation and improve the bearing capacity of two sides [13]. Thus the magnitude of abutment pressure transmitted by two sides will decrease.

We take a gob-side entry retaining in Xinyuan coal mine as an example to analysis control effect of strengthening support of roof and two sides. The original support parameters of roof are follows: the anchor bolts with 20 mm in diameter, and 2400 mm in length are used to support roof and coal side, the column space is 800 mm, the row space is 900 mm and the anchored mode is extensible-length anchorage. The anchor cables with 21.6 mm in diameter, and 10300 mm in length are also used to support roof, the column space is 1600 mm, and the row space was 900 mm. The calculation result of the vertical stress based on the above mechanical calculation model (k_2 = 2.18, k_3 = 0.74) is illustrated in Fig. 11. If three anchor cables are added between two anchor bolts for roof (k_2 = 2.06, k_3 = 0.65, see Fig. 12), the magnitude of vertical stress and plastic zone decrease. In addition, if one anchor cable are also added between two anchor bolts for coal side (k_2 = 2.026, k_3 = 0.67, see Fig. 13), the magnitude of vertical stress and plastic zone decrease again. So the technology of strengthening support of roof and two sides can reduce floor heave in gob-side entry retaining.

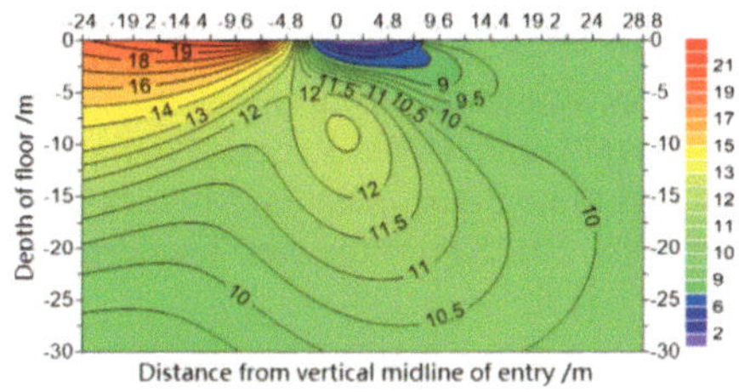

Fig. 11. Distribution of vertical stress in floor of entry with original support (MPa)

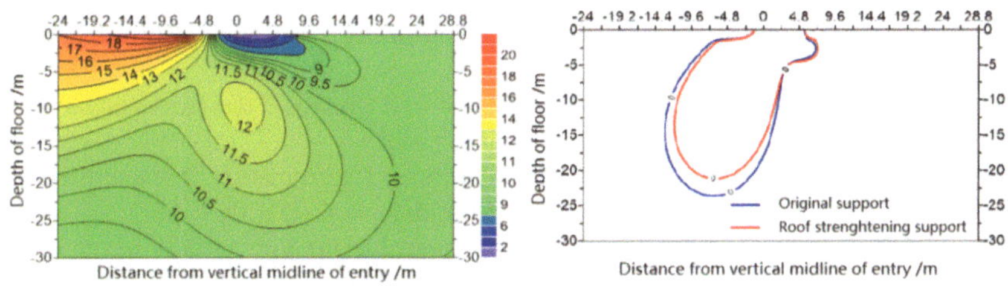

Fig. 12. Distribution of vertical stress in floor of entry with roof strengthening support (MPa)

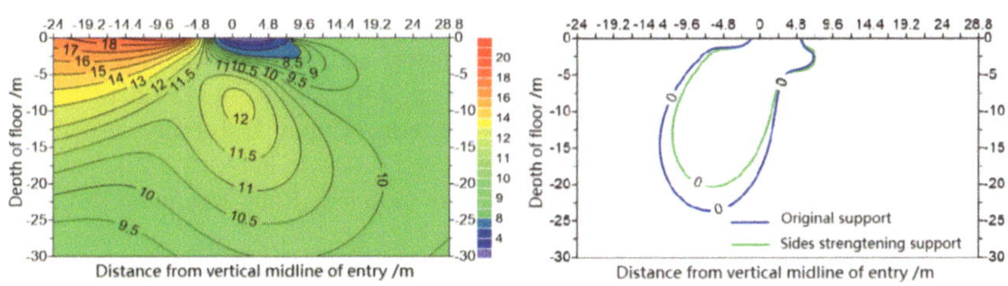

Fig. 13. Distribution of vertical stress in floor of entry with side strengthening support (MPa)

4 Conclusion

(1) The established elastoplastic model can be used to analysis the effect of mining abutment pressure on floor of gob-side entry retaining.
(2) The vertical stress and plastic zone in the floor increase with the advancing of working face at the stage of leading influence during first mining, increase first and then decrease at the stage of entry retaining during first mining, and increase at the stage of leading influence during second mining.
(3) The abutment pressure transmitted by two entry sides with different properties cause the asymmetrical characteristic of the vertical stress distribution and plastic zone.
(4) The technology of strengthening support of roof and two sides can reduce floor heave in gob-side entry retaining.

Acknowledgments. The authors would like to acknowledge National Natural Science Foundation of China (No. 511204167) and Open Project Program Foundation of Engineering Research Center of Underground Mine Construction of Ministry of Education (Anhui University of Science and Technology)) (No. 2015KF03) for their financial support.

References

1. Lu, S.L.: Strata Behaviors of Roadway Without Coal Pillar, 1st edn. China Coal Industry Publishing House, Beijing (1982)
2. Kang, H.P., Niu, D.L., Zhang, Z., et al.: Deformation characteristics of surrounding rock and supporting technology of gob-side entry retaining in deep coal mine. Chin. J. Rock Mech. Eng. **29**(10), 1977–1987 (2010)
3. Wang, W.J., Hou, C.J., Feng, T.: Floor Heave of Dynamic Pressure Roadway, 1st edn. China Coal Industry Publishing House, Beijing (2003)
4. Wei, X.H., Lin, D.C., Xu, C.Z.: The floor stress and deformation characteristics of deep roadway along goaf. Industr. Saf. Environ. Prot. **3**(1), 4–6 (2013)
5. Hua, X.Z., Lu, X.Y., Li, Y.F.: Prevention and control technology of floor heave in gob-side entry retaining with large section of deep mine. Coal Sci. Technol. **41**(9), 100–104 (2013)
6. Xu, Y., Chen, J., Bai, J.B.: Control of floor heaves with steel pile in gob-side entry retaining. Int. J. Min. Sci. Technol. **26**(3), 527–534 (2016)
7. Xu, Z.L.: Elastic Mechanics Brief Tutorial, 4th edn. Higher Education Press, Beijing (2013)
8. Cai, M.F.: Rock Mechanics and Engineering, 1st edn. Science Press, Beijing (2002)
9. Xu, Y., Zhou, H., Bai, J.B., Chen, J.: Characteristics and control method of floor heave in gob-side entry retaining. Chin. J. Rock Mech. Eng. **34**(s2), 4235–4243 (2015)
10. Hou, C.J.: Ground Control of Roadway, 1st edn. China University of Mining and Technology Press, Xuzhou (2012)
11. Sun, H.H., Zhao, B.L.: The Theory and Practice of Gob-side Entry Retaining, 1st edn. China Coal Industry Publishing House, Beijing (1993)
12. Deng, Y.H., Tang, J.X., Zhu, X.K.: Analysis and application in controlling surrounding rock of support reinforced roadway in gob-side entry with full mechanized mining. Min. Sci. Technol. **20**(6), 839–845 (2010)
13. Chen, Y., Bai, J.B., Yan, S., Xu, Y., Wang, X.Y., Ma, S.Q.: Control mechanism and technique of floor heave with reinforcing solid coal side and floor corner in gob-side coal entry retaining. Int. J. Min. Sci. Technol. **22**(6), 841–845 (2012)

Author Index

L. Zhang et al. (Eds.): GSIC 2018, , pp. 409–411, 2018.
https://doi.org/10.1007/978-981-13-0113-1

Z